NITRIC OXIDE

Principles and Actions

NITRIC OXIDE
Principles and Actions

EDITED BY
Jack Lancaster, Jr.
Departments of Physiology and Medicine
Louisiana State University Medical Center
New Orleans, Louisiana

ACADEMIC PRESS

San Diego   New York   Boston   London   Sydney   Tokyo   Toronto

This book is printed on acid-free paper. ∞

Copyright © 1996 by ACADEMIC PRESS, INC.

All Rights Reserved.
No part of this publication may be reproduced or transmitted in any form or by any means, electronic or mechanical, including photocopy, recording, or any information storage and retrieval system, without permission in writing from the publisher.

Academic Press, Inc.
A Division of Harcourt Brace & Company
525 B Street, Suite 1900, San Diego, California 92101-4495

United Kingdom Edition published by
Academic Press Limited
24-28 Oval Road, London NW1 7DX

Library of Congress Cataloging-in-Publication Data

Nitric oxide : principles and actions / edited by Jack Lancaster, Jr.
     p.   cm.
  Includes index.
  ISBN 0-12-435555-2
  1. Nitric oxide--Physiological effect. I. Lancaster, Jack R.
QP535.N1N556 1996
574.19'214--dc20   96-2094
                                                CIP

PRINTED IN THE UNITED STATES OF AMERICA
96 97 98 99 00 01 MM 9 8 7 6 5 4 3 2 1

To Judith

Contents

Contributors xi
Preface xiii

1
The Physiological and Pathological Chemistry of Nitric Oxide
Joseph S. Beckman

I. Introduction 1
II. Why Is the Half-Life of Nitric Oxide So Short *in Vivo* 10
III. Chemistry of Nitric Oxide and Related Nitrogen Oxides 18
IV. Handling and Detection of Nitric Oxide 32
V. Chemistry of Nitric Oxide Reacting with Superoxide 39
VI. Conclusions 69
References 71

2
Nitric Oxide Complexes of Metalloproteins: An Introductory Overview
J. C. Salerno

I. Introduction 83
II. Nitric Oxide as a Paramagnetic Ligand 84
III. Iron Model Complexes 86
IV. Nitric Oxide Complexes of Ferrohemes in Proteins 88
V. Nitric Oxide Complexes of Iron–Sulfur Proteins 91
VI. Nitric Oxide Complexes of Other Nonheme Iron Proteins 95
VII. Copper Complexes 97
VIII. Conclusion 98
IX. Appendix: Electron Paramagnetic Resonance 99
References 107

3

Nitric Oxide as a Communication Signal in Vascular and Neuronal Cells
Louis J. Ignarro

I. Introduction   111
II. Evidence That EDRF Is Nitric Oxide or a Labile Nitroso Precursor   113
III. Biosynthesis and Metabolism of Endothelium-Derived Nitric Oxide   115
IV. Unique Chemical Properties for a Ubiquitous and General Physiological Mediator   119
V. Signal Transduction Mechanisms for Transcellular Communication   120
VI. Nitric Oxide as a Neuronal Messenger   124
VII. Conclusions   129
References   130

4

The Intracellular Reactions of Nitric Oxide in the Immune System and Its Enzymatic Synthesis
Jack Lancaster, Jr.
Dennis J. Stuehr

I. Introduction   139
II. Intracellular Reactions of Nitric Oxide in the Immune System   140
III. Enzymology of Nitric Oxide Synthesis   148
References   164

5

The Role of Nitric Oxide in Autoimmune Diabetes
John A. Corbett
Michael L. McDaniel

I. Introduction   177
II. Effects of Cytokines on β-Cell Function   181
III. What Mediates Interleukin 1-Induced Islet Dysfunction?   185
IV. Mechanism of Nitric Oxide Mediated Islet Dysfunction and Destruction   187
V. Cellular Source of Interleukin 1-Induced Nitric Oxide Formation by Islets of Langerhans   191
VI. Signaling Mechanism of Interleukin 1-Induced Expression of Nitric Oxide Synthase   195

VII. Constitutive Nitric Oxide Synthase and Insulin Secretion   196
VIII. Effects of Activated Macrophages on Islet Function   198
IX. Proposed Mechanism by Which Interleukin 1 Induces β-Cell Dysfunction and Destruction   198
X. Animal Models of Insulin-Dependent Diabetes Mellitus   200
XI. Nitric Oxide and Islet Inflammation   201
XII. Nitric Oxide Production by Human Islets   202
XIII. Interventions   204
XIV. Conclusions   206
References   207

6

A Role for Nitric Oxide in Liver Inflammation and Infection
Mauricio Di Silvio
Andreas K. Nussler
David A. Geller
Timothy R. Billiar

I. Introduction   219
II. Nitric Oxide and the Liver in Sepsis and Inflammation   221
III. The Biochemistry of Hepatocyte Nitric Oxide Synthesis   225
IV. Studies on *in Vitro* Actions of Nitric Oxide in Liver Cells   227
V. Studies on *in Vivo* Actions of Nitric Oxide in the Liver   231
VI. Summary   233
References   233

7

Role of Nitric Oxide in Allograft Rejection
Rosemary A. Hoffman
Jan M. Langrehr
Richard L. Simmons

I. Introduction   237
II. Nitric Oxide Production in Rat Splenocyte Mixed Lymphocyte Reaction   238
III. *In Vitro* Nitric Oxide Synthesis by Cells Recovered from Rat Sponge Matrix Allografts   239
IV. *In Vivo* Nitric Oxide Synthesis during Allograft Rejection   244
V. Effect of Nitric Oxide Synthesis in Defined Macrophage–Lymphocyte Cocultures   245
VI. Effect of Authentic Nitric Oxide on Lymphocyte Function   246
VII. Possible Mechanisms of Nitric Oxide-Induced Inhibition of Lymphocyte Activation   248

VIII. Conclusion   252
References   253

8
Role of Nitric Oxide in Treatment of Foods
Daren Cornforth

I. Introduction   259
II. Historical Use of Nitrate and Nitrite in Cured Meats   260
III. Nitric Oxide in Meat Curing: An Overview   261
IV. Nitric Oxide and Cured Meat Color   261
V. Nitric Oxide as an Antioxidant in Cured Meats   266
VI. Nitrite and Nitric Oxide as Antimicrobial Agents   269
VII. Direct Application of Nitric Oxide to Meats   277
VIII. Nitrite, Nitric Oxide, and Nitrosamine Formation   278
IX. Nitrate Prevention of "Late Gas" Defect in Cheese   279
X. Nitric Oxide in Swollen Cans of Green Beans and Spinach   279
XI. Summary   279
References   280

9
The Enzymology and Occurrence of Nitric Oxide in the Biological Nitrogen Cycle
Thomas C. Hollocher

I. Global Nitrogen Cycle   289
II. Denitrifying Bacteria   292
III. Evidence For and Against Nitric Oxide as an Intermediate in Denitrification Pathway   297
IV. Nitric Oxide Reductase, Nitric Oxide-Consuming Enzyme of Denitrification Pathway   307
V. Dissimilatory Nitrite Reductases, Enzymes That Generate Nitric Oxide in Denitrifying Bacteria   312
VI. Genome for Denitrification   320
VII. Nitric Oxide-Reducing Nitrite Reductases   321
VIII. Denitrification by Chemolithotrophic Ammonia Oxidizers   321
IX. Denitrification by Eukaryotic Microorganisms   323
X. Nitric Oxide Production from Nitrite by Enteric and Related Bacteria   324
XI. Nitric Oxide Production by Heterotrophic Nitrifiers   326

XII. Toxicity of Nitric Oxide toward Bacteria, a Topic Joining Bacteria with the Nitric Oxide Synthase of Animal Cells   328
References   329

Index   345

Contributors

Numbers in parentheses indicate the pages on which the authors' contributions begin.

Joseph S. Beckman (1) Departments of Anesthesiology and Biochemistry, The University of Alabama at Birmingham, Birmingham, Alabama 35233.

Timothy R. Billiar (219) Department of Surgery, University of Pittsburgh, Pittsburgh, Pennsylvania 15261.

John A. Corbett (177) Department of Pathology, Washington University School of Medicine, St. Louis, Missouri 63110.

Daren Cornforth (259) Department of Nutrition and Food Sciences, Utah State University, Logan, Utah 84322.

Mauricio Di Silvio (219) Department of Surgery, University of Pittsburgh, Pittsburgh, Pennsylvania 15261.

David A. Geller (219) Department of Surgery, University of Pittsburgh, Pittsburgh, Pennsylvania 15261.

Rosemary A. Hoffman (237) Department of Surgery, University of Pittsburgh, Pittsburgh, Pennsylvania 15261.

Thomas C. Hollocher (289) Department of Biochemistry, Brandeis University, Waltham, Massachusetts 02254.

Louis J. Ignarro (111) Department of Pharmacology, UCLA School of Medicine, Center for the Health Sciences, Los Angeles, California 90024.

Jack Lancaster, Jr. (139) Departments of Physiology and Medicine, Louisiana State University Medical Center, New Orleans, Louisiana 70112.

Jan M. Langrehr (237) Department of Surgery, University of Pittsburgh, Pittsburgh, Pennsylvania 15261.[1]

Michael L. McDaniel (177) Department of Pathology, Washington University School of Medicine, St. Louis, Missouri 63110.

---

[1] Present Address: Department of Surgery, Free University of Berlin, D-13353 Berlin, Germany.

Andreas K. Nussler (219) Department of Surgery, University of Pittsburgh, Pittsburgh, Pennsylvania 15261.

J. C. Salerno (83) Center for Biochemistry and Biophysics and Biology Department, Rensselaer Polytechnic Institute, Troy, New York 12181.

Richard L. Simmons (237) Department of Surgery, University of Pittsburgh, Pittsburgh, Pennsylvania 15261.

Dennis J. Stuehr (139) The Cleveland Clinic, Immunology Section NN-1, Cleveland, Ohio 44195.

Preface

Nitric oxide (NO) is one of the 10 smallest stable molecules in nature and has been a favorite subject of research for chemists, beginning with its discoverer, Joseph Priestly, more than 200 years ago. Biologists have long known of the importance of the inorganic nitrogen oxides in the biological nitrogen cycle (microbes and plants), and they also have been utilized extensively for many decades in the food preservation field. The chemistry and toxicology of the nitrogen oxides is an important area in atmospheric pollution. In addition, due to its unique properties NO has been used since 1865 as a probe to study the ligand environment in metalloproteins. In spite of this intense interest in the biological roles of the nitrogen oxides, it came as a great surprise when in the 1980s a series of investigators demonstrated that not only is NO synthesized by mammalian cells, but it is involved in an astonishing array of critically important physiological and pathophysiological phenomena.

This book is not meant to provide a comprehensive description of the multiple roles of NO in biology, but is intended to be a sourcebook of overall principles which apply to nitrogen oxides in all of its biological actions. In addition to several chapters which provide an overview of the basic chemistry/biochemistry of NO, there are individual chapters on its actions in several specific physiological/pathophysiological phenomena. These specific phenomena have been selected because they have been some of the most thoroughly studied at the molecular level and have provided insight into general principles. There are also chapters devoted to NO in the microbial nitrogen cycle and in food preservation, in hopes for closer communication between all fields united by a common biochemistry.

I thank the contributors for conscientiously composing their chapters, and Academic Press for substantial editorial help.

Jack Lancaster, Jr.

1

The Physiological and Pathological Chemistry of Nitric Oxide

Joseph S. Beckman
*Departments of Anesthesiology
and Biochemistry
The University of Alabama at Birmingham
Birmingham, Alabama 35233*

I. INTRODUCTION

Mammalian cells were discovered to produce nitric oxide (·NO) as a short-lived intercellular messenger. The physiological implications continue to rapidly expand and have so far have implicated nitric oxide in the regulation of blood pressure, platelet adhesion, neutrophil aggregation, as well as synaptic plasticity in brain (Feldman *et al.*, 1993; Gally *et al.*, 1990; Moncada *et al.*, 1991). Nitric oxide and its secondary oxidants are also major cytotoxic agents produced by activated macrophages and neutrophils. Nitric oxide is a simple molecule, consisting of a single oxygen bonded to one nitrogen atom, which makes its chemistry accessible to study in great detail. It is a remarkably stable free radical and has been kept in the gas phase for at least 40 years without evidence of decomposition. The thermodynamics, quantum mechanics, and gas phase reactions of nitric oxide and related nitrogen oxides are covered by an enormous amount of chemical literature. Excellent general references concerning nitrogen chemistry are written by Jones (1973) and Colburn (1973). However, much of the chemical literature is not easily read by biologists. The wide range of possible oxides involving nitrogen can be daunting. Many of the reactions commonly proposed to occur in biological systems were derived from accounts in inorganic chemistry textbooks, which describe the nitrogen oxides from the viewpoint of industrial applications and air pollutants. However, the general chemistry is often not directly applicable to biological systems because the reactions may have been studied in the gas phase at relatively high concentrations of nitrogen oxides and at temperatures relevant either to the interior of automobile engines or to the surface of Neptune.

My goal here is to provide an intuitive introduction to the chemistry of nitrogen oxides while emphasizing physical properties and the reactions with oxygen that are more likely to predominate in biological systems. Because nitric oxide is dilute in biological systems (typically under 10–400 nM), the physiological chemistry is both simpler and subtly different than commonly inferred. A major theme developed in this chapter is that the slow reaction of nitric oxide with oxygen to form the strongly oxidizing nitrogen dioxide ($\cdot NO_2$) cannot account for the short biological half-life of nitric oxide. The rate of nitrogen dioxide formation falls off with the square of nitric oxide concentration. Thus, the toxicity of nitric oxide *per se* at the concentrations used for signal transduction is much lower than commonly assumed, and one cannot extrapolate from experiments conducted at concentrations of nitric oxide above 1 μM. We can generally ignore the nitrogen oxides, N_2O_2, N_2O_3, N_2O_4, and N_2O_5, which are formed by weak nitrogen–nitrogen bonding of nitric oxide and nitrogen dioxide at higher concentrations. At the extreme dilutions encountered *in vivo*, the odds of two nitrogen oxides combining with each other are for the most part too small to be considered when compared to reactions with other biological molecules. The important exception occurs in the immediate vicinity of activated macrophages and neutrophils, which can produce substantially greater quantities of nitric oxide than endothelium (Hibbs et al., 1988).

This review will focus on three main reactions of dilute nitric oxide in physiological solutions. The first reaction is the binding of nitric oxide to ferrous heme iron of guanylate cyclase or other proteins, which is important for the activation of signal transduction pathways.

$$\text{Heme}-Fe^{2+} + \cdot NO \longrightarrow \text{heme}-Fe^{2+}-NO \qquad \text{(Reaction 1)}$$

The second reaction and certainly the major route for the destruction of nitric oxide *in vivo* is the fast and irreversible reaction with oxyhemoglobin (Hb) or oxymyoglobin to produce nitrate.

$$Hb-Fe^{2+}-O_2 + \cdot NO \longrightarrow Hb-Fe^{2+}OONO \longrightarrow Hb-Fe^{3+} + NO_3^- \qquad \text{(Reaction 2)}$$

Red blood cells typically contain 20 mM oxyhemoglobin and thus will destroy nitric oxide diffusing into the vascular stream. Nitric oxide can diffuse over

100 μm, the average distance to a capillary, in a few seconds. The resulting methemoglobin will be reduced by NADPH-dependent mechanisms in the red blood cell.

Superoxide ($O_2 \cdot ^-$) is the one electron reduced form of molecular oxygen. It reacts irreversibly and at close to the diffusion limit with nitric oxide (Huie and Padmaja, 1993) to form the powerful oxidant peroxynitrite anion ($ONOO^-$).

$$\cdot NO + \cdot O\text{-}O\text{:}^- \longrightarrow ONOO^- \quad \text{(Reaction 3)}$$

It is particularly toxic because peroxynitrite is stable enough to diffuse over a cell diameter, but is selective in its reactivity. Peroxynitrite can be produced in substantial concentrations by activated macrophages (Ischiropoulos *et al.*, 1992a) and neutrophils (Carreras *et al.*, 1994a, b), and may also be a major damaging species produced after cerebral and myocardial ischemia as well as by inflammation, sepsis, and many other pathological conditions (Matheis *et al.*, 1992; Mulligan *et al.*, 1991, 1992; Nowicki *et al.*, 1991). The rapid reaction with superoxide may also explain the short half-life of nitric oxide in perfusion cascades. Peroxynitrite has generally been overlooked in the chemical literature because it is unstable in the gas phase and has limited commercial usefulness.

A. Why Does Nitric Oxide Have an Unpaired Electron?

The unusual properties of nitric oxide result from an unpaired electron in the highest occupied molecular orbital (HOMO). Nitrogen contains five valence electrons (electrons in the outermost shell with the greatest influence on bonding), while oxygen contains six valence electrons (Fig. 1). Therefore, nitric oxide contains a total of eleven valence electrons. Because orbitals can hold only two electrons with each electron possessing an opposite spin, there must be a single

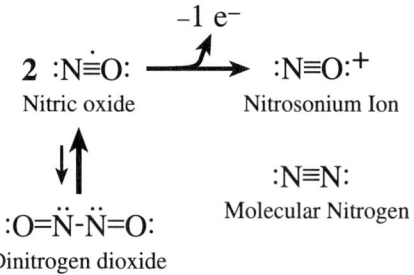

FIGURE 1

Lewis dot diagrams of nitric oxide compared to the nitrosonium ion and molecular nitrogen. Each bond contains one electron from each atom. These simple diagrams fail to properly account for the effective bond order of 2.5 predicted by molecular orbital theory and must be only considered as illustrative. The dimer of two nitric oxide molecules has five bonds, which is the same as two individual molecules. Thus, nitric oxide remains dissociated at room temperatures.

electron left alone in an orbital of nitric oxide. According to molecular orbital theory, nitric oxide has three fully occupied bonding orbitals with the unpaired electron residing in a fourth antibonding π orbital (Jones, 1973). The unpaired electron in the antibonding orbital weakens the overall bonding of the nitrogen to oxygen by one-half of a bond. Thus, the nitrogen and oxygen in nitric oxide are effectively held together by 2.5 bonds (3.0 bonding–0.5 antibonding orbitals). The bond between nitrogen and oxygen becomes stronger if one electron is removed from nitric oxide (Fig. 1). The resulting nitrosonium ion ($^+N\equiv O$) has the same number and distribution of electrons (i.e., it is isoelectronic) as molecular nitrogen (N_2). Both are triply bonded. The triple bond stabilizes the nitrosonium ion as reflected in its shorter N–O bond distance of 1.14 Å and the substantially stronger bond energy of 250.5 kcal/mol, compared to the bond distance and energy of 1.2 Å and 149.9 kcal/mol for nitric oxide (Jones, 1973).

Why do the unpaired electrons on two molecules of nitric oxide not combine to form a bond producing the dimer, dinitrogen dioxide (N_2O_2)? Curiously, no net bonds are formed when nitric oxide dimerizes, so the decrease in Gibbs energy from dimerization is negligible. The dimer O=N—N=O has a total of 5 bonds, which is the same as for 2 separate molecules of nitric oxide with 2.5 bonds each. Consequently, the enthalpy (the amount of heat released) for dimerization is only -2.6 kcal/mol (Ragsdale, 1973). This slight energy decrease is offset by the decrease in entropy because the dimer is more ordered than two separate monomers. At 300°C and 1 atm, the entropic energy ($-T\Delta S$) is $+4.3$ kcal/mol. Thus, the N_2O_2 dimer can exist with high concentrations of nitric oxide in the solid state at liquid nitrogen temperatures where the temperature-dependent entropic contribution is minimized. The remarkably weak N–N bond of N_2O_2 will be readily broken by the thermal energy available at room temperature to form two nitric oxide radicals.

Although nitric oxide is a free radical and a modest one electron oxidant, it does not readily react with most organic molecules. Ground states for the vast majority of organic molecules possess orbitals that are filled with two electrons of opposite spins. Therefore, a reaction involving nitric oxide will leave the organic molecule with an unpaired electron, resulting in an energetic organic radical intermediate. Such reactions tend to have high activation energies and are generally slow. However, nitric oxide can rapidly and directly react with the unpaired electron on both organic and oxygen-centered radicals to yield a variety of highly reactive intermediates.

B. Oxygen Reactions

The interactions of nitric oxide with oxygen help explain the short half-life of nitric oxide as a messenger and implicate nitric oxide as a major participant in free radical injury. Oxygen is required for the synthesis of nitric oxide and therefore must be present in tissue in substantially higher concentrations than

1 Physiological and Pathological Chemistry of Nitric Oxide

nitric oxide, potentially a 1000- to 10,000-fold higher on the arterial side of a vascular bed. Nitric oxide readily reacts with oxygen because molecular oxygen is in a triplet ground state, containing two unpaired electrons with parallel spins in separate orbitals, one on each oxygen atom (Fig. 2). Thus the unpaired electron on nitric oxide can form a weak bond with one of the unpaired electrons on oxygen to form nitrosyldioxyl radical (ONOO·).

The unpaired electrons on molecular oxygen create a pivotal role for oxygen in propagating free radical damage to tissues. As a consequence, the reduction of oxygen to water proceeds most readily by one electron steps producing a series of partially reduced oxygen species that are cytotoxic. In the order of sequential one electron additions, these are superoxide ($O_2^{·-}$), hydrogen peroxide (H_2O_2), hydroxyl radical (HO·), and finally water (Fig. 2). Superoxide is scavenged in cells by several distinct types of superoxide dismutases, while hydrogen peroxide is scavenged by catalase and glutathione peroxidase. No specific scavenger of hydroxyl radical exists because hydroxyl radical is so reactive that it attacks on collision with virtually every organic molecule.

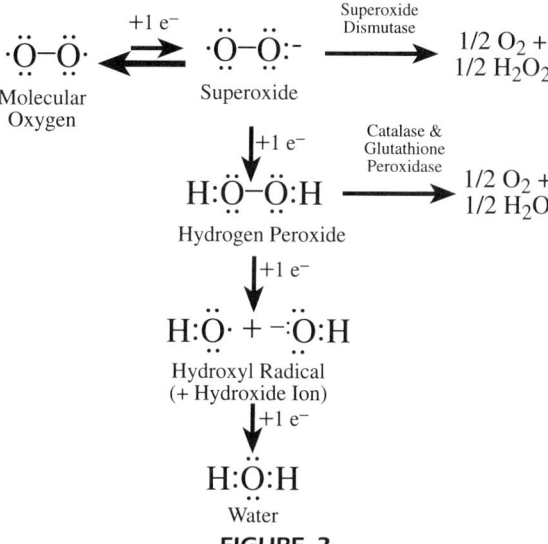

FIGURE 2

The reductive chemistry of oxygen illustrated by Lewis dot diagrams. However, the representation is not accurate for oxygen, failing to account for the two net bonds between the oxygen atoms, and should be only viewed as a illustrative teaching aid. Because molecular oxygen contains two unpaired electrons, its reduction to water proceeds most readily through a series of four single electron reduction steps. The lifetimes of superoxide and hydrogen peroxide are sufficiently long that they can be scavenged by the specific antioxidant enzymes, superoxide dismutase, catalase, and glutathione peroxidase.

The initial reduction of oxygen to superoxide is reversible and requires a substantial activation energy. Consequently, the reactivity of oxygen is limited by a high kinetic barrier, even though oxygen can oxidize virtually all biological molecules from a strictly thermodynamic point of view. The next two partially reduced species of oxygen, hydrogen peroxide and hydroxyl radical, are strong oxidants and damaging to biological tissue. Molecular oxygen also reacts with organic free radicals (R· + ·O—O· → R—O—O·) and therefore can propagate many forms of free radical-mediated injury. The potential interactions of biological sources of nitric oxide have only more recently begun to be explored. For example, organic dioxyl radicals (R—O—O·) may react with nitric oxide to form organic peroxynitrites (ROONO) that are strong oxidants in their own right capable of further propagating free radical injury (Pryor et al., 1985).

C. Metal Reactions

Nitric oxide rapidly reacts with transition metals, which have stable oxidation states differing by one electron (see Chapters 2 and 3). Nitric oxide is unusual in that it reacts with both the ferric (Fe^{3+}) and ferrous forms (Fe^{2+}) of iron. The unpaired electron of nitric oxide is partially transferred to the metal forming a principally ionic bond. Complexes of ferric iron with nitric oxide are called nitrosyl compounds and will nitrosate (add an NO^+ group) many compounds, while reducing the iron to the ferrous state (Wade and Castro, 1990).

The binding of nitric oxide with ferrous iron is reversible and occurs with a remarkably high affinity. For example, the affinity constant for nitric oxide binding to deoxyhemoglobin is about 10,000 times higher than molecular oxygen (Sharma et al., 1987; Traylor and Sharma, 1992). The major cellular signal transduction mechanism of nitric oxide is mediated by activation of soluble guanylate cyclase, which elevates intracellular concentrations of cyclic guanosine monophosphate (cGMP). This enzyme contains the same heme protoporphyrin IX as hemoglobin with iron in the ferrous form.

D. Why Is Nitric Oxide Used as an Intercellular Messenger?

Nitric oxide is a well known toxic gas, so its role as a biological messenger seems maladaptive. Yet, the same chemical properties that make nitric oxide toxic also explain why it is so useful as a rapid, locally acting messenger. As a small hydrophobic gas, nitric oxide crosses cell membranes as readily as molecular oxygen and carbon dioxide, without mediation of channels or receptors. Thus, nitric oxide will diffuse isotropically to surrounding tissue. The diffusion coefficient of nitric oxide in water is higher than oxygen, carbon dioxide, or carbon monoxide at 37°C (Wise and Houghton, 1968), which is ideal for carrying information. The signal is inherently short-lived and localized because nitric oxide decomposes spontaneously by reaction with oxygen as well as with heme proteins (Furchgott and Vanhoutte, 1989). A gradient is spontaneously formed because the biological activity of nitric oxide continuously declines by reaction with oxygen as it diffuses radially from its source. The local increase in nitric oxide can be immediately translated to a cellular signal because the unpaired electron of nitric oxide binds to transition metals with great affinity. Interaction with the heme iron activates guanylate cyclase and thereby increases the synthesis of the intracellular messenger cGMP. Nitric oxide or more likely a species derived from nitric oxide may also have other targets such as a soluble ADP-ribosylating enzyme (Brune and Lapetina, 1990), which has been identified as glyceraldehyde-3-phosphate dehydrogenase (Dimmeler et al., 1992; Kots et al., 1992; Zhang and Snyder, 1992). The binding of nitric oxide to ferrous heme groups is reversible, which allows the guanylate cyclase to turn off immediately after the nitric oxide gradient has dissipated.

E. How Is Information Communicated by Nitric Oxide?

Most messenger molecules encode information within their shape, which is recognized by a specific receptor. Nitric oxide is the smallest of biological messenger molecules, with the possible exception of carbon monoxide. Because of its chemical simplicity, nitric oxide must convey information by its concentration, which is interpretable by the spatial proximity of the source and target cells and the short duration of nitric oxide. Thus, the short half-life and limited diffusion distance of nitric oxide confers specificity, allowing the target tissue to derive information based solely on nitric oxide concentration.

For example, release of nitric oxide from endothelium causes a local relaxation only of the underlying vascular smooth muscle. The arterial bed consists of a series of branching smaller and smaller blood vessels, which constrict and relax rapidly according to local tissue needs (Fig. 3). If a blood vessel supplying one region should suddenly constrict, the upstream pressure will immediately rise and could potentially cause turbulent flow in other branches should the Reynolds number for laminar flow be exceeded (Griffith et al., 1987). Turbulent flow can disrupt perfusion to other vascular branches by increasing shear stress, which will constrict upstream blood vessels by myogenic responses. Further constriction can amplify turbulent flow leading to a catastrophe collapse of blood flow to distal vascular beds. However, shear-induced stress from turbulence induces endothelium to synthesize nitric oxide. The local relaxation of the underlying smooth muscle counterbalances myogenic responses to turbulent flow and thereby insures a laminar distribution of blood between vessels (Griffith et al., 1987). Part of the vasospasm associated with subarachnoid hemorrhage may be simply due to the scavenging of nitric oxide by hemoglobin released around the hemorrhage site.

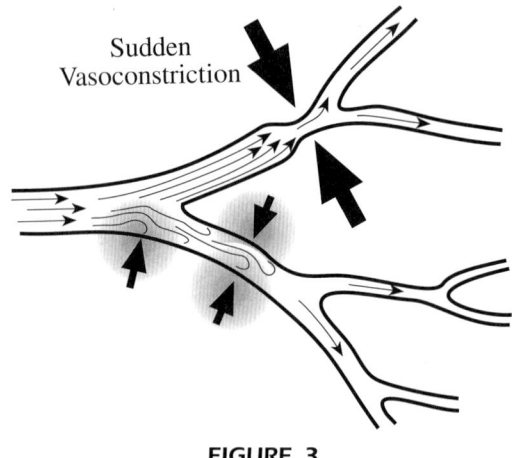

FIGURE 3

The role of nitric oxide in maintaining laminar blood flow. A sudden vasoconstriction in one branch of a vascular bed will cause an immediate increase in the upstream pressure that could cause turbulent flow. The myogenic constriction of the upstream branches responding to the turbulent flow will be counterbalanced by increased synthesis of nitric oxide, shown as a radial gradiant, due to shear stress on endothelium.

Oxyhemoglobin rapidly oxidizes nitric oxide to nitrate and is converted to methemoglobin (Doyle and Hoekstra, 1981; Goretski and Hollocher, 1988).[1]

Gradients of nitric oxide can convey information by a more subtle means related to its half-life. The lifetime of a local gradient of nitric oxide is relatively long compared to the firing of a nerve impulse or even to the activation of a neural network and thus could serve to integrate neuronal activity in a small volume of brain. Neural circuits fire on a millisecond time scale and the shortest perceptible time interval consciously resolvable in humans is about one-fortieth of a second. Gally et al. (1990) hypothesized that nitric oxide could be an important signal averaging mechanism controlling synaptic plasticity in the central nervous system. Much of the spatial organization on a local scale of the nervous system is determined by temporally correlated activation of neurons (Kandel and Hawkins, 1992). A general principle controlling the organization of the brain is that groups of neurons activated in synchrony will localize into the same region, whereas axons from neurons with uncorrelated activity project into spatially different regions (Yuste et al., 1992). Nitric oxide may be a key determinant underlying this principle because it carries information retrograde to the normal

[1] This is a reaction where both nitric oxide and hemoglobin are oxidized. The oxygen bound to the hemoglobin acts as the oxidizing agent and is consumed to produce nitrate. The reaction appears to proceed through an iron bound peroxynitrite intermediate (Doyle and Hoekstra, 1981).

mode of neurotransmission (Fig. 4). Classical neurotransmitters such as acetylcholine only affect receptors on the adjacent dendritic spine. Nitric oxide is produced from dendrites and diffuses radially to affect surrounding synapses, in the opposite direction of normal neurotransmission. Furthermore, nitric oxide does not distinguish between synapses that are in direct contact with the dendrite versus those that are localized in the same region.

The synthesis of nitric oxide by neurons is carefully regulated to occur only under special circumstances, and is initiated by extracellular calcium entry controlled by a particular glutamate receptor (the N-methyl-D-aspartate or NMDA receptor) (Garthwaite et al., 1988; Garthwaite, 1991) that has been strongly

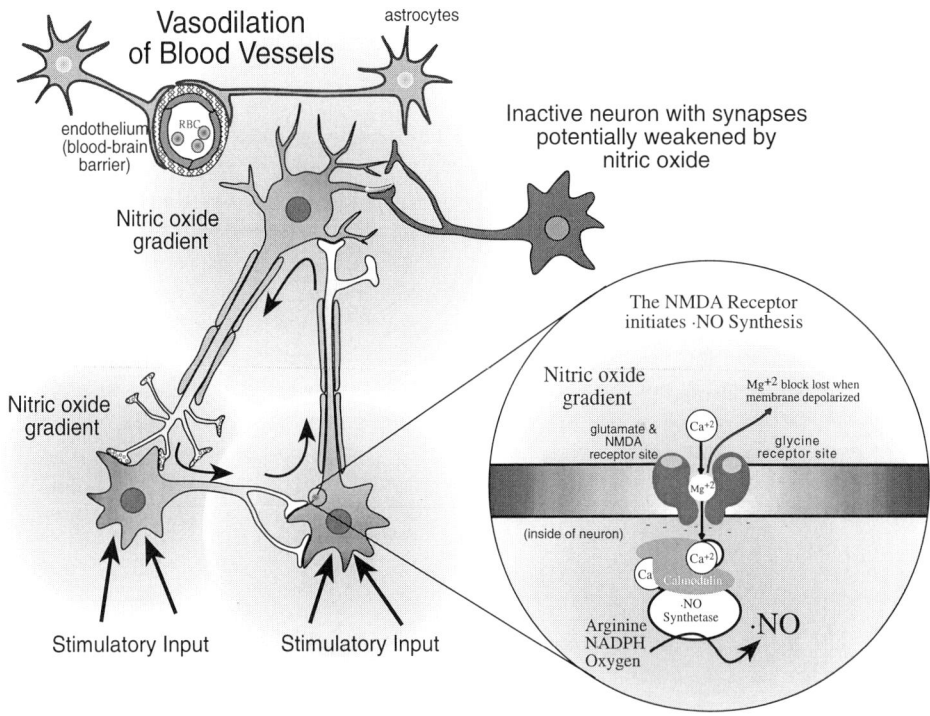

FIGURE 4

Role of nitric oxide in controlling synaptic plasticity. Nitric oxide synthesis is initiated by opening of the N-methyl-D-aspartate receptor/channel, and therefore will only occur when neurons have been actively firing and are partially depolarized. Nitric oxide can diffuse to surrounding synapses, carrying information in the opposite direction of neural transmission. It may affect synaptic plasticity by a combination of strengthening synapses that have been firing actively and weakening synapses that have not been active. In this manner, temporally correlated activity can be translated into spatial patterns of functionally related synapses mapping into the same brain regions.

implicated in learning and development. The NMDA receptor opens only when a neuron is partially depolarized, as occurs when a neuron has been firing rapidly. A local cluster of neurons repeatedly firing for even a second is sufficient to cause a local increase in nitric oxide concentrations (Shibuki and Okada, 1991). Nitric oxide has been shown to play an essential role in controlling synaptic plasticity in the cerebellum (Shibuki and Okada, 1991) and for the induction of long-term potentiation in the hippocampus, the most widely studied neuronal equivalent of learning (Böhme et al., 1991; East and Garthwaite, 1991; Nowak, 1992; Shulman and Madison, 1991; Zorumski and Izumi, 1993). Integration of neuronal activity occurs because nitric oxide will only be synthesized when localized groups of neurons have been partially depolarized by repeated activation necessary to open NMDA receptors. By selectively enhancing the synaptic efficiency of surrounding axonal arbors of neurons that have been firing frequently and weakening arbors from neurons that have not been active (Izumi et al., 1992), nitric oxide can control the remodeling of neuronal circuits important for learning and development (Gally et al., 1990).

The role of nitric oxide is analogous to the function of the reference light beam necessary for the recording of holograms. The reference beam, which allows for the constructive and destructive interference of light reflected from the object being photographed, allows the phases as well as the amplitudes to be recorded. Just as long-term memories are relatively unaffected by the selective loss of neurons, the entire holographic image is still present when the holographic film is cut in half. Only the resolution of the holographic image is decreased.

Clearly, the short half-life of nitric oxide is an important determinant for its biological function, but the chemical basis for this short half-life is still unknown. It cannot be due to the most commonly accepted mechanism: reaction with oxygen to form nitrogen dioxide.

II. WHY IS THE HALF-LIFE OF NITRIC OXIDE SO SHORT *IN VIVO*?

The major sink for nitric oxide produced by endothelium is almost certainly oxyhemoglobin in red blood cells and to a lesser extent myoglobin. For example, the shortest measured half-life of nitric oxide in blood-free perfused guinea pig heart is about 1 sec (Kelm and Schrader, 1988), which contains substantial amounts of myoglobin that rapidly scavenges nitric oxide. Even in blood-free cascade perfusion systems that assay the relaxation of vascular smooth muscle, the half-life of the endothelium-derived relaxing factor (EDRF) is surprisingly short: approximately 6–8 sec in buffers saturated with 95% oxygen and 30 sec at normal oxygen tensions (Furchgott and Vanhoutte, 1989). When these data are replotted as a function of oxygen concentration (Fig. 5), the biological in-

1 Physiological and Pathological Chemistry of Nitric Oxide 11

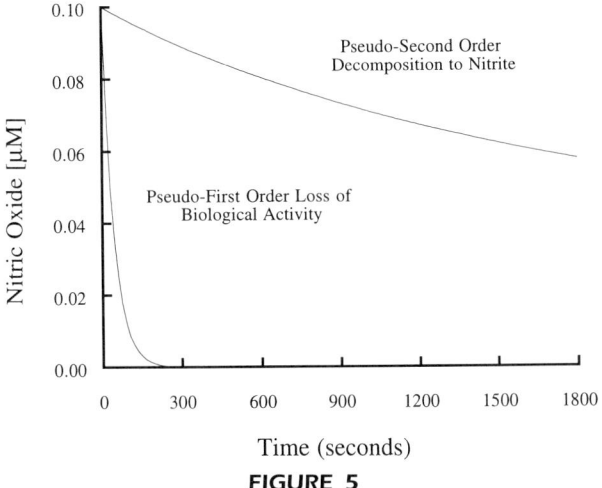

Time (seconds)

FIGURE 5

Rate of biological inactivation of nitric oxide. The apparent first-order rate constants were calculated as the natural logarithm of two divided by the half-life of nitric oxide measured in perfusion cascades. The half-lives were 30 sec with room air and 6–8 sec with 95% oxygen (Furchgott and Vanhoutte, 1989). Under anaerobic conditions, nitric oxide is indefinitely stable. The slope of the line give a second-order rate constant for the inactivation of nitric oxide by oxygen as approximately 100 M^{-1} sec^{-1}.

activation by oxygen proceeds with a second-order rate constant of approximately 100 M^{-1} sec^{-1}. This is a relatively slow reaction compared to diffusion-limited reactions that occur at rates of 10^9-10^{11} M^{-1} sec^{-1}.

The decomposition of nitric oxide in oxygenated perfused organ baths is commonly assumed to occur by the third-order reaction[2] of

$$2 \cdot NO + O_2 \xrightarrow{k_3} 2 \cdot NO_2. \quad \text{(Reaction 4)}$$

If one opens the stopcock on a gas sampling cylinder containing gaseous nitric oxide, the characteristic orange-brown color of nitrogen dioxide forms immediately. However, the local concentration of nitric oxide in the sampling cylinder is at least 10,000 times higher than tissue levels of nitric oxide, based on *in vivo* measurements (Shibuki and Okada, 1991). Because the overall rate of Reaction 4 is given by the third-order rate equation

$$\frac{d[NO]}{dt} = k_3[\cdot NO]^2[O_2], \quad (1)$$

[2] The third-order rate constant k_3 depends on the simultaneous collision of three reactants and has different units (M^{-2} sec^{-1}) from the second-order rate of biological inactivation estimated from Fig. 6. As described below, Reaction 4 actually occurs in two separate steps.

the effect of reducing the nitric oxide concentration by 10,000-fold will be to slow the overall reaction by a 100,000,000-fold. Thus, the low probability of any two nitric oxide molecules encountering each other in a biological system makes the formation of nitrogen dioxide extremely slow under physiological conditions.

How long would the decomposition of physiological concentrations of nitric oxide take if decomposition is due only to Reaction 4? Shibuki and Okado (1991) have measured up to 100 nM nitric oxide being produced in cerebellum following brief electrical stimulation and similar concentrations were measured more recently by Malinski and Taha (1992). The minimal concentration of nitric oxide required to stimulate synthesis of cGMP in vascular smooth muscle is approximately 10 nM (Hutchinson et al., 1987; Liu et al., 1994; Tracey et al., 1991). The overall third-order rate constant for Reaction 4 is about 6×10^6 M^{-2} sec^{-1} in aqueous solutions (Wink et al., 1993; Zafiriou and McFarland, 1980). The rate constant is slightly slower in the gas phase. The time for a given concentration to decrease by one-half is given by $1/(k_3[O_2][\cdot NO])$. Thus, the half-life depends on nitric oxide concentration and becomes progressively longer as nitric oxide decomposes (Fig. 6). In a simple chemical system, the loss of nitric oxide is in fact twice as fast as predicted by this rate constant because the product nitrogen dioxide will react with a second nitric oxide to form dinitrogen trioxide (N_2O_3) and then water to give two nitrites (2 NO_2^- + 2 H$^+$; see Section III,D).

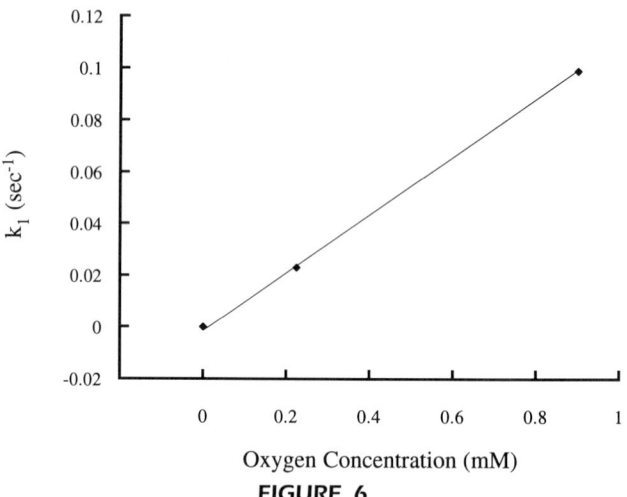

FIGURE 6

Comparison of the observed pseudo-first-order decay of biological activity with a half-life of 30 sec at normal oxygen tensions versus decomposition via nitrogen dioxide by pseudo-second-order kinetics predicted by Reaction 4. The loss of nitric oxide through formation of nitrogen oxide is twice as fast as calculated by Reaction 4 because each nitrogen dioxide formed rapidly attacks a second nitric oxide to form nitrite.

Thus, the predicted time for 100 nM nitric oxide at physiological oxygen tensions to decrease to 50 nM by Reaction 4 would take 1 hr, then 2 hr to decrease to 25 nM, and an additional 4 hr to reach 12.5 nM. The overall time for 100 nM nitric oxide to decompose to 10 nM purely by Reaction 4 is thus about 7 hr. Clearly, the third-order kinetics for the formation of nitrogen dioxide is far too slow to account for the biological inactivation of nitric oxide (Fig. 6).

The short biological half-life of nitric oxide measured in perfusion cascades is surprising to atmospheric chemists because nitric oxide persists for hours even on the most polluted days over Los Angeles. In atmospheric processes, the reaction of oxygen with two nitric oxides is considered to be a negligible source of nitrogen dioxide. Instead, the slow conversion of nitric oxide to nitrogen dioxide in the atmosphere is attributed to reaction with either the hydrogen dioxyl radical ($HO_2\cdot$; the conjugate acid of superoxide anion) or ozone (O_3). The slow formation of nitrogen dioxide and weak oxidative potential of nitric oxide makes nitric oxide less toxic than is commonly assumed in the biological literature. Low concentrations of nitric oxide can be safely administered in the breathing circuit to patients for the treatment of pulmonary hypertension with only trace conversion to nitrogen dioxide (Frostell et al., 1991).

Destruction of nitric oxide by superoxide in the buffers is more likely to account for the short half-life of nitric oxide *in vitro*. Superoxide dismutase (15–100 U/ml) substantially increased the apparent half-life of EDRF, strongly suggesting that superoxide contributes to the short biological half-life of nitric oxide. In the perfusion cascade bioassay system, the buffers are bubbled with 95% oxygen, contain 11 mM glucose as well as trace iron plus copper contamination and are incubated under the weak ultraviolet (UV) radiation of fluorescent lights. These are prime conditions for the autoxidation of glucose to form small amounts of superoxide in sufficient amounts to account for the short half-life of nitric oxide in nanomolar concentrations. The rate of reaction between superoxide and nitric oxide is 6.7×10^9 M^{-1} sec^{-1}. The shortest half-life of nitric oxide measured is approximately 6 sec. To achieve a half-life of 6 sec, the steady state concentration of superoxide would only need to be 17 pM, calculated as $\ln(2)/(6 \text{ sec} \times 6.7 \times 10^9 \text{ } M^{-1} \text{ } sec^{-1})$.

Another possibility that deserves further investigation is the formation of a transient and reversible complex of nitric oxide and oxygen to yield the nitrosyldioxyl radical ($ONOO\cdot$). The nitrosyldioxyl radical may be stabilized by hydrogen bonding to water, which may prevent it from activating guanylate cyclase (Beckman and Koppenol, 1992).

When nitric oxide is present in much lower concentrations than oxygen, the formation of nitrogen dioxide shown in Reaction 4 is initiated by the reversible reaction of nitric oxide with molecular oxygen to form nitrosyldioxyl radical.

$$\cdot NO + O_2 \rightleftharpoons ONOO\cdot \qquad \text{(Reaction 5)}$$

14 Joseph S. Beckman

The subsequent rate-limiting or slowest step is then the addition of a second nitric oxide to nitrosyldioxyl radical

$$\cdot NO + ONOO\cdot \rightleftharpoons ONOONO \qquad \text{(Reaction 6)}$$

which rapidly decomposes by homolytic cleavage of the O–O bond to nitrogen dioxide

$$ONOONO \longrightarrow 2\ \cdot NO_2 \qquad \text{(Reaction 7)}$$

The nitrosyldioxyl radical has largely been ignored in the chemical literature because it is relatively unstable in air. Nitrosyldioxyl radical is approximately 4.8 kcal/mol less stable than nitric oxide and oxygen in the gas phase; less than 0.1% of the nitric oxide will combine with oxygen under standard conditions in the gas phase. Although present in low concentrations, the infrared spectrum of nitrosyldioxyl radical has been reported in the gas phase (Guillory and Johnston, 1965) and *ab initio* quantum mechanics calculations have been performed (Boehm and Lohr, 1989).

However, thermodynamic and quantum mechanical calculations indicate that the stability of nitrosyldioxyl radical will be greatly increased by hydrogen bonding with water (Beckman and Koppenol, 1992). The reason for this increased stability can be readily visualized in Fig. 7. Nitrosyldioxyl radical is most stable when bent into the cis (or **C**-shaped) conformation (Boehm and Lohr, 1989). In this conformation, the lobes of the highest occupied molecular orbital from

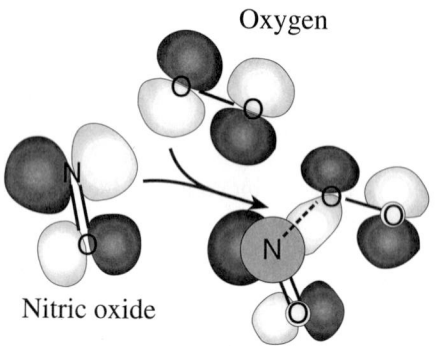

Oxygen

Nitric oxide

Nitrosyldioxyl Radical

FIGURE 7

The highest occupied molecular orbital and partial charges on each atom of oxygen, nitric oxide, and nitrosyldioxyl radical in the cis conformation. The positive and negative phases of each of the π-like orbitals are shown. Thus, overlap of orbitals with the same shading or phase produce a net bond between the two molecules. The partial charge on each atom is proportional to the circle around each atom, with the nitrogen being positive and the oxygen negative. The results were calculated with the MOPAC 6.0 of the Tetronix CAChe workstation, using the AM1 parameter set and a unrestricted Hartree–Foch approximation.

the two terminal oxygens are oriented towards each other and overlap to form a weak partial bond. In the cis conformation, the two distal oxygens, each with one greater nuclear charge than nitrogen, collaborate to pull electron density from the nitrogen and thus create a net dipole (separation of charge) across the molecule (Fig. 7). Calculated electrostatic fields of nitrosyldioxyl radical compared to nitric oxide and molecular oxygen are shown in Fig. 8. The electrostatic fields of nitric oxide and molecular oxygen repel hydrogen ions, accounting for the hydrophobic properties of these molecules. In contrast, the partial separation of charge (dipole moment) makes nitrosyldioxyl radical unstable in the gas phase, but favors hydrogen bonding with water. The hydrogen bonding stabilizes nitrosyldioxyl radical in solution and makes it hydrophilic. The stabilization may be substantial: Nitrosyldioxyl radical in solution is estimated to be about 8.8 kcal/mol lower than nitric oxide and molecular oxygen under standard conditions, approximately the same energy as released by the hydrolysis of ATP under standard conditions (Beckman and Koppenol, 1992).

A major unanswered question is what causes the inactivation of guanylate cyclase after nitric oxide has bound. The enzyme stops producing cGMP immediately after removal of nitric oxide (Garthwaite *et al.*, 1988). Nitric oxide potentially could be either oxidized or reduced by guanylate cyclase, though no such electron-accepting or donating cofactors have been identified yet as necessary for guanylate cyclase inactivation. The heme in guanylate cyclase is in the ferrous state, which binds nitric oxide tightly and reversibly to ferrous heme proteins. Thus, we will assume that guanylate cyclase is activated by a reversible equilibrium of nitric oxide with the ferrous heme group and propose the fol-

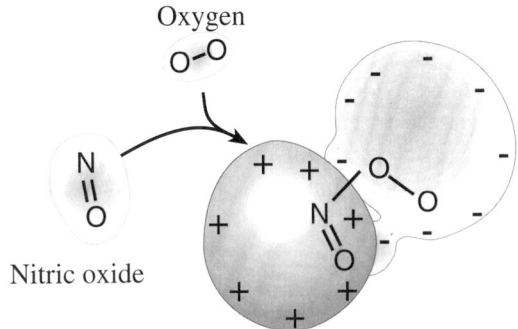

FIGURE 8

Electrostatic fields of nitric oxide, molecular oxygen, and nitrosyldioxyl radical. The size of the field is an index of the magnitude of the electrostatic field. A negatively charged field that will attract hydrogen bonding. Electrostatic fields were calculated using the analytical method with the CAChe system using the results from the MOPAC calculation described for Fig. 7.

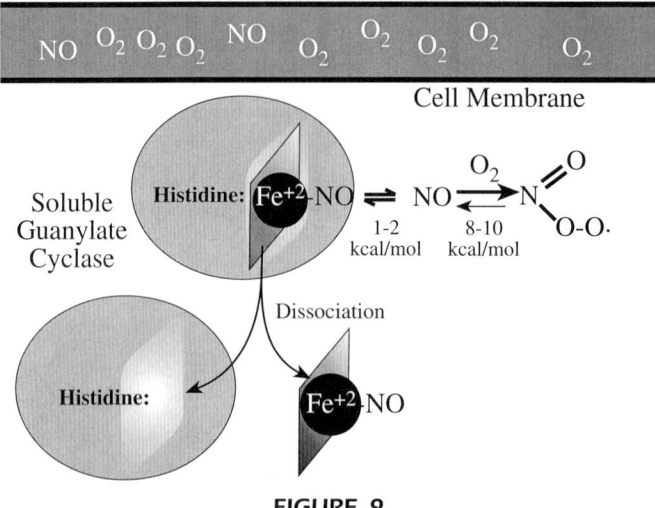

FIGURE 9

Possible equilibrium involved in the rapid activation of soluble guanylate cyclase and the slower inactivation by reaction of nitric oxide with oxygen. Nitric oxide dissolved in membranes may be more stable than in solution, because the nitrosyldioxyl radical cannot be stabilized by hydrogen bonding to water.

lowing model for the short half-life of nitric oxide in isolated organ baths perfused with oxygen-saturated buffers (Fig. 9).

$$GC-Fe^{2+}-NO \rightleftharpoons GC-Fe^{2+} + NO + O_2 \rightleftharpoons ONOO\cdot \qquad \text{(Reaction 8)}$$

The reaction rate for nitric oxide binding to guanylate cyclase is much faster (typically about 10^7 M^{-1} sec^{-1}) than the formation of nitrosyldioxyl radical by reaction with oxygen (assumed to be about 10^2 M^{-1} sec^{-1} to account for the observed half-life of nitric oxide). When nitric oxide synthesis is activated, nitric oxide will first associate with guanylate cyclase due to the faster reaction rate with ferrous heme. Nitric oxide will react more slowly with oxygen to form nitrosyldioxyl radical and nitric oxide will gradually be bound as nitrosyldioxyl radical because of high oxygen concentrations relative to guanylate cyclase. In perfusion cascades bubbled with 95% oxygen, the oxygen concentration is close to 1 mM, approximately a millionfold higher than the concentration of guanylate cyclase. Furthermore, the loss of biological activity according to this model follows the observed simple exponential kinetics as nitric oxide forms nitrosyldioxyl radical, because the reaction rate varies directly with nitric oxide concentration and not with its square.

The affinity of nitric oxide for ferrous heme proteins is exceptionally high, on the order of 10^{11}–10^{12} M^{-1} for deoxyhemoglobin (Traylor and Sharma, 1992).

However, the high affinity is necessary for guanylate cyclase to bind dilute concentrations of nitric oxide. In any equilibrium, the fraction associated as a complex becomes smaller as the concentrations of reactants decrease.

$$K_a = \frac{[AB]}{[A][B]} \quad (2)$$

The physiological concentrations of nitric oxide are in the range of 10^{-8} to 10^{-7} M, while guanylate cyclase concentrations are low and likely to be in the range of 10^{-8} to 10^{-7} M in cells. Thus, the product of the nitric oxide and guanylate cyclase concentrations is slightly smaller than the affinity constants reported for ferrous heme proteins for nitric oxide. Because both nitric oxide and guanylate cyclase are dilute *in vivo*, the binding energy for the association of nitric oxide with guanylate cyclase will be small under physiological conditions. The ratio of oxygen to guanylate cyclase concentrations determines whether the formation of nitroxyldioxyl radical can inactivate guanylate cyclase. In a purified solution of guanylate cyclase at the high concentrations needed for spectroscopy, the complex of nitric oxide and the ferrous heme will be stable even in the presence of oxygen. It is only when the guanylate cyclase–nitric oxide complex is diluted to more than a thousandfold lower than oxygen on a molar basis that the formation of nitrosyldioxyl radical can dissociate nitric oxide from the enzyme. At these dilute concentrations, the greater concentrations of oxygen combined with hydrogen bonding of water to nitrosyldioxyl radical should favor the dissociation of nitric oxide from activated guanylate cyclase. Thus, a continuous synthesis of nitric oxide would be necessary to keep guanylate cyclase activated because the nitric oxide will gradually complex with oxygen.

Oxygen is slightly more hydrophobic than nitric oxide and is concentrated up to eightfold within the core of membranes compared to cytoplasm. Nitric oxide will also be expected to partition within lipid membranes. However, the nitrosyldioxyl radical in the membrane cannot be stabilized by hydrogen bonding with water, making the effective lifetime of nitric oxide longer in the core of membranes. The nitric oxide-activated guanylate cyclase is the only soluble member of a large family of receptor-activated guanylate cyclases that are otherwise all membrane bound. If guanylate cyclase was anchored to a membrane, it would be activated for a much longer duration than the soluble enzyme because of the greater stability of nitric oxide in the hydrophobic membrane core. Thus, the cytoplasmic localization of the nitric oxide-activated guanylate cyclase may be an important determinant of the short-lived physiological effects of nitric oxide. In contrast, the predominant form of the constitutive nitric oxide synthase in endothelial cells appears to be membrane associated due to the addition of a myristate group to the amino terminus (Pollock *et al.*, 1991). This would favor the partitioning of nitric oxide directly into the plasma membrane, minimizing inactivation within the cytoplasm of endothelial cells, while allowing nitric oxide to escape to surrounding smooth muscle and platelets. Lateral diffusion of nitric oxide within membranes could be a significant contribution to its distribution in the central nervous system.

In blood-containing vascular beds, the inactivation of nitric oxide by oxygen is of minor importance because of the rapid and irreversible reactions of nitric oxide with oxyhemoglobin in red blood cells. Any nitric oxide that diffuses into the vascular lumen will be quickly destroyed, making blood vessels effective sinks for nitric oxide. The half-life of nitric oxide is sufficiently long that nitric oxide diffusing into the vascular smooth muscle could also diffuse back out to the lumin to be inactivated by hemoglobin in red blood cells.

Heme groups are typically bound to proteins only by coordination with histidine, but the affinity of nitric oxide for ferrous heme groups is so high that it can dislodge heme groups from proteins (Traylor and Sharma, 1992). Guanylate cyclase is often purified as an apoprotein, which is readily activated by porphyrins (Ignarro, 1991). This helps explain how transfer of the nitrosyl heme from nitric oxide-treated catalase can activate guanylate cyclase (Craven et al., 1979). Removal of heme inactivates guanylate cyclase, which may provide a safety valve to limit cGMP synthesis by prolonged exposure to elevated nitric oxide concentrations. The dissociation of heme from guanylate cyclase may be one of several contributing factors to the development of nitroglycerin tolerance (Axelsson and Karlsson, 1984).

III. CHEMISTRY OF NITRIC OXIDE AND RELATED NITROGEN OXIDES

Most of biological chemistry can be understood in terms of simple ball and stick models. The chemistry of nitric oxide and related oxides is more intimidating because its patterns of bonding depend strongly on quantum mechanics and molecular orbital theory. But the basics can be grasped by comparison to other molecules and a simple consideration of where nitrogen sits in the periodic table.

Nitrogen has an atomic number of seven, placing it between carbon with an atomic number of six and oxygen with an atomic number of eight. Because nitrogen carries one more nuclear (positive) charge than carbon, it is more electronegative and pulls electron density away from carbon in organic molecules (Table 1). Electronegativity describes the relative ability of an atom to draw electron density towards itself in a chemical bond. In most organic molecules, nitrogen readily forms hybrid orbitals directly analogous with carbon, forming the familiar sp^3 (with a pyramidal shape) as in ammonia and amines, sp^2 (planar) found in histidine, purines, and pyrimidines, and sp^1 (linear) found in nitrogen

TABLE 1
Electronegativity of Nitrogen Compared to Oxygen and Carbon[a]

Atom	Atomic number	Electron configuration	Electronegativity (nuclear charge)
Hydrogen	1	$1s$	2.20
Carbon	6	$1s^2\ 2s^22p^2$	2.55
Nitrogen	7	$1s^2\ 2s^22p^3$	3.04
Oxygen	8	$1s^2\ 2s^22p^4$	3.44
Fluorine	9	$1s^2\ 2s^22p^5$	3.98
Chlorine	17	— $3s^23p^5$	3.16

[a] The values are based on the Pauling scale (DeKock and Gray, 1989)

gas (N_2) and cyanide (HCN). This branch of nitrogen chemistry is familiar to biochemists because it resembles the chemistry of carbon.

However, nitrogen bound to oxygen behaves differently from its better known organic relatives. Oxygen is even more electronegative than nitrogen, due to an additional nuclear charge, and withdraws electron density away from nitrogen. The strong electronegativity of oxygen and nitrogen makes most nitrogen oxides strongly oxidizing species that will readily attack organic molecules. Nitric oxide per se is among the least reactive of the nitrogen oxides.

The dynamic range of the compounds formed by nitrogen can be understood by organizing the compounds according to the oxidation state of nitrogen (Table 2). Two of the seven electrons in nitrogen occupy the $1s$ level and thus can be ignored because they do not participate in bonding. That leaves five valence electrons to be placed in the $2s$ and $2p$ levels. According to the octet rule, the

TABLE 2
The Oxidation States of Some Common Nitrogen Compounds

Oxidation state	Examples	Common names
−3	NH_3, R—NH_2	ammonium, amines
−2	R=NH_2, N_2H_2	amides, hydrazine
−1	NH_2OH	hydroxylamine
0	N_2	molecular nitrogen
+1	N_2O	nitrous oxide
+2	NO	nitric oxide
+3	NO_2^-	nitrite
+4	NO_2	nitrogen dioxide
+5	$ONOO^-$, NO_3^-	peroxynitrite, nitrate
+6	$ONOO\cdot$, $NO_3\cdot$	nitrosyldioxyl radical, nitrate radical

outer shell of electrons is most stable when a total of eight electrons are available through bonding with other atoms. Nitrogen can donate up to five electrons by binding to oxygen and other more electronegative elements. In fact, the oxidation states of the nitrogen oxides is defined by assuming that electrons are formally transferred to the oxygen. The overwhelming diversity of nitrogen-containing compounds arises from nine possible oxidation states because nitrogen can gain up to three electrons by bonding with weaker electronegative atoms like carbon or hydrogen, as occurs in most biological molecules, or can lose up to five electrons (Fig. 10).

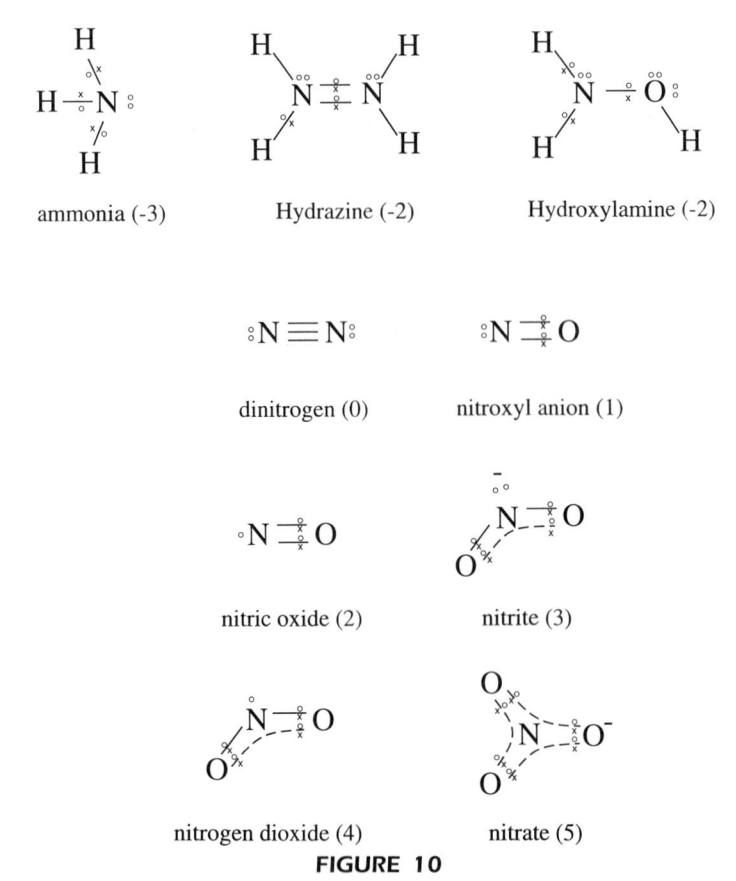

ammonia (-3)   Hydrazine (-2)   Hydroxylamine (-2)

dinitrogen (0)   nitroxyl anion (1)

nitric oxide (2)   nitrite (3)

nitrogen dioxide (4)   nitrate (5)

FIGURE 10

Oxidation states illustrated by Lewis dot diagrams for some simple nitrogen containing molecules. The formal oxidation states are given in parentheses. Open circles are used to arbitrarily show electrons donated by the nitrogen and "x" shows electrons from hydrogen or oxygen. In nitrogen–oxygen bonds, the oxidation state is determined by assuming that the electron from nitrogen is transferred to oxygen.

1 Physiological and Pathological Chemistry of Nitric Oxide

The chemistry of the nitrogen oxides dates back to the days of the Reverend Joseph Priestley, who used the reaction of nitric oxide to measure the concentration of oxygen in air. As a consequence, many of the recommended IUPAC names for nitrogen species have common names. As a general rule, common names are used when they have been widely utilized in the biological literature and IUPAC names for less well-known chemical species. Table 3 should help facilitate translation among the different names.

To better characterize the chemistry of nitric oxide, we will first consider the chemistry of the one-electron-oxidized and -reduced species of nitric oxide and then of the equivalent states of nitrogen dioxide. The distinction is artificial because many of these species are readily interconverted to each other. The interconversion is controlled by thermodynamics, and an extremely useful graphical summary is given by an oxidation state diagram (Fig. 11). The electrochemical reduction potential between any two nitrogen species is given by the slope of a line connecting the two species. Equivalent but less complete diagrams at pH 0 and 14 are presented by Jones (1973).

TABLE 3
Common Names and Recommended IUPAC Names for Nitrogen Oxides and Related Oxygen Radical Species[a]

Formula	Common name	IUPAC name
NO	nitric oxide	nitrogen monoxide
NO$^-$	nitroxyl anion	oxonitrate(1−)
HNO	nitroxyl	oxonitric acid
NO$^+$	nitrosonium ion	nitrosyl cation
N$_2$O	nitrous oxide	dinitrogen monoxide
NO$_2$	nitrogen dioxide	nitrogen dioxide
NO$_2^-$	nitrite	dioxonitrate(1−)
HNO$_2$	nitrous acid	dioxonitric acid
NO$_2^+$	nitronium ion	nitryl cation
ONOO·	peroxynitrite radical	nitrosyldioxyl radical
NO$_3^-$	nitrate	nitrate
HNO$_3$	nitric acid	nitric acid
ONOO$^-$	peroxynitrite	oxoperoxonitrate(1−)
ONOOH	peroxynitrous acid	oxoperoxonitric acid
N$_2$O$_3$	dinitrogen trioxide	dinitrogen trioxide
N$_2$O$_4$	dinitrogen tetroxide	dinitrogen tetroxide
N$_2$O$_5$	dinitrogen pentoxide	dinitrogen pentoxide

[a] Prepared with the assistance of Dr. Willem Koppenol, ETH, Zürich, Switzerland.

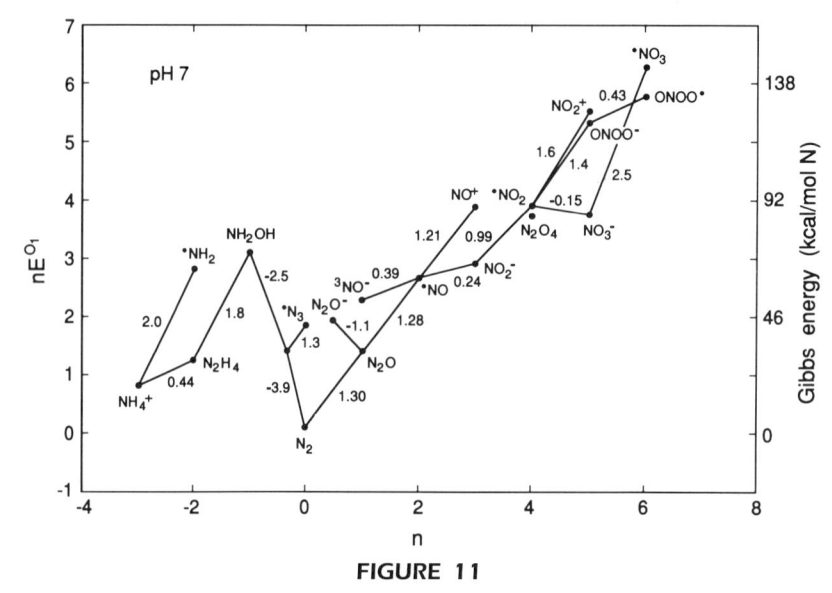

FIGURE 11
Oxidation state diagram of nitrogen at pH 7.0. The numbers refer to the slopes of the lines and give the reduction potentials. Conditions are for pH 7 at 298°C and 1-molar concentrations for all compounds including gases. The abscissa is the oxidation state of nitrogen and the left ordinate is the reduction potential in volts relative to the normal hydrogen electrode. The right ordinate is the Gibbs energy for reaction of one mole of nitrogen to form the compound of interest from nitrogen, water, and hydrogen. Reprinted with permission from Koppenol et al. (1992).

A. Nitric Oxide, Nitrosonium Ion, and Nitroxyl Anion

Nitric oxide may be oxidized by one electron to give nitrosonium ion (NO$^+$) or reduced by one electron to form nitroxyl anion (NO$^-$), which are important intermediates in the chemistry of nitric oxide (Stamler et al., 1992a).

$$^+N\equiv O \xrightarrow[1.2\text{ V}]{+1e^-} \cdot N=O \xrightarrow[0.39\text{ V}]{+1e^-} N=O^-$$

Although much of the biological literature focuses on nitrosating reactions of nitric oxide, chemically nitric oxide is a moderate one-electron oxidant, making formation of nitroxyl anion feasible under physiological conditions. The reduction potential to reduce nitric oxide to nitroxyl anion is +0.39 V, whereas it requires +1.2 V to oxidize nitric oxide to nitrosonium ion. Nitrosating reactions of nitric oxide are often mediated by conversion of nitric oxide to another nitrogen oxide species or by direct reaction with transition metals (Wade and Castro, 1990).

1 Physiological and Pathological Chemistry of Nitric Oxide

Chemical reduction of nitric oxide will generally produce nitroxyl anion (NO^-), which is isoelectronic with molecular oxygen. Both molecular oxygen and nitroxyl anion have two unpaired electrons in separate orbitals, where electrostatic repulsion is minimized. In the lowest energy configuration, the two unpaired electrons have parallel spins, which is called the triplet state. The two electrons could have opposite spins resulting in the singlet state, though this state is significantly higher in energy (about 21 kcal/mol) (Stanbury, 1989).

$$\uparrow \cdot N{=}O \cdot \uparrow \ ^- \qquad\qquad \uparrow \cdot N{=}O \cdot \downarrow \ ^-$$

Triplet nitroxyl anion   Singlet nitroxyl anion

The directions of the spins of the two unpaired electrons are indicated by the arrows. Molecular oxygen also exists in a triplet ground state, whereas most organic molecules are in the singlet state. Thus, molecular oxygen is relatively unreactive even though the oxidation of organic molecules is thermodynamically favored (Fig. 12). Photosensitizing agents like methylene blue, rose bengal, or the antibiotic tetracycline can transfer visible light energy to convert triplet oxygen to the more reactive singlet state (+23 kcal/mol). Singlet oxygen rapidly attacks organic compounds with double bonds because its reactivity is not limited by the spin restriction of triplet molecular oxygen. Consequently, tetracycline makes patients hypersensitive to sunlight. Singlet nitroxyl anion is produced in photochemical processes and reacts differently from triplet nitroxyl anion (Donald et al., 1986). Similar reactions might occur with singlet nitroxyl anion, though there is a lack of information about its biological effects. Sunburn is accompanied by vasodilation, so the potential biological reactions of singlet nitroxyl anion will be an interesting area for future research.

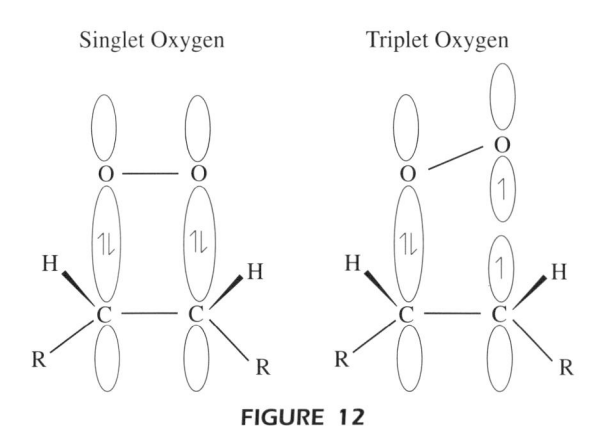

FIGURE 12
Reaction of singlet versus triplet oxygen with a carbon double bond. The arrows correspond to electron spin.

Nitroxyl anion in the triplet state can be detected by the conversion of methemoglobin (Fe^{3+}) to nitrosohemoglobin ($Fe^{2+}-NO$). Nitroxyl anion in the triplet state also reacts reversibly with Cu,Zn superoxide dismutase (Murphy and Sies, 1991), which is reasonable because nitroxyl anion has the approximate shape and charge of superoxide anion. Superoxide dismutase in the Cu^{2+} state (the resting enzyme) will be reduced by nitroxyl anion to form cuprous (Cu^{1+}) superoxide dismutase and nitric oxide (Fig. 13). Conversely, cuprous (Cu^{1+}) superoxide dismutase can also reduce nitric oxide to nitroxyl anion. Cuprous superoxide dismutase is an important intermediate formed during the first step in the catalytic cycle of resting superoxide dismutase (Fig. 13). Cupric superoxide dismutase oxidizes superoxide to release molecular oxygen. The reduction potential for copper in superoxide dismutase is close to that of nitroxyl anion (NO^-), so the equilibrium constant for this reaction should be approximately one. Although the rate constant for the reaction of nitroxyl anion with superoxide dismutase is not known at present, the positively charged electrostatic field near the active site of superoxide dismutase will make the trapping of nitroxyl anion rapid. The reaction of superoxide dismutase with superoxide is among the fastest known for any enzyme (Getzoff et al., 1983).

Nitroxyl anion has been suggested to be a potential EDRF (Murphy and Sies, 1991). However, the triplet state of nitroxyl anion reacts with molecular oxygen (also in a triplet state) to form peroxynitrite anion (Donald et al., 1986; Hughes et al., 1971; Yagil and Anbar, 1964).

$$NO^- + O_2 \longrightarrow ONOO^- \qquad \text{(Reaction 9)}$$

The rate of Reaction 9 may be as fast as 3.4×10^7 M^{-1} sec^{-1} (Huie and Padmaja,

FIGURE 13

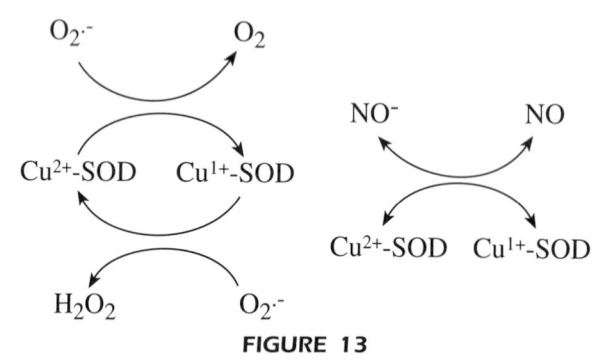

The reaction of nitric oxide with superoxide dismutase is a simple reversible equilibrium, whereas the catalytic cycle with superoxide involves a two step sequence. Consequently, superoxide dismutase may be reduced by superoxide and then react with nitric oxide to form nitroxyl anion. Nitroxyl anion may react with molecular oxygen to form peroxynitrite anion ($ONOO^-$).

1993), which would mean that the half-life of nitroxyl anion would only be 80 μsec at normal oxygen tensions. Peroxynitrite is a powerful and long-lived oxidant, which will be described in detail below. Peroxynitrite will cause vessel relaxation and increase cGMP in smooth muscle in concentrations that are 50- to 1000-fold higher than nitric oxide (Liu et al., 1994; Tarpey et al., 1992). However, it is unlikely that either nitroxyl anion or peroxynitrite would be produced as a biological EDRF, given their reactivity, toxicity, and modest efficacy as vasorelaxing agents.

B. Nitrosating Reactions

Nitrosonium ion is an important intermediate in the formation of carcinogenic nitrosamines and to the deamination of DNA and amine-containing amino acids residues in proteins (Wink et al., 1991). It readily adds to the lone electron pair of amine as follows

$$R-\ddot{N}H_2 + N{=}O^+ \longrightarrow R-NH_2{}^+{-}N{=}O \longrightarrow$$
$$R-NH-N{=}O + H^+ \quad \text{(Reaction 10)}$$

When the R is a secondary amine (such as in dimethylamine or morpholine), the resulting nitrosamine is relatively stable. However, primary amines such as on lysine or cytosine will undergo further reaction, giving rise to diazonium intermediates ($R-N{\equiv}N^+$). First, the hydrogen on the amine nitrogen will migrate to the more electronegative oxygen and then leave as hydroxide ion.

$$R-NH-N{=}O \longrightarrow R-N{=}N-OH \longrightarrow R-N{\equiv}N^+ + {}^-OH$$
$$\text{(Reaction 11)}$$

The diazonium intermediate releases molecular nitrogen (N_2) to form reactive carbonium ions. If the R group is aromatic, the diazonium intermediates can be stabilized as a salt, and are widely used as intermediates in synthetic chemistry. The formation of a diazonium intermediate with sulfanilic acid by acidified nitrite, a source of nitrosonium ion, is the basis for measuring nitrite by the Griess reaction.

Because of the high 1.2 V potential for oxidation, nitric oxide cannot directly nitrosate organic molecules without a cofactor to accept the unpaired electron. Clearly, the following reaction to form a nitrosothiol from nitric oxide is unbalanced unless there is a strong electron acceptor.

$$\cdot NO + RSH \longrightarrow RSNO + H^+ + e^- \quad \text{(Reaction 12)}$$

Common electron acceptors that will participate in nitrosating reactions include nitrogen dioxide, transition metals, and possibly oxygen. For example, nitric oxide bound to ferric heme iron nitrosates phenolics, thiols, and secondary amines (Wade and Castro, 1990).

$$\cdot NO + Fe^{3+} \longrightarrow Fe^{2+}NO \qquad \text{(Reaction 13)}$$

$$Fe^{2+}NO + RSH \longrightarrow Fe^{2+} + RSNO + H^+ \qquad \text{(Reaction 14)}$$

In the process, the iron is reduced to the ferrous form. Ferric cytochrome c is reduced by nitric oxide through a nitrosyl intermediate to produce ferrous cytochrome c and nitrite (Orii and Shimada, 1978). The nitrosyl cytochrome c absorbs at 560 nm, which is slightly higher than the 550-nm peak observed for reduced cytochrome c. Nitric oxide may be an interference in the assay of superoxide from cultured cells by the cytochrome c method. When nitric oxide reacts with cytochrome c, there is an initial decrease in absorbance at 550 nm as the nitrosyl complex is formed followed by a rise in absorbance as the complex decomposes to nitrite and reduced cytochrome c. This is a potential artifact in studies measuring the release of superoxide from cultured endothelial cells or other cells that make nitric oxide.

C. Nitrogen Dioxide, Nitronium Cation, and Nitrite

Like nitric oxide, nitrogen dioxide has an unpaired electron and can either be oxidized by one electron to form nitronium ion (NO_2^+) or reduced by one electron to nitrite (NO_2^-) as shown in Fig. 14.

The positively charged nitronium ion is isoelectronic with carbon dioxide and is a linear molecule. It is a strong oxidant with a reduction potential of +1.6 V at pH 7.0, favoring electrophilic attack. Nitronium ion is the active agent in fuming nitric acid (a caustic mixture of sulfuric and nitric acids), where the nitric acid is effectively dehydrated to give nitronium ion.

$$H_2SO_4 + HNO_3 \rightleftharpoons HSO_4^- + H_2O + NO_2^+ \qquad \text{(Reaction 15)}$$

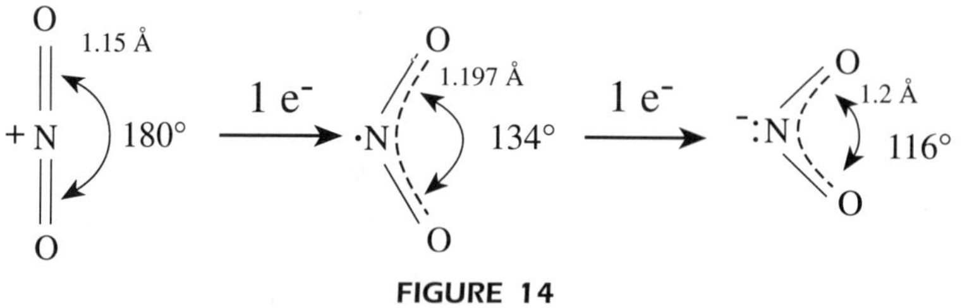

FIGURE 14

Structures of NO_2^+, NO_2, and NO_2^-. In nitrogen dioxide and nitrite, the presence of electrons in the lone orbital on nitrogen repels the two oxygen atoms to cause an asymmetric bent structure. The bending is greater in nitrite than nitrogen dioxide because the lone pair contains two electrons rather than one.

Such strongly acidic solutions are important synthetic routes for the preparation of nitrophenols including TNT (2,4,6-trinitrotoluene). Protonation of nitronium ion to give HNO_2^{2+} in highly acidic media may be important in nitration of deactivated aromatic rings in strongly acidic media (Olah et al., 1992ab). Nitronium ion is usually considered to be a significant species in highly acidic media. We have shown that a nitronium-like species is produced from peroxynitrite reacting with metal ions at physiological pH (Beckman et al., 1992; Ischiropoulos et al., 1992b; Koppenol et al., 1992). Macrophages activated in the presence of either superoxide dismutase or Fe^{3+} EDTA will nitrate low molecular weight phenolic compounds via the nitronium-like species (Ischiropoulos et al., 1992a). Its chemistry is described in greater detail with the chemistry of peroxynitrite in Section V.

Nitrogen dioxide is a common pollutant from internal combustion engines, accounting for the orange–brown tinge of smog. It is slightly denser than air, tending to cling to the ground in polluted air, making nitrogen dioxide a serious irritant of mucus membranes and lungs. Nitrogen dioxide is a strong one electron oxidant ($E^{o\prime} = +0.99$ V) and at low concentrations efficiently initiates free radical oxidation of unsaturated lipids, thiols, and proteins (Pryor et al., 1982; Pryor and Lightsey, 1981).

$$R-H + \cdot NO_2 \longrightarrow R\cdot + NO_2^- + H^+ \qquad \text{(Reaction 16)}$$

At higher concentrations, nitrogen dioxide will further react with most organic radicals at near diffusion limited rates to form a nitroderivative ($R-NO_2$).

$$R\cdot + \cdot NO_2 \longrightarrow R-NO_2 \qquad \text{(Reaction 17)}$$

For example, tyrosine residues on proteins are readily nitrated by a free radical mechanism (Prütz et al., 1985), where one nitrogen dioxide oxidizes the tyrosine to a phenyl radical that reacts with a second nitrogen dioxide to give nitrotyrosine.

Nitrite is the product from the one electron reduction of nitrogen dioxide. Nitrite is relatively stable in hemoglobin-free solutions and is frequently assayed as an indirect measure of nitric oxide production. Nitrite converts oxyhemoglobin to methemoglobin, and can be toxic at moderate dosages by causing methemoglobemia. Deaths have resulted from nitrite in cans of spinach. Nitrite readily complexes with metals and prevents their slow autooxidation. Thus, it is commonly added to cure meats such as bacon and sausage and may be a significant source of carcinogenic nitrosamines in the diet.

The chemistry of nitrite at acidic pH is closely related to that of nitrosonium ion described earlier. The pK_a of nitrous acid is around 3.4–3.6, though precise measurement is difficult because of its rapid secondary reactions. Acidification of nitrite produces nitrous acid, which is in reversible equilibrium with nitrosonium ion and hydroxide ion (Turney and Wright, 1959), although in aqueous solutions, the nitrosonium ion-like intermediate is more reasonably considered to exist as the hydrated form, $H_2O—N^+≡O$.

$$NO_2^- + H^+ \rightleftharpoons HO—NO \longleftrightarrow HO^-\cdots N≡O^+ \overset{H^+}{\rightleftharpoons} H_2O\cdots N≡O^+$$
(Reaction 18)

Acidic nitrite solutions readily donate nitrosonium ion and are important synthetic reactions to form diazonium salts.

The UV spectrum of nitrite at acidic pH is unusual. Nitrite at neutral to alkaline pH has a single peak at 356 nm, which is converted to a series of multiple peaks at acidic pH (Fig. 15). Normally, a UV peak is broadened by a combination of vibrational and more closely spaced rotational quantum states. However, light in the near UV region can separate nitrous acid into hydroxyl radical and nitric oxide.

$$ON—OH \xrightarrow{h\nu} ON\cdot + \cdot OH \quad \text{(Reaction 19)}$$

This reduces the rotational broadening and makes the vibrational subspectra apparent as a series of multiple peaks. If hydroxyl radical scavengers such as buffer components, the glass or plastic wall of test tubes, or trace metal contaminants are present, exposure to light will result in the formation of nitric oxide from

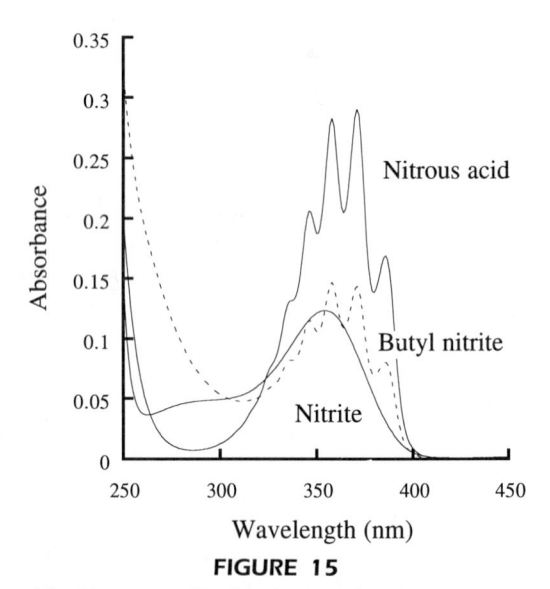

FIGURE 15

Ultraviolet spectra of 5 mM nitrite in 50 mM potassium phosphate at pH 7.5, nitrous acid at pH 1.9, and butyl nitrite at pH 7.5.

solutions of nitrite at neutral to acidic pH. Nitrite stored in Hamilton syringes under ordinary fluorescent lights has been shown to be able to relax vascular smooth muscle, which may be due to the trapping of hydroxyl radical with the glass walls releasing nitric oxide (Matsunaga and Furchgott, 1989, 1991). The ability of organic nitrites to release nitric oxide helps explain their biological activity. The nitrovasodilators have the structure of R—O—N=O, whereas isomers with the structure R—NO$_2$ have little biological activity. It is far easier for nitrites to release nitric oxide than to rupture the stronger carbon–nitrogen bond of a nitroderivative. The nitrosyl moiety may also be readily transferred to thiols groups to form nitrosothiols, explaining why thiols are important for the vasoactivity of some nitrites (Kowaluk and Fung, 1990).

D. Dimerization Reactions between Nitric Oxide and Nitrogen Dioxide

Although these reactions are too slow to account for the half-life of nitric oxide in signal transduction, they could be important in the phagosomes of activated macrophages, where substantial concentrations of nitric oxide are produced as part of the microbicidal mechanism of the macrophage. The slowest or rate-limiting elementary step occurring in the formation of nitrogen dioxide from nitric oxide and oxygen (Reaction 4) is ONOO· combining with a second nitric oxide.

$$ONOO\cdot + \cdot NO \rightleftharpoons ONOONO \longrightarrow 2 \cdot NO_2 \qquad \text{(Reaction 20)}$$

Nitrogen dioxide contains an unpaired electron. It reversibly dimerizes to form dinitrogen tetraoxide, which can exist in several isomeric forms shown in decreasing order of stability (Jones, 1973).

$$2 \cdot NO_2 \rightleftharpoons O_2N-NO_2 \rightleftharpoons NO^+ - NO_3^- \rightleftharpoons$$
$$ONONO_2 \rightleftharpoons ONOONO \qquad \text{(Reaction 21)}$$

Dinitrogen tetraoxide is a strong oxidant as well as a nitrosylating agent and a nitrating agent (adding an NO$_2$ group). Dinitrogen tetraoxide tends to be more reactive than either nitrogen dioxide or nitric oxide because it is a two electron oxidant. It readily attacks water, an energetically expensive reaction, to give the stable decomposition products nitrite and nitrate.

$$N_2O_4 + H_2O \longrightarrow NO_2^- + NO_3^- + 2 H^+ \qquad \text{(Reaction 22)}$$

Under physiological conditions, Reaction 22 is neglectable, because of the slow rate of nitrogen dioxide formation *in vivo*. Nitrogen dioxide is more likely to bond to residual nitric oxide rather than a second nitrogen dioxide. The dimer of nitric oxide and nitrogen dioxide is dinitrogen trioxide (N_2O_3), which like N_2O_4 also exists in several isomeric forms.

$$\cdot NO + \cdot NO_2 \rightleftharpoons ON-NO_2 \rightleftharpoons ON-O-NO \rightleftharpoons NO_2^- -NO^+$$
(Reaction 23)

Dinitrogen trioxide will rapidly react with water to form two nitrites (Marletta et al., 1988).

$$N_2O_3 + HO^- \longrightarrow 2\ NO_2^- + 2\ H^+ \quad \text{(Reaction 24)}$$

This contributes to the propensity of nitric oxide in dilute solutions to principally form nitrite rather than an equamolar amount of nitrite and nitrate as predicted by Reaction 22 (Ignarro, 1990; Ignarro et al., 1993). In a biological system, nitrogen dioxide may also directly extract an electron from lipids or other compounds to form nitrite directly (Pryor and Lightsey, 1981).

$$R_1-CH=CH-R_2 + \cdot NO_2 + H^+ \longrightarrow$$
$$R_1-\cdot CH-CH_2-R_2 + NO_2^- \quad \text{(Reaction 25)}$$

Dinitrogen trioxide is a strong oxidant that may be an important nitrosylating agent produced by activated macrophages. When a macrophage phagocytizes a foreign body, the potential local concentration of nitric oxide will be the highest produced in a biological tissue, potentially allowing secondary reactions between two nitrogen oxide species (e.g., Reactions 22 and 24) to occur. Wink et al. (1993) have proposed that another distinct oxidant besides dinitrogen trioxide may be formed in dilute aqueous solutions of nitric oxide. The nature of this species remains to be determined.

Still, the importance of dimers of nitric oxide and nitrogen dioxide is difficult to estimate. The dissociation constant for N_2O_3, defined as the ratio of $[NO][NO_2]/[N_2O_3]$ is 2×10^{-5} M (Treinin and Hayon, 1970). As the total concentration of nitrogen oxides decreases, the percentage of dimer rapidly falls off. If the total concentration of nitric oxide and nitrogen dioxide are both 1 μM, then the concentration of N_2O_3 will be 50 nM, but only 0.5 nM if the monomer concentrations are each 0.1 μM. The dissociation constant for N_2O_4 is 1.3×10^{-5} M (Treinin and Hayon, 1970), so with 0.1 μM nitrogen dioxide the concentration of N_2O_4 will be approximately 0.7 nM. While the great reactivity of N_2O_3 and N_2O_4 as two electron oxidants will drive such reactions, other direct reactions of nitrogen dioxide with biological molecules may compete at the necessarily dilute concentrations produced in vivo. In addition, activated macrophages and neutrophils produce large amounts of superoxide, which rapidly reacts with nitric oxide to form peroxynitrite (Section V,Q).

E. Thiols, Nitric Oxide, Nitrosothiols, and Endothelium-Derived Relaxing Factor

Controversy still exists about whether the EDRF is identical with nitric oxide. In some bioassay methods, EDRF and nitric oxide behave differently in bioassay cascades (Long et al., 1987; Marshall and Kontos, 1990). The biological activity of EDRF can be mimicked by nitrosothiols (R—S—N=O), which can decompose to release nitric oxide and a disulfide (R—S—S—R) (Myers et al., 1989, 1990). Furthermore, nitrosothiols are detected as nitric oxide in chemiluminescent assays that use acid reflux chambers as originally used to detect nitric oxide. The lifetime and concentration dependence of nitrosocysteine closely mimic the biological activity of nitric oxide in isolated vascular strips and in cerebral vessels (Marshall and Kontos, 1990). The lifetime of nitrosoglutathione is slightly longer lived than cysteine and the bovine serum albumin derivative may be stable for a day (Stamler et al., 1992b). Nitrosoalbumin has been detected in micromolar concentrations as a natural product *in vivo* (Stamler et al., 1992a). It has been suggested that nitrosothiols could be storage forms of nitric oxide and could also be carriers of nitric oxide across cell membranes.

The issue of whether nitrosothiols are EDRFs is difficult to resolve. Nitrosothiols lack the unique properties of nitric oxide as a messenger. Nitric oxide readily diffuses across cell membranes, so there is no need for a carrier system as postulated for nitrosothiols. The short half-life of nitric oxide is an important determinant of how it transmits information, so long-acting forms would undermine the usefulness of nitric oxide as a signaling mechanism. Moreover, nitric oxide appears to be the direct product of purified preparations of nitric oxide synthase (Marletta et al., 1988). Although many assay systems include a thiol, typically cysteine, glutathione, or dithiothreitol, these are added to protect the biopterin cofactor and the enzyme from oxidative inactivation. The initial velocity of nitric oxide production from the macrophage enzyme is the same in the absence of glutathione as in its presence, but the enzyme is apparently inactivated more rapidly without thiols (Stuehr et al., 1990). The ability of either cysteine, glutathione, or dithiothreitol (DTT) to prevent the loss of nitric oxide synthase activity also suggests that the thiol requirement is nonspecific.

Nitrosothiols should not bind directly to the heme of guanylate cyclase, but rather to activate guanylate cyclase by releasing nitric oxide. The simple cleavage of nitrosothiols into nitric oxide and a thiol radical is energetically expensive, so the activation may involve an oxidation with a second thiol group.

$$R-S-N=O + RSH \longrightarrow RS^{\cdot-}-SR + \cdot NO + H^+ \qquad \text{(Reaction 26)}$$

The resulting disulfide radical is strongly reducing and readily reduces oxygen to form superoxide (Koppenol, 1993).

$$RS^{\cdot-}-SR + O_2 \longrightarrow RS-SR + O_2^{\cdot-} \qquad \text{(Reaction 27)}$$

The thiol-dependent release of nitric oxide may well involve metal catalysis from trace contamination of copper or iron in buffers. The disulfide radical may also reduce another nitrosothiol to produce more nitric oxide.

$$RS^{\cdot-}-SR + R-S-N=O \longrightarrow$$
$$RS-SR + \cdot NO + RSH \qquad \text{(Reaction 28)}$$

Superoxide is also a moderately good reducing agent and may reduce another nitrosothiol to release additional nitric oxide.

$$R-S-N=O + O_2^{\cdot-} \longrightarrow \cdot NO + RSH + O_2 \qquad \text{(Reaction 29)}$$

The spontaneous reaction of nitric oxide with thiols is slow at physiological pH and the final product under anaerobic conditions is not a nitrosothiol (Pryor et al., 1982). The reaction is slow because it involves the conjugate base of the thiol ($R-S^-$). At pH 7.0, the oxidation of cysteine by nitric oxide required 6 hr to reach completion and yields RSSR and N_2O as the products. The synthetic preparation of nitrosothiols usually involves the addition of nitrosonium ion from acidified nitrite to the thiol, or oxidation of the thiol with nitrogen dioxide under anaerobic conditions in organic solvents. Nitric oxide will form nitrosothiols by reaction with ferric heme groups, such as found in metmyoglobin or methemoglobin (Wade and Castro, 1990). It is also possible that nitrosyldioxyl radical also reacts with thiols to form a nitrosothiol.

$$ONOO\cdot + R-SH \longrightarrow R-S-N=O + O_2^{\cdot-} + H^+ \qquad \text{(Reaction 30)}$$

Possibly, cellular thiols may be oxidized by the inactive adduct of nitric oxide and oxygen to regenerate a nitrosothiol or related species with EDRF activity. Some of the inconsistent results observed in bioassay systems may be due to the secondary and nonenzymatic formation of a nitrosothiols or other species capable of regenerating nitric oxide, which are leached into perfusion cascades. Consequently, bioassay systems should not be the gold standard to distinguish whether nitric oxide is the EDRF, because secondary reactions of nitric oxide decomposition products may regenerate nitric oxide.

IV. HANDLING AND DETECTION OF NITRIC OXIDE

Understanding the role of nitric oxide in vascular physiology and pathology has been hindered by the difficulties in both measurement and handling of dilute concentrations of nitric oxide. This has led to continuing and still unsettled uncertainty as to whether EDRF is identical with nitric oxide *in vivo*. Special care is required for the preparation of nitric oxide solutions because of its reactions with oxygen and trace metals. Both oxygen and nitric oxide are surprisingly permeable to most types of plastic tubing, necessitating the avoidance of any significant lengths of plastic or rubber tubing. Nitric oxide will corrode plastic and rubbers in a few days or weeks, causing leaks and contamination. Commercial cylinders of high purity nitrogen, helium, and argon contain up to a few parts per billion (ppb) of oxygen, so that a small amount of oxygen will still be present even after extensive bubbling. For strict anaerobic conditions, the inert gases should first be passed over oxygen-scrubbing columns available from gas chromatography suppliers. Argon may be advantageous over helium or nitrogen because it is heavier than air, so that less oxygen will settle in vessels as they are disconnected from the gas supply or during repeated needle punctures to withdraw samples. On the other hand, helium has a lower solubility in water. Regular grades of nitrogen contain significant backgrounds of nitric oxide and should be avoided. Contamination with oxygen is also a major problem when transferring nitric oxide in syringes. Gas-tight syringes with Teflon plungers are essential, and the syringe should be washed at least five times with the nitric oxide solution before drawing the sample for injection.

A. Reactions with Trace Metal Contaminants

Buffers are typically contaminated with $1-10$ μM iron, copper, and other transition metals due to a combination of impurities in the deionized water, phosphate salts, microscopic dust, and leaching from the walls of glass containers. When dilute solutions of nitric oxide (>1 μM) are prepared using such buffers, the majority of nitric oxide may be bound by the trace transition metal contaminants. Metal contaminants can be controlled by metal chelators, such as diethylenetriaminepentaacetic acid. The most commonly used metal chelator, EDTA, should be avoided because it leaves a reactive ligand open in the complex with ferric iron (Stezowski et al., 1973) and can strongly promote many types of metal-catalyzed reactions. In addition, UV light from fluorescent bulbs can photooxidize EDTA to form superoxide. These subtle effects may explain why different laboratories find substantial differences in the biological activity of nitric oxide (Tracey et al., 1991). Therefore, dilute solutions of nitric oxide should be prepared using deionized water that has been passed over metal chelator columns and stored for only short periods of time in carefully cleaned containers. In addition, the headspace above any dilute solution of nitric oxide must be minimized as nitric oxide will preferentially partition into the gas phase.

The saturating concentration of nitric oxide in water at 25°C is 1.87 mM. Because nitric oxide is a hydrophobic gas, it becomes less soluble as either temperature or the ionic strength increase (Fig. 16). The solubility of nitric oxide will be decreased in solutions containing physiological concentrations of salts, because salts reduce the solubility of hydrophobic gases. Solutions with nitric oxide bubbled into a gas sampling cylinder half filled with water are quite stable for many months. Any oxygen that diffuse through gas sampling ports quickly reacts with nitric oxide and forms nitrous acid in the water phase. Consequently, the water can become quite acidic.

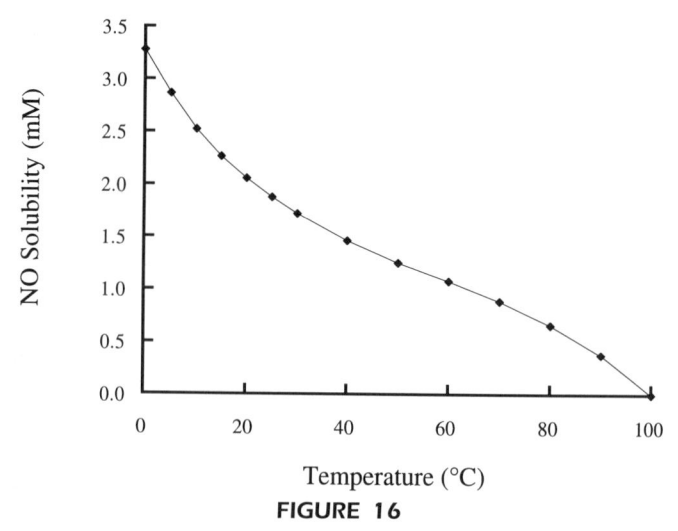

FIGURE 16
Solubility of nitric oxide in water as a function of temperature (Seidell, 1919).

A particularly sensitive and simple assay to quantify the concentration of stock nitric oxide solutions is the conversion of oxyhemoglobin to a mixture of methemoglobin and an as yet uncharacterized higher oxidized state, monitored by the decrease in absorbance of the hemoglobin Soret band at 401 nm (Feelish and Noack, 1987). The assay can be used to follow the kinetics of nitric oxide formation *in vitro*. However, the assay does not distinguish between nitric oxide, ONOO·, nitrogen dioxide, or nitrosothiols. Other oxidants including superoxide and hydrogen peroxide can oxidize oxyhemoglobin as well. Thus, it is not definitive proof of nitric oxide being present. Nitrite will also oxidize oxyhemoglobin, but at a significantly lower rate than nitric oxide and with an initial lag phase (Kosaka *et al.*, 1979, 1981). Care must also be taken in the preparation of hemoglobin, which is largely in the met form as purchased from commercial sources and must be converted to the oxy form by titrated reduction with sodium hydrosulfite (dithionite), desalting over a size exclusion column, and then reoxygenation (Di Iorio, 1981). Oxyhemoglobin is also relatively easy to purify in bulk from blood (Huisman and Dozy, 1965).

The most common assay of nitric oxide or EDRF is through bioassay, measuring either relaxation of isolated vessels denuded of endothelium, inhibition of platelet aggregation, or by the assay of cGMP production from crude preparations of guanylate cyclase. Such assays are extremely sensitive and by definition detect only physiologically active forms. However, such assays are difficult to quantify and do not distinguish between the endogenous EDRF and a large variety of compounds which can generate nitric oxide. Furthermore, bioassay does not directly measure concentration but rather the net exposure measured as concentration × time.

Other assay methods rely on measuring nitrite and nitrate, the breakdown products of nitric oxide. A relatively sensitive and simple colorimetric assay for nitrite is the Griess reaction. Acidified nitrite forms nitrous acid which reacts by separating into nitrosonium ion (N≡O$^+$) plus hydroxide ion. The Griess reaction uses sulfanilic acid as a trap for nitrosonium ion. The resulting diazonium salt is coupled with naphthalene ethyldiamine to form a strongly absorbing purple-red product, with a lower limit of sensitivity of 0.1–0.5 μM nitrite in the final assay solution. Related assays are based on the formation of fluorescent products and are roughly 10-fold more sensitive (Ohta, 1986).

Nitrate does not react in the Griess reaction. Because nitrite is rapidly converted to nitrate by hemoglobin in blood and more gradually by autooxidation, the Griess reaction cannot be directly used with plasma and urine samples unless nitrate is first reduced to nitrite. A variety of methods are available. Reduction by cadmium columns lends itself to an automated HPLC (high-performance liquid chromatography) method, though considerable caution is required because cadmium is highly toxic and poses a difficult and hazardous waste disposal problem. Reduction with purified bacterial nitrate reductases works well, but is expensive and the assays must be kept anaerobic with sodium hydrosulfite. A simpler procedure is to use a frozen suspension of either *Escherichia coli* or *Pseudomonas oleovorans* to metabolize nitrate to nitrite, and then remove the bacteria by centrifugation (Hibbs *et al.*, 1992). *Escherichia coli* must be grown under anaerobic conditions with nitrate to induce nitrate reductase and then incubated with ammonium formate as a source of reducing equivalents.

Nitrite and nitrate may also be determined by HPLC using anion-exchange columns with either UV absorption or electrochemical detection. These methods are generally used for environmental samples and urine where fewer interfering compounds exist. The more extensive sample preparation and analysis time have limited its use for biological samples.

Chemiluminescent detection in the gas phase is an extremely sensitive physical technique for measuring nitric oxide (Fontijn *et al.*, 1970), which was originally developed to monitor air pollution. The detector is based on the reaction of nitric oxide with ozone (O_3) to give nitrogen dioxide in an excited state plus molecular oxygen. The excited state of nitrogen dioxide decays to give a weak infrared chemiluminescense above 600 nm. Photons are counted by a red sensitive photomultiplier tube with a cutoff filter below 600 nm. An inlet gas stream containing the nitric oxide to be measured is mixed with ozone directly in front of the photomultiplier, while the reaction chamber is kept under moderate vacuum to minimize collisional deactivation of excited nitrogen dioxide. The reaction is highly specific for nitric oxide since most radical species that are chemiluminescent would decay long before entering the detector. The detector is insensitive to nitrogen dioxide, unless an 600°C oven or strong UV light is placed between the nitric oxide source and the detector. Both heat and UV radiation convert nitrogen dioxide to nitric oxide.

Nitric oxide release from blood vessels was first detected by chemiluminescence (Palmer et al., 1987). In the original adaptations of the nitric oxide detector, perfusates from isolated vessels were directly mixed in a reflux chamber containing acetic acid and iodine. The iodine in the reflux chamber served to reduce any nitrites or nitroso-containing groups to nitric oxide, which was stripped from the chamber by a continuous stream of nitrogen or helium that flowed to the chemiluminescent detector. Replacement of the acetic acid with the less volatile trichloroacetic acid reduces problems with contamination of the nitric oxide detector (Dr. D. Harrison, Emory University, Atlanta, Georgia, personal communication, 1991). While extremely sensitive, the use of the acid reflux chamber also reduces the specificity of the assay, raising questions as to whether nitric oxide or a nitrosothiol is EDRF (Myers et al., 1990).

The acidic reflux chamber using iodine will only reduce nitrite and nitrosocompounds to nitric oxide, but will not reduce nitrate. Nitrate may be analyzed by first converting it to nitrite with bacteria or nitrate reductase as described above for the Griess nitrosylating reaction. Nitrate can also be quantified in reflux chambers with heating to approximately 100°C and with the use of stronger reducing agents. A mixture of ferrous ammonium sulfate and ammonium molybdate was used by Cox (1980) as the reducing agent. Some caution must be used when examining samples treated with the nitric oxide synthase inhibitor, nitroarginine, because the nitro group will hydrolyze when heated under acidic conditions to give a quantitative nitric oxide signal.

Because nitric oxide is a hydrophobic gas, it will partition into the gas phase. Nitric oxide release from biological samples has been detected by sampling the headspace above solutions in sealed vials. After an incubation of several minutes, a sample of the headspace is collected with a gas-tight syringe and injected directly into the nitric oxide detector (Kowaluk and Fung, 1990). These experiments also show that the lifetime of nitric oxide is much longer than its biological activity would suggest.

We have confirmed the apparent long duration of nitric oxide in physiological solutions by a newly developed technique to measure nitric oxide and nitrogen dioxide continuously in dilute solutions (Fig. 17). Because nitric oxide is a hydrophobic gas, it crosses hydrophobic membranes much like oxygen or carbon dioxide. We have submerged U-shaped lengths of thin microporous tubing made of polypropylene (Celgard tubing, Hoechst-Celanese, Inc., Charlotte, NC) into well-stirred solutions of nitric oxide. One end of the tubing was connected to the vacuum system of a sensitive gas-phase chemiluminescent detector for measuring nitric oxide (Antek Inc., Houston, TX). A controlled flow of helium or other inert carrier gas is allowed to sweep through the polypropylene tubing (Fig. 18). The apparent half-life of nitric oxide was hundreds of seconds long even in the presence of 95% oxygen. Furthermore, no nitrogen dioxide was detectable

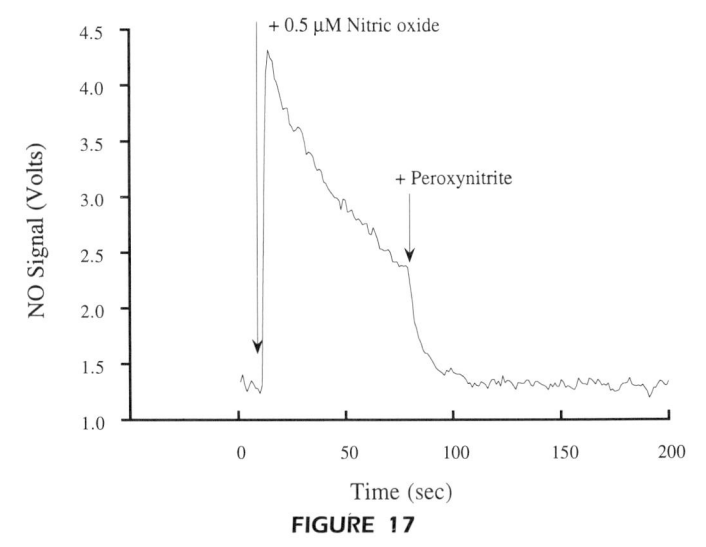

1 Physiological and Pathological Chemistry of Nitric Oxide

FIGURE 17

Measurement of nitric oxide by absorption into hydrophobic tubing. The apparent disappearance of nitric oxide is due to uptake into the hydrophobic tubing and is independent of the oxygen concentration in the solution. Peroxynitrite (ONOO$^-$) rapidly destroys nitric oxide as shown by the rapid decrease in the signal. The details of the basic apparatus are shown in Fig. 18.

when a nitrogen dioxide converter was interposed in the gas stream between the measurement chamber and the nitric oxide detector. The nitric oxide signal was immediately lost when either oxyhemoglobin or methylene blue were added.

Nitrogen dioxide can also be measured by passing the gas stream from the fibers through an oven heated to 600°C just before entering the nitric oxide detector (Fig. 18). The detector will then register the total concentration of nitric oxide plus nitrogen dioxide. One can also verify that the signal is due to nitric oxide by adding ozone to the inlet stream (Zafiriou and True, 1986). Ozone will destroy the nitric oxide by converting it to nitrogen dioxide before it enters the detector. If the ozone-treated gas stream is then passed through the 600°C oven, the nitric oxide signal will be regenerated. Only nitric oxide can be destroyed by treatment with ozone and regenerated with heating in an oven or exposure to UV light.

B. Electrochemical Detection

Nitric oxide can be assayed directly in tissues by its electrochemical oxidation on electrode surfaces (Shibuki, 1990). The technique was successfully used by Shibuki in cerebellar slices. However, the probe was fabricated in a glass micropipet coated with a thin hydrophobic chloroneoprene membrane, which makes the technique experimentally difficult. More recently, a carbon fiber electrode

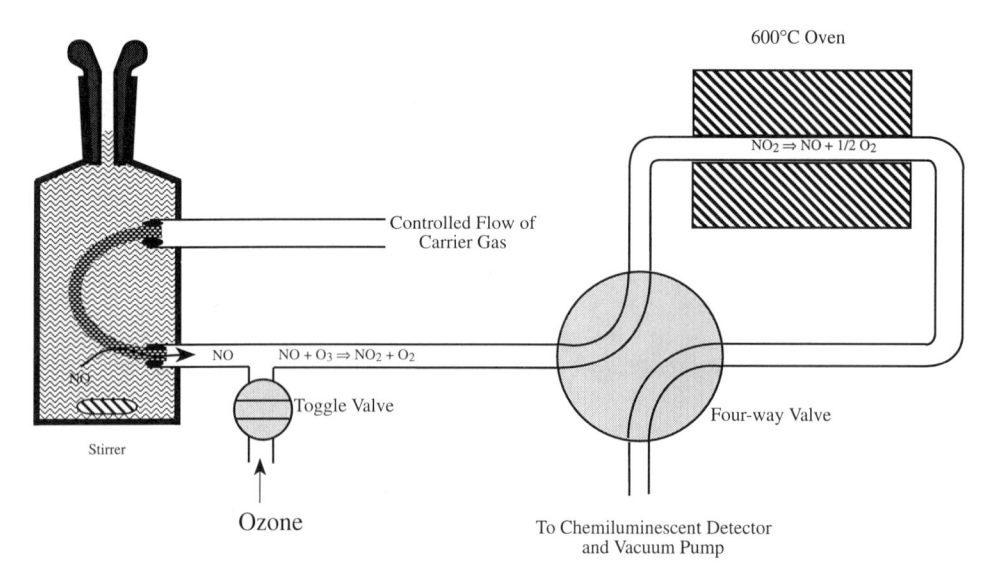

FIGURE 18

Measurement of nitric oxide by chemiluminescense. A slow flow of carrier gas is pulled through microporous hydrophobic tubing (Celgard, Celanese-Hoechst Inc.). The nitric oxide will diffuse through the wall of the tubing and into the gas stream leading to the chemiluminescent detector. If the gas stream is diverted through the oven, nitrogen dioxide will be converted to nitric oxide plus oxygen and then be detectable by the analyzer. It is possible to definitively prove that the signal is from nitric oxide by first adding a small amount of ozone to the gas stream, converting nitric oxide to nitrogen dioxide. This will destroy the signal from nitric oxide. Then, nitrogen dioxide so formed can be converted back to nitric oxide by switching a valve to divert the gas stream through an oven heated to 600°C. No other compound should be destroyed by ozone and regenerated by the oven.

with nickel–porphyrin plated on the surface of a glassy carbon electrode and then covered with a Nafion membrane (Aldrich, Milwaukee, WI) has been used to measure nitric oxide in a single endothelial cell (Malinski and Taha, 1992). These techniques hold great promise for measuring nitric oxide produced under physiological conditions, and commercial instruments are now available (e.g., World Precision Instruments, Sarasota, FL).

C. Electron Paramagnetic Spin Resonance

Although nitric oxide has an unpaired electron, it is difficult to detect directly by electron paramagnetic resonance. In addition to the low concentration of nitric oxide *in vivo*, the angular momentum of the unpaired electron can couple with the angular momentum of the nitric oxide molecule to obscure the paramagnetic properties of nitric oxide (Jones, 1973). However, nitric oxide can be detected as a complex with heme groups, which has been used to show the

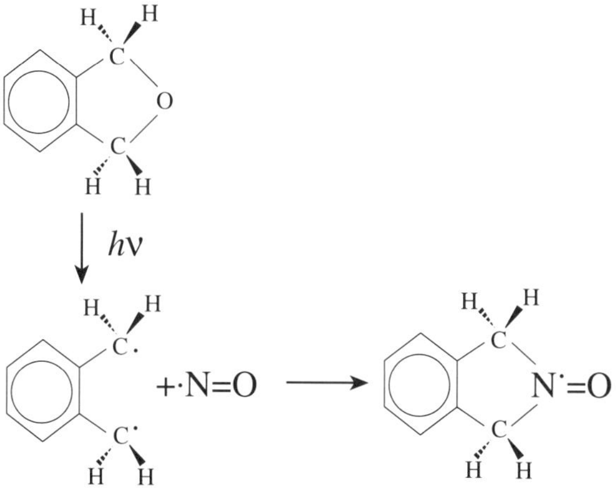

1 Physiological and Pathological Chemistry of Nitric Oxide 39

FIGURE 19
Cheletrophic spin trapping of nitric oxide. Ultraviolet light converts the phenolic compound to a biradical, which rapidly reacts with nitric oxide to form a stable nitroxyl radical that is readily observable by electron paramagnetic resonance.

presence of nitric oxide in immunologically rejected rat hearts (Lancaster et al., 1992) and from tumor cells exposed to activated macrophages (Drapier et al., 1991).

Nitric oxide has been difficult to detect with conventional spin-trapping agents. However, a new approach has been to use stable biradicals to trap nitric oxide (Fig. 19). The cheletrophic trap has two carbon centered radicals spaced the correct distance to catch nitric oxide and form a new ring (Korth et al., 1992). The trapped nitric oxide becomes a stable nitroxide radical that is the same chemical moiety present in most common spin-trapping reagents. Nitric oxide production from activated macrophages has been directly assayed by this method (Korth et al., 1992).

V. CHEMISTRY OF NITRIC OXIDE REACTING WITH SUPEROXIDE

Among the primary lines of evidence for demonstrating the action of EDRF is enhanced biological activity in the presence of superoxide dismutase (Moncada et al., 1991). Furthermore, superoxide-generating compounds are well known to inactivate EDRF. When nitric oxide was proposed to be the principal form of EDRF, the reason for its inactivation by superoxide was obvious. Both superoxide and nitric oxide contain unpaired electrons and react at near-diffusion-limited rates to form peroxynitrite anion (ONOO$^-$).

$$^-{:}O{-}O{\cdot} + {\cdot}N{=}O \longrightarrow {}^-{:}O{-}O{-}N{=}O \qquad \text{(Reaction 31)}$$

The reaction rate has been determined to be 6.7 ($\pm$0.9) $\times$ 10^9 M^{-1} sec^{-1}, which is approximately six times faster than the reaction of superoxide with one monomer of Cu,Zn–superoxide dismutase. The reaction is essentially irreversible because of a 22 kcal/mol decrease in Gibbs energy by forming peroxynitrite (Koppenol, 1993). Peroxynitrite anion is not a free radical because the unpaired electrons on superoxide and nitric oxide combine to form a new bond. Peroxynitrite is stable at alkaline pH, but rapidly decomposes to nitrate at neutral pH when it is protonated to form peroxynitrous acid (ONOOH). Originally, nitric oxide was proposed to be a scavenger of superoxide that would detoxify oxygen radicals (Feigl, 1988). While this view still persists (Kanner et al., 1991; Rubanyi et al., 1991), it has become clear that peroxynitrite is a powerful, relatively long-lived, and toxic oxidant that can be produced *in vivo* (Beckman et al., 1990; Ischiropoulos et al., 1992b; Wang et al., 1991). The oxidative chemistry of peroxynitrite is particularly complex at neutral pH and is controlled in part by peroxynitrite conformation and in part by the type of molecule being oxidized. The long life-time of peroxynitrite anion results from its being locked in the stable cis conformation, but the conjugate acid, peroxynitrous acid, can more readily isomerize to the reactive trans configuration. This peculiar behavior allows *cis*-peroxynitrite to diffuse to critical cellular targets and even cross cell membranes before becoming activated and fragmenting into extremely potent oxidants, which include species with the reactivity of hydroxyl radical, nitronium ion, and nitrogen dioxide.

Many pathological conditions, including ischemia/reperfusion, inflammation, and sepsis may induce tissues to simultaneously produce both superoxide and nitric oxide. For example, ischemia allows intracellular calcium to accumulate in endothelium (Fig. 20). If the tissue is reperfused, the readmission of oxygen will allow nitric oxide as well as superoxide to be produced (Beckman, 1990). For each 10-fold increase in the concentration of nitric oxide and superoxide, the rate of peroxynitrite formation will increase by 100-fold. Sepsis causes the induction of a second nitric oxide synthase in many tissues, which can produce a thousand times more nitric oxide than the normal levels of the constitutive enzyme (Moncada et al., 1991). Nitric oxide and indirectly peroxynitrite have been implicated in several important disease states. Blockade of nitric oxide synthesis with N-methyl or N-nitroarginine reduces glutamate-induced neuronal degeneration in primary cortical cultures (Dawson et al., 1991). Nitroarginine also decreases cortical infarct volume by 70% in mice subjected to middle cerebral artery occlusion (Nowicki et al., 1991). Myocardial injury from a combined hy-

FIGURE 20

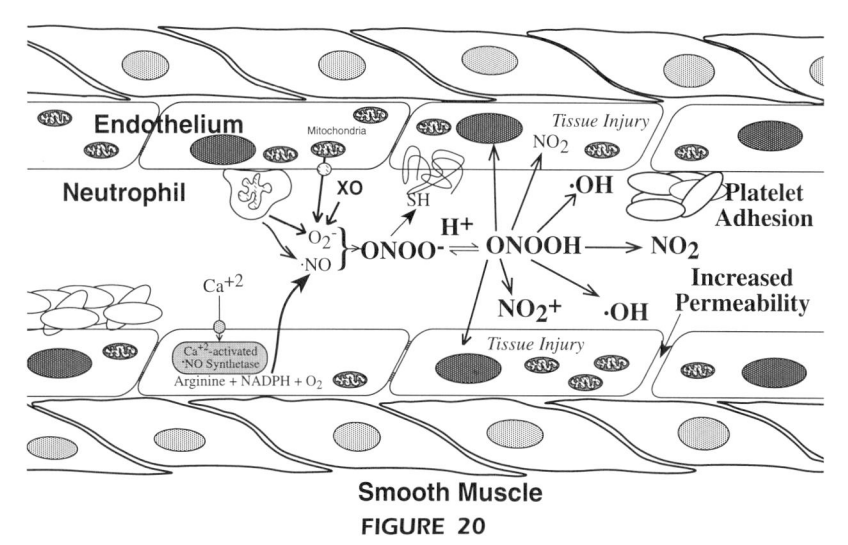

Generation of peroxynitrite in the vascular compartment as the result of ischemia/reperfusion. The introduction of oxygen following ischemia will initiate the simultaneous production of superoxide and nitric oxide. Neutrophils and macrophages may also generate nitric oxide and peroxynitrite directly. The rate of forming peroxynitrite will increase as the product of nitric oxide and superoxide concentration, and thus will increase rapidly under conditions when both are produced simultaneously.

poxic/ischemic insult in newborn pigs is substantially reduced by N-methylarginine (Matheis et al., 1992). Immune-complex induced pulmonary edema in rats mediated by either IgG or IgA can be largely blocked by N-methylarginine (Mulligan et al., 1991, 1992a). In many of these disease models, superoxide dismutase is also protective, suggesting that a connection may exist between nitric oxide and superoxide-dependent toxicity.

Nitric oxide itself is not particularly reactive with most organic molecules and its rate of conversion to more damaging species like NO_2, N_2O_3, or N_2O_4 can be rate limiting at the low concentrations produced *in vivo*. The direct reactions of nitric oxide with transition metals result in the inactivation of iron–sulfur centers in mitochondrial proteins (Stuehr and Nathan, 1989) and in the formation of a nitrosylating agent (Wade and Castro, 1990). However, the oxidizing potential and toxicity of nitric oxide are enormously increased if it reacts with superoxide to form peroxynitrite.

A. How Does Superoxide Dismutase Reduce Tissue Injury?

Vascular injury resulting from ischemia–reperfusion, inflammation, xenobiotic metabolism, hyperoxic exposure, and other diseases, causes loss of endothelial barrier function, adhesion of platelets, and abnormal vasoregulation (Moncada et al., 1991). Superoxide dismutase can reduce such injury, which indirectly implicates superoxide in many pathological processes (Flaherty and Weisfeldt, 1988). Although superoxide dismutase has been shown to reduce ischemia–reperfusion injury in many tissues, the mechanism of protection remains puzzling. For example, intravenously administered superoxide dismutase is protective when injected immediately before reperfusion and is most likely confined to the intravascular compartment (Beckman et al., 1988). Thus, superoxide scavenged by intravenously administered superoxide dismutase either is primarily produced in the extracellular compartment or else escapes through anion exchange proteins in cell membranes (Kontos and Wei, 1986; Lynch and Fridovich, 1978).

Superoxide can be directly toxic (Fridovich, 1986), particularly to iron–sulfur containing dehydratases in *E. coli* and aconitase (Gardner and Boveris, 1990; Gardner and Fridovich, 1991). However, the toxicity does not generally arise from the oxidative strength of superoxide at neutral pH (Sawyer and Valentine, 1981). At neutral pH, superoxide is generally a mild reductant, preferring to give up an electron to form molecular oxygen and is not a strongly oxidizing species. The limited reactivity of superoxide anion with many biological molecules has raised questions about the toxicity of superoxide per se (Baum, 1984; Sawyer and Valentine, 1981).

To account for the toxicity of superoxide in ischemic pathology, the secondary production of the far more reactive hydroxyl radical ($\cdot OH$) by the iron-catalyzed Haber–Weiss reaction has been frequently proposed to occur via the following two step reaction mechanism.

$$O_2^{\cdot -} + Fe^{3+} \longrightarrow O_2 + Fe^{2+} \quad \text{(Reaction 32)}$$

$$Fe^{2+} + H_2O_2 \longrightarrow \cdot OH + {}^-OH + Fe^{3+} \quad \text{(Reaction 33)}$$

Although the iron chelator, desferrioxamine, and low molecular weight scavengers of hydroxyl radical, such as mannitol, dimethylthiourea, and dimethylsulfoxide (DMSO), may reduce ischemic injury, the Haber–Weiss reaction is not a satisfactory explanation for oxygen radical injury to ischemic tissue. Hydroxyl radical reacts within a few nanometers of its site of formation (Hutchinson, 1957), which prevents hydroxyl radical from reaching a critical cellular target unless it is formed by a metal complexed directly to the target. Furthermore, the iron-catalyzed Haber–Weiss reaction requires the interaction of three species, making the rate of hydroxyl radical formation dependent on the product of minuscule reactant concentrations *in vivo* due to efficient scavenging and sequestering mechanisms. Millimolar concentrations of hydrogen peroxide are required for toxicity in most *in vitro* studies (Beckman and Siedow, 1985), which are far higher than anyone has measured *in vivo*. Furthermore, the catalytic iron must be both reduced and suitably chelated to react with hydrogen peroxide. Most biological sources of iron are not directly reactive because they are bound in proteins as inert complexes. Furthermore, the rate of Fe^{2+} reacting with H_2O_2 is relatively slow (only 10^3-10^4 M^{-1} sec^{-1}; (Rush and Koppenol, 1990) while the rate of Fe^{3+} reduction by superoxide is 10^6 M^{-1} sec^{-1}. Ascorbate, which also reduces Fe^{3+}, is present in much higher concentrations than superoxide. Because ascorbate or other reductants may readily substitute for superoxide as a reductant for iron, the ability of superoxide dismutase to decrease hydroxyl radical formation via the iron-catalyzed Haber–Weiss reaction may be limited *in vivo* (Winterbourn, 1979). These inconsistencies suggest that other reactions may be important for understanding superoxide toxicity to vascular targets. Nitric oxide may be a pathologically important target of superoxide both by removing an important intercellular messenger and by forming a potent cytotoxic agent.

B. Diffusion Distances of Hydroxyl Radical and Peroxynitrite

The potentially greater toxicity of peroxynitrite can be readily visualized by comparing the mean diffusion distances that various nitrogen and oxygen-centered species may traverse in one lifetime. The definition of lifetime (τ) is the time required for 67% of the initial concentration to decompose, and is readily calculated as the reciprocal of the pseudo-first-order rate constant for the disappearance of the species in question. Distances were calculated from the following equation, which is readily derived from the Fick's laws of diffusion (Nobel, 1983; Pryor, 1992).

$$x = (4D\tau)^{1/2} \tag{3}$$

where D is the diffusion coefficient. A value of 4.8×10^{-5} $cm^2 \cdot sec^{-1}$ for nitric oxide at 37°C was used for the following calculations (Wise and Houghton, 1968). The half-life of nitric oxide in normoxic physiological buffers is about 30 sec, but decreases to under 1 sec for nitric oxide in blood-free perfused guinea pig heart (Kelm and Schrader, 1988), most likely due to reactions with myoglobin. From Eq. (3), the estimated diffusion distances of nitric oxide are approximately 700 μm in normoxic buffer and 130 μm in blood-free isolated perfused heart (Fig. 21). Assuming the average intracellular concentration of superoxide dismutase is about 1 to 10 μM, superoxide could diffuse for approximately 1 to 4 μm. In the vascular compartment, the concentration of superoxide dismutase is typically about 3 U/ml based on the standard cytochrome c reduction assay of McCord and Fridovich (1969). This low superoxide dismutase activity increases the diffusion of superoxide in the vascular compartment to approximately 15 μm. Hydroxyl radical can react with essentially all cellular components at diffusion-limited rates, ranging between 10^9 and 10^{10} M^{-1} sec^{-1}. Radiation chemists have estimated the diffusion distance of hydroxyl radical to be only 3 nanometers (Hutchinson, 1957), about the average diameter of a typical protein. To

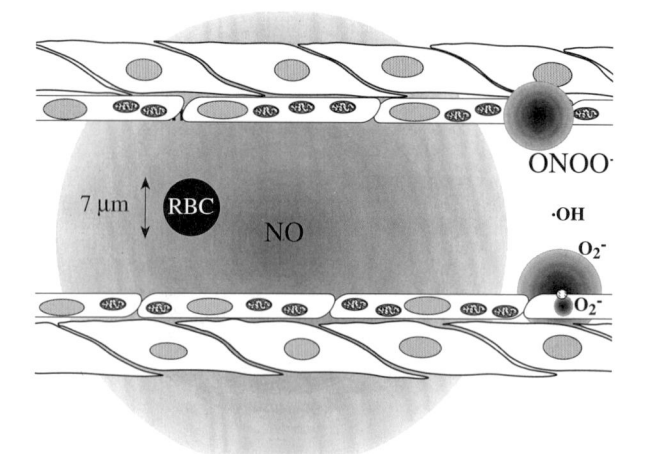

FIGURE 21

The mean diffusion distance of various nitrogen and oxygen based species within one estimated lifetime. The diameter of the dot for hydroxyl radical is still 100-fold larger than the actual diffusion distance (Hutchinson, 1957).

place this in perspective, the tiny dot shown for the diffusion distance of hydroxyl radical in Fig. 21 is still a hundred times larger than its potential diffusion distance.

The half-life of peroxynitrite in phosphate buffer at 37°C and pH 7.4 is about 1 sec, which gives it a diffusion distance of approximately 100 μm. However, reactions with cellular thiols (Radi et al., 1991a,b) and bicarbonate (Zhu et al., 1992) in a cellular milieu may decrease the half-life to 9 millisec. Thus, the diffusion distance of peroxynitrite would be roughly 9 μm. Many approximations make these estimated diffusion distances only a rough estimate that could be in error by as much as 5- to 10-fold, but my main point is the dramatic difference in diffusion distances between hydroxyl radical and peroxynitrite. Peroxynitrite is formed by a reaction that is at least a million times faster than the reaction of hydrogen peroxide with ferrous iron and can diffuse ten thousand times further than hydroxyl radical or ferryl iron. For hydroxyl radical to be toxic, it must be generated at the immediate site of a critical target. Peroxynitrite not only has the reactivity of hydroxyl radical, but will also directly attack critical cellular targets by other oxidative pathways that are of potentially greater pathological significance.

C. Historical Studies on Peroxynitrite

The formation of peroxynitrite in acidic solutions of nitrite and hydrogen peroxide was first proposed by Baeyer and Villiger (1901). However, subsequent studies erroneously proposed the structure of the oxidizing intermediate to be HNO_4, a view that persisted into the 1920s. Gleu and Roell (1929) conducted extensive studies on the reaction of ozone with alkaline azide, and concluded that peroxynitrite was the strongly oxidizing intermediate formed by this reaction (Gleu and Hubold, 1935). They deduced the correct formula for peroxynitrite and provided a wealth of information concerning its reactivity. Halfpenny and Robinson (1952a,b) showed that acidic solutions of nitrite and hydrogen peroxide catalyzed the formation of phenols, nitrophenols, and biphenol products from benzene. Such solutions also initiated the free radical-mediated polymerization of methylmethacrylate, leading Halfpenny and Robinson (1952a,b) to propose that peroxynitrous acid was decomposing to form hydroxyl radical and nitrogen dioxide. Ray (1962) showed that the enthalpy of peroxynitrite decomposition to nitrate was 38 kcal/mol. Hughes and Nicklin (1968) prepared stable solutions of peroxynitrite anion at alkaline pH. They estimated that peroxynitrite had a pK_a near 8.0 from extrapolation of its rate of decomposition in varying concentrations of NaOH. They subsequently showed that copper ions catalyze a variety of reactions with peroxynitrite, providing the first evidence for an alternative oxidative mechanism via peroxynitrite involving metal catalysis (Hughes and Nicklin, 1970; Hughes et al., 1971). Keith and Powell (1969) determined the pK_a of peroxynitrite to be approximately 6.6 at 0°C and reported that most buffers react directly with peroxynitrite at neutral to alkaline pH. Hughes and Nicklin (1968) also noted that complex reactions of peroxynitrite occurred with bicarbonate and borate buffers at alkaline pH. Mahoney provided further evidence that peroxynitrous acid decomposes to form a hydroxyl radical-like species by examining the oxidation of hydrogen peroxide to oxygen by varying concentrations of nitrite at pH 2 (Mahoney, 1970). Blough and Zafiriou (1985) showed that superoxide and nitric oxide react to give peroxynitrite in high yield by a reaction that was too rapid to permit determination of its rate.

D. Radiation Damage to Nitrate and the Viking Mars Mission

Photochemists have known for many years that irradiation of potassium nitrate crystals results in the formation of a stable yellow product, which turned out to be peroxynitrite. Continued UV exposure leads to the further decomposition to nitrite and oxygen. Addition of irradiated nitrate crystals containing about 0.3% peroxynitrite to biological molecules nicks DNA and cleaves polypeptide chains (King et al., 1992). The initially promising signs of life observed during Viking missions to Mars during the mid 1970s may have resulted from the formation of peroxynitrite from nitrate in the soil samples (Plumb et al., 1989) caused by UV radiation from the sun. Two key tests for life on the Viking missions were the evolution of carbon dioxide and oxygen. Peroxynitrite oxidatively decarboxylates amino acids to release carbon dioxide and also forms oxygen by kinetics that could explain the data obtained from the Viking missions (Plumb et al., 1989). Only the failure to detect the formation of complex carbohydrates from simple substrates prevented the conclusion that life was present on Mars. It remains to be determined whether sufficient nitrate is present in the Martian soil for the peroxynitrite mechanism to operate.

E. Complex Reactivity of Peroxynitrite

A major contributory factor to the toxicity of peroxynitrite is its unusual stability as an anion. Peroxynitrite can be stored for weeks in the freezer in 0.1 M NaOH. Even at physiological pH, the stability of peroxynitrite allows it to diffuse for a considerable distance on a cellular scale and to even cross cell membranes. But it becomes highly reactive at physiological pH by at least three distinct oxidative pathways (Fig. 22): hydrogen ion-catalyzed decomposition to form an intermediate with the reactivity of hydroxyl radical and nitrogen dioxide, direct reaction with sulfhydryl groups, and reaction with metal ions to form a potent nitrating agent resembling nitronium ion. In effect, peroxynitrite can be compared to an explosive mine with a proximity fuse that fragments into highly reactive species when in contact with critical cellular components, such as thiols and metal cofactors in enzymes. The rapid and specific reactions of peroxynitrite with sulfhydryls and metals increase its likelihood of inactivating a key cellular target. The importance of thiol and metal reactions can be readily visualized by comparing the reaction rates versus concentration of substrate (Fig. 23). The hydroxyl radical like species is formed randomly and reacts with what ever biological molecule it comes in contact with.

The different oxidative pathways are possible because the O–O bond of peroxynitrite can react as if it was either cleaved homolytically into HO· + ·NO$_2$ or heterolytically into HO$^-$ and NO$_2^+$. The third possibility of separating peroxynitrite into HO$^+$ and NO$_2^-$ is too energetically costly to be considered, requiring 105 ± 20 kcal/mol. Koppenol et al. (1992) have calculated the energetic cost of separating peroxynitrite at pH 7.0 homolytically to be 21 ± 3 kcal/mol, which compares favorably with the experimental activation energy for spontaneous decomposition of 18 ± 2 kcal/mol. The complete separation of peroxy-

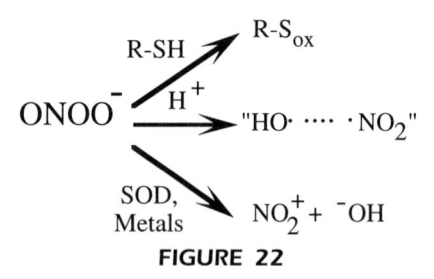

$$ONOO^- \begin{array}{c} \xrightarrow{R\text{-}SH} R\text{-}S_{ox} \\ \xrightarrow{H^+} \text{"HO·} \cdots \text{·NO}_2\text{"} \\ \xrightarrow{SOD, Metals} NO_2^+ + {}^-OH \end{array}$$

FIGURE 22

Three distinct oxidative pathways for peroxynitrite. In addition, peroxynitrite can directly rearrange to nitrate at slightly alkaline pH without acting as an oxidant (~pH 8) (Crow et al., 1994).

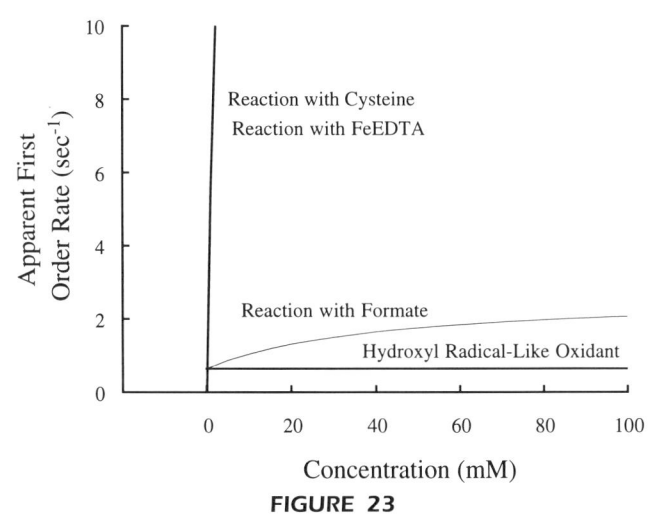

1 Physiological and Pathological Chemistry of Nitric Oxide   47

[Figure: plot of Apparent First Order Rate (sec^{-1}) vs Concentration (mM), showing curves labeled "Reaction with Cysteine", "Reaction with FeEDTA", "Reaction with Formate", and "Hydroxyl Radical-Like Oxidant"]

FIGURE 23

The effects of reactant concentrations on the rate decomposition of peroxynitrite. Peroxynitrite reacts directly with thiols and with Fe^{3+}EDTA, so the rate of decomposition increases linearly as the concentrations of thiol or Fe^{3+}EDTA are increased. The hydroxyl radical-like reactivity is controlled by peroxynitrite reaching a high energy state, so the rate of reaction is independent of substrate concentration. In a third type of reaction, the decomposition of peroxynitrite is accelerated by certain buffer anions. The rate initially rises with formate concentration but then becomes independent of buffer concentration. This behavior may be due the effects of buffer ions on water structure affecting the decomposition of trans configuration of peroxynitrite. At higher buffer concentrations, the rate of isomerization from cis to trans peroxynitrite becomes rate limiting, resulting in the plateau in reaction rate. All reaction conditions were adjusted to give pseudo-first-order kinetics and the temperature was 37°C (Beckman et al., 1992; Koppenol et al., 1992; Radi et al., 1991a,b).

nitrite into hydroxide ion (HO$^-$) and nitronium ion (NO$_2^+$) requires only 13 kcal/mol (Koppenol et al., 1992). Although the energetic cost to form nitronium ion is lower, a substantial energy barrier of as great as 45 kcal/mol exists to initially form the charge pair HO$^{\delta-}$ $\cdots$ $^{\delta+}$NO$_2$ in water. Thus, a kinetic barrier favors decomposition to the hydroxyl radical-like pathway unless metal ions are present. Metal ions such as Fe^{3+}EDTA circumvent this barrier and catalyze a nitronium-like pathway with an activation energy of 12 kcal/mol. Consequently, metal-catalyzed reactions are likely to play a significant role in peroxynitrite mediated toxicity.

The redox potentials of various oxidants derived from nitric oxide and peroxynitrite are summarized in Table 4. Clearly, as the adducts of molecular oxygen and nitric oxide become more reduced, they form substantially stronger oxidizing agents. In effect, addition of one electron makes these nitrogen oxides more ready to accept the next. The precise pathway of decomposition followed is influenced by what types of target molecules come in contact with peroxynitrite and is

TABLE 4
Reductions Potentials at pH 7.0 of Nitric Oxide, Peroxynitrite, and Other Secondary Species Derived from Peroxynitrite[a]

NO	+1 e^-	$\longrightarrow$ NO$^-$	0.39 V
ONOO·	+1 e^-	$\longrightarrow$ ONOO$^-$	0.43 V
ONOO$^-$ + 2 H$^+$	+1 e^-	$\longrightarrow$ NO$_2$· + H$_2$O	1.4 V
NO$_2$·	+1 e^-	$\longrightarrow$ NO$_2^-$	0.99 V
NO$_2^+$	+1 e^-	$\longrightarrow$ NO$_2$·	1.6 V
·OH	+1 e^-	$\longrightarrow$ HO$^-$	2.3 V

[a] These reactions are summarized from a more extensive list in Koppenol et al. (1992) and from references cited therein.

strongly dependent on pH in the region of physiological interest. The multiple mechanisms of peroxynitrite reactions will be described individually in greater detail before attempting to describe a model with supporting physical evidence for the complex reactivity of peroxynitrite.

F. Hydroxyl Radical-like Oxidative Pathway

Peroxynitrite is capable of initiating many of the reactions commonly attributed to hydroxyl radical, particularly under acidic conditions. Halfpenny and Robinson (1952a,b) showed that nitrous acid plus hydrogen peroxide in aqueous solutions at pH 2, which generates peroxynitrous acid, initiated the polymerization of methylmethacrylate (the precurser to Plexiglas) as well as the hydroxylation, nitration, and polymerization of benzene.

We initially investigated the apparent formation of hydroxyl radical from peroxynitrite by looking at the formation of formaldehyde from the oxidation of DMSO (Beckman et al., 1990). Only 25–30% of added peroxynitrite is detected as hydroxyl radical adducts (Beckman et al., 1990; Mahoney, 1970). DMSO does not accelerate the decomposition of peroxynitrite, showing that peroxynitrite does not directly react with DMSO by a second-order reaction. The yield of formaldehyde increased by a hyperbolic relationship with DMSO, requiring approximately 5 mM for 50% yield and 50–100 mM concentrations to become saturated (Fig. 24). Such high concentrations are necessary to successfully trap a species as reactive as hydroxyl radical. Other hydroxyl radical scavengers competitively inhibited the yield of formaldehyde with relative efficiency as would be predicted for reactions with hydroxyl radical generated by pulse radiolysis (Beckman et al., 1990). DMSO also increases the amount of nitrogen dioxide detectable from the decomposition of peroxynitrite from 0.2% in phosphate buffer to 20% of added peroxynitrite (Zhu et al., 1992).

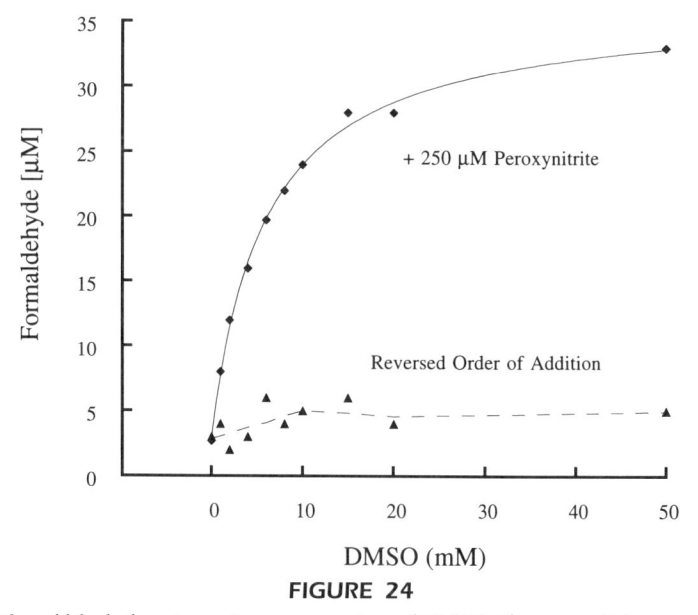

1 Physiological and Pathological Chemistry of Nitric Oxide 49

FIGURE 24

Yield of formaldehyde from increasing concentrations of DMSO. The peroxynitrite concentration was 250 μM added to 50 mM potassium phosphate, pH 7.4 at 37°C. In the reverse order of addition experiment (♦), peroxynitrite was added to phosphate buffer 3 min before the DMSO, where it would rapidly decompose.

Based on these data, we proposed that peroxynitrous acid in the trans configuration could spontaneously rupture the O–O bond homolytically into free hydroxyl radical and nitrogen dioxide (Beckman et al., 1990). The experimental enthalpy for spontaneous decomposition of peroxynitrous acid of 18 ± 2 kcal/mol is close to the theoretical energy of 21 ± 3 kcal/mol for separating peroxynitrous acid homolytically into hydroxyl radical and nitrogen dioxide (Koppenol et al., 1992). However, the assumption of complete homolytic cleavage of peroxynitrous acid has proved to be an over simplification based on more subtle kinetic and thermodynamic arguments (Koppenol et al., 1992). For example, the activation entropy is only 3 cal/mol/°K, which is substantially lower than the entropy of 13–20 cal/mol/°K typically required for homolytic radical cleavage (Benson, 1976). Homolytic cleavage generally involves a loose and floppy transition state, whereas the low activation energy for peroxynitrous acid suggests that the transition state involving peroxynitrous acid is a rigid and relatively symmetric structure (higher entropy implies greater disorder). From the activation energy and the forward rate of reaction for the hydroxyl radical-like pathway, it is possible to calculate that the rate of hydroxyl radical recombination with nitrogen dioxide will be about 10^{14} M^{-1} sec^{-1}, at least 10^2–10^3 times faster than the diffusion limit (Koppenol et al., 1992). The low activation entropy and the impossibly rapid reverse reaction suggest that the hydroxyl radical-like reactivity of peroxynitrous acid is mediated by a vibrationally excited intermediate that does not physically separate into free hydroxyl radical and nitrogen dioxide (Koppenol et al., 1992).

Instead, peroxynitrous acid appears to distort into higher energy conformations, which have the reactivity of hydroxyl radical but do not physically separate into free hydroxyl radical and nitrogen dioxide (Fig. 25). This high energy conformations result from vibrations of the N—O—O bond angle and lengthening the O—O bond angle. Semiempirical quantum mechanics calculations indicate that the lowest unoccupied molecular orbital becomes a sigma antibonding orbital[3] that is substantially lower in energy than for the corresponding orbital in *trans*-peroxynitrous acid. Thus, when the high energy intermediate oxidizes a target molecule, the electron enters the sigma antibonding orbital and promotes the homolytic cleavage of peroxynitrous acid to give ROH· + ·NO$_2$. The vibrationally excited intermediate also provides a direct pathway for isomerization of *trans*-peroxynitrous acid to nitric acid, accounting for the upper 20–30% limit of hydroxyl radical products formed by peroxynitrous acid. Augusto and Radi (1994) have trapped a hydroxyl radical adduct with DMPO from peroxynitrite by using a carefully titrated amount of glutathione to remove the nitrogen dioxide that is also formed. Nitrogen dioxide would otherwise rapidly destroy the DMPO–OH adduct.

A perplexing phenomenon has been the decrease in oxidative reactions initiated by peroxynitrite at alkaline pH. For example, the yield of formaldehyde as well as nitrogen dioxide from DMSO decreased at more alkaline pH with an apparent pK_a of about 7.9 (Fig. 26; Beckman et al., 1990). These results cannot be explained by the secondary reaction of formaldehyde with peroxynitrite as proposed by Yang et al. (1992), since formaldehyde standards were not consumed by peroxynitrite under the reaction conditions utilized (Fig. 26). A similar decrease in yield was observed with the oxidation of deoxyribose to malonyldialdehyde (Beckman et al., 1990) and with hydroxylation of phenol (H. Ischiropoulos and J. S. Beckman, unpublished data). Isomerization of the trans-geometry of peroxynitrite anion directly to nitrate by the bending of the N—O—O bond angle (the same pathway described for *trans*-peroxynitrous acid) may account for the decreasing yield of hydroxyl radical products at more alkaline pH (Fig. 27). Peroxynitrite anion, already carrying one negative charge, would not be a good oxidant because the oxidative intermediate would be required to pull a second negative charge from the reductant molecule being attacked by peroxynitrite. Thus, electrostatic repulsion would inhibit oxidation by

---

[3] A sigma bond is a molecular orbital that looks like an *s*-type atomic orbital when viewed down its axis and has cylindrical symmetry. A π bond looks like a *p*-type of atomic orbital from the same view.

1 Physiological and Pathological Chemistry of Nitric Oxide   51

$$\text{H-O-O-N=O} \rightleftharpoons \text{O-N=O (with OH)} \rightarrow \text{H-O---N=O}$$

$$\downarrow R$$

$$\text{R-O-H} \cdot + \cdot NO_2$$

FIGURE 25

The lowest unoccupied molecular orbital (LUMO) for *trans*-peroxynitrous acid and for putative hydroxyl radical-like intermediates. The LUMO is the orbital that electrons must enter when peroxynitrous acid oxidizes another molecule.

peroxynitrite anion. The apparent pK_a of 7.9 for the decrease in oxidative yield can be explained by assuming that it corresponds to the pK_a of *trans*-peroxynitrous acid. This pK_a is slightly higher than the apparent pK_a of 6.8 measured by spontaneous decomposition in phosphate buffer (Keith and Powell, 1969; Radi *et al.*, 1991a,b; Yang *et al.*, 1992), which we attribute to the pK_a of the cis anion. Further experimental and theoretical evidence for the existence of two pK_a values will be presented below.

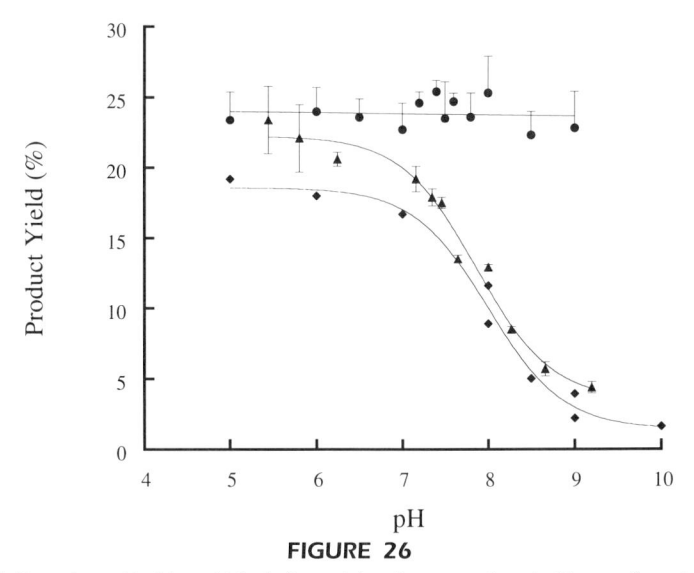

FIGURE 26

Effect of pH on the yield of formaldehyde (triangle) and nitrogen dioxide (diamond) produced from peroxynitrous acid attack upon 100 mM DMSO. All reactions were conducted at 37°C in 50 mM potassium phosphate buffer. The pH was measured after the addition of peroxynitrite to account for the alkaline peroxynitrite addition. Peroxynitrite did not attack 24 μM formaldehyde (circle) under the reactions conditions utilized.

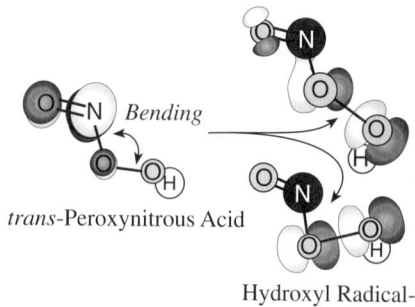

FIGURE 27
Isomerization of *trans*-peroxynitrite anion and *trans*-peroxynitrous acid to nitrate through the putative hydroxyl radical-like intermediates.

G. Formation of a Nitronium-like Species Catalyzed by Metals

Although hydroxyl radical is commonly assumed to be the most toxic of the oxygen radicals (with little direct evidence), other direct reactions are more likely to be important for understanding the cytotoxicity of peroxynitrite. A second oxidative pathway involves the heterolytic cleavage of peroxynitrite to form a nitronium-like species (NO_2^+), which is catalyzed by transition metals (Beckman et al., 1992). Low molecular weight metal complexes as well as metals bound in superoxide dismutase and other proteins catalyze the nitration of a wide range of phenolics, including tyrosine residues in most proteins (Beckman et al., 1992).

Transition metals like $Fe^{3+}EDTA$ apparently catalyze the formation of a nitronium ion-like species from peroxynitrite by the following reactions

$$ONOO^- + Fe^{3+}EDTA \longrightarrow NO_2^{\delta+}—O^{\delta-}—Fe^{3+}EDTA \quad \text{(Reaction 34)}$$

$$phenol + NO_2^{\delta+}—O^{\delta-}—Fe^{2+}EDTA \longrightarrow$$
$$NO_2\text{-phenol} + H^+ + O^-—Fe^{2+}EDTA \quad \text{(Reaction 35)}$$

$$2H^+ + O^-—Fe^{2+}EDTA \longrightarrow H_2O + Fe^{3+}EDTA \quad \text{(Reaction 36)}$$

The δ symbol is used to describe that only a partial separation of charge has occurred. In the first step, the negatively charged peroxynitrite anion will be electrostatically attracted to $Fe^{3+}EDTA$ to form an intermediate complex. Electron density in the peroxynitrite–$Fe^{3+}EDTA$ complex will be pulled away from the nitrogen toward the iron, favoring heterolytic cleavage to give a nitronium-like species that attacks phenols. The peroxynitrite–$Fe^{3+}EDTA$ complex may directly react with phenol without nitronium ion being physically separated from the complex. The O^-–$Fe^{2+}EDTA$ will rapidly add two hydrogen ions from the solvent to release water and regenerate Fe^{3+}EDTA (Reaction 36). The energy calculated from thermodynamics for heterolytic cleavage of peroxynitrite into nitronium ion and hydroxide ion is estimated to be about 13 kcal/mol in water at pH 7.0 (Koppenol et al., 1992).

$$ONOOH \longrightarrow NO_2^+ + OH^- \quad 13 \pm 2 \text{ kcal/mol} \quad \text{(Reaction 37)}$$

This compares favorably with the 12 kcal/mol activation energy measured experimentally for Fe^{3+}EDTA-catalyzed nitration (Beckman et al., 1992).

Metal-catalyzed nitration by peroxynitrite also provides an alternative explanation to the Haber–Weiss reaction for the role of transition metals in oxidative tissue injury. The rate of peroxynitrite reaction with Fe^{3+}EDTA is 5700 M^{-1} sec^{-1}, which is in the same range as the rate of hydrogen peroxide reacting with Fe^{2+}EDTA (7000 M^{-1} sec^{-1}) (Beckman et al., 1992). Moreover, the reaction with peroxynitrite does not require that iron first be reduced by superoxide or another reductant to be toxic.

H. Pathological Implications of Tyrosine Nitration

Nitration of tyrosine residues by tetranitromethane is well known to alter protein function, including cytochrome P450 (Janing et al., 1987), α-thrombin (Lundblad et al., 1988), and mitochondrial ATPase (Guerrieri et al., 1984). Phosphorylation of tyrosines plays a critical role in cell regulation and is an important target damaged by nitration (Martin et al., 1990). Treatment with tetranitromethane inactivates the complement component C1q binding capacity of human IgG (McCall and Easterbrook-Smith, 1989), abolishes the inhibitory activity of human α_1-protease inhibitor for elastase (Mierzwa and Chan, 1987), and inhibits the binding of human high-density lipoprotein (HDL3) to liver plasma membranes (Chacko, 1985). Nitration of polycyclic aromatics found in cigarette tar makes them some of the most mutagenic compounds known. Coupling of dinitrophenol to proteins is commonly used to increase the antigenicity of proteins. Thus, endogenous nitration of cellular proteins might induce inflammatory reactions characteristic of autoimmune diseases. Because activated macrophages and potentially many other cell types can produce peroxynitrite (Ischiropoulos et al., 1992a), metal-catalyzed nitration could be a major pathological mechanism of tissue injury. Furthermore, nitrotyrosine has been detected in human urine (Ohshima et al., 1990), indicating that such nitrating reactions occur in vivo. We have developed monoclonal and polyclonal antibodies to nitrotyrosine and shown extensive nitration within human atherosclerotic lesions, particularly around foam cells (Beckman et al., 1994). Extensive nitration was also found around alveolar inflammatory cells in patients with respiratory distress syndrome, pneumonia, and sepsis (Kooy et al., 1994).

I. Sulfhydryl Oxidation

Peroxynitrite oxidizes directly both low molecular weight sulfhydryl groups and sulfhydryls in proteins (Radi et al., 1991b). The rate of peroxynitrite decomposition increases in direct proportion to sulfhydryl concentration, indicating that the principal pool of peroxynitrite, which is in the cisgeometry, directly attacks the sulfhydryl group by a simple second-order reaction. The only other direct reaction of cisperoxynitrite anion so far identified is with low molecular weight metal complexes. Since no rate-limiting isomerization must take place before peroxynitrite can oxidize a sulfhydryl group, peroxynitrite is a relatively selective oxidant of cellular sulfhydryls. Sulfhydryl agents such as cysteine or glutathione are scavengers of peroxynitrite, though moderately high concentrations (0.1–10 mM) of thiol are necessary because the reactions rates with peroxynitrite are relatively slow ($2-6 \times 10^3$ M^{-1} sec^{-1}).

J. Reaction with Buffer Anions

Several investigators have noted the direct reaction of peroxynitrite with many buffer anions at neutral to alkaline pH (Hughes and Nicklin, 1970; Keith and Powell, 1969). We have found that many common buffer anions, such as formate, will accelerate the decomposition of peroxynitrite (Fig. 28). The mechanism is unknown, but may involve perturbations of water structure by anions.

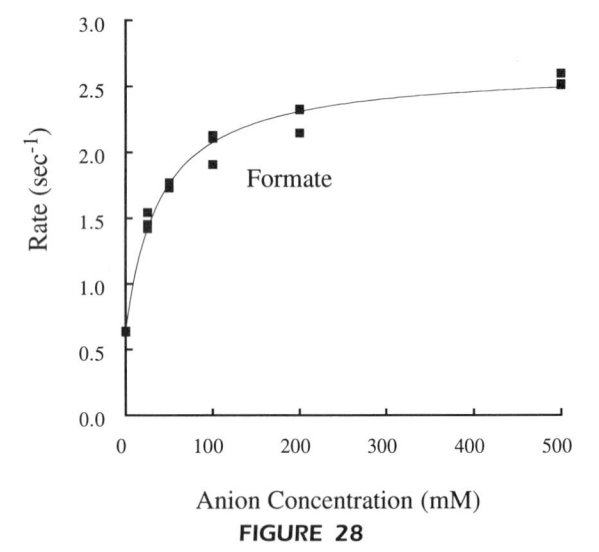

FIGURE 28

The effect of buffer anions on peroxynitrite decomposition. As the concentration of formate is increased, the rate of peroxynitrite decomposition increases but reaches a plateau at high concentrations. The maximum rate depends on the pH.

Curiously, the reaction rate becomes zero order with respect to buffer anion at higher buffer concentrations. If the zero-order rate constants for a variety of buffers are plotted versus pH, an apparent pK_a is observed at pH 7.9 to 8.0 (Fig. 29). In contrast, the apparent pK_a observed in phosphate buffer is at 6.8. As described below, we propose that the difference in pK_a values is due to a slight difference in the Gibbs energies of the cis and trans peroxynitrite anions relative to the corresponding conjugate acid and has important consequences concerning the reactivity of peroxynitrite.

K. pH Dependence of Peroxynitrite Oxidation

The yield of products by the different oxidative mechanisms from peroxynitrite are strongly dependent on pH (Fig. 30). For example, the yield of expected hydroxyl radical products from dimethylsufoxide decreases with an apparent pK_a of approximately 7.8. The rate of sulfhydryl oxidation decreases with an apparent

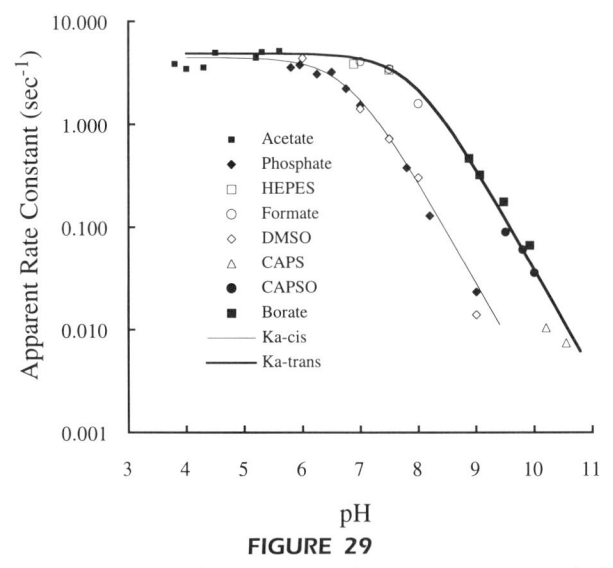

FIGURE 29

Apparent first-order rate constants for peroxynitrite decomposition in various buffers versus pH. When peroxynitrite is fully protonated at acidic pH, the decomposition rate is constant. The breakpoint in the curve identifies the pK_a of peroxynitrite since a larger fraction present as an anion slows the rate of decomposition. In 50 mM potassium phosphate, the apparent pK_a is at 6.8 and is not affected by temperature (Koppenol, 1993). The rate of decomposition is not affected by DMSO, mannitol, or ethanol. As shown in Fig. 28, many buffers can slightly accelerate the decomposition of peroxynitrite and the rate of decomposition reaches a maximum at high buffer concentrations. When these maximal rates are plotted as a function of pH, peroxynitrite exhibits a second pK_a of approximately 8.0.

FIGURE 30

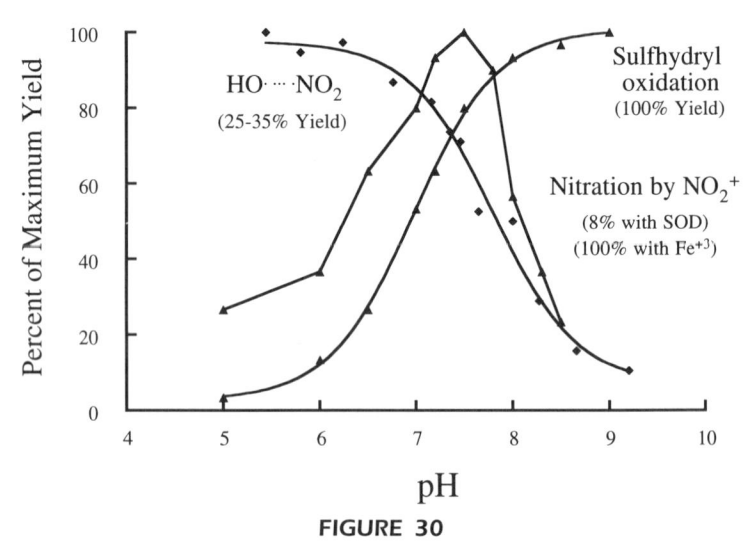

The oxidative reactivity of peroxynitrite as a function of pH. The data are plotted as the percentage of the maximal yield under optimal conditions. Data were originally reported in Beckman et al. (1990, 1992); Radi et al. (1991a,b).

pK_a of 6.8, consistent with the pK_a of peroxynitrite determined by its spontaneous decomposition rate (Radi et al., 1991b). The yields of lipid peroxidation products as well superoxide dismutase-catalyzed nitration were maximal at pH 7.5 and decreased with apparent pK_a values at 6.8 and at 7.9–8.0 (Radi et al., 1991b). All three oxidative pathways occur at physiological pH and are important for understanding the toxicity of peroxynitrite. We have proposed that the complex oxidative behavior appears to be controlled by subtle differences in the reactivity for differing conformations of peroxynitrite (Beckman et al., 1992).

L. cis Conformation of Peroxynitrite

Peroxynitrite anion appears to be present only in the cis conformation in strongly alkaline solution, which is important for understanding why such a strong oxidant can be stored for weeks. Cis-peroxynitrite is 36 kcal/mol higher in energy than its isomer nitrate. In the cis conformation, the negative charge on the terminal oxygen is partially delocalized over the entire peroxynitrite molecule, forming weak partial bonds with the distal nitrogen and oxygen with the strength of a typical hydrogen bond (Fig. 31). This stabilization results in peroxynitrite having the lowest pK_a of any hydrogen peroxide-containing compound (Edwards and Plumb, 1994). More importantly, the delocalization of the negative charge also partially locks peroxynitrite into the cis conformation by inhibiting

1 Physiological and Pathological Chemistry of Nitric Oxide   57

cis-Peroxynitrite

trans-Peroxynitrite

FIGURE 31
Highest occupied molecular orbitals for *cis*- and *trans*-peroxynitrite anion. The orbitals were calculated with the AM1 parameter set using MOPAC 6.0 and the CACHe interface (Textronix Inc., Beaverton, OR).

the terminal oxygen from rotating around the OO—NO bond (torsion) or bending of the N—O—O⁻ bond angle to the trans configuration. The terminal oxygen cannot directly attack the nitrogen to form nitrate, even though this is extremely favorable (−36 kcal/mol). Contracting of the O—O—N bond angle produces a strong repulsion between the terminal peroxide oxygen and the two oxygens bound to the nitrogen (Fig. 32).

In the trans configuration, shortening of the O—O—N bond angle coupled with a slight lengthening of the O—O bond will allow the terminal peroxide oxygen to directly attack the nitrogen to give nitrate (Fig. 27). Consequently, any *cis*-peroxynitrite that isomerizes to the trans configuration can decompose rapidly to nitrate.

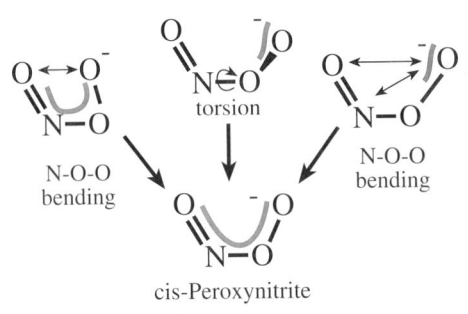

cis-Peroxynitrite
FIGURE 32
Stability of the cis conformation. There is no vibration which can allow *cis*-peroxynitrite to directly isomerize to nitrate. Each vibration that distorts the cis conformation is resisted because the delocalization of the negative charge is disrupted.

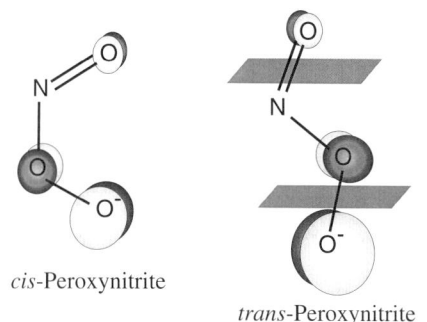

Physical studies of peroxynitrite in alkaline solution indicate only one geometry is present, which appears to be the cis conformation (Tsai, 1991). The ^{15}N-NMR spectra shows only a single peak from peroxynitrite in 0.3 M NaOH (Fig. 33), whereas two peaks should be observed if peroxynitrite was present in both the cis and trans forms. These results were supported by the Raman spectra of peroxynitrite (Fig. 34). Raman spectroscopy measures the slight shift in wavelength of laser light scattered by internal vibrations of the peroxynitrite molecule. Raman spectra are analogous to infrared spectra, but can be recorded in aqueous solutions. Because peroxynitrite has four atoms, each conformation generates only six possible vibrational modes (Fig. 35). For a nonlinear atom, there are $3N - 6$ possible bands where N is the number of atoms. The slight shifts in vibrational frequencies cause by using ^{15}N- and ^{18}O-substituted peroxynitrite appear to be better predicted by assuming that peroxynitrite in alkaline solution is present in only the cis geometry (Tsai et al., 1994; Tsai, 1991).

Usually, Raman peaks are quite narrow as observed for the nitrate peak at 1000 cm^{-1}. The peroxynitrite peaks are all broadened by strong interactions with water. The Raman peak for torsion around the OO–NO bond at 640 cm^{-1} is exceptionally broad and strong, which appears to result from the combination of the high energy required to rotate from the cis to the trans geometry and solvent interactions (Tsai, 1991). The negative charge on the terminal oxygen appears to be partially delocalized over all four atoms in peroxynitrite. As the dihedral angle is changed by the torsion around the OO–NO bond, the negative charge

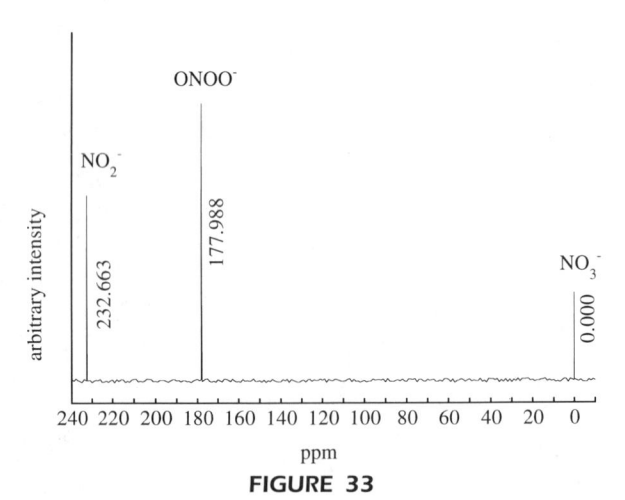

FIGURE 33

Nuclear magnetic resonance spectrum of ^{15}N-labeled peroxynitrite in 0.1 M NaOH relative to nitrite and nitrate.

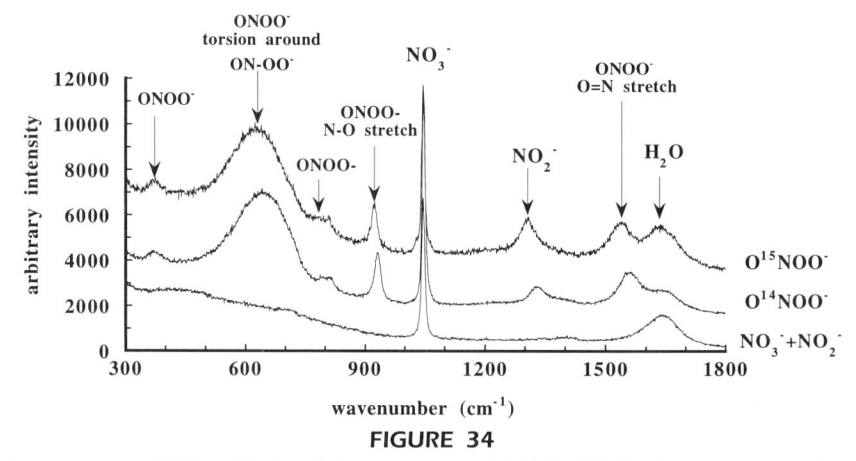

1 Physiological and Pathological Chemistry of Nitric Oxide   59

FIGURE 34

Raman spectra of ^{14}N and ^{15}N-labeled peroxynitrite in 0.3 M NaOH. The bottom spectra is for a sample of ^{14}N-labeled peroxynitrite that has decomposed after adding a small amount of hydrochloric acid. The O[14]NOO spectra revealed only five bands with a sixth band apparently obscured by the predominant nitrate peak. The sixth band can be resolved in solid state spectra (Tsai et al., 1994).

is forced onto the terminal oxygen, greatly changing the polarizability of the molecule. This contributes to the large Raman peak at 640 cm^{-1}, because the intensity of a Raman peak is proportional to a change in polarizability of the molecule. Strong interactions with water also broaden the torsional band, because it is progressively sharpened when water is replaced by methanol, ethanol, or propanol.

Quantum mechanical calculations suggest that the cis conformation is lower in energy because of interactions between the terminal peroxide oxygen with the distal oxygen (Fig. 31). In the trans anion, these interactions are weaker because

FIGURE 35

Vibrational modes of peroxynitrite anion observable by Raman spectroscopy. A nonlinear molecule with four atoms can have only six independent modes. These consist of three bond stretching modes, two bending modes, and torsion of the OO–NO bond, which will make peroxynitrite nonplanar.

the terminal oxygens cannot interact. The implications are that peroxynitrite anion is locked into a relatively stable cis conformation. The energy barrier for rotation from the cis to the trans configuration may be as high as 25 kcal/mol, based on semiempirical quantum mechanical calculations (Fig. 36). Such calculations further indicate that the barrier for rotation about the O–O bond is significantly smaller for peroxynitrous acid compared to the anion because the negative charge on the anion limiting rotation is neutralized by the hydrogen (Fig. 36).

Nevertheless, peroxynitrous acid (ONOOH), the conjugate acid of peroxynitrite, is also more stable in the cis form by 1–2 kcal/mol (Cheng et al., 1991; McGrath et al., 1988).[4] Infrared spectroscopy of peroxynitrous acid trapped in an argon matrix also indicated that peroxynitrous acid is also present only in the cis conformation (Cheng et al., 1991). The trans geometry may be more reactive than the cis geometry, since the hydroxyl radical-like oxidant appears to be derived from *trans*-peroxynitrous acid (Fig. 25) (Koppenol et al., 1992). Because of its higher pK_a, *trans*-peroxynitrous acid is relatively stabilized at neutral pH with respect to *cis*-peroxynitrous acid. If the attacking species is derived by *trans*-peroxynitrous acid, the slight difference in pK_a values between the cis and trans geometries would explain the increase in nitration (Fig. 30) and lipid peroxidation (Radi et al., 1991a) observed at pH 7.5. Thus, subtle shifts in pH in the physiological range greatly influences the reactivity of peroxynitrite.

M. Nitric Oxide As a Scavenger of Superoxide and Hydroxyl Radical

In the preceding discussion, the oxidative potential of peroxynitrite has been emphasized. Yet, several authors have claimed that nitric oxide may also serve as an antioxidant. Clearly, the reaction of superoxide with nitric oxide leads to the formation of a much stronger oxidant. However, nitric oxide does react at the diffusion limit with hydroxyl radical to form nitrous acid. The hydroxyl radical has an extremely short diffusion distance, on the order of 3 nanometers, as shown in Fig. 21. The concentration of nitric oxide would have to be extremely high, in the millimolar range, to effectively compete for hydroxyl radical. In the experimental paradigm used by Kanner et al. (1991), the nitric oxide concentration was 200 μM and kept anaerobic during the assays. Clearly, the reactions of this amount of nitric oxide with oxygen in a biological setting would lead to cytotoxicity from the formation of nitrogen dioxide. Nitric oxide can also bind to transition metals and slow their catalytic role in driving the Haber–Weiss

[4] There is a 3 kcal difference in energy between the energy difference shown in Fig. 38 and the McGrath results. The calculations used for Fig. 38 are based on a lower level, semiempirical quantum mechanics calculation, while the McGarth calculations utilize computationally more expensive *ab initio* methods.

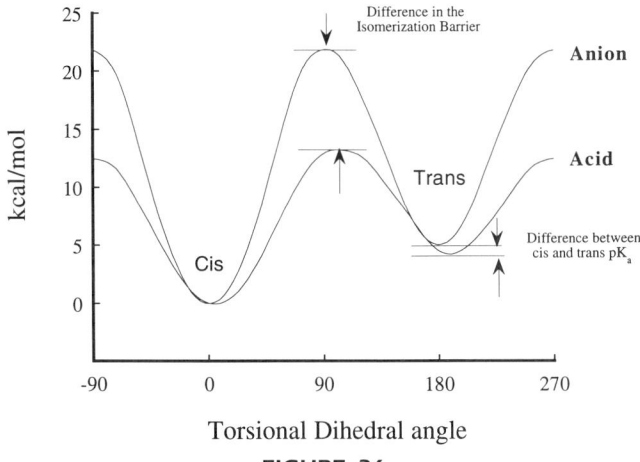

1 Physiological and Pathological Chemistry of Nitric Oxide 61

Torsional Dihedral angle

FIGURE 36

Energy curve for the isomerization of the cis and trans geometries of peroxynitrite anion and peroxynitrous acid. The energies were calculated with MOPAC 6.0, using the AM1 parameter set. For comparison, the energy of the cis anion and acid are set to be equal as would occur at the pK_a of 6.8. The barrier for isomerization is substantially lower for the acid form compared for the anion. In the trans geometry, the cis anion is slightly higher in energy than the acid. Thus, ionization of *trans*-peroxynitrous acid will require slightly greater energy than the cis geometry, resulting in a slightly greater pK_a for the trans form. The energy difference of 1.0 kcal/mol predicts the difference in pK_a values should be 0.7 pH units, while experimentally the difference is approximately 1.2 pH units. The energy minima for peroxynitrous acid is not precisely at 0° or 180° because the ON–OO atoms do not lie exactly in the same plane. The proton projects at a 90° angle from the plane, which causes a slight distortion of about 5° in the torsion angle between ON–OO (McGrath et al., 1988). For the anion, all four atoms of $ONOO^-$ lie in the same plane.

formation of hydroxyl radical in the test tube. However, nitric oxide also reduces ferric iron, and therefore could replace superoxide in the first step in the Haber–Weiss reaction. Furthermore, nitric oxide is converted to a nitrosonium ion-like species, forming nitrosamines and nitrosophenolics (Wade and Castro, 1990). Thus, the conclusions about whether nitric oxide is an oxidant or antioxidant may be strongly influenced by the assays utilized. For example, the effects of nitric oxide on oxidation of low density lipoproteins was markedly influenced by experimental conditions (Dee et al., 1991). When nitric oxide is present, the yield of thiobarbiturate products might be reduced, while the yield of cytotoxic nitrosation and nitration reactions may be increased.

The term free radical scavenger is often loosely used with little consideration of subsequent reactions. Virtually all organic molecules are scavengers of reactive species like hydroxyl radical. To be useful at reducing oxidative injury, a scavenger must react rapidly with longer lived oxidants and should not be converted to a more reactive species after trapping an oxidant. Thus, nitric oxide would not be a good candidate as a scavenger of superoxide.

N. Reaction of Peroxynitrite with Superoxide Dismutase

Peroxynitrite reacts with the active site of superoxide dismutase (SOD) to form a nitronium-like species (Fig. 37), analogous to the Fe^{3+}EDTA reactions described earlier. However, copper in the active site of superoxide dismutase was necessary for the formation of the adduct. Removing copper from the active site by reduction with borohydride and dialysis against 50 mM KCN resulted in no adduct being formed, while restoration of copper to the active site gave back full enzyme activity. To account for the essential role of copper in the active site and the subsequent formation of 3-nitrotyrosine located 18–21 Å distal from the active site, we proposed that peroxynitrite is attracted by the same electrostatic force field that draws superoxide into the active site (Beckman et al., 1992; Ischiropoulos et al., 1992b). Peroxynitrite appears to bind to copper in the active site to form a transient cuprous adduct as shown.

$SOD-Cu^{2+}--^-O-O-N=O \longrightarrow$

$\qquad SOD-Cu^{1+}O^-\!-\!O=N^+\!=\!O$ (Reaction 38)

$SOD-Cu^{1+}O^-\!-\!O=N^+\!=\!O + phenol \longrightarrow$

$\qquad SOD-Cu^{2+} + NO_2-phenol + HO^-$ (Reaction 39)

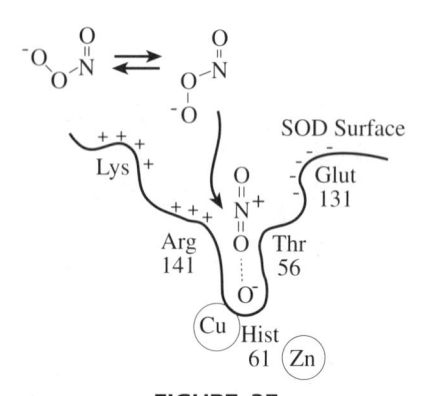

FIGURE 37

Reaction of *trans*-peroxynitrite with superoxide dismutase. The placement of positively charged amino acids around the active site facilitates the attraction of the negatively charged peroxynitrite anion. Because the copper of superoxide dismutase is buried in a pocket shaped to accommodate superoxide, only peroxynitrite in the trans configuration should be able to fit in the active site. Because the predominant form of peroxynitrite is the cis form, the isomerization from the cis to trans geometry limits the reaction of peroxynitrite with superoxide dismutase.

The peroxynitrite bound in superoxide dismutase will then attack phenol via a nitronium-like intermediate to form a nitrophenol. In the absence of exogenous phenolics, peroxynitrite bound to superoxide dismutase will slowly nitrate tyrosine residue 108 on a second superoxide dismutase molecule (Ischiropoulos et al., 1992b; Smith et al., 1992). Cu,Zn-Superoxide dismutase is not inactivated by this reaction and continues to act catalytically after modification. Both manganese and iron superoxide dismutase is slowly inactivated by reaction with superoxide dismutase, which may be due to nitration of a tyrosine in the active site. The overall rate of nitration catalyzed by superoxide dismutase of low molecular weight phenolics is among the most rapid reaction we have observed with peroxynitrite, occurring at about 10^5 M^{-1} sec^{-1} (Beckman et al., 1992). However, the reaction rate becomes saturated at higher superoxide dismutase concentrations with only 8-9% of added peroxynitrite resulting in detectable nitration of phenols. Apparently, superoxide dismutase can only react with peroxynitrite in the trans configuration, so the reaction can become limited by the rate of cis-to-trans isomerization (Fig. 37) (Beckman et al., 1992).

O. Superoxide Dismutase and Amyotrophic Lateral Sclerosis

A series of autosomal dominant mutations in the Cu, Zn-superoxide dismutase gene in humans has been linked with familial amyotrophic lateral sclerosis (Rosen et al., 1993). A total of 25 mutations at 16 distinct positions have now been identified. All of the mutations are missense and none are nonsense, arguing that the mutations have a gained function. We have proposed that these mutations act by increasing nitration rather than simply by diminishing superoxide scavenging (Beckman et al., 1993). The disease is characterized by the selective loss of motor neurons, which are the largest neurons in the body. The selectivity of the disease may result from selective nitration of either a key tyrosine kinase coupled to a growth factor receptor or by nitration of a key cytoskeletal element such as neurofilaments.

P. Superoxide Dismutase as a Probe for Peroxynitrite

The reaction of superoxide dismutase with peroxynitrite can be a useful assay for peroxynitrite *in vivo*, even though it traps only 8-9% of the available peroxynitrite. However, superoxide dismutase can also inhibit the formation of peroxynitrite by directly scavenging superoxide before it reacts with superoxide dismutase. Cu,Zn-Superoxide dismutase can be chemically modified to circumvent this competing reaction. Greater than 99% of the superoxide scavenging activity of superoxide dismutase can be inactivated by the combination of cleaving one histidine in the active site with H_2O_2 and coupling phenylglyoxal to the arginine 141. Both of these modifications inhibit superoxide dismutation (Beyer et al.,

1987), but do not significantly affect the reaction with peroxynitrite (Ischiropoulos et al., 1992b).

Q. Macrophages Produce Peroxynitrite

Inflammatory mediators such as lipopolysaccharide, interferon-γ, and tissue necrosis factor activate macrophages to synthesize nitric oxide from arginine via the inducible nitric oxide synthase (Nathan and Hibbs, 1991). Inhibition of nitric oxide synthesis with the competitive inhibitor N-methylarginine limits the microbicidal and tumoricidal activities of macrophages (Albina et al., 1989). Activated macrophages also generate superoxide by a membrane-bound NADPH oxidase making it reasonable to expect that activated inflammatory cells could be producing peroxynitrite. N-Methylarginine increases detectable superoxide, suggesting that nitric oxide is reacting with superoxide to produce peroxynitrite (Albina et al., 1989).

We have used the nitration of a tyrosine analog with superoxide dismutase to measure the peroxynitrite production from activated rat alveolar macrophages (Fig. 38). The estimated rate of peroxynitrite synthesis was estimated to be 0.1 nmol/10^6 cells/min (Ischiropoulos et al., 1992a). The rate of nitration was the same whether native Cu,Zn-superoxide dismutase or the phenylglyoxyl–H_2O_2 modified superoxide dismutase (which is >99% inhibited with respect to its superoxide scavenging activity) was used (Fig. 39). Three other independent but indirect estimates of peroxynitrite formation were consistent with the superoxide

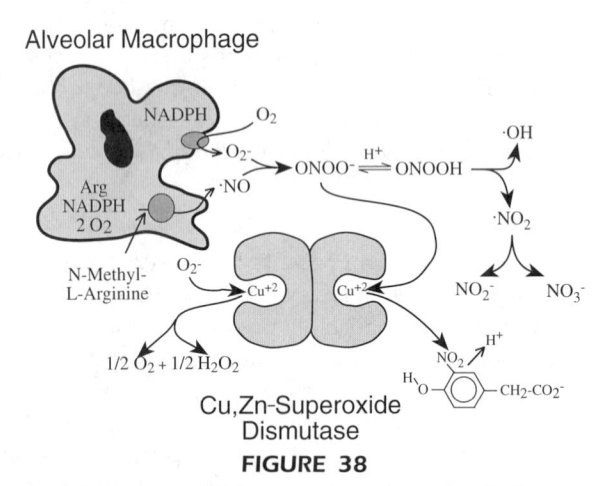

FIGURE 38

Trapping of peroxynitrite from rat alveolar macrophages by superoxide dismutase. Although the formation of peroxynitrite is drawn as superoxide reacting with nitric oxide in the extracellular space, the actual reactions may be a combination of the pathways shown in Fig. 40.

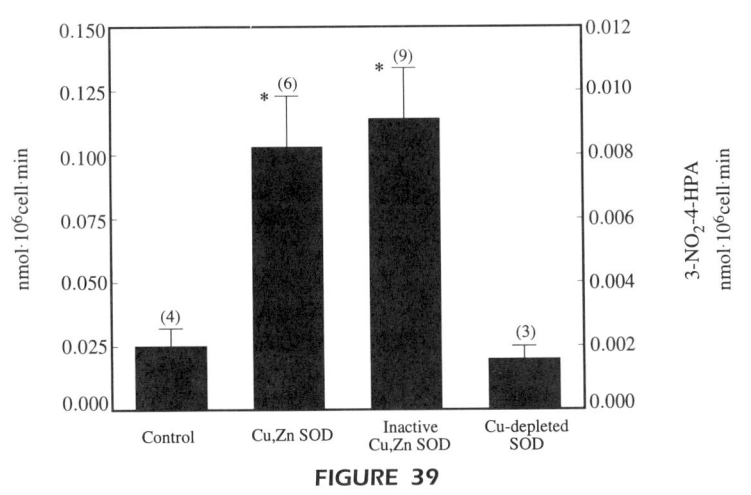

1 Physiological and Pathological Chemistry of Nitric Oxide 65

FIGURE 39

Superoxide dismutase-catalyzed nitration of 4-hydroxyphenylacetate by alveolar rat macrophages. Reprinted from Ischiropoulos et al. (1992a).

dismutase-based measurement (Table 5). Inhibition of nitric oxide synthesis with N-methylarginine increased the amount of superoxide detected by the superoxide dismutase-inhibitable cytochrome c reduction by 0.12 ± 0.02 nmol/10^6 cells/min. Under the experimental conditions utilized for these studies, the macrophages produced four times more superoxide than nitric oxide. Methylarginine also depressed phorbol ester-stimulated oxygen consumption by 0.24 ± 0.03 nmol/10^6 cells/min, but had no effect on nonstimulated oxygen consumption.

Inhibition of nitric oxide synthesis with methylarginine would be expected to decrease oxygen consumption due to peroxynitrite formation by a factor of 2.5, because two moles of oxygen are apparently consumed per mole of nitric oxide formed (Stuehr et al., 1991) while an additional oxygen is consumed to produce superoxide. If superoxide does not react with nitric oxide, it will regenerate half a mole of oxygen by dismutation. A further indication of peroxynitrite formation was the increase percentage of nitrate relative to nitrite after exposure to phorbol myristate ester (before 30% versus 67% after phorbol treatment). Nitric oxide in dilute solution without a source of superoxide predominantly decays to give nitrite (Ignarro, 1990; Ignarro et al., 1993). Peroxynitrite decomposes principally to nitrate in buffer, though it can also yield nitrite when oxidizing other molecules (Hughes et al., 1971). The stable decomposition products of nitric oxide, nitrite (NO_2^-) and nitrate (NO_3^-), accumulated at a rate of 0.10 ± 0.01 nmol/10^6 cells/min in activated macrophages. Thus, apparently all of the nitric oxide produced by rat alveolar macrophages stimulated with phorbol esters is converted into peroxynitrite.

TABLE 5
Production of Nitrite, Nitrate, Superoxide, and Oxygen Consumption by Macrophages[a]

	Control	+PMA −NMA	+PMA +1 mM NMA	Difference with and without NMA
$NO_2^- + NO_3^-$	0.06 ± 0.01	0.10 ± 0.01	0.003 ± 0.001	0.10[b]
% NO_3^-	35 ± 11	62 ± 7	—	—
Superoxide production	0	0.37 ± 0.04	0.49 ± 0.04	0.12[b]
Oxygen consumption	1.28 ± 0.11	1.96 ± 0.20	1.71 ± 0.20	0.25 ± 0.05[c]

[a] Macrophages were simulated with phorbol 12-myristate 13-acetate, and nitric oxide synthesis was inhibited with N-methyl-L-arginine. Percent NO_3^- represents the percentage of NO_3^- compared to $NO_2^- + NO_3^-$. Results are expressed as nmol/10⁶ cells/min ± STD (standard deviation), n = 3–7 from at least three separate macrophage preparations pooled from 2–4 rats.
[b] Significantly different (p < 0.05) by one-way analysis of variance (ANOVA) with the Least Significant Difference *post hoc* test using JMP program from SAS, Cary, NC on a MacIntosh. Reproduced with permission from (Ischiropoulos et al., 1992a).
[c] The standard deviation was calculated from averaging the differences from the same macrophage preparations from experiments conducted on three separate days.

R. Four Routes That Yield Peroxynitrite

The direct reaction of superoxide with nitric oxide is only one of at least four possible pathways that can form peroxynitrite (Fig. 40). For example, superoxide should also efficiently reduce nitrosyldioxyl radical to peroxynitrite. Alternatively, nitric oxide may be reduced to nitroxyl anion, which reacts with oxygen to form peroxynitrite. Superoxide dismutase could even catalyze the formation of peroxynitrite, since reduced (Cu^{1+} or cuprous) superoxide dismutase can reduce nitric oxide to nitroxyl anion (Murphy and Sies, 1991). Thus, superoxide might first reduce superoxide dismutase to the cuprous form, with nitric oxide reacting with reduced superoxide dismutase to produce nitroxyl anion. A fourth pathway to form peroxynitrite is by the rapid reaction of nitrosonium ion (NO^+) with hydrogen peroxide. This is a convenient synthetic route for experimental studies (Reed et al., 1974), but not likely to be physiologically relevant due to the low concentrations of hydrogen peroxide and the difficulty of oxidizing nitric oxide to nitrosonium ion.

S. Toxicity of Peroxynitrite

Peroxynitrite is a potent bactericidal agent for *E. coli*, with an LD_{50} of 250 μM at pH 7.4 and 37°C (Zhu et al., 1992). Neither 1 mM H_2O_2 nor 10 mU/ml of xanthine oxidase plus xanthine were toxic in the same system. The toxicity was not due to the hydroxyl radical-like reactivity because lipid soluble radical scavengers either had no effect or slightly increased toxicity of peroxynitrite (Zhu

FIGURE 40

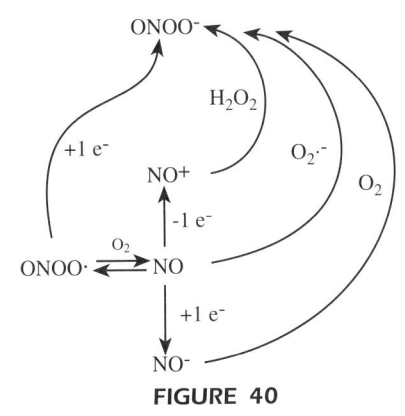

Four routes to form peroxynitrite from nitric oxide. The reaction of nitric oxide with superoxide is only one mechanism leading to the formation of peroxynitrite. Superoxide could also reduce the nitrosyldioxyl radical. If nitric oxide is directly reduced to nitroxyl anion, it will react with molecular oxygen to form peroxynitrite. At acidic pH, nitrite may form nitrous acid and nitrosonium ion, which reacts with hydrogen peroxide to form peroxynitrite.

et al., 1992). At first glance, this concentration of peroxynitrite may seem high, but one must consider toxicity in terms of concentration × time (Fig. 41). If spontaneous decomposition of peroxynitrite is the principal reaction and proceeds at a rate of k_1, then the concentration of peroxynitrite at time t is $[ONOO^-]_0\, e^{-k_1 t}$. If this expression is integrated from time zero to infinity, the exposure to peroxynitrite is $[ONOO^-]_0/k_1$, which has units of concentration × time. At pH 7.4 and 37°C, we have measured k_1 to be 0.65 sec^{-1}, which implies that exposure to a 250 μM concentration was equivalent to exposure to a steady-state concentration of 1 μM peroxynitrite for 6 min (Fig. 41). Thus, exposure to low steady-state concentrations of peroxynitrite for a few minutes could produce substantial toxicity.

T. Recent Observations with Peroxynitrite

Moreno and Pryor (1992) have shown that peroxynitrite inactivates α_1-antiprotease by oxidizing a critical methionine. This protease inhibitor is important for protecting the lung from neutrophil dependent proteases. Peroxynitrite damages both the lipid and protein components of pulmonary surfactant (Haddad et al., 1994a,b), which is essential for proper gas exchange in the lung. Peroxynitrite also can damage amelioride-sensitive sodium channels (Bauer et al., 1992), which are important for preventing pulmonary edema. Peroxynitrite also selectively damages oxygen uptake in pulmonary type II cells before affecting trypan blue exclusion (Hu et al., 1994). We have already demonstrated that activated rat macrophages can produce peroxynitrite (Ischiropoulos et al., 1992a).

FIGURE 41

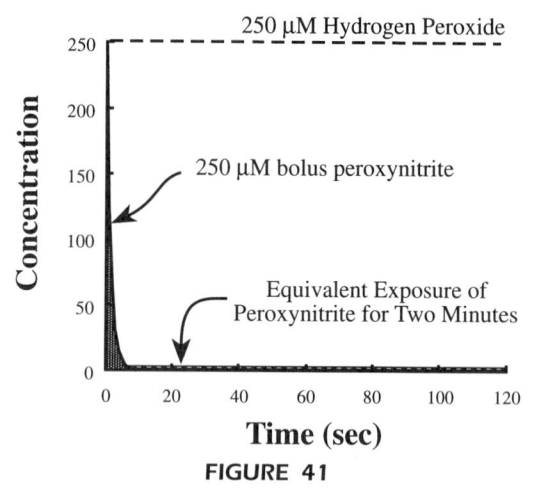

Lifetime of peroxynitrite versus hydrogen peroxide. Peroxynitrite decomposes rapidly at physiological pH with a half-life of less than a second in phosphate buffer. In biological media, the half-life is even shorter. A better comparison is to use the area under each curve as a measure of exposure.

Human neutrophils have also been shown to produce peroxynitrite (Carreras et al., 1994a,b). We have observed extensive nitration in autopsy samples from septic and adult respiratory distress syndrome (ARDS) patients using antibodies specific for nitrotyrosine. These results strongly suggest that peroxynitrite is an important mediator of pulmonary inflammation (Mulligan et al., 1991, 1992a,b). Peroxynitrite can also induce a massive colitis in rat colon (Rachmilewitz et al., 1993). Blockade of nitric oxide synthesis also is known to greatly attenuate colitis resulting from chronic inflammation (Miller et al., 1993).

Kooy and Royall (1994) have shown that cultured endothelial cells produce peroxynitrite when stimulated by bradykinen. They used an ultrasensitive chemiluminescent assay based on luminol developed by Radi et al. (1993). Peroxynitrite is a potent toxin to trypanosomes, attacking both sulfhydryl dependent enzymes and respiratory enzymes (Rubbo et al., 1994). Radi et al. (1994) have also shown that it is far more damaging to mitochondria than nitric oxide.

Hogg et al. (1992) have shown that high concentrations of the nitrovasodilator 3-morpholine sydnonimime (SIN-1) actually produces peroxynitrite and initiates many free radical-like reactions. They have also shown that peroxynitrite oxidizes low density lipoproteins to a form that is rapidly taken up by activated macrophages (Darley-Usmar et al., 1992; Graham et al., 1993). Peroxynitrite also depletes lipoproteins of α-tocopherol (Hogg et al., 1993) and may be the biological oxidant responsible for oxidizing low density lipoprotein *in vivo*. Peroxynitrite added to plasma oxidizes sulfhydryls and nitrates tyrosines in proteins, even before it depletes stores of tocopherol and ascorbate (van der Vliet et al., 1994).

Additional physical studies and calculations have been performed on peroxynitrite (Koppenol and Klasinc, 1993; Kraus, 1994). Part of the infrared spectrum of peroxynitrite anion in a solid state matrix has been reported (Plumb and Edwards, 1991).

VI. CONCLUSIONS

The identification of nitric oxide as an endogenous vasodilator and second messenger has helped to clarify many physiological processes, but relatively little is known about the reactivity of nitric oxide in dilute oxygenated solutions. Some of these chemical reactions are summarized in Fig. 42. Many of the widely cited reactions in the physiological literature are based on chemical studies done with either high concentrations of nitric oxide or nitric oxide donors that produce other complicating by-products. The dependence of reaction rates on the square of nitrogen oxide concentration makes direct extrapolation from studies using high concentrations of nitric oxide inappropriate and misleading. However, nitric oxide formed *in vivo* is far more likely to diffuse out of tissue and react with hemoglobin in red blood cells than it is to react with oxygen and a second nitric

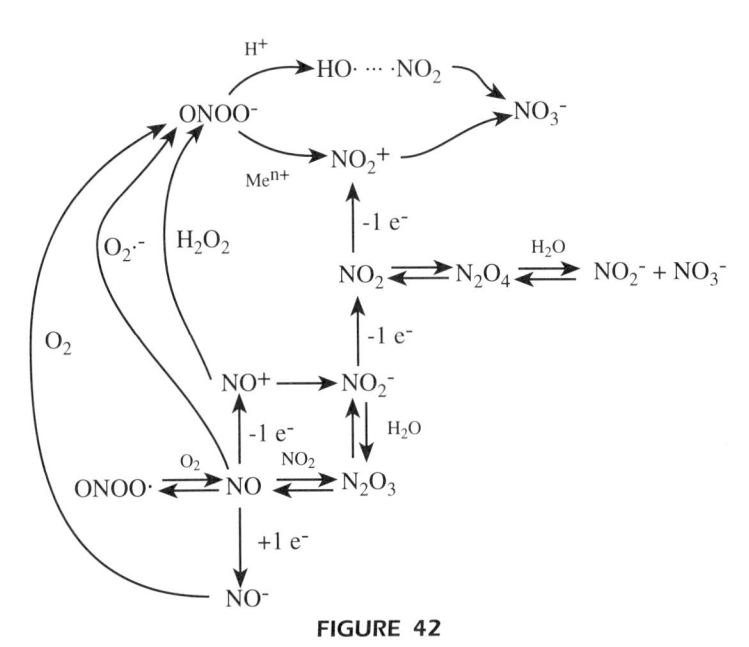

FIGURE 42

Summary of reactions of nitrogen oxides, water, and oxygen.

oxide to produce nitrogen dioxide. Nitric oxide will initiate different reactions when added in 10–1000 μM concentrations compared to the 10–400 nM concentrations measured *in vivo* (Malinski and Taha, 1992; Shibuki and Okada, 1991). We have focused on the reactions of both oxygen and superoxide with nitric oxide as being more important than dimerization of the nitrogen oxides. Thus, the physiological chemistry of nitric oxide is more likely to involve the relatively unstable and poorly understood species, nitrosyldioxyl radical (ONOO·) and peroxynitrite (ONOO⁻). The chemistry of both nitrosyldioxyl radical and peroxynitrite appear to be determined by the interactions of the two terminal oxygens in cis conformation.

The short biological half-life is a critical element for the biological usefulness of nitric oxide as a cellular messenger. The rate of decomposition to form nitrogen dioxide is too slow to account for the short half-life observed in isolated perfusion cascades. However, the stabilization of ONOO· in water makes this a reasonable candidate for the biologically inactive form of nitric oxide in perfusion cascades. In blood-containing vascular beds, the major route of inactivation is the reaction with hemoglobin and to a lesser extent myoglobin. Even a half-life of 6 sec is much longer than the 1 sec required for nitric oxide to diffuse 50–100 μm through a blood capillary.

Reaction with oxygen also can greatly increase the toxicological reactivity of nitric oxide. Peroxynitrite, the one electron reduced form of ONOO·, is a powerful oxidant that is implicated in stroke, heart disease, inflammation, and many other pathological processes. During its decomposition at physiological pH, peroxynitrite can produce some of the strongest oxidants known in a biological system, initiating reactions characteristic of hydroxyl radical, nitronium ion, and nitrogen dioxide. Its unusual stability in the cis conformation contributes to its toxicity by allowing peroxynitrite to diffuse far from its site of formation and to be selectively reactive with cellular targets. Continued progress in understanding the chemistry of these reactive species will help unravel the many conflicting studies on the role of nitric oxide in physiological and pathological processes.

ACKNOWLEDGMENTS

I am grateful for many helpful discussions with Drs. Rafael Radi (University of the Republic, Montevideo, Uruguay), Willem H. Koppenol (ETH, Zürich, Switzerland), Harry Ischiropoulos (University of Pennsylvania), John Crow, Sadi Matalon, Meg Tarpey, and James Royall (the University of Alabama at Birmingham). Quantum mechanical calculations were performed by Mark Van der Woerd, Dr. Joseph Harrison, and Dr. Tracy P. Hamilton of the Center for Macromolecular Crystallography, the Department of Physics and the Department of Chemistry respectively at the University of Alabama at Birmingham. Laser Raman spectroscopy and normal mode calculations on *cis*-peroxynitrite were conducted by Michael Tsai and James C. Martin of the Department of Physics, The University of Alabama at Birmingham. Michael Jabonski performed the ^{15}N-NMR experiments. This work was supported by Grants HL 46407, NS24338, and HL48676 from the National Institutes of Health and a Grant-in-Aid from the American Heart Association. Joseph S. Beckman is an Established Investigator of the American Heart Association.

REFERENCES

Albina, J. E., Mills, C. D., Henry, W. L., Jr., and Caldwell M. D. (1989). Regulation of macrophage physiology by L-arginine. Role of the oxidative L-arginine deaminase pathway. *J. Immunol.* **143,** 3641–3646.

Augusto, O., Gatti, R. M., and Radi, R. (1994). Spin-trapping studies of peroxynitrite decomposition and of 3-morpholinosydnonimine N-ethylcarbamide autooxidation. *Arch. Biochem. Biophys.* **310,** 118–125.

Axelsson, K. I., and Karlsson, J.-O. G. (1984). Nitroglycerin tolerance *in vitro:* Effect on cGMP turnover in vascular smooth muscle. *Acta Pharmacol. Toxicol.* **55,** 203–210.

Baeyer, A., and Villiger, V. (1901). Uber die salpetrige Säure. *Berichte* **34,** 755–763.

Bauer, M., Beckman, J. S., Bridges, R., and Matalon, S. (1992). Peroxynitrite inhibits sodium transport in rat colonic membrane vesicles. *Biochim. Biophys. Acta* **1104,** 84–87.

Baum, R. M. (1984). Superoxide theory of oxygen toxicity is center of heated debate. *Chem. Eng. News* **April 9,** 20–28.

Beckman, J. S. (1990). Ischaemic injury mediator. *Nature (London)* **345,** 27–28.

Beckman, J. S., and Siedow, J. N. (1985). Bactericidal agents generated by the peroxynitrite-catalyzed oxidation of *para*-hydroquinones. *J. Biol. Chem.* **260,** 14604–14609.

Beckman, J. S., and Koppenol, W. H. (1992). Why is the half-life of nitric oxide so short? "Biology of Nitric Oxide. 2. Enzymology, Biochemistry and Immunology." Portland Press Proceedings, London.

Beckman, J. S., Minor, R. M., Jr., White, C. J., Repine, J., Rosen, G. M., and Freeman, B. A. (1988). Superoxide dismutase and catalase conjugated to polyethylene glycol increases endothelial enzyme activity and oxidant resistance. *J. Biol. Chem.* **263,** 6584–6802.

Beckman, J. S., Beckman, T. W., Chen, J., Marshall, P. M., and Freeman, B. A. (1990). Apparent hydroxyl radical production from peroxynitrite: Implications for endothelial injury by nitric oxide and superoxide. *Proc. Natl. Acad. Sci. U.S.A.* **87,** 1620–1624.

Beckman, J. S., Ischiropoulos, H., Zhu, L., van der Woerd, M., Smith, C., Chen, J., Harrison, J., Martin, J. C., and Tsai, M. (1992). Kinetics of superoxide dismutase and iron catalyzed nitration of phenolics by peroxynitrite. *Arch. Biochem. Biophys.* **298,** 438–445.

Beckman, J. S., Carson, M., Smith, C. D., and Koppenol, W. H. (1993). ALS, SOD and Peroxynitrite. *Nature (London)* **364,** 584.

Beckman, J. S., Ye, Y. Z., Anderson, P., Chen, J., Accavetti, M. A., Tarpey, M. M., and White, C. R. (1994). Extensive nitration of protein tyrosines observed in human atherosclerosis detected by immunohistochemistry. *Biol. Chem. Hoppe-Seyler* **375**, 81–88.

Benson, S. W. (1976). "Thermodynamic Kinetics. Methods for the Estimation of Thermodynamic Data and Rate Parameters." Wiley, New York.

Beyer, W. F., Jr., Fridovich, I., Mullenbach, G. T., and Hallewell, R. (1987). Examination of the role of arginine-143 in the human copper and zinc superoxide dismutase by site-specific mutagenesis. *J. Biol. Chem.* **262**, 11182–11187.

Blough, N. V., and Zafiriou, O. C. (1985). Reaction of superoxide with nitric oxide to form peroxonitrite in alkaline aqueous solution. *Inorg. Chem.* **24**, 3504–3505.

Boehm, R. C., and Lohr, L. L. (1989). An *ab initio* characterization of nitrogen trioxide electronic states. *J. Phys. Chem.* **93**, 3430–3433.

Böhme, G. A., Bon, C., Stutzmann, J. M., Doble, A., and Blanchard, J. C. (1991). Possible involvement of nitric oxide in long-term potentiation. *Eur. J. Pharmacol.* **199**, 379–381.

Brune, B., and Lapetina, E. G. (1990). Properties of a novel nitric oxide-stimulated ADP-ribosyltransferase. *Arch. Biochem. Biophys.* **279**, 286–290.

Carreras, M. C., Catz, S. D., Pargament, G. A., Del Bosco, C. G., and Poderoso, J. J. (1994a). Decreased production of nitric oxide by human neutrophils during septic multiple organ dysfunction syndrome. *Inflammation* **18**, 151–161.

Carreras, M. C., Pargament, G. A., Catz, S. D., Poderoso, J. J., and Boveris, A. (1994b). Kinetics of nitric oxide and hydrogen peroxide production and formation of peroxynitrite during the respiratory burst of human neutrophils. *FEBS Lett.* **341**, 65–68.

Chacko, G. K. (1985). Modification of high density lipoprotein (HDL3) with tetranitromethane and the effect on its binding to isolated rat liver plasma membranes. *J. Lipid Res.* **26**, 745–754.

Cheng, B. M., Lee, J. W., and Lee, Y. P. (1991). Photolysis of nitric acid in solid argon: The infrared absorption of peroxynitrous acid (ONOOH). *J. Phys. Chem.* **95**, 2814–2817.

Colburn, C. B. (1973). "Developments in Inorganic Nitrogen Chemistry." Elsevier, Amsterdam.

Cox, R. D. (1980). Development of analytical methodologies for parts per billion level determination of nitrate, nitrite, and N-nitroso group content. Ph.D. dissertation, University of Iowa.

Craven, P. A., DeRuberts, F. R., and Pratt, D. W. (1979). Electron spin resonance study of the role of NO · catalase in the activation of guanylate cyclase by NaN$_3$ and NH$_2$OH. Modulation of enzyme responses by heme proteins and their nitrosyl derivatives. *J. Biol. Chem.* **254**, 8213–8222.

Crow, J. P., Spruell, C., Chen, J., Gunn, C., Ischiropoulos, H., Tsai, M., Smith, C. D., Radi, R., Koppenol, W. H., and Beckman, J. S. (1994). On the pH-dependent yield of hydroxyl radical products from peroxynitrite. *Free Radical Biol. Med.* **16**, 331–338.

Darley-Usmar, V. M., Hogg, H., O'Leary, V. J., Wilson, M. T., and Moncada, S. (1992). The simultaneous generation of superoxide and nitric oxide can initiate lipid peroxidation in human low density lipoprotein. *Free Radicals Res. Commun.* **17**, 9–20.

Dawson, V. L., Dawson, T. M., London, E. D., Bredt, D. S., and Snyder, S. H. (1991). Nitric oxide mediates glutamate neurotoxicity in primary cortical cultures. *Proc. Natl. Acad. Sci. U.S.A.* **88**, 6368–6371.

Dee, G., Rice-Evans, C., Obeyesekera, S., Meraji, S., Jacobs, M., and Bruckdorfer, K. R. (1991). The modulation of ferryl myoglobin formation and its oxidative effects on low density lipoproteins by nitric oxide. *FEBS Lett.* **294,** 38–42.

DeKock, R. L., and Gray, H. B. (1989). "Chemical Structure and Bonding." University Science Books, Mill Valley, California.

Di Iorio, E. E. (1981). Preparation of derivatives of ferrous and ferric hemoglobin. In "Methods in Enzymology" (E. Antonini, L. Rossi-Bernardi, and E. Chiancone, eds.), Vol. 76, pp. 57–72. Academic Press, New York.

Dimmeler, S., Lottspeich, F., and Brüne, B. (1992). Nitric oxide causes ADP-ribosylation and inhibition of glyceraldehyde-3-phosphate dehydrogenase. *J. Biol. Chem.* **267,** 16771–16774.

Donald, C. E., Hughes, M. N., Thompson, J. M., and Bonner, F. T. (1986). Photolysis of the N=N bond in trioxodinitrate: Reaction between triplet NO- and O_2 to form peroxonitrite. *Inorg. Chem.* **25,** 2676–2677.

Doyle, M. P., and Hoekstra, J. W. (1981). Oxidation of nitrogen oxides by bound dioxygen in hemoproteins. *J. Inorg. Biochem.* **14,** 351–358.

Drapier, J. C., Pellat, C., and Henry, Y. (1991). Generation of EPR-detectable nitrosyl–iron complexes in tumor target cells cocultured with activated macrophages. *J. Biol. Chem.* **266,** 10162–10167.

East, S. J., and Garthwaite, J. (1991). NMDA receptor activation in the rat hippocampus induces cyclic GMP formation through the L-arginine–nitric oxide pathway. *Neurosci. Lett.* **123,** 17–19.

Edwards, J. O., and Plumb, R. C. (1994). The chemistry of peroxonitrites. *Prog. Inorg. Chem.* **41,** 599–635.

Feelish, M., and Noack, E. (1987). Correlation between nitric oxide formation during degradation of organic nitrates and activation of guanylate cyclase. *Eur. J. Pharmacol.* **139,** 19–30.

Feigl, E. O. (1988). EDRF—a protective factor. *Nature (London)* **331,** 490–491.

Feldman, P. L., Griffith, O. W., and Stuehr, D. J. (1993). The surprising life of nitric oxide. *Chem. Eng. News.* **December 20,** 26–38.

Flaherty, J. T., and Weisfeldt, M. L. (1988). Reperfusion injury. *Free Radical Biol. Med.* **5,** 409–419.

Fontijn, A., Sadadell, A. J., and Ronco, R. J. (1970). Homogeneous chemiluminescent measurement of nitric oxide with ozone. *Anal. Chem.* **42,** 575–578.

Fridovich, I. (1986). Biological effects of the superoxide radical. *Arch. Biochem. Biophys.* **247,** 1–11.

Frostell, D., Fratacci, M. D., Wain, J. C., Jones, R., and Zapol, W. M. (1991). Inhaled nitric oxide. A selective pulmonary vasodilator reversing hypoxic pulmonary vasoconstriction. *Circulation* **83,** 2038–2047.

Furchgott, R. E., and Vanhoutte, P. M. (1989). Endothelium-derived relaxing and contracting factors. *FASEB J.* **3,** 2007–2018.

Gally, J. A., Montague, P. R., Reeke, G. N., Jr., and Edelman, G. M. (1990). The NO hypothesis: Possible effects of a short-lived, rapidly diffusible signal in the development and function of the nervous system. *Proc. Natl Acad. Sci. U.S.A.* **87,** 3547–3551.

Gardner, J. F., and Boveris, A. (1990). Generation of superoxide anion by the NADH dehydrogenase of bovine heart mitochondria. *Biochem. J.* **191,** 421–427.

Gardner, P. R., and Fridovich, I. (1991). Superoxide sensitivity of the *Escherichia coli* 6-phosphogluconate dehydratase. *J. Biol. Chem.* **266,** 1478–1483.

Garthwaite, J. (1991). Glutamate, nitric oxide and cell–cell signalling in the nervous system. *Trends Neurosci.* **14,** 75–82.

Garthwaite, J., Charles, S. L., and Chess-Williams, R. (1988). Endothelium-derived relaxing factor release on activation of NMDA receptors suggests role as intercellular messenger in the brain. *Nature (London)* **336,** 385–388.

Getzoff, E. D., Tainer, J. A., Weiner, P. K., Kollman, P. A., Richardson, J. S., and Richardson, D. C. (1983). Electrostatic recognition between superoxide and copper, zinc superoxide dismutase. *Nature (London)* **306,** 287–290.

Gleu, K., and Hubold, R. (1935). Die Einwirkung von Wasserstoffsuperoxyd auf salpetrige Säure. Persalpetrige Säure. *Z. Anorg. Allg. Chem.* **223,** 305–317.

Gleu, K., and Roell, E. Z. (1929). Die Einwirkung von Ozon auf Alkaliazid. Persalpetrige Säure I. *Z. Anorg. Allg. Chem.* **179,** 233–266.

Goretski, J., and Hollocher, T. C. (1988). Trapping of nitric oxide produced during denitrification by extracellular hemoglobin. *J. Biol. Chem.* **263,** 2316–2323.

Graham, A., Hogg, N., Kalyanaraman, B., O'Leary, V., Darley-Usmar, V., and Moncada, S. (1993). Peroxynitrite modification of low-density lipoprotein leads to recognition by the macrophage scavenger receptor. *FEBS Lett.* **330,** 181–185.

Griffith, T. M., Edwards, D. H., Davies, R. L., Harrison, T. J., and Evans, K. T. (1987). EDRF coordinates the behaviour of vascular resistance vessels. *Nature (London)* **329,** 442–445.

Guerrieri, F., Yagi, T., and Papa, S. (1984). On the mechanism of H^+ translocation by mitochondrial H^+-ATPase. Studies with chemical modifier of tyrosine residues. *J. Bioenerg. Biomembr.* **16,** 251–262.

Guillory, W. A., and Johnston, H. S. (1965). Infrared absorption by peroxy-nitrogen trioxide free radical in the gas phase. *J. Chem. Phys.* **42,** 2457–2461.

Haddad, I. Y., Crow, J. P., Hu, P., Ye, Y. Z., Beckman, J. S., and Matalon, S. (1994a). Concurrent generation of nitric oxide and superoxide damages surfactant protein A (SP-A). *Am. J. Phys.* **265,** L555–L564.

Haddad, I. Y., Ischiropoulos, H., Holm, B. A., Beckman, J. S., Baker, J. R., and Matalon, S. (1994b). Mechanisms of peroxynitrite induced injury to pulmonary surfactants. *Am. J. Physiol.* **265,** L555–L564.

Halfpenny, E., and Robinson, P. L. (1952a). The nitration and hydroxylation of aromatic compounds by pernitrous acid. *J. Chem. Soc.*, 939–946.

Halfpenny, E., and Robinson, P. L. (1952b). Pernitrous acid. The reaction between hydrogen peroxide and nitrous acid, and the properties of an intermediate product. *J. Chem. Soc.*, 928–938.

Hibbs, J. B., Jr., Taintor, R. R., Vavrin, Z., and Rachlin, E. M. (1988). Nitric oxide: A cytotoxic activated macrophage effector molecule. *Biochem. Biophys. Res. Commun.* **157,** 87–94.

Hibbs, J. B., Jr., Westenfelder, C., Taintor, R., Vavrin, Z., Kablitx, C., Baranowski, R. L., Ward, J. H., Menlove, R. L., McMurry, M. P., Kushner, J. P., and Samlowski, W. E. (1992). Evidence for cytokine-inducible nitric oxide synthesis from L-arginine in patients receiving interleukin-2 therapy. *J. Clin. Invest.* **89,** 867–877.

Hogg, N., Darley-Usmar, V. M., Wilson, M. T., and Moncada, S. (1992). Production of hydroxyl radicals from the simultaneous generation of superoxide and nitric oxide. *Biochem. J.* **281**, 419–424.

Hogg, N., Darley-Usmar, V. M., Graham, A., and Moncada, S. (1993). Peroxynitrite and atherosclerosis. *Biochem. Soc. Trans.* **21**, 358–362.

Hu, P., Ischiropoulos, H., Beckman, J. S., and Matalon, S. (1994). Peroxynitrite inhibition of oxygen consumption and sodium transport in alveolar type II cells. *Am. J. Physiol.* **266**, L628–L634.

Hughes, M. N., and Nicklin, H. G. (1968). The chemistry of pernitrites. Part I. Kinetics of decomposition of pernitrous acid. *J. Chem. Soc. (A)*, 450–452.

Hughes, M. N., and Nicklin, H. G. (1970). The chemistry of peroxonitrites. Part II. Copper (II)-catalysed reaction between hydroxylamine and peroxonitrite in alkali. *J. Chem. Soc. (A)*, 925–928.

Hughes, M. N., Nicklin, H. G., and Sackrule, W. A. C. (1971). The chemistry of peroxonitrites. Part III. The reaction of peroxynitrite with nucleophiles in alkali, and other nitrite producing reactions. *J. Chem Soc. (A)*, 3722–3725.

Huie, R. E., and Padmaja, S. (1993). The reaction rate of nitric oxide with superoxide. *Free Radicals Res. Commun.* **18**, 195–199.

Huisman, T. H. J., and Dozy, A. M. (1965). Studies on the heterogeneity of hemoglobin. IX. The use of Tris(hydroxylmethyl)aminomethane-HCl buffers in the anion-exchange chromatography of hemoglobins. *J. Chromatogr.* **19**, 160–169.

Hutchinson, F. (1957). The distance that a radical formed by ionizing radiation and diffuse in a yeast cell. *Radiat. Res.* **7**, 473–483.

Hutchinson, P. J. A., Palmer, R. M. J., and Moncada, S. (1987). Comparative pharmacology of EDRF and nitric oxide on vascular strips. *Eur. J. Pharmacol.* **141**, 445–451.

Ignarro, L. J. (1990). Biosynthesis and metabolism of endothelium-derived nitric oxide. *Annu. Rev. Pharmacol. Toxicol.* **30**, 535–560.

Ignarro, L. J. (1991). Signal transduction mechanisms involving nitric oxide. *Biochem. Pharmacol.* **41**, 485–490.

Ignarro, L. J., Fukuto, J. M., Griscavage, J. M., Rogers, N. E., and Burns, R. E. (1993). Oxidation of nitric oxide in aqueous solution to nitrite but not nitrate: Comparison with enzymatically formed nitric oxide from L-arginine. *Proc. Natl. Acad. Sci. U.S.A.* **90**, 8103–8107.

Ischiropoulos, H., Zhu, L., and Beckman, J. S. (1992a). Peroxynitrite formation from activated rat alveolar macrophages. *Arch. Biochem. Biophys.* **298**, 446–451.

Ischiropoulos, H., Zhu, L., Chen, J., Tsai, H. M., Martin, J. C., Smith, C. D., and Beckman, J. S. (1992b). Peroxynitrite-mediated tyrosine nitration catalyzed by superoxide dismutase. *Arch. Biochem. Biophys.* **298**, 431–437.

Izumi, Y., Clifford, D. B., and Zorumski, C. F. (1992). Inhibition of long-term potentiation by NMDA-mediated nitric oxide release. *Science* **257**, 1273–1276.

Janing, G. R., Kraft, R., Blanck, J., Rabe, H., and Ruckpaul, K. (1987). Chemical modification of cytochrome P-450 LM4. Identification of functionally linked tyrosine residues. *Biochim. Biophys. Acta.* **916**, 512–523.

Jones, K. (1973). The Chemistry of Nitrogen. "Comprehensive Inorganic Chemistry." Pergamon, Oxford.

Kandel, E. R., and Hawkins, R. D. (1992). The biological basis of learning and individuality. *Sci. Am.* **267,** 78–86.
Kanner, J., Harel, S., and Granit, R. (1991). Nitric oxide as an antioxidant. *Arch. Biochem. Biophys.* **289,** 130–136.
Keith, W. G., and Powell, P. E. (1969). Kinetics of decomposition of peroxynitrous acid. *J. Chem. Soc. (A)*, 453.
Kelm, M., and Schrader, J. (1988). Nitric oxide release from the isolated guinea pig heart. *Eur. J. Pharmacol.* **155,** 317–321.
King, P. A., Anderson, V. E., Edwards, J. O., Gustafson, G., Plumb, R. C., and Suggs, J. W. (1992). A stable solid that generates hydroxyl radical on dissolution in aqueous solutions: Reaction with proteins and nucleic acid. *J. Am. Chem. Soc.* **114,** 5430–5432.
Kontos, H. A., and Wei, E. P. (1986). Superoxide production in experimental brain injury. *J. Neurosurg.* **64,** 803–807.
Kooy, N. W., and Royall, J. A. (1994). Agonist-induced peroxynitrite production by endothelial cells. *Arch. Biochem. Biophys.* **310,** 353–359.
Kooy, N. W., Royall, J. A., Ye, Y. Z., Kelley, D. R., and Beckman, J. S. (1994). Evidence for *in vivo* peroxynitrite production in human acute lung injury. *Am. Rev. Respir. Dis.* **151,** 1250–1254.
Koppenol, W. H. (1993). A thermodynamic appraisal of the radical sink hypothesis. *Free Radical Biol. Med.* **14,** 91–94.
Koppenol, W. H., and Klasinc, L. (1993). *Ab initio* calculations on ONOOH and ONOO$^-$. *Int. J. Quantum Chem.: Quantum Biol. Symp.* **20,** 1–6.
Koppenol, W. H., Moreno, J. J., Pryor, W. A., Ischiropoulos, H., and Beckman, J. S. (1992). Peroxynitrite: A cloaked oxidant from superoxide and nitric oxide. *Chem. Res. Toxicol.* **5,** 834–842.
Korth, H. G., Ingold, K. U., Sustmann, R., de Groot, H., and Sies, H. (1992). Tetramethyl-*ortho*-chinodiethan (NOCT-1), das erste Mitglied einer Familie massgescheiderter chletroper Sinfänger für Stickstoffmonoxide. *Angew. Chem.* **104,** 915–917.
Kosaka, H., Imaizumi, K., Imai, K., and Tyuma, I. (1979). Stoichiometry of the reaction of oxyhemoglobin with nitrite. *Biochim. Biophys. Acta* **581,** 184–188.
Kosaka, H., Imaizumi, K., and Tyuma, I. (1981). Mechanism of autocatalytic oxidation of oxyhemoglobin by nitrite an intermediate detected by electron spin resonance. *Biochim. Biophys. Acta* **702,** 237–241.
Kots, A. Y., Skurat, A. V., Serienko, E. A., Bulargina, T. V., and Severin, E. S. (1992). Nitric oxide stimulates the cysteine-specific mono(ADP-ribosylation) of glyceraldehyde-3-phosphate dehydrogenase from human erythrocytes. *FEBS Lett.* **300,** 9–12.
Kowaluk, E. A., and Fung, H. L. (1990). Spontaneous liberation of nitric oxide cannot account for *in vitro* vascular relaxation by S-nitrosothiols. *J. Pharmacol. Exp. Ther.* **255,** 1256–1264.
Kraus, M. (1994). Electronic structure and spectra of the peroxynitrite anion. *Chem. Phys. Lett.* **222,** 513–516.
Lancaster, J. R., Jr., Langrehr, J. M., Bergonia, H. A., Murase, N., Simmons, R. L., and Hoffman, R. A. (1992). EPR detection of heme and nonheme iron-containing protein nitrosylation by nitric oxide during rejection of rat heart allograft. *J. Biol. Chem.* **267,** 10994–10998.

Liu, S. Y., Beckman, J. S., and Ku, D. D. (1994). Peroxynitrite, a product of superoxide and nitric oxide, produces coronary vasorelaxation in dogs. *J. Pharmacol. Exp. Ther.* **268,** 1114–1120.

Long, C. J., Shikano, K., and Berkowitz, B. A. (1987). Anion exchange resins discriminate between nitric oxide and EDRF. *Eur. J. Pharmacol.* **142,** 317–318.

Lundblad, R. L., Noyes, C. M., Featherstone, G. L., Harrison, J. H., and Jenzano, J. W. (1988). The reaction of alpha-thrombin with tetranitromethane. *J. Biol. Chem.* **263,** 3729–3734.

Lynch, R. E., and Fridovich, I. (1978). Permeation of erythrocyte stroma by superoxide radical. *J. Biol. Chem.* **253,** 4697–4699.

McCall, M. N., and Easterbrook-Smith, S. B. (1989). Comparison of the role of tyrosine residues in human IgG and rabbit IgG in binding of complement subcomponet C1q. *Biochem. J.* **257,** 845–851.

McCord, J. M., and Fridovich, I. (1969). Superoxide dismutase: An enzymic function for erythrocuprein (hemocuprein). *J. Biol. Chem.* **244,** 6049–6055.

McGrath, M. P., Francl, M. M., Rowland, F. S., and Hehre, W. J. (1988). Isomers of nitric acid and chlorine nitrate. *J. Phys. Chem.* **92,** 5352–5357.

Mahoney, L. R. (1970). Evidence for the formation of hydroxyl radical in the isomerization of pernitrous acid to nitric acid in aqueous solution. *J. Am. Chem. Soc.* **92,** 5262–5263.

Malinski, T., and Taha, Z. (1992). Nitric oxide release from a single cell measured *in situ* by a porphyrinic-based microsensor. *Nature (London)* **358,** 676–678.

Marletta, M. A., Yoon, P. S., Iyengar, R., Leaf, C. D., and Wishnok, J. S. (1988). Macrophage oxidation of L-arginine to nitrite and nitrate: Nitric oxide is an intermediate. *Biochemistry* **27,** 8706–8711.

Marshall, J. J., and Kontos, H. A. (1990). Endothelium-derived relaxing factors. A perspective from *in vivo* data. *Hypertension (Dallas)* **16,** 371–386.

Martin, B. L., Wu, D., Jakes, S., and Graves, D. J. (1990). Chemical influences on the specificity of tyrosine phosphorylation. *J. Biol. Chem.* **265,** 7108–7111.

Matheis, G., Sherman, M. P., Buckberg, G. D., Haybron, D. M., Young, H. H., and Ignarro, L. J. (1992). Role of L-arginine-nitric oxide pathway in myocardial reoxygenation injury. *Am. J. Physiol.* **262,** H616–H620.

Matsunaga, K., and Furchgott, R. F. (1989). Interactions of light and sodium nitrite in producing relaxation in rabbit aorta. *J. Pharmacol. Exp. Ther.* **248,** 687–695.

Matsunaga, K., and Furchgott, R. F. (1991). Responses of rabbit aorta to nitric oxide and superoxide generated by ultraviolet irradiation of solutions containing inorganic nitrite. *J. Pharmacol. Exp. Ther.* **259,** 1140–1146.

Mierzwa, S., and Chan, S. K. (1987). Chemical modification of human alpha 1 proteinase inhibitor by tertanitromethane. Structure–function relationship. *Biochem. J.* **246,** 37–42.

Miller, M. J. S., Chotinaruemol, S., Sadowska-Krowicka, H., Kakkis, J. L., Munshi, U. K., Zhang, X. J., and Clark, D. A. (1993). Nitric oxide: The Jekyll and Hyde of gut inflammation. *Agents Actions* **39,** C180–C182.

Moncada, S., Palmer, R. M. J., and Higgs, E. A. (1991). Nitric oxide: Physiology, pathophysiology, and pharmacology. *Pharmacol. Rev.* **43,** 109–142.

Moreno, J. J., and Pryor, W. A. (1992). Inactivation of α-1-proteinase inhibitor by peroxynitrite. *Chem. Res. Toxicol.* **5**, 425–431.
Mulligan, M. S., Hevel, J. M., Marletta, M. A., and Ward, P. A. (1991). Tissue injury caused by deposition of immune complexes is L-arginine dependent. *Proc. Natl. Acad. Sci. U.S.A.* **88**, 6338–6342.
Mulligan, M. S., Moncada, S., and Ward, P. A. (1992a). Protective effects of inhibitors of nitric oxide synthase in immune complex-induced vasculitis. *Br. J. Pharmacol.* **107**, 1159–1162.
Mulligan, M. S., Warren, J. S., Smith, C. W., Anderson, D. C., Yeh, C. G., Rudolph, A. R., and Ward, P. A. (1992b). Lung injury after deposition of IgA immune complexes. Requirements for CD18 and L-arginine. *J. Immunol.* **148**, 3086–3092.
Murphy, M. E., and Sies, H. (1991). Reversible conversion of nitroxyl anion to nitric oxide by superoxide dismutase. *Proc. Natl. Acad. Sci. U.S.A.* **88**, 10860–10864.
Myers, P. R., Guerra, R., Jr., Bates, J. N., and Harrison, D. G. (1989). Release of NO and EDRF from cultured bovine aortic endothelial cells. *Am J. Physiol.* **256**, H1030–H1037.
Myers, P. R., Minor, R. L., Jr., Guerra, R., Jr., Bates, J. N., and Harrison, D. G. (1990). Vasorelaxant properties of the endothelium-derived relaxing factor more closely resemble S-nitrosocysteine than nitric oxide. *Nature (London)* **345**, 161–163.
Nathan, C. F., and Hibbs, J. B. J. (1991). Role of nitric oxide synthesis in macrophage antimicrobial activity. *Curr. Opin. Immunol.* **3**, 65–70.
Nobel, P. S. (1983). "Biophysical Plant Physiology and Ecology." Freeman, New York.
Nowak, R. (1992). Corners of the mind: The cellular basis of memory and learning. *J. NIH Res.* **4**, 49–55.
Nowicki, J. P., Duval, D., Poignet, H., and Scatton, B. (1991). Nitric oxide mediates neuronal death after focal cerebral ischemia in the mouse. *Eur. J. Pharmacol.* **204**, 339–340.
Ohshima, H., Friesen, M., Brouet, I., and Bartsch, H. (1990). Nitrotyrosine as a new marker for endogenous nitrosation and nitration of proteins. *Food Chem. Toxicol.* **28**, 647–652.
Ohta, T. (1986). Fluorometric determination of nitrite with 4-hydroxycoumarin. *Anal. Chem.* **58**, 3132–3135.
Olah, G. A., Laali, K. K., and Sandford, G. (1992a). Comparison of the nitration of polyfluoronitrobenzenes by nitronium salts in superacidic and aprotic media: Activation of the nitronium ion by protosolvation. *Proc. Natl. Acad. Sci. U.S.A.* **89**, 6670–6672.
Olah, G. A., Rasul, G., Aniszfeld, R., and Surya-Prakash, G. K. (1992b). Protonitronium dication (NO_2H^+). *J. Am. Chem. Soc.* **114**, 5608–5609.
Orii, Y., and Shimada, H. (1978). Reaction of cytochrome c with nitrite and nitric oxide. A model of dissimulatory nitrite reductase. *J. Biochem. (Tokyo)* **84**, 1543–1552.
Palmer, R. M. J., Ferrige, A. G., and Moncada, S. (1987). Nitric oxide release accounts for the biological activity of endothelium-derived relaxing factor. *Nature (London)* **327**, 523–526.
Plumb, R. C., and Edwards, J. O. (1991). Color centers in UV-irradiated nitrates. *J. Phys. Chem.* **96**, 3245–3247.
Plumb, R. C., Tantayanon, R., Libby, M., and Xu, W. W. (1989). Chemical model for Viking biology experiments: Implications for the composition of the martian regolith. *Nature (London)* **338**, 633–635.

Pollock, J. Förstermann, U., Mitchell, J. A., Warner, T. D., Schmidt, H. H. H. W., Nakane, M., and Murad, F. (1991). Purification and characterization of particulate endothelium-derived relaxing factor synthase from cultured and native bovine aortic endothelial cells. *Proc. Natl. Acad. Sci. U.S.A.* **88,** 10480–10484.

Prütz, W. A., Mönig, H., Butler, J., and Land, E. J. (1985). Reactions of nitrogen dioxide in aqueous model systems: Oxidation of tyrosine units in peptides and proteins. *Arch. Biochem. Biophys.* **243,** 125–134.

Pryor, W. (1992). How far does ozone penetrate into the pulmonary air/tissue boundary before it reacts. *Free Radical Biol. Med.* **12,** 83–88.

Pryor, W. A., and Lightsey, J. W. (1981). Mechanisms of nitrogen dioxide reactions: Initiation of lipid peroxidation and the production of nitrous acid. *Science* **214,** 435–437.

Pryor, W. A., Church, D. F., Govindan, C. K., and Crank, G. (1982). Oxidation of thiols by nitric oxide and nitrogen dioxide: Synthetic utility and toxicological implications. *J. Org. Chem.* **147,** 156–158.

Pryor, W. A., Castle, L., and Church, D. F. (1985). Nitrosation of organic hydroperoxides by nitrogen dioxide/dinitrogen tetraoxide. *J. Am. Chem. Soc.* **107,** 211–217.

Rachmilewitz, D., Stamler, J. S., Karmeli, F., Mullins, M. E., Singel, D. J., Loscalzo, J., Xavier, R. J., and Podolsky, D. K. (1993). Peroxynitrite-induced rat colitis—a new model of colonic inflammation. *Gastroenterology* **105,** 1681–1688.

Radi, R., Beckman, J. S., Bush, K. M., and Freeman, B. A. (1991a). Peroxynitrite-induced membrane lipid peroxidation. The cytotoxic potential of superoxide and nitric oxide. *Arch. Biochem. Biophys.* **288,** 481–487.

Radi, R., Beckman, J. S., Bush, K. M., and Freeman, B. A. (1991b). Peroxynitrite-mediated sulfhydryl oxidation: The cytotoxic potential of superoxide and nitric oxide. *J. Biol. Chem.* **266,** 4244–4250.

Radi, R., Cosgrove, T. P., Beckman, J. S., and Freeman, B. A. (1993). Peroxynitrite-induced luminol chemiluminescence. *Biochem. J.* **290,** 51–57.

Radi, R., Rodriguez, M., Castro, L., and Telleri, R. (1994). Inhibition of mitochondrial electron transport by peroxynitrite. *Arch. Biochem. Biophys.* **308,** 89–95.

Ragsdale, R. O. (1973). Reactions of nitrogen (II) oxide. "Developments in Inorganic Nitrogen Chemistry." Elsevier, Amsterdam.

Ray, J. D. (1962). Heat of isomerization of peroxynitrite to nitrate and kinetics of isomerization of peroxynitrous acid to nitric acid. *J. Inorg. Nucleotide Chem.* **24,** 1159–1162.

Reed, J. W., Ho, H. H., and Jolly, W. L. (1974). Chemical syntheses with a quenched flow reactor. Hyroxytrihydroborate and peroxynitrite. *J. Am. Chem. Soc.* **96,** 1248–1249.

Rosen, D. R., Siddique, T., Patterson, D., Figlewicz, D. A., Sapp, P., Hentati, A., Donaldson, D., Goto, J., O'Regan, J. P., Deng, H.-X., Rahmani, Z., Krizus, A., McKenna-Yasek, D., Cayabyab, A., Gaston, S. M., Berger, R., Tanszi, R. E., Halperin, J. J., Herzfeldt, B., Van den Bergh, R., Hung, W.-Y., Bird, T., Deng, G., Mulder, D. W., Smyth, C., Lang, N. G., Soriana, E., Pericak-Vance, M. A., Haines, J., Rouleau, G. A., Gusella, J. S., Horvitz, H. R., and Brown, R. H. J. (1993). Mutations in Cu/Zn superoxide dismutase gene are associated with familial amyotrophic lateral sclerosis. *Nature (London)* **362,** 59–62.

Rubanyi, G., Ho, E. H., Cantor, E. H., Lumma, W. C., and Parker Botelho, L. H. (1991). Cytoprotective function of nitric oxide: Inactivation of superoxide radicals produced by human leukocytes. *Biochem. Biophys. Res. Commun.* **181,** 1392–1397.

Rubbo, H., Denacola, A., and Radi, R. (1994). Peroxynitrite inactivates thiol-containing enzymes of *Trypanosoma cruzi* energetic metabolism and inhibits cell respiration. *Arch. Biochem. Biophys.* **308,** 96–102.

Rush, J. D., and Koppenol, W. H. (1990). Reactions of Fe(II)–ATP and Fe(II)–citrate complexes with *tert*-butyl hydroperoxide and cumylhydroperoxide. *FEBS Lett.* **275,** 114–116.

Sawyer, D. T., and Valentine, J. (1981). How super is superoxide? *Acc. Chem. Res.* **14,** 393–400.

Seidell, A. (1919). "Solubilities of Inorganic and Organic Compounds." Van Nostrand, New York.

Sharma, V. S., Traylor, T. G., Gardiner, R., and Mizukami, H. (1987). Reaction of nitric oxide with heme proteins and model compounds of hemoglobin. *Biochemistry* **26,** 3837–3843.

Shibuki, K. (1990). An electrochemical microprobe for detecting nitric oxide release in brain tissue. *Neurosci. Res.* **9,** 69–76.

Shibuki, K., and Okada, D. (1991). Endogenous nitric oxide release required for long-term synaptic depression in the cerebellum. *Nature (London).* **349,** 326–329.

Shulman, E. M., and Madison, D. V. (1991). A requirement for the intercellular messenger nitric oxide in long-term potentiation. *Science* **254,** 1503–1506.

Smith, C. D., Carson, M., Van der Woerd, M., Chen, J., Ischiropoulos, H., and Beckman, J. S. (1992). Crystal structure of peroxynitrite-modified bovine Cu,Zn superoxide dismutase. *Arch. Biochem. Biophys.* **299,** 350–355.

Stamler, J. S., Singel, D. J., and Loscalzo, J. (1992a). Biochemistry of nitric oxide and its redox-activated forms. *Science* **258,** 1898–1902.

Stamler, J. S., Simon, D. I., Osborne, J. A., Mullins, M. E., Jaraki, O., Michel, T., Singel, D. J., and Loscalzo, J. (1992b). S-Nitrosylation of proteins with nitric oxide: Synthesis and characterization of biologically active compounds. *Proc. Natl. Acad. Sci. U.S.A.* **89,** 444–448.

Stanbury, D. M. (1989). Reduction potentials involving inorganic free radicals in aqueous solution. *Adv. Inorg. Chem.* **33,** 69–138.

Stezowski, J. J., Countryman, R., and Hoard, J. L. (1973). Structure of the ethylenediaminetetraacetatoaquomangensate(II) ion in a crystallin sodium salt. Comparative stereochemistry of the seven-coordinate chelates of magnesium(II), manganese(II), and iron(II). *Inorg. Chem.* **12,** 1749–1754.

Stuehr, D. J., and Nathan, C. F. (1989). Nitric oxide. A macrophage product responsible for cytostasis and respiratory inhibition in tumor cells. *J. Exp. Med.* **169,** 1543–1555.

Stuehr, D. J., Kwon, N. S., and Nathan, C. F. (1990). FAD and GSH participate in macrophage synthesis of nitric oxide. *Biochem. Biophys. Res. Commun.* **168,** 558–565.

Stuehr, D. J., Kwon, N. S., Nathan, C. F., Griffin, O. W., Feldman, P. L., and Wiseman, J. (1991). N^W-hydroxy-L-arginine is an intermediate in the biosynthesis of nitric oxide from L-arginine. *J. Biol. Chem.* **266,** 6259–6263.

Tarpey, M. M., Ischiropoulos, H., Gore, J. S., Beckman, J. S., and Brock, T. A. (1992). Peroxynitrite stimulates cGMP synthesis by vascular smooth muscle. *Am. Rev. Respir. Dis.* **145,** A714 (abstract).

Tracey, W. R., Linden, J., Peach, M. J., and Johns, R. A. (1991). Comparison of spectrophotometric and biological assays for nitric oxide (NO) and endothelium-derived relaxing factor (EDRF): Nonspecificity of the diazotization reaction for NO and failure to detect EDRF. *J. Pharmacol. Exp. Ther.* **252**, 922–928.

Traylor, T. G., and Sharma, V. S. (1992). Why NO? *Biochemistry* **31**, 2847–2849.

Treinin, A., and Hayon, E. (1970). Absorption spectra and reaction kinetics of NO_2, N_2O_3 and N_2O_4 in aqueous solution. *J. Am. Chem. Soc.* **92**, 5281–5828.

Tsai, J.-H. M., Hamilton, T. P., Harrison, J. G., Jablowski, M., van der Woerd, M., Martin, J. C., and Beckman, J. S. (1994). Role of peroxynitrite conformation with its stability and toxicity. *J. Am. Chem. Soc.* **116**, 4115–4116.

Tsai, M. (1991). Raman spectra of peroxynitrite anion. Master's Thesis, University of Alabama at Birmingham.

Turnery, T. A., and Wright, G. A. (1959). Nitrous acid and nitrosation. *Chem. Rev.* **59**, 497–513.

van der Vliet, A., O'Neill, C. A., Halliwell, B., Cross, C. E., and Kaur, H. (1994). Aromatic hydroxylation and nitration of phenylalanine and tyrosine by peroxynitrite. *FEBS Lett.*, 96–102.

Wade, R., and Castro, C. (1990). Redox reactivity of iron(III) porphyrins and heme proteins with nitric oxide. Nitrosyl transfer to carbon, oxygen, nitrogen and sulfur. *Chem. Res. Toxicol.* **3**, 289–291.

Wang, J. F., Komarov, P., Sies, H., and DeGroot, H. (1991). Contribution of nitric oxide synthase to luminol-dependent chemiluminescence generated by phorbol-ester-activated Kupffer cells. *Biochem. J.* **279**, 311–314.

Wink, D., Kasprzak, K., Maragos, C., Elespuru, R., Misra, M., Dunams, T., Cebula, T., Koch, W., Andrews, A., Allen, J., and Keefer, L. (1991). DNA deaminating ability and genotoxicity of nitric oxide and its progenitors. *Science* **254**, 1001–1003.

Wink, D. A., Darbyshire, J. F., Nims, R. W., Saavedra, J. E., and Ford, P. C. (1993). Reactions of the bioregulatory agent nitric oxide in oxygenated aqueous media: Determination of the kinetics for oxidation and nitrosation by intermediates generated in the NO/O_2 reaction. *Chem. Res. Toxicol.* **6**, 23–27.

Winterbourn, C. C. (1979). Comparison of superoxide with other reducing agents in the biological production of hydroxyl radicals. *Biochem. J.* **182**, 625–628.

Wise, D. L., and Houghton, G. (1968). Diffusion of nitric oxide. *Chem. Eng. Sci.* **23**, 1211–1216.

Yagil, G., and Anbar, M. (1964). The formation of peroxynitrite by oxidation of chloramine, hydroxylamine and nitrohydroxamate. *J. Inorg. Nucleotide Chem.* **26**, 453–460.

Yang, G., Candy, T., Boaro, M., Wilkin, H., Jones, P., Nazhat, N., Saadalla-Nazhat, R., and Blake, D. (1992). Free radical yields for the homolysis of peroxynitrous acid. *Free Radical Biol. Med.* **12**, 327–330.

Yuste, R., Peinado, A., and Katz, L. C. (1992). Neuronal domains in developing neocortex. *Science* **257**, 665–669.

Zafiriou, O., and McFarland, M. (1980). Determination of trace levels of nitric oxide in aqueous solution. *Anal. Chem.* **52**, 1662–1667.

Zafiriou, O. C., and True, M. B. (1986). Interferences in environmental analysis of NO by NO plus O_3 detectors: A rapid screening technique. *Environ. Sci. Technol.* **20**, 594–596.

Zhang, J., and Snyder, S. H. (1992). Nitric oxide stimulates auto-ADP-ribosylation of glyceraldehyde-3-phosphate dehydrogenase. *Proc. Natl. Acad. Sci. U.S.A.* **89,** 9382–9385.

Zhu, L., Gunn, C., and Beckman, J. S. (1992). Bactericidal activity of peroxynitrite. *Arch. Biochem. Biophys.* **298,** 452–457.

Zorumski, C. F., and Izumi, Y. (1993). Nitric oxide and hippochampal synaptic plasicity. *Biochem. Pharmacol.* **44,** 777–785.

2
Nitric Oxide Complexes of Metalloproteins: An Introductory Overview

J. C. Salerno
Center for Biochemistry and Biophysics and Biology Department
Rensselaer Polytechnic Institute
Troy, New York 12181

I. INTRODUCTION

It is now clear that nitric oxide (NO) has numerous roles in biological systems. These include functions as a signal molecule in smooth muscle and nerve (Palmer et al., 1988; Garthwaite et al., 1988; Knowles et al., 1989) and as a killing agent directed against target cells by the immune system (Hibbs et al., 1988; Marletta et al., 1988; Steuhr and Nathan, 1989; Steuhr et al., 1989; Lancaster, and Hibbs, in press). Research in this area has experienced explosive growth after the relatively recent discovery of the physiological importance of NO; this has led to the participation of biomedical scientists with a wide range of expertise. It is not surprising, therefore, that investigators on the cutting edge of exciting biomedical developments are apparently often unaware of previous work that has great potential relevance to their fields of interest.

It has long been known that NO forms colored, paramagnetic complexes with many transition metals (Kon, 1968; Yonetoni et al., 1972; Chevion et al., 1977; Hille et al., 1979; Morse and Chen, 1980; Hori et al., 1981; LoBrutto et al., 1983; Kon, 1975; Kon and Katakoa, 1969); the inorganic chemistry literature goes back at least to the early years of the twentieth century. Because NO has an odd number of electrons, binding of NO to an even electron metal ion produces a species amenable to study by electron paramagnetic resonance (EPR). This has led to the extensive use of NO as a spin probe for the study of the active sites of metalloenzymes in biochemical research.

The close relationship between O_2, CO, and NO binding by metal centers, particularly ferrous hemes, has led to a special interest in the use of NO to probe the environment of the prosthetic group in heme proteins. Nitric oxide may be a catalytic intermediate in nitrite reductases (Cammack et al., 1978; LeGall et al., 1979). While the usefulness of NO as a paramagnetic O_2 and CO analog has made it especially valuable in the study of heme-containing oxygenases, oxidases, and oxygen carriers, NO binds to a wide variety of metal-containing proteins and metal complexes.

The literature is rich with examples of metal–nitrosyl complexes, and it would be surprising if the generation of NO by the immune system did not result in the formation of many such adducts. Previous articles have presented summaries of metal proteins that form NO complexes (Butler et al., 1985; Henry et al., 1993), and more recently evidence has mounted that generation of NO by the immune system and by endothelial cells produces a variety of iron–nitrosyl complexes (Mulsch et al., 1993; Vanin et al., 1993; Lancaster et al., 1994). It is unclear which of the potential products will prove to be of physiological relevance, but because the enzymes that may be involved range from the central focus of oxidative cellular metabolism (LoBrutto et al., 1983) to the enzymes of DNA repair (Asahara et al., 1989), the list of potential targets is long and varied.

Because this chapter includes material that may be difficult for interested workers with different backgrounds, an attempt has been made to organize sections so that some material may be skipped without loss of continuity. The boxed sections provide some insight into aspects of the physical chemistry of metal nitrosyl complexes, but the chapter can be read without them.

II. NITRIC OXIDE AS A PARAMAGNETIC LIGAND

The ground state electronic configuration of the oxygen atom is $1s^2 2s^2 2p^4$. The ground states of atomic nitrogen and carbon are $1s^2 2s^2 2p^3$ and $1s^2 2s^2 2p^2$, respectively; nitrogen has an odd number of electrons (seven), while oxygen and carbon have even numbers of electrons (eight and six, respectively).

Dioxygen has 16 electrons. In a naive valence bond model, a double bond between the two oxygen atoms would result in a spin of 0. In fact, the ground state of free dioxygen has a net spin of 1, which can be rationalized by assuming some triple bond character. This allows the remaining single electrons associated with each oxygen atom to assume a parallel spin configuration.

The singlet (S = 0) state lies about 1000 cm^{-1} above the ground state triplet (S = 1) in the EPR spectrum of free dioxygen. Transitions associated with triplet oxygen in solution are detectable by EPR at low temperatures, but dioxygen complexes with even electron metal centers (e.g., ferroheme) are not generally observable by this method. Usually, only odd electron systems (Kramers' systems) are detectable by magnetic resonance.

Kramers' theorem (Kramers, 1930; Griffiths, 1961; Tinkham, 1964) states that in the absence of external fields the ground state of an odd electron system must be at least twofold degenerate. This means that in the absence of such a field two states always have the same energy; the Kramers' doublet they form is split in energy by application of a magnetic field. It is only necessary to increase the magnetic field so that the resulting Zeeman splitting is equal to the energy of the exciting radiation for transitions to be induced.

On the other hand, the ground state of an even (non-Kramers') spin system is usually a singlet even in the absence of an applied field. This is obviously the case for S = 0. Even when S = 1 or a larger integer, the spin multiplet is, in general, subject to zero field splittings. Transitions between states differing in M_s by +1 are allowed, but they can be observed only if the zero field splittings are not large compared to the energy of a microwave quantum. Because these splittings are typically on the order of a few wave numbers for iron group transition metal complexes, and because the energy of an X-band (the most widely used EPR frequency, around 9 GHz) microwave quantum is about 0.3 cm^{-1}, integral spin systems are not usually observed in conventional EPR experiments. In some cases, small zero field splittings allow the observation of multiple transitions within the spin multiplet (Hagen, 1982).

Carbon monoxide has 14 electrons, which pair to give a net spin of zero. Carbon monoxide complexes of transition metals, like oxygen complexes, cannot convert an even electron system to an odd electron system. In the case of iron, CO usually binds only to ferrous ions, which have six 3d electrons. As a consequence, CO complexes and O_2 complexes with iron-containing proteins are generally not detectable by EPR.

Carbon monoxide binds end on to ferroheme with the carbon atom facing the iron (Scheidt and Piciulo, 1976; Jameson et al., 1981; Peng and Ibers, 1976). It is in this regard a good analog for O_2, which is constrained to bind only end on in many heme proteins, but CO differs from dioxygen in important details. It is a much more tightly binding ligand than dioxygen, and it binds with a different geometry than O_2. Carbon monoxide is a potent inhibitor of heme-containing enzymes.

Nitrous oxide has 15 electrons. Transition metal complexes of NO may be pictured as having spins of $S_m + \frac{1}{2}$, where S_m is the spin of the metal center alone. Since the spin of $\frac{1}{2}$ contributed by the NO may add or subtract from the spin of the metal center, a variety of possibilities are open.

A characteristic of the magnetic resonance spectra of nitrosyl complexes is the presence of hyperfine interactions between the unpaired electrons and the nuclear spin of the nitrogen. The strength and orientation dependence of these interactions are determined by the geometry of the complex and the localization of unpaired electrons in ligand and metal orbitals.

The nuclear spin I of ^{14}N, the most abundant naturally occurring isotope of nitrogen, is 1. The three allowed orientations of I with respect to the electron spin S give rise to $2S + 1$ states which differ in energy through the term $\mathbf{I} \cdot \mathbf{A} \cdot \mathbf{S}$ in the absence of an applied magnetic field. Each electron spin resonance transition is split into three resonances by the effects of this hyperfine interaction. The hyperfine splitting is resolved in many but not all nitrosyl complexes; if it is not resolved the hyperfine coupling will contribute to the line width and can be detected by other methods, for example, ENDOR (electron nuclear double resonance) and PFS (pulsed field sweep) EPR (Scholes, 1979; Falkowski et al., 1986).

Nitrogen-15, a readily available isotope of nitrogen, has a nuclear spin of $\frac{1}{2}$ and a somewhat larger magnetic moment. The two allowed orientations of I with respect to S produce a doublet hyperfine splitting. It is often easier to resolve the doublet produced by ^{15}N than the triplet produced by ^{14}N, and the resulting spectra are simpler and easier to interpret.

Nitric oxide binds in the end on position to ferrous hemes, with the nitrogen atom facing the iron (Scheidt and Piciulo, 1976; Jameson et al., 1981; Peng and Ibers, 1976). It binds more tightly to ferrous iron, but has a greater ability than CO to bind to iron in the ferric state, producing a diamagnetic complex. Ferrous–NO complexes are reasonably stable compared to ferrous–O_2 complexes, but in many cases they can be oxidized or reduced. Nitric oxide is much more likely than CO to react at a metal site rather than to bind reversibly.

In addition to the widely known NO complexes of ferrous hemes, NO forms complexes with iron–sulfur proteins (Salerno et al., 1976; Woolum et al., 1968) other nonheme iron proteins (Galpin et al., 1978; Salerno and Siedow, 1979; Lipscomb et al., 1979), as well as copper proteins (Verplaest et al., 1981; Spira and Solomon, 1983; Himmelwright et al., 1980). The properties of NO as a paramagnetic ligand and as a relatively stable dioxygen analog have led to wide application of the molecule by enzymologists and spectroscopists interested in enzyme structure and mechanism. The remainder of this chapter provides information on some of the nitrosyl complexes that may be relevant to the study of NO as a physiologically important molecule.

III. IRON MODEL COMPLEXES

Ferroheme–NO complexes are the best known and most thoroughly studied NO adducts with biochemical relevance. Some of the vast body of existing work on heme protein–nitrosyl complexes is discussed later in Section IV. A number of ferroheme–nitrosyl model complexes have been studied that provide insight into the binding of NO by ferrohemes and allow the role of the protein in modifying the resulting adduct to be assessed.

The nitrosyl complex of ferrous tetraphenylporphyrin 1-methylimidazole [F(II) TPP 1-Me-Im] has been crystallized and its structure determined by X-ray crystallography (Scheidt and Piciulo, 1976; Jameson et al., 1981). The NO group binds end on to the iron with an Fe–N–O bond angle of 140°. This bent structure is reflected in the NO adducts of heme proteins. In contrast, CO model complexes tend to have linear Fe–C–O bond geometries; for example, the bond angle for ferrous tetraphenylporphyrin pyridine [Fe(II) TPP Py] CO is 179° (Peng and Ibers, 1976). This is the result of differences in hybridization in the ligand orbitals. Dioxygen binds in a bent configuration similar to that of NO; the Fe–O–O bond angles for similar model complexes of dioxygen are about 130°, close to the value of 120° expected purely on the grounds of sp^2 hybridization. As discussed later, in proteins the polypeptide can impose other geometries on the complexes. In addition to the bond angle, interatomic distances and the projection of the NO vector on the porphyrin plane can be affected.

Simple complexes of ferrous iron in solution often readily bind NO, forming paramagnetic adducts that can readily be observed in EPR experiments. A solution of ferrous iron and cysteine will react with NO to form a complex in which each iron has two cysteine ligands, an NO ligand, and an NO$^-$ ligand. This brownish complex has a broad absorbance in the visible region and an approximately axial electron paramagnetic resonance spectrum with $g_x \approx g_y \approx 2.04$ and $g_z \approx 1.99$ (Woolum et al., 1968). The deviation from the free electron value is caused by orbital angular momentum contributions from the iron d orbitals. The signals can be observed from very low temperatures (at least to 10 K), where microwave power saturation is difficult to avoid, up to at least the freezing point. No hyperfine splittings can be resolved at X-band, but unresolved hyperfine couplings to the two nitrogen nuclei almost certainly are responsible for the relatively broad signals (considering the modest g tensor anisotropy, slow relaxation rate, and $S = \frac{1}{2}$ character of the complex).

Similar complexes can be prepared using other thiol-containing reagents and/or by starting with hydroxylamine or with nitrite and a reductant such as sodium dithionite. Formation of the complex is favorable enough so that it can be quantitatively prepared. The reaction may be used as an assay for free iron (Salerno et al., (1976).

In the absence of thiols or other strong ligands, iron complexes may be formed in solution with ionic ligands including phosphates and carboxylic acid groups (McDonald et al., 1965). Treatment of such a solution of ferrous iron with NO results in the formation of minority $S = \frac{1}{2}$ species with g_{av} slightly greater than 2.0 (owing to orbital angular momentum contributions from the iron d orbitals) and resolved hyperfine splitting from the nitrogen nucleus. Unlike the cysteinyl–iron–NO complexes, these species have never been quantitatively prepared.

The original studies (McDonald et al., 1965) that reported the formation of these species were limited by the available technology to temperatures above 77 K. At those temperatures, the minority $S = \frac{1}{2}$ species are the only ones detectable by EPR, but at temperatures below 25 K, attained by the use of liquid helium as a refrigerant, other species can be observed (Salerno and Siedow, 1979; Rich et al., 1978). These species have much faster relaxation rates because they are not $S = \frac{1}{2}$ systems.

When NO is passed through a solution of ferrous iron in which only ionic ligands are available, the solution becomes brown or green because of the formation of nitrosyl complexes. The description of a brown ring compound in aqueous solution was an early example of the formation of an iron–nitrosyl complex. Under many conditions, a bright green color may be obtained; addition of NO to a solution of ferrous ethylenediaminetetraacetate [FE(II)EDTA], for example, produces a bright green solution due to the presence of a nitrosyl adduct with charge transfer bands near 400 and 620 nm (J. C. Salerno, unpublished observations). The green color probably indicates an anionic complex.

In both brown and green solutions, an intense EPR spectrum can be observed at low temperatures with $g_x \approx g_y \approx 4$ and $g_z \approx 2$. This set of g values is characteristic of an $S = \frac{3}{2}$ species with a strong axial zero field splitting (see Section IX,D) (Salerno and Siedow, 1979; Rich et al., (1978). Rapid relaxation, probably due to Orbach processes involving low-lying excited states, causes the signals to broaden rapidly as the temperature is raised. Spectroscopically similar species can be observed in aqueous solutions in the presence of a wide variety of ionic ligands; the coordination of iron after formation of the NO adduct is unknown.

Ferrous tetramethylcyclam forms an NO adduct for which a crystal structure is available (Hodges et al., 1979). The iron atom is out of plane with respect to its four nitrogen ligands; the single NO ligand occupies a position at the vertex of the pyramid. Although this complex was originally described as having $S = \frac{1}{2}$, the EPR spectrum of the adduct has features near $g = 4$ and $g = 2$. Hodges et al. attributed this anisotropy to orbital angular momentum effects, but the magnetic resonance spectra strongly suggest that the system is $S = \frac{3}{2}$. Spin $\frac{3}{2}$ nitrosyl complexes can thus be formed in which the other ligands are either oxygen or nitrogen atoms.

IV. NITRIC OXIDE COMPLEXES OF FERROHEMES IN PROTEINS

Heme–NO complexes are the best known and most thoroughly studied of the biological metal–nitrosyl complexes. So many heme enzymes have been studied using NO as a probe that it is neither practical nor desirable to mention them all here; instead, a few illustrative examples are given.

The functions of hemoglobin and myoglobin depend on their unusual ability to reversibly bind molecular dioxygen without significant catalysis of reduction to superoxide, peroxide, or water. In addition, both proteins have a greatly decreased affinity for CO relative to dioxygen in comparison to heme model complexes; the cooperative behavior of hemoglobin is one of the most thoroughly studied problems in biochemistry (Fermi and Perutz, 1981). Despite these and other unusual specializations, these two proteins are commonly used as the standard for comparison for other heme proteins because of the existence of numerous classic studies devoted to their structure and function. The availability of single crystal EPR data in addition to crystal structures has provided the opportunity for unusually complete understanding of these systems to be developed.

Compared to model complexes such as fe(II) TPP 1-Me-Im and Fe(II) TPP Py, hemoglobin and myoglobin both form complexes with ligands with altered geometries (Scheidt and Piciulo, 1976; Jameson *et al.*, 1981; Peng and Ibers, 1976; *The Porphyrins*, 1978; Gouterman *et al.*, 1986). This is particularly true for ligands such as CO which bind in a linear configuration normal to the heme plane in model systems; the structure of the heme pocket apparently effectively prevents such binding in these proteins. The ligands instead bind in a bent configuration (the Fe–C–O bond angle in myoglobin is 137° as compared to 179° in the TPP model complex), or the CO ligand is tilted with respect to the heme normal. This is the result of interactions with histidine and valine residues adjacent to the heme iron.

Binding of dioxygen is less constrained; values as low as 115° have been reported for the Fe–O–O bond angle, but derivatives with virtually the same Fe–O–O bond angle, but derivatives with virtually the same Fe–O–O bond angle as model complexes (~130°) have also been observed. Nitrosyl complexes of hemoglobin have similar Fe–N–O bond angles (145° versus 140°) compared to the model systems. The plane defined by the positions of these three atoms is constrained with respect to rotation about the heme normal, however, in both the dioxygen and nitrosyl adducts. In model complexes the orientation of this plane with respect to the plane of the axial base minimizes steric interactions between the axial ligands and the porphyrin, but this is not the case in the proteins.

Electron paramagnetic resonance has been widely used to study the nitrosyl complexes of hemoglobin and myoglobin (Kon, 1975; Kon and Katakoa, 1969). Modest g tensor anisotropy arises from delocalization of the unpaired electron into the empty e_g orbitals of the ferrous iron; the largest value of **g** is about 2.07 while the other principal values are about 2.0 and 1.98. Hyperfine coupling is fully resolved only in the central region of the spectrum. The maximum value hyperfine coupling tensor is about 60 MHz, corresponding to slightly more than

20 gauss in an X-band EPR experiment; the other principal values of the **A** tensor are about half this large. Because of the bent Fe−N−O configuration, the **g** and **A** tensors are not colinear. The **A** tensor orientation is dominated by the direction of the N−O bond, while the largest value of **g** lies approximately along the orientation in the heme plane perpendicular to the Fe−N−O plane.

The α and β chains of hemoglobin form slightly different nitrosyl complexes (Shiga et al., 1969; Henry and Banerjee, 1973). The sensitivity of the α chain−NO complex to the state of the β chain heme has been used to study cooperativity in hemoglobin. Organic phosphates such as 2,3-diphosphoglycerate change the EPR spectrum of hemoglobin NO; this may correspond to an R/T transition, and probably implies the weakening or breaking of the metal−distal ligand bond.

Yonetoni and co-workers (1972) have shown that hemoglobin and myoglobin form nitrosyl complexes with different bond angles at 77 K and at room temperature. The high-temperature species has less **g** tensor anisotropy (g_z = 2.03, g_y = 1.98–1.99) and poorly resolved hyperfine splitting. Addition of glycerol at high concentrations prevented the transition between these forms.

Horseradish peroxidase (HRP) is another enzyme which has been extensively studied and which forms well-defined heme−nitrosyl complexes. Hori et al. (1981) reported that the nitrosyl complex of HRP had g_x = 2.08, g_y = 2.004, and g_z = 1.95. The **A** tensor produced a maximal splitting near g_z; with [14]NO this was about 22 gauss. A smaller splitting, of about 6 gauss, was produced by hyperfine coupling to a histidyl nitrogen. Substitution of [57]Fe introduced a third hyperfine coupling with a maximum splitting of about 6 gauss. Use of [15]NO produced a doublet of triplets, because ^{15}N is an $I = \frac{1}{2}$ nucleus and the histidyl nitrogen contributing the weaker interaction was still [14]N.

The NO complex of fully reduced cytochrome oxidase has g values of 2.08, 2.005, and 1.98 (LoBrutto et al., 1983). Hyperfine coupling to [14]NO gives rise to a maximum splitting of 21 gauss; the resulting triplet is in turn split into a triplet by interaction with a histidinie ligand. If care is not taken in preparing the sample, a different EPR spectrum lacking the hyperfine splitting from the axial histidine results. The heme iron in this case is believed to be five coordinate.

Addition of NO to oxidized cytochrome oxidase produces a state in which NO binds to the copper center rather than to the heme (Brudvig et al., 1980. The Cu(II)−NO complex is diamagnetic; EPR signals can be observed at $g = 6$ which probably result from the ferric heme a_3, now uncoupled from Cu(II). It is also possible to assign these signals to some $S = \frac{5}{2}$ coupled state involving both iron and cooper, but this is much less likely.

In the presence of NO and azide, cytochrome oxidase forms a complex with integral spin EPR spectra that have been assigned to a triplet state formed by coupling of $S = \frac{1}{2}$ heme and copper centers (Brudvig et al., 1980). This explanation is possible, but other net integral spin possibilities could also explain the observed spectra. It seems probable that under the conditions of the experiment the enzyme is capable of catalyzing redox reactions involving NO and azide.

Nitrite reductase and sulfite reductase are enzymes found in choroplasts and in prokaryotes that reduce nitrite to ammonia and sulfite to sulfide (Scott et al., 1978). Sulfite reductase also catalyzes reduction of nitrite at a lower rate. Both enzymes contain a siroheme prosthetic group linked to an iron–sulfur cluster. In siroheme, the porphyrinoid moiety is present in the more reduced chlorin form. Because NO lies between nitrite and ammonia in oxidation state, it is a potential intermediate.

Both siroheme enzymes form ferroheme–NO complexes in which the g value anisotropy appears somewhat smaller than in the corresponding complexes of most other enzymes. The EPR spectra of the complexes somewhat resemble the spectra of the high-temperature myoglobin–NO complexes. The hyperfine splitting from the NO nitrogen nucleus is evident at intermediate g values but is not well resolved. These enzymes are capable of reducing NO to ammonia if supplied with low potential reducing equivalents. Other heme proteins also catalyze oxidation reduction reactions with NO.

Cytochromes P450 form a very large group of heme enzymes that catalyze the hydroxylation of a variety of substrates. They are important in drug metabolism, in cholesterol and steroid hormone biosynthesis, and in numerous other pathways. They have been found to participate in reactions other than hydroxylations.

Nitrosyl complexes of cytochrome P450 have poorly resolved **g** tensor anisotropy; the principal g values are around 2.07, 2.004, and 1.98 (Stern and Peisach, 1976; O'Keefe et al., 1978). The lack of resolution is probably due at least in part to the presence of multiple species. Well-resolved hyperfine splitting from the NO is present at intermediate g values. No hyperfine splitting from the proximal axial ligand can be observed; in P450, this ligand is a cysteinyl sulfur. Conversion of P450 to the inactive P420 form shifts the EPR spectrum of the nitrosyl complex in a manner consistent with the loss of the proximal ligand; no additional hyperfine splittings appear.

V. NITRIC OXIDE COMPLEXES OF IRON–SULFUR PROTEINS

The iron–sulfur proteins include small proteins that function as remarkably simple electron carriers and large multisubunit complexes with multiple activities. Although many of these enzymes function in electron transfer in bioenergetic or biosynthetic pathways, it has become clear that iron–sulfur proteins catalyze a broad array of reactions not always involving electron transfer.

The prosthetic groups of iron–sulfur proteins fall into several classes (Lovenberg, 1977; Spiro 1982). Rubredoxins bind single iron atoms with four cysteinyl sulfur ligands; they function as electron carriers in some bacterial systems. Rubredoxins generally have two such centers per molecule; in the ferric state each iron center is EPR detectable.

The 2Fe2S* (S*, acid-labile sulfur) ferredoxins have a redox active binuclear center, with each of the two iron atoms attached to the protein by two cysteinyl sulfur ligands and connected by two inorganic acid-labile sulfur ligands. At cryogenic temperatures these clusters are EPR detectable, with characteristic features in the vicinity of $g = 1.94$. Spinach ferredoxin has principal g values of 2.03, 1.96, and 1.88 and a broad absorbance spectrum with a weak maximum around 420 nm, giving these proteins a reddish brown color which bleaches on reduction. Ferredoxins are low potential electron carriers; chloroplast ferredoxins function in photosynthetic electron transfer, but related proteins such as adrenal ferredoxin are involved in steroidogenic electron transfer in mitochondria in tissues which produce steroid hormones.

Ferredoxins with 4Fe4S* clusters are small, low potential electron carriers that function in bacterial electron transfer. Like the binuclear clusters, each tetranuclear cluster can reversibly accept a single electron. The tetranuclear prosthetic group is a cubane with iron and acid-labile inorganic sulfide groups at alternate vertices; the four iron atoms and the four sulfides form interpenetrating tetrahedra of slightly different sizes.

Many bacterial ferredoxins have two such clusters, each of which can be reduced to a paramagnetic state. In other proteins, 4Fe4S* clusters can be oxidized to a paramagnetic state. The tetranuclear clusters in these two types of proteins are similar in structure, but they functionally shuttle between different reduction states. In both cases, each iron atom is additionally coordinated by four cysteinyl sulfur ligands.

The optical and EPR spectra of the clusters which are paramagnetic when reduced are only slightly different than those of the 2Fe2S* proteins; the EPR spectra of the 4Fe4S* clusters tend to be most easily studied at lower temperatures (10–15 K versus 20–30 K) and to have numerically larger values of g_z (2.04–2.12 versus 2.01–2.03) compared to the binuclear clusters. The clusters which are paramagnetic when oxidized have less anisotropic EPR spectra with an average g value of greater than 2.

In addition, 3Fe4S* clusters also function as electron carriers. The structure of these complexes has been determined by X-ray diffraction; 3Fe4S* sites can resemble 4Fe4S* sites with a missing cysteinyl residue (Ghoush et al., 1982). They are paramagnetic when oxidized and have EPR spectra that are more isotropic than those of the other clusters (Emptage et al., 1983). All three g values may be within 0.02 of 2.00, but relaxation is rapid compared to free radicals because of the presence of low-lying excited states.

More complex iron–sulfur clusters are also known to exist. These include the iron–molybdenum cofactor of nitrogenase (Thornely and Lowe, 1984) and probably larger clusters in which the only metal is iron (Hagen, 1987). They are characterized by highly anisotropic EPR spectra from $S > \frac{1}{2}$ ground states; the nitrogenase cluster, for example is $S = \frac{3}{2}$ and has EPR features near $g = 4$ and $g = 2$.

Many complex enzymes contain multiple iron–sulfur clusters (Hatefi, 1985; Ohnishi, 1987). Mitochondrial succinate dehydrogenase contains 2Fe2S*, 3Fe4S*, and 4Fe4S* clusters. NADH dehydrogenase contains at least seven iron–sulfur clusters of the 2Fe2S* and 4Fe4S* classes. In addition to the involvement of these and other iron–sulfur enzymes in energy-related electron transfer, iron–sulfur proteins such as adrenal ferredoxin function in biosynthetic electron transfer pathways (Omura et al., 1967).

The classic picture of iron–sulfur clusters included the binding of the iron to the polypeptide through cysteinyl sulfur ligands. It has become clear, however, that amines and carboxylic acid groups are sometimes substituted for one or more cysteinyl ligands; this alters the spectroscopic and thermodynamic properties of the iron–sulfur center. The Rieske iron–sulfur center, which functions as the electron acceptor from quinol in the quinol cytochrome c reductase of the mitochondrial respiratory chain, has two histidyl nitrogen and two cysteinyl sulfur ligands (Gabriel et al., 1989a,b). It has a high potential ($E^{\circ\prime} \approx 260$ mV) and an unusually anisotropic EPR spectrum with features near g values of 2.02, 1.90, and 1.78.

It is now clear that in addition to their widespread involvement in electron transfer pathways, iron–sulfur clusters function as catalytic centers in a wide variety of enzymes. The first example of such an enzyme is aconitase. It was at first thought that the role of the iron–sulfur group was regulatory, but it is now clear that in this enzyme the iron–sulfur group is part of the catalytic site. One of the iron atoms can coordinate water or hydroxyl and plays a key role in the isomerization catalyzed by the enzyme (Emptage et al., 1983).

Aconitase, although not an electron transfer protein, is involved in cellular energy production in the tricarboxylic acid (TCA) cycle. Other iron–sulfur enzymes are now known to catalyze reactions that are related neither to bioenergetics nor to electron transfer. In particular, some endonucleases have been shown to have iron–sulfur prosthetic groups. Endonuclease III, a DNA repair enzyme from *Escherichia coli*, has a tetranuclear iron–sulfur cluster similar to that found in aconitase (Asahara et al., 1989). This enzyme catalyzes excision reactions involving hydrolytic cleavage of bonds. Reviews have described numerous examples or iron–sulfur-containing enzymes catalyzing a wide variety of reactions not directly related to electron transfer (Beinert, 1990).

In general, reaction of iron–sulfur proteins with NO produces cysteinyl$_2$–Fe–(NO)$_2^-$ complexes similar to those described earlier (section III; see also Butler et al., 1985; Henry et al., 1993; Mulsch et al., 1993; Vanin et al., 1993; Lancaster et al., 1994). The g values and overall line shape can be very slightly different because of the effects of the polypeptide on the geometry of the complex; free dithiol–dinitrosyl complexes are also possible. This suggests that the iron–nitrosyl complex not only remains associated with the protein in some cases, but that conformational or steric factors allow the polypeptide to influence the geometry of the complex. Similar complexes have been observed after the reaction of succinate dehydrogenase, spinach chloroplast ferredoxin, NADH dehydrogenase, adrenal ferredoxin, and Rieske iron–sulfur protein with NO, hydroxylamine, or nitrite and a reductant. Destruction of bacterial iron–sulfur centers with NO was also accompanied by formation of the complex (Reddy et al., 1983).

Quantitative conversion of the iron in succinate dehydrogenase to this form is possible if additional cysteine is added to the reaction mixture. It is probable that not enough cysteinyl sulfur ligands are available for complex formation without addition of the extra cysteine; some of the nitrosyl complex does form without any cysteine addition in these systems.

Although 2Fe2S* clusters such as the one in spinach chloroplast ferredoxin have enough cysteinyl residues to allow quantitative formation of the $Cys_2-Fe-(NO)_2^-$ complex with no added cysteine, trinuclear and tetranuclear clusters require only one cysteine per iron atom. Substitution of other residues (histidine, glutamic acid, aspartic acid) or water as a ligand for the iron can further reduce the content of cysteine in the vicinity of the iron in the native protein. It is to be expected that addition of cysteine would be needed to quantitatively form the $Cys_2-Fe-(NO)_2^-$ complex in all but the classic binuclear complexes, or in iron–sulfur clusters which at least had access to neighboring cysteinyl residues.

Formation of this nitrosyl complex requires destruction of the iron–sulfur group, probably involving the liberation of the acid-labile sulfur and disruption of the local three-dimensional (3D) polypeptide folding. Intact succinate dehydrogenase buffered at pH 8.0 forms nitrosyl complexes slowly in the presence of nitrite and reductant, while lowering the solution below pH 6.0 allows rapid complex formation. Replacement of NO with argon or nitrogen does not regenerate the intact iron–sulfur cluster, although the EPR signal of the iron–nitrosyl complex decreases with time.

In the absence of added cysteine, the EPR signals of other nitrosyl complexes can sometimes be observed when iron–sulfur-containing enzymes are exposed to NO. In particular, EPR signals near $g = 4$ are often observed at low temperature, indicating the formation of a spin $\frac{3}{2}$ complex. It is likely that this represents NO complexes of iron from dissociated iron–sulfur clusters for which there are insufficient thiol ligands. It is possible, however, that NO can be coordinated by some iron–sulfur clusters without dissociation of the cluster and that some such nitrosyl complexes can give rise to $S = \frac{3}{2}$ states. In particular, it is possible that under some conditions NO and other ligands may be able to bind to those iron–

sulfur clusters that catalyze group transfers rather than electron transfers and hence have positions in their coordination sphere which are filled by weak ligands such as water. Such complexes could be very different from other NO complexes.

VI. NITRIC OXIDE COMPLEXES OF OTHER NONHEME IRON PROTEINS

In addition to the large number of enzymes with iron–sulfur prosthetic groups, many other enzymes are nonheme iron proteins in which the iron is coordinated by different ligands. A large class of such nonheme iron proteins is the dioxygenases, which catalyze the incorporation of dioxygen into a variety of substrates. These enzymes are often employed as ring cleavers in degradation of substrates by microorganisms (e.g., the catechuate dioxygenases; Palmer, 1980) but may also be involved in biosynthetic pathways (e.g., soybean lipoxygenase, which catalyzes lipid peroxidation; Que et al., 1976).

Many of these enzymes have EPR spectra characteristic of ferric high-spin iron in a rhombic environment (E/D close to $\frac{1}{3}$, where E and D are the rhombic and axial zero field-splitting parameters), or can be oxidized to such a state. The most readily observed EPR features from such a system are the middle Kramers' doublet resonances near $g = 4.2$, although features from the ground state can often be observed near $g = 9$; the other features from the ground state are at very high field and difficult to detect. Addition of substrate or inhibitor typically changes the geometry of the system to produce a more axial $S = \frac{5}{2}$ system with g values around 6. This system differs from the axial high-spin heme in that the sign of D in the dioxygenase system is apparently negative; the ground state is $S_z = +\frac{5}{2}$, and the EPR spectrum is from an excited state doublet.

Treatment of soybean lipoxygenase with NO produces a nitrosyl complex very different from the product of the reaction of NO with iron–sulfur proteins. The EPR spectrum of the complex, with features near 4.1, 3.9, and 2.0, is indicative of an $S = \frac{3}{2}$ system (Galpin et al., 1978; Salerno and Siedow, 1979). This suggests that the ligands of the iron are nitrogen and/or oxygen rather than sulfur. Evidence has been presented that the lipoxygenase pathway in platelets is inhibited by NO generated either enzymatically from arginine or from nitroprusside, but no connection was made with the early studies on soybean lipoxygenase (Nakatsuka and Osawa, 1994).

The protocatechuate 3,4- and 4,5-dioxygenases also react with NO to form $S = \frac{3}{2}$ complexes (Lipscomb et al., 1979). The EPR spectrum of the 3,4-dioxygenase–nitrosyl adduct has g values of 4.3, 3.7, and 2.0, whereas the EPR spectrum of the nitrosyl adduct of the 4,5-dioxygenase is similar to that of the lipoxygenase NO adduct. The differences in the EPR spectra are the result of the

E/D ratios. The differences in coordination geometry implied by the EPR spectra need not be great.

In addition to the EPR spectra features at $g = 4$ and $g = 2$ assigned to $S = \frac{3}{2}$ complexes of the enzymes, a broad signal near $g = 1.9$ is often observed in the presence of NO. Similar signals have been attributed to matrix-associated NO radicals. However, the presence of strong chelators such as EDTA sometimes reduces the appearance of the $g = 1.9$ signals; also, the signals are visible at low temperature (~ 10 K), and the shape and position of the signals are variable. They clearly represent paramagnetic NO in some form, but the species responsible for the signals is not yet well understood.

Other enzymes that are not obviously related to the dioxygenases have at least superficially similar metal sites. The fatty acid desaturase of the endoplasmic reticuluum is a nonheme iron protein and requires both oxygen and reducing equivalents for activity (Strittmatter and Enoch, 1978). It is not known whether this enzyme forms a nitroxyl complex, but rat liver microsomes containing the enzyme form an $S = \frac{3}{2}$ nitroxyl adduct when treated with nitrite and dithionite.

Iron-containing superoxide dismutases are present in many species of bacteria (Hassan and Fridovitch, 1978). These nonheme iron proteins have a characteristic set of EPR lines split about $g = 4.2$ in the ferric state, arising from the middle Kramers' doublet of a rhombic high-spin site. Ferrous iron superoxide dismutase forms an $S = \frac{3}{2}$ complex with NO that resembles the lipoxygenase–NO adduct by EPR criteria (I. Fridovich, T. Kirby, and J. C. Salerno, (1978) unpublished observations).

Metal storage proteins also form a variety of nitrosyl complexes. Metallothionein has been shown to form a low-spin complex with Fe^{2+} and NO (Kennedy et al., 1993). Ferritin contains multiple iron binding sites capable of forming spectroscopically distinguishable nitrosyl complexes (Lee et al., 1994; LeBrun et al., 1993).

The iron responsive element, a critical factor in the control of proteins involved in iron utilization, has been identified as the cytoplasmic form of the iron–sulfur protein aconitase (Kennedy et al., 1992). Activated macrophages have been shown to activate this element, presumably by attack of the iron–sulfur cluster by NO (Drapier et al., 1993). It has been claimed that this attack is mediated by peroxynitrite (Castro et al., 1994; Hausladen and Fridovich, 1994, but this conclusion is not universally accepted.

The nonheme iron enzymes discussed so far in this section either utilize oxygen as a substrate or form it as a product. Other nonheme iron sites that do not bind O_2 as part of their catalytic function have similar ligand environments. An example of such a system is the QFe site associated with the reaction centers of photosynthetic bacteria and with photosystem II of chloroplasts (Feher et al., 1989).

VII. COPPER COMPLEXES

The most widely studied nitrosyl complexes of biological significance are those of Fe(II); these compounds are widely distributed, paramagnetic, and of proven value as probes of metal components of biomolecules. Copper is also a component of many proteins; it is present at the active site in a large number of enzymes, and in some organisms copper proteins such as hemocyanin fill the oxygen transport role taken by hemoglobin vertebrates. In particular, copper is the site of oxygen binding in many oxygenases and oxidases. As in enzymes in which that role is filled by iron, NO is an analog for O_2. The copper centers of these proteins thus also tend to form nitrosyl complexes.

Copper in aqueous solution, and in proteins, is usually present in the Cu(I) or Cu(II) states. Cu(I) has 10 3d electrons, constituting a filled shell. All the electrons are paired and $S = 0$. Cu(II) has 9 3d electrons; Cu(II) complexes are paramagnetic with $S = \frac{1}{2}$. Because the 3d orbitals are more than half-filled, orbital angular momentum contributions from the d orbitals cause the average g value to be over 2. The EPR spectra are also affected by hyperfine coupling to the copper nucleus; both of the common natural isotopes have $I = \frac{3}{2}$.

In electron carrying cuproproteins such as plastocyanin (Ryden, 1984), the copper shuttles reversibly between the Cu(I) and Cu(II) states. Ligand binding and the catalysis of additional reactions by the metal center is inhibited by the polypeptide, which provides the copper center with a filled coordination sphere. In oxygenases and oxidases, an open ligand position enables copper to bind a variety of exogenous ligands. Copper, like iron, binds dioxygen in the reduced form, which in its case is Cu(I). Whereas NO binds preferentially to Cu(I), forming a paramagnetic complex, Cu(II)NO, which may produce Cu(I)NO$^+$, is also observed. Like Fe(III)NO, this is an even electron complex not detectable by EPR.

The binuclear copper center of molluscan hemocyanin binds a single NO molecule (Verplaest et al., 1981; Spira and Solomon, 1983). In the resting state both metal ions are cupric; the Cu(II)–Cu(II) spin-coupled cluster gives rise to broad, relatively weak signals near $g = 2$ and at the half-field position (apparent g value near 4). Binding of a single NO appears to yield a Cu(II)–Cu(I)–NO$^-$ state, with an EPR spectrum characteristic of a single copper in the cupric state.

Binding of NO to the spin-coupled copper pair of tyrosinase (Schoot Uterkamp and Mason, 1973; Malmstrom, 1978) yields a dipolar coupled pair in which both copper ions bind NO; the EPR spectra of the NO-complexed enzyme exhibits both broadened $S_z = 1$ transitions and $S_z = 2$ transitions.

Nitric oxide binds to the Cu(II) ion in the binuclear center of fully oxidized cytochrome oxidase (Brudvig et al., 1980). The binding of NO creates an even spin copper center and effectively breaks the spin coupling between the heme and copper metal ions. As a result, the high-spin heme EPR signal is visible at $g = 6$.

VIII. CONCLUSION

A large number of iron-containing proteins form nitrosyl complexes. Heme proteins, iron–sulfur proteins, and other iron proteins such as nonheme iron dioxygenases all form characteristic nitrosyl complexes. In enzymes in which the metal center has an open coordination position, NO often can be bound without severe disruption of the site. This introduces the possibility of reversibility of inhibition.

There are at least three mechanisms for the reversal of NO inhibition of an enzyme. The most obvious is that the NO can be released as the concentration of NO in solution falls or the temperature rises. This depends on the (kinetic) off constant for NO in the system in question as well as the (equilibrium) dissociation constant; for most enzymes that form nitrosyl complexes, this has not been measured. Nitric oxide bound to a metal center could also undergo oxidation or reduction followed by the release of product. Finally, NO could be photodissociated from a metal center by light of the correct wavelength.

Iron–sulfur centers form NO complexes that require the disruption of the complex and almost certainly at least some unfolding of the polypeptide. Reversibility of NO binding in these proteins is not enough to regain function; it would be necessary to reform the clusters and partially refold the proteins. Although this is not impossible, it is clearly a more difficult process to imagine.

It appears likely that the range of iron–nitrosyl complexes in proteins is not exhausted. Of particular interest is the admittedly speculative possibility that NO may bind to open ligand positions in iron–sulfur clusters which have open ligand positions and catalyze reactions other than oxidation/reduction chemistry. Such putative nitrosyl complexes might be spectroscopically very different than those which are more familiar. If they were $S = 0$ (or integral spin), they could be undetectable by EPR. Metal centers with open ligand positions may not be irreversibly deactivated by NO binding; on the other hand, they may represent sites of high sensitivity to NO because the metal center is unprotected by a filled coordination sphere of endogenous ligands. It appears that the rapid expansion of our knowledge of metal centers is timely since it appears to be important to understand the effects of biologically produced NO on critical enzymes.

The existence of copper–nitrosyl complexes of biological significance has been briefly discussed here (Section VII). It is worth pointing out that nitrosyl complexes of other metal-containing proteins may form, and that these may be important in understanding the effects of NO on living cells. Nitrosyl complexes of many other metals are well documented (e.g., Werner and Karrer, 1918; Moeller, 1952) and include complexes of nickel, cobalt, and ruthenium. Some such complexes may be less obvious than the paramagnetic and often colorful iron–NO complexes. Certainly, the demonstrated role of NO in biology gives the study of nitrosyl complexes of metalloproteins importance extending beyond their previous use as probes of structure and function.

IX. APPENDIX: ELECTRON PARAMAGNETIC RESONANCE

In electron paramagnetic resonance (EPR) experiments transitions between states characterized by the orientation of unpaired electron spins in a magnetic field can be induced by microwave radiation. This is analogous to the transitions between states characterized by the orientation of nuclear spins in a magnetic field in NMR (nuclear magnetic resonance) experiments. Electron paramagnetic resonance experiments provide information about the environment of the unpaired electrons. Numerous books are available that cover the subject at levels ranging from the introductory (Carrington and McLachlan, 1967) to the advanced (Abragam and Bleaney, 1970).

A. Zeeman Splitting and g Tensor

An electron has a spin S of $\frac{1}{2}$; a single unpaired electron in an atomic or molecular orbital has two possible orientations (parallel or antiparallel) with respect to an applied magnetic field. These are characterized by the spin orientation quantum numbers S_z (sometimes referred to as M_s). For $S = \frac{1}{2}$, S_z can be $+\frac{1}{2}$ or $-\frac{1}{2}$.

A free electron in a magnetic field has two such states; the Zeeman energy term that describes the strength of their interaction with the magnetic field H is $g_e \beta \mathbf{H} \cdot \mathbf{S}$, where g_e is the spectroscopic splitting factor of the free electron (about 2.0023) and β is the Bohr magneton; together, these two constants define the magnetic moment of the free electron. The Zeeman energies of the two states are $g_e \beta \mathbf{H} \cdot \mathbf{S}$ and $-g_e \beta \mathbf{H} \cdot \mathbf{S}$; the separation between the states is $g_e \beta H$.

In an EPR experiment, transitions can be induced between the two spin states by microwave photons of the appropriate polarization. When $h\nu$, the energy of the photon, is equal to $g_e \beta H$, the Zeeman splitting, the system is in resonance and microwave energy can be absorbed by the sample. In all conventional EPR spectrometers, the magnetic field is scanned while the frequency is held constant. The most common frequency region used in EPR is the X-band, around 9 to 9.5 GHz. A free electron would give rise to a single EPR line between 3200 and 3400 gauss at X-band, depending on the exact frequency used. This line would usually appear as the first derivative of an absorption line (often Gaussian or Lorentzian in shape) because EPR spectrometers are usually field modulated as well as field scanned.

In an orbital, the orbital angular momentum L as well as the electron spin must be considered. In a free ion, L and S are tightly coupled to give a total angular momentum quantum number J which interacts with an applied field. In molecules, interactions with neighboring atoms affect the energies and form of the orbitals with the result that **L** and **S** are no longer tightly coupled; this effect is called quenching of orbital angular momentum.

In systems such as ionic complexes of paramagnetic ions the unpaired electrons can be productively treated as if they were completely localized in the metal orbitals. Effects of **L** on the interaction of the electron with a magnetic field can be described by a formalism in which orbitals appropriate to the ligand symmetry of the site are mixed by the spin–orbit coupling energy $\lambda \mathbf{L} \cdot \mathbf{S}$. This has the effect of restoring some of the effect of the orbital angular momentum that had been quenched by the ligand field. Some of the effects of covalent character can be approximated by a reduction in the effective size of the spin–orbit coupling constant.

When the unpaired electron is delocalized over a number of atoms, molecular orbital theory must be applied to obtain a molecular description of the resulting magnetic species. In this situation there is less opportunity for substantial contributions from **L**, and in general the more delocalized the electron the more like a free electron it appears. In some cases, the electron is delocalized over only a few atoms, and in these cases modest contributions from **L** are expected, especially if one of the atoms is a transition metal. If more extensive delocalization is present, or if all the atoms involved are light, only small contributions (e.g., from $2p$ orbitals) may be observed.

Rather than dealing with **S** and **L** separately in these situations, we attribute the entire interaction to a fictitious spin (sometimes distinguished from the true spin by writing it **S′**; here we refer to it merely as **S**). To do so we must assign the spin a value of g different than g_e; since the contributions to the magnetic interactions of $\mathbf{L}_z$, $\mathbf{L}_x$, and $\mathbf{L}_y$ will in general be different g will be direction dependent and can conveniently be represented by a second-order third-rank tensor. The Zeeman interaction for an electron in a molecule is written as $-\beta \mathbf{H} \cdot \mathbf{g} \cdot \mathbf{S}$. It is possible to choose a molecular coordinate system so that the **g** tensor is diagonal. The form of **g** is then

$$\begin{pmatrix} g_x & 0 & 0 \\ 0 & g_y & 0 \\ 0 & 0 & g_z \end{pmatrix}$$

where g_x, g_y, and g_z are the principal values of the **g** tensor. When the magnetic field is aligned with one of the principal axes of the coordinate system in which **g** is diagonal, the Zeeman energy is determined by the corresponding principal value of g. For example, with the magnetic field along the z axis the Zeeman splitting is $g_z \beta H$.

At intermediate orientations the effective g values will also be intermediate. Consider a molecule oriented so that the magnetic field makes an angle θ with the z axis, and the projection of the magnetic field on the xy plane makes an angle ϕ with the x axis. The effective value of g is the $(g_z^2 \cos^2 \theta + g_x^2 \sin^2 \theta \cos^2 \phi + g_y^2 \sin^2 \theta \sin^2 \phi)^{1/2}$, and the Zeeman splitting is $g\beta H$. Each g value can be positive or negative. Only the sign of the product of the three g values is physically meaningful, however; the individual signs change with choice of coordinate systems. The sign of the product does not affect the spectrum in normal EPR experiments; it can be determined by special experiments using circularly polarized microwaves.

In a frozen solution, all possible orientations of the molecule with respect to the magnetic field are always present. Each orientation gives rise to microwave absorption at resonance. If the **g** tensor is isotropic, $g_z = g_x = g_y$ and only a single EPR line will be observed. If the **g** tensor is axial, one of the three g values will be different. If this value ($g_{||}$) is larger than the other two, a positive peak will result at $H = h\nu/g_{max}$. A derivative shaped peak will be present at the resonant field corresponding to g_{min}; the negative excursion will be greater than the positive excursion. If $g_{||}$ is the minimum g value, a negative signal will be observed at g_{min} and the derivative feature at g_{max} will have a dominant positive lobe.

In a rhombic spectrum, the three g values are all different and result in three lines in the EPR spectrum. Here, g_{max} and g_{min} produce positive and negative features, respectively; the intermediate g value produces a derivative feature. Absorption of microwaves by intermediate orientations is strong between these features, but because this absorption does not vary rapidly with field it does not produce much apparent intensity in the derivative presentation.

B. Hyperfine Coupling

In addition to **g** tensor anisotropy, EPR spectra are often strongly affected by hyperfine interactions between the nuclear spin **I** and the electron spin **S**. These interactions take the form $\mathbf{I} \cdot \mathbf{A} \cdot \mathbf{S}$, where **A** is the hyperfine coupling tensor. Like the **g** tensor, the **A** tensor is a second-order third-rank tensor that expresses orientation dependence, in this case, of the hyperfine coupling. The **A** and **g** tensors need not be colinear; in other words, **A** is not necessarily diagonal in the coordinate systems which diagonalize **g**.

The magnitude of **A** is affected by the magnetic moment associated with the nuclear spin and by the electron distribution on the molecule. The tensor **A** has a dipolar contribution resulting from through space magnetic interactions between the nuclear and electron spins. Such interactions vary as $1/R^3$ and have a $1 - \cos^2 \theta$ angular dependence even if both spins are isotropic. In addition, **A** has an isotropic contact term resulting from unpaired electron density at the nucleus in question. Knowledge of the **A** tensor thus provides information about the electron distribution on the molecule.

A nuclear spin of $\frac{1}{2}$ has two possible orientations with respect to the electron spin to which it is coupled. The antiparallel and parallel states differ in energy by $\frac{1}{2}A = \frac{1}{4}A - (-\frac{1}{4}A)$, where A is the effective value of the hyperfine coupling at a particular field orientation. Each allowed EPR transition corresponds to flipping an electron spin with the nuclear spin orientation unchanged. The hyperfine coupling thus results in the splitting of each EPR line into two lines separated by A. Since the hyperfine coupling is anisotropic, the splittings may be resolved at some orientations but be smaller than the line width at others.

Just as an interaction with a nuclear spin of $\frac{1}{2}$ splits an EPR line into a doublet, hyperfine coupling to $I = 1$ produces a triplet and coupling to $I = \frac{3}{2}$ produces a quartet. In general, hyperfine coupling to a nucleus of spin I produces $2I + 1$ lines. Coupling to two or more nuclei produces more complex spectra. At most, $(2I + 1)(2I_2 + 1)$ lines can result from coupling to two nuclei. The triplet produced by coupling to two identical $I = \frac{1}{2}$ can be distinguished from the triplet produced by coupling to $I = 1$ because the three lines in the former case have a $1:2:1$ intensity ratio, whereas in the latter case the three lines have a $1:1:1$ intensity ratio.

Protons have a spin $I = \frac{1}{2}$ and therefore often give rise to doublets. Nitrogen-14 has $I = 1$ while ^{15}N has $I = \frac{1}{2}$; use of ^{15}N can therefore simplify EPR spectra when hyperfine coupling to nitrogen is important. Both of the naturally abundant isotopes of copper have $I = \frac{3}{2}$, and their magnetic moments are similar. Numerous other elements can give rise to hyperfine splitting either in naturally abundant isotopic forms or in less common isotopes after enrichment.

C. Relaxation, Saturation, and Temperature

Because the coefficients of absorbance and emission are identical, a microwave photon of the appropriate frequency would be equally likely to cause emission of an identical photon from an excited molecule as to be absorbed by a molecule in the ground state if equal numbers of molecules were present in each state. Therefore, net absorbance of energy depends on the difference in populations between the ground state and the excited state.

Because the population ratio is determined by the appropriate Boltzmann expression ($e^{-\Delta/kT}$, where Δ is the energy difference between states and k is the Boltzmann constant), lowering the temperature often results in an increase in signal in EPR experiments. X-band microwave photons have an energy of around $\frac{1}{3}$ cm^{-1}; because kT is much larger than the difference in energy between the Zeeman states, the population difference and signal intensity often vary approximately as $\frac{1}{T}$ over a significant range of temperature. This is refered to as Curie law behavior.

As the temperature is lowered, the line width may significantly decrease and then approach a lower limit. This sharpening also increases the apparent intensity of the signal. The increased line width at high temperature is usually attributable to lifetime broadening. If many thermally accessible states are present, lowering the temperature will increase EPR signals from the ground state Kramers' doublet more rapidly then Curie law behavior can account for; EPR transitions from excited doublets may eventually become weakened at low temperatures by this mechanism.

As the temperature is further lowered, the natural processes that maintain the Boltzmann distribution (relaxation processes) may be no longer able to keep up with the rate of transitions induced by the microwave radiation. Power saturation leads to a decrease in signal at low temperatures and high levels of microwave power. Because the rate and temperature dependence of relaxation processes is very different in different systems, different paramagnetic species saturate at different levels of power and are best observed at different temperatures. Organic radicals are best observed at relatively high temperature and low levels of power; transition metals, especially in systems in which $S > \frac{1}{2}$, are usually observed at cryogenic temperatures because of their rapid relaxation rates.

D. $S = \frac{1}{2}$ and $S > \frac{1}{2}$ Systems

Neglecting orbital angular momentum, a single unpaired electron has a spin of $\frac{1}{2}$ and a g value just over 2. In many cases, more than one unpaired electron is present in a molecule; these may be coupled to give spins of greater than $\frac{1}{2}$. We first briefly discuss some odd electron cases.

Ferric iron has five $3d$ electrons. In a strong octahedral ligand field (e.g., as in bisimidazole ferriheme), the five ferric $3d$ orbitals are split in energy into a low-lying triplet (the t_{2g} set) and a higher lying doublet (the e_g set). The five electrons may be pictured as occupying the three lowest d orbitals (neglecting covalent participation of the d orbitals); the Pauli principle forces the net spin to be $\frac{1}{2}$. In this low-spin complex, deviations from the free electron value of d are brought about by contributions from the spin–orbit coupling. The extent of these contributions depends on the ability of the spin–orbit coupling term (λ is about 400 cm^{-1} for ionic iron complexes and may be effectively reduced by a significant amount by covalency) to mix the t_{2g} orbitals. These orbitals would be degenerate in pure octahedral symmetry, but are separated by a few hundred or a few thousand wave numbers by axial and rhombic distortions in the ligand field. The contributions to the **g** tensor from the orbitals angular momentum term may be substantial; bisimidazole–ferriheme complexes typically have **g** tensors with principal values (neglecting signs) near 2.9, 2.2, and 1.5.

Cupric ions are $3d^9$ and are therefore always $S = \frac{1}{2}$. They show smaller deviations from $g = 2$ due to admixture of the d orbitals by spin–orbit coupling. The anisotropy is not as great as for ferric heme because the orbitals are further apart. Because the d orbitals are more than half-filled, the sign of the spin–orbit coupling causes the g values to be greater than 2. As mentioned earlier, copper complexes usually display resolved copper hyperfine splitting.

If the ligand field of a ferric complex is weak, it will not be large enough to compensate for the exchange energy, which tends to align the spins of the electrons. Ferric iron in high-spin complexes has one electron in each $3d$ orbital. This happens when electrostatic repulsion between d electrons is more important than interactions between the electrons and the ligands. In a high-spin heme, the net spin of $\frac{5}{2}$ is quantized along a molecular axis that roughly corresponds to the heme normal. There are six states corresponding to the six allowed orientations of S along this axis: $S_z = \frac{5}{2}, \frac{3}{2}, \ldots -\frac{1}{2})$.

High-spin hemes have a strong axial zero field splitting which splits these states into three doublets, and because the value of the zero field splitting parameter D is positive (typically 5–10 cm^{-1}) for hemes the doublet consisting of the $S_z = +\frac{1}{2}, -\frac{1}{2}$ states is lowest lying. There is usually also a smaller rhombic term E which cannot split the Kramers' doublets but mixes states differing in S_z by 2. The main net effect of E is to remove the equivalence of the x and y (in heme plane) directions so that g_x and g_y are unequal.

When the magnetic field is along the axis of quantization, the Zeeman interaction splits the $S_z = \frac{1}{2}$ and $S_z = \frac{1}{2}$ states but does not mix them. The splitting is just $\beta \mathbf{H} \cdot \mathbf{g} \cdot \mathbf{S}$; there is little contribution from orbital angular momentum as there are no low-lying excited states to be mixed in by the spin–orbit coupling. Therefore, along the z direction the effective g value is about 2.

When the magnetic field is in the plane of the heme, the Zeeman term does not commute with the zero field terms; the Zeeman term mixes the previously degenerate $S_z = \frac{1}{2}$ and $S_z = \frac{1}{2}$ states. Unlike the situation for $S = \frac{1}{2}$, for $S > \frac{1}{2}$ the value of the integral $\langle \frac{1}{2} | S_z | -\frac{1}{2} \rangle$ is not equal to $\langle \frac{1}{2} | S_x | -\frac{1}{2} \rangle$ or $\langle \frac{1}{2} | S_y | -\frac{1}{2} \rangle$. For an axial $S = \frac{5}{2}$ system, the latter two terms are three times as large as the first. This means that the interaction of the ground state of the axial $S = \frac{5}{2}$ with a magnetic field is the same as the interaction of an $S = \frac{1}{2}$ free electron when the field is along the axis of quantization, but is three times as large when the field is perpendicular to the z axis. We express this anisotropy by assigning the $S = \frac{5}{2}$ system a g value of 6 along the directions perpendicular to the axis of quantization, as if it had $S = \frac{1}{2}$.

Nonzero values of E split g_x and g_y [$g_x - g_y = 48(E/D)$ for $g_x - g_y < 2$] to first order and decrease all three g values to second order [most importantly, $g_z = 2-34(E^2/D^2)$ when $g_x - g_y < 2$] (Palmer 1985).

Nitrosyl complexes of both $S = \frac{1}{2}$ and $S = \frac{3}{2}$ are common. The $S = \frac{1}{2}$ nitrosyl complexes or iron and copper have slightly anisotropic $\mathbf{g}$ tensors ($\Delta g/g < 2$) with the anisotropy provided primarily by contributions of orbital angular momentum from the metal d orbitals. As an example, we consider the model proposed by Kon and Katakoa (1969) for ferroprotoheme–NO complexes, which is also applicable to the ferrous–nitrosyl complexes of heme proteins. In the complexes studied by these workers, the axial ligands are a nitrogenous base and NO. They proposed that the unpaired electron resides primarily in the metal d_{z^2} orbital. The spin–orbit coupling would then mix contributions from the d_{xz} and d_{yz} orbitals.

The expected g values would be

$$g_z = g_e$$
$$g_x = g_e(1 - 3\lambda/\Delta_{xy})$$
$$g_y = g_e(1 - 3\lambda/\Delta_{yx}).$$

This picture can qualitatively account for the **g** tensor anisotropy of nitrosyl complexes in which $g_x \approx 2.08$, $g_y \approx 2.01$, and $g_z \approx 2.00$. However, g_y is often less than 2 and is as small as 1.95 in proteins such as horseradish peroxidase. To explain the reduction in g from the free electron value along the y axis, it is necessary to postulate delocalization of the electron over the molecule. This can best be done by a complete molecular orbital description, but it is instructive to consider the formation of bonding and antibonding orbitals with d_{yz} character from the metal d_{xy} orbital and a p orbital from the nitrogen. The filled orbital would then contribute positively to the g value while admixture of the empty orbital would decrease the g value. Thus, the value of g_y could be quite variable. The delocalization of the electron into ligand orbitals reduces the occupancy of the metal d_z^2 orbital. This effectively reduces the coefficients of the wavefunction components which account for the **g** tensor anisotropy; hence, the anisotropy is an order of magnitude less than might be expected for a pure ionic d^7 complex in which the unpaired electron resides in the d_z^2 orbital.

The magnitude of the hyperfine coupling to the NO nitrogen suggests, in fact, that the unpaired electron spin is primarily (about two-thirds) associated with the nitrogen orbitals. The contribution of nitrogen orbitals to the **g** tensor anisotropy is negligible with respect to the contribution of the iron orbitals because of the much smaller size of the spin–orbit coupling matrix elements and the greater separation of orbitals in nitrogen and oxygen. Just as the iron orbitals dominate the g tensor, the nitrogen orbitals dominate the A tensor. Since the NO usually binds at an angle to the heme plane, the g and A tensors are not colinear. In cytochrome oxidase, the ferrocytochrome a_3–nitrosyl complex has a major splitting of about 22 gauss along the direction in which **A** is maximum; this is contributed by the ^{14}N nucleus of the nitroxyl group. In addition, a small splitting (6–7 gauss) along the same axis is contributed by the ligating nitrogen of the axial histidine ligand. The relative sizes of the splitting reflect the displacement of the unpaired electron toward the NO side of the heme. The resulting triplet of triplets can be seen in many heme–nitrosyl complexes in which the other axial ligand is nitrogenous.

In other $S = \frac{1}{2}$ iron–nitrosyl complexes, such as the Cys_2–Fe(II)–$(NO)_2^-$ complex, delocalization of the electron in extended molecular orbitals results in reduced orbital angular momentum contributions from the iron d orbitals. In this complex, delocalization apparently reduces the association of the unpaired electron with each nitrogen nucleus so that the hyperfine splittings from them are not resolved but instead contribute to line width.

In $S = \frac{3}{2}$ nitrosyl complexes, the spin in quantized along a molecular axis. There are four allowed orientations of S along z; these correspond to four states with $S = \frac{3}{2}, \frac{1}{2}, -\frac{1}{2}$, and $-\frac{3}{2}$. The $S_z = \frac{1}{2}$ and $S_z = -\frac{1}{2}$ states are separated from the other two states by zero field splitting terms. These are of the form $D(S_z^2 - \frac{5}{4}) + E(S_x^2 - S_y^2)$, where D and E are the axial and rhombic zero field splitting parameters. Including both the zero field splitting terms and the Zeeman terms appropriate for a magnetic filed along z, the matrix of the spin Hamiltonian is

S_z	$\frac{3}{2}$	$\frac{1}{2}$	$-\frac{1}{2}$	$-\frac{3}{2}$
$\frac{3}{2}$	$D + \frac{3}{2}gH$	0	$3E$	0
$\frac{1}{2}$	0	$-D + \frac{1}{2}gH$	0	$3E$
$-\frac{1}{2}$	$3E$	0	$-D - \frac{1}{2}gH$	0
$-\frac{3}{2}$	0	$3E$	0	$D - \frac{3}{2}gH$

where the rows and columns are labeled in S_z. In the absence of a magnetic field, the ground state Kramers' doublet is $2(D^2 + 3E^2)^{1/2}$ below the excited doublet; this reduces to $2D$ in the axial case when $E = 0$.

When the magnetic field is along z, the axis of quantization, the terms on the diagonal split the lowest doublet by $g\beta H$, just as if the spin of the system were $\frac{1}{2}$. A magnetic field along x or y, however, would give rise instead to off-diagonal matrix elements. The matrix element connecting the states with $S_z = \frac{1}{2}$ and $S_z = -\frac{1}{2}$, which makes up the lowest doublet, is $[(S_z + \frac{1}{2})(S - \frac{1}{2} + 1)]^{1/2}$ or $S + \frac{1}{2}$. This mixes the degenerate states completely, giving rise to two new states separated by $g_e(2S + 1)\beta H$. For $S = \frac{3}{2}$, this is the only Zeeman matrix element of significance when $E = 0$ and D is much larger than the Zeeman terms. In an $S = \frac{3}{2}$ system with a strong axial zero field splitting, we account for the large numerical value of the off-diagonal matrix element by assigning g_x and g_y values near 4, much as we assigned g_x and g_y values near 6 for high-spin heme.

The introduction of modest rhombic terms ($E < < D$) mixes the doublets because of the off-diagonal matrix elements in E. This causes g_x and g_y to separate; second-order effects slightly reduce the average value of g_x and g_y and cause g_z to fall below 2. Unlike the $S = \frac{5}{2}$ system discussed earlier, $S = \frac{3}{2}$ states are not orbital singlets, and there is the possibility of angular momentum contributions from admixture of d orbitals. This effect will be stronger if the separation between orbitals is small and if the unpaired electron density is localized on the metal ion.

An $S = \frac{3}{2}$ iron–nitrosyl complex could result from several possibilities, including $Fe^{2+}NO$, in which $S = 2$ Fe^{2+} is coupled to $S = \frac{1}{2}$ NO, and $Fe^{3+}NO$, in which the ferric ion is intermediate spin. All known $S = \frac{3}{2}$ iron–nitrosyl complexes are dominated by spin-only effects, although contributions from orbital angular momentum are responsible for small shift in the g values.

REFERENCES

Abragam, A., and Bleaney, B. (1970). "Electron Paramagnetic Resonance of Transition Ions." Clarendon Press, Oxford.

Asahara, H., Bank, T., Cunningham, R., Scholes, C. P., Salerno, J. C., Munck, E., McCracken, J., Peisach, J., and Emptage, M. (1989). *Biochemistry* **28**, 4450–4454.

Bastian, N. R., Yim, C.-Y., Hibbs, J. B., and Samlowski, W. E. (1994). Induction of iron-derived epr signals in murine cancers by nitric oxide—Evidence for multiple intracellular targets. *J. Biol. Chem.* **269**, 5127–5131.

Beinert, H. (1990). *FASEB J.* **4**, 2483–2491.

Brudvig, G. W., Stevens, T. H., and Chan, S. I. (1980). *Biochemistry* **19**, 5275–5285.

Butler, A. R., Glidewell, C., Hyde, A. R., and Walton, J. C. (1985). *Polyhedron* **4**, 797–809.

Cammack, R., Hucklesby, D. P., and Hewitt, E. J. (1978). *Biochem. J.* **171**, 519–526.

Carrington, A., and McLachlan, A. D. (1967). "Introduction to Magnetic Resonance." Harper and Row, New York.

Castro, L., Rodriguez, M., and Radi, R. (1994). *J. Biol. Chem.* **269**, 29409–29415.

Chevion, M., Peisach, J., and Blumberg, W. E. (1977). *J. Biol. Chem.* **252**, 3637–3645.

Dolphin, D. (1978). "The Porphyrins," Vol. 5. Academic Press, N. Y.

Drapier, J. C., Hirling, H., Wietzerbin, J., Kaldy, P., and Kuhn, L. C. (1993). Biosynthesis of nitric oxide activates iron regulatory factor in macrophages. *EMBO J.* **12**, 3643–3649.

Emptage, M. H., Dryer, J. L., Kennedy, M. C., and Beinert, H. (1983). *J. Biol. Chem.* **258**, 11106–11111.

Falkowski, K. M., Scholes, C. P., and Taylor, H. (1986). *J. Mag. Res.* **68**, 453–468.

Feher, G., Allen, J. P., Okamura, M. Y., and Reis, D. C. (1989). *Nature* **339**, 111–116.

Fermi, G., and Perutz, M. F. (1981). "Hemoglobin and Myoglobin." Clarendon Press, Oxford.

Gabriel, R. J., Batie, C. J., Sivarja, M., True, A. E., Fee, J. A., Hoffman, B. M., and Ballou, D. (1989). *Biochemistry* **28**, 4861–4871.

Gabriel, R. J., Ohnishi, T., Robertson, D. E., Daldal, F., and Hoffman, B. M. (1989). *Biochemistry* **30**, 11579–11584.

Galpin, J. R., Veldink, G. A., Vliegenthardt, J. F. G., and Boldingh, J. (1978). *Biochem. Biophys. Acta* **536**, 356–362.

Garthwaite, J., Charles, S. L., and Chess-Williams, R. (1988). *Nature* **336**, 385–388.

Ghoush, D., O'Donnell, W., Furey, W., Robbins, A., and Stout, D. C. (1982). *J. Mol. Biol.* **158**, 73–109.

Gouterman, M., Rentzepis, P., and Straub, K. (eds.) (1986). "Porphyrins: Excited States and Dynamics." American Chemical Society, Washington, DC.

Griffiths, J. S. (1961). "Theory of Transition Metal Ions," p. 205. Cambridge Univ. Press, London.
Hagen, W. R. (1982). *Biochim. Biophys. Acta* **708,** 82–98.
Hagen, W. R. (1987). *In* "Cytochrome Systems: Molecular Biology and Bioenergetics" (S. Papa, B. Chance, and L. Ernster, eds.). Plenum, New York.
Hassan, H. M., and Fridovitch, I. (1978). *J. Biol. Chem.* **253,** 8143–8148.
Hatefi, Y. (1985). *Ann. Rev. Biochem.* **54,** 1015–1069.
Hausladen, A., and Fridovich, I. (1994). Vol. 269, 29405–29408.
Henry, Y., and Banerjee, R. (1973). *J. Mol. Biol.* **73,** 469–482.
Henry, Y., Lepoivre, M., Drapier, J. C. Ducrocq, C., Boucher, J. L., and Guissani, A. (1993). Epr characterization of molecular targets for NO in mammalian cells and organelles. *FASEB J.* **7,** 1124–1134.
Hibbs, J. B., Taintor, R. R., Vavrin, Z., and Rachlin, C. M. (1988). *Biochem. Biophys. Res. Commun.* **157,** 87–94.
Hille, R., Olson, J. S., and Palmer, G. (1979). *J. Biol. Chem.* **254,** 12110–12120.
Himmelwright, R. S., Lubien, C. D., Lerch, K., and Solomon, E. I. (1980). *J. Am. Chem. Soc.* **102,** 7339–7344.
Hodges, K. D., Wollman, R. G., Kessel, S. L., Hendrickson, D. N., VanDerveen, D. G., and Barefield, E. K. (1979). *J. Am. Chem. Soc.* **101,** 905–917.
Hori, H., Ikeda-Saito, M., and Yonetoni, T. (1981). *J. Biol. Chem.* **256,** 7849–7855.
Jameson, G. B., Robinson, W. I., and Ibers, J. A. (1981).
Kennedy, M. C., Gan, T., Antholine, W. E., and Petering, D. H. (1993). *Biochem. Biophys. Res. Commun.* **196,** 632–635.
Kennedy, M. C., Mendemueller, L., Blondin, G. A., and Beinert, H. (1992). Purification and characterization of cytosolic aconitase from beef liver and its relationship to the iron-responsive element binding protein. *Proc. Natl. Acad. Sci., USA* **89,** 11730.
Knowles, R. G., Palacios, M., Palmer, R. M. J., and Moncada, S. (1989). *Proc. Natl. Acad. Sci. USA* **86,** 5159–5162.
Kon, H. (1968) *J. Biol. Chem.*
Kon, H. (1975). *Biochim. Biophys. Acta* **379,** 103–113.
Kon, H., and Katakoa, N. (1969). *Biochemistry* **8,** 4759–4762.
Kramers, K. (1930). *Acad. van Wetenschappen* (Amsterdam) **33,** 959.
Lancaster, J. R., and Hibbs, J. B. *Proc. Natl. Acad. Sci. USA, in press.*
Lancaster, J. R., Wernerfelmayer, G., and Wachter, H. (1994). Coinduction of nitric oxide synthesis and intracellular nonheme iron–nitrosyl complexes in murine cytokine-treated fibroblasts. *Free Radical Biol. Med.* **16,** 869–870.
LeBrun, N. F., Cheeseman, M. A., Thomson, A. J., Moore, G. L., Anders, S. C., Guest, J. R., and Harrison, P. M. (1993). An epr investigation of non-haem iron sites in *Escherichia coli* bacterioferritin and their interaction with phosphate—A study using nitric oxide as a spin probe. *FEBS Lett.* **323,** 261–266.
Lee, M. H., Arosio, P., Cozzi, A., and Chasteen, N. D. (1994). *Biochemistry* **33,** 3679–3687.
LeGall, J., Payne, W. J., Morgan, T. V., and DerVartanian, D. (1979). *Biochem. Biophys. Res. Commun.* **87,** 355–362.
Lipscomb, J. D., Huyuh, B.-H., and Munck, E. (1979). *Fed. Proc. Am. Soc. Med.* **38,** 731.

LoBrutto, R., Wei, Y.-H., Mascarenhas, R., Scholes, C. P., and King, T. E. (1983). *J. Biol. Chem.* **258,** 7437–7448.
Lovenberg, W. (ed.) (1977). "The Iron Sulfur Proteins." Academic Press, NY.
Malmstrom, B. G. (1978). In "New Trends in Bioinorganic Chemistry" (R. J. P. Williams and J. R. DaSilva, eds.), p. 59. Academic Press, London.
Marletta, M. A., Yoon, P. S., Inyeagar, R. M., Leaf, C. D., and Wishnok, J. S. (1988). *Biochemistry* **27,** 8706–8711.
McDonald, C. C., Phillips, W. D., and Mower, H. F. (1965). *J. Am. Chem. Soc.* **87,** 3319–3329.
Moeller, T. (1952). "Inorganic Chemistry: An Advanced Textbook." Wiley, New York.
Morse, R., and Chen, S. I. (1980). *J. Biol. Chem.* **255,** 7876–7882.
Mulsch A., Mordvintcev, P. I., Vanin, A. F., and Busse, R. (1993). Formation and release of dinitrosyl iron complexes by endothelial cells. *Biochem. Biophys. Res. Commun.* **196,** 1303–1308.
Nakatsuka, M., and Osawa, Y. (1994). *Biochem. Biophys. Res. Commun.* **200,** 1630–1634.
Ohnishi, T. (1987). *Curr. Topics Bioenerg.* **15,** 37–65.
O'Keefe, D. H., Ebal, R. E., and Petersen, J. A. (1978). *J. Biol. Chem.* **253,** 3509–3516.
Omura, I., Sato, R., Cooper, D. Y., Rosenthal, O., and Estabrook, R. W. (1967). In "Methods in Enzymology," Vol. 10, pp. 362–367. Academic Press, New York.
Palmer, G. (1980). In "Methods for Determining Metal Ion Environments in Proteins" (D. Darnal, and R. Wilkins, eds.), pp. 153–181. Elsevier, N. Holland and New York.
Palmer, G. (1985). *Biochem. Soc. Trans.* **13,** 548–560.
Palmer, R. M. J., Ashton, D. S., and Moncada, S. (1988). *Nature* **333,** 664–666.
Peng, S. M., and Ibers, J. A. (1976). *J. Am. Chem. Soc.* **98,** 8032.
Que, L., Lipscomb, J. D., Zimmermann, R., Munck, E., Orme-Johnson, N. R., and Orme-Johnson, W. H. (1976). *Biochim. Biophys. Acta* **452,** 320–324.
Reddy, D., Lancaster, J. R., and Cornforth, D. P. (1983). *Science* **221,** 769–770.
Rich, P. R., Salerno, J. C., Leigh, J. S., and Bonner, W. D. (1978). *FEBS Lett.* **93,** 323.
Ryden, L. (1984). In "Copper Proteins and Copper Enzymes" (R. Lontie, ed.), Vol. 1, pp. 157–182. CRC Press, Boca Raton, FL.
Salerno, J. C., Ohnishi, T., Lim, J., and King, T. E. (1976). *Biochem. Biophys. Res. Commun.* **73,** 833–890.
Salerno, J. C., and Siedow, J. N. (1979). *Biochim. Biophys. Acta* **579,** 246–251.
Scheidt, W. R., and Piciulo, P. L. (1976). *J. Am. Chem. Soc.* **98,** 1913.
Scholes, C. P. (1979). In "Multile Electron Resonsance Spectroscopy" (M. Durio and J. H. Freed, eds.), pp. 297–329. Plenum, New York.
Schoot Uterkamp, A. J. M., and Mason, H. S. (1973). *Proc. Natl. Acad. Sci. USA* **70,** 993–996.
Scott, A. I., Irwin, A. J., Siegel, L. M., and Schoolery, J. A. (1978). *J. Am. Chem. Soc.* **100,** 316–318.
Shiga, T., Hwang, K.-J., and Tyuma, I. (1969). *Biochemistry* **8,** 378–392.
Spira, D. J., and Solomon E. I. (1983). *Biochem. Biophys. Res. Commun.* **112,** 729–736.
Spiro, T. G. (ed.) (1982). "Iron Sulfur Proteins, Metal Ions in Biology Series," Vol. 4. Wiley, New York.

Stern, J. O., and Peisach, J. (1976). *FEBS Lett.* **62,** 364–368.
Steuhr, D. J., Gross, S. S., Sakuma, I., Levi, R., and Nathan, C. F. (1989). *J. Expl. Med.* **169,** 1011–1020.
Steuhr, D. J., and Nathan, C. F. (1989). *J. Exp. Med.* **169,** 1543–1555.
Strittmatter, P., and Enoch, H. G. (1978). *Methods Enzymol.* **52,** 188–193.
Tinkham, M. (1964). "Group Theory and Quantum Mechanics," p. 143. McGraw–Hill, New York.
Thornely, R. N. F., and Lowe, D. J. (1984). *In* "Metal Ions in Biology" (T. Spiro, ed.) Vol. 7, chapter 5. Wiley, New York.
Vanin, A. F., Mordvincev, P. I., Huischildt, S., and Mulsch, A. (1993). The relationship between L-arginine-dependent nitric oxide synthesis, nitrite release and dinitrosyl–iron complex formation by activated macrophages. *Biochim. Biophys. Acta* **1177,** 37–42.
Verplaest, J., Vanturnout, P., DeFreyn, G., Witters, R., and Lontie, R. (1981). *Eur. J. Biohem.* **95,** 327–331.
Werner, A., and Karrer, P. (1918). *Helv. Chim. Acta* **1,** 54.
Woolum, J. C., Tiezzi, E., and Commoner, B. (1968). *Biochim. Biophys. Acta* **160,** 311–320.
Yonetoni, T., Yamamato, H., Herman, J. E., Leigh, J. S., and Reed, G. H. (1972). *J. Biol. Chem.* **247,** 2447–2455.

3
Nitric Oxide as a Communication Signal in Vascular and Neuronal Cells

Louis J. Ignarro
Department of Pharmacology
UCLA School of Medicine, Center for the Health Sciences
Los Angeles, California 90024

I. INTRODUCTION

In 1979 we made our initial discovery that bubbling nitric oxide (NO) gas into Krebs bicarbonate solution bathing isolated precontracted strips of coronary artery caused a marked relaxation response (Gruetter et al., 1979). The NO activated cytosolic guanylate cyclase prepared from coronary artery, and hemoproteins and methylene blue inhibited the NO-elicited guanylate cyclase activation and vascular smooth muscle relaxation. On the basis of these observations, we forwarded the hypothesis that NO is a potent vasodilator whose biological actions are mediated by cyclic GMP. These studies were completed prior to the discovery of endothelium-derived relaxing factor (EDRF) (Furchgott and Zawadzki, 1980).

Additional studies conducted in this laboratory revealed that NO was responsible for the vascular smooth muscle relaxant effects of several different nitrovasodilators, and that S-nitrosothiols were intermediates in the intracellular formation of NO (Ignarro et al., 1981). We showed also that NO was a potent inhibitor of human platelet aggregation, and that NO elicited such effects by stimulating platelet cyclic GMP formation (Mellion et al., 1981). We started studying endothelium-dependent vascular smooth muscle relaxation in 1983 and found that EDRF formation was associated with cyclic GMP formation in the smooth muscle, and that both relaxation and cyclic GMP formation were blocked by methylene blue (Ignarro et al., 1984b). These observations together with the knowledge that EDRF is unstable suggested that EDRF might be related to NO.

Further studies revealed that EDRF, like NO, could activate purified cytosolic guanylate cyclase (Ignarro et al., 1986b). Based on this and other observations, we forwarded the hypothesis in 1986 that EDRF is NO or a labile nitroso precursor (Fourth Symposium on Mechanisms of Vasodilatation, Rochester, MN, July, 1986). One year later we (Ignarro et al., 1987a) and others (Palmer et al., 1987) provided chemical evidence to support the view that EDRF is NO. As was first discovered for NO (Mellion et al., 1981), EDRF was shown to inhibit platelet aggregation and adhesion (Radomski et al., 1987).

Research on NO increased dramatically during the late 1980s. NO was clearly identified as an important cytotoxic product of cytokine-activated macrophages (Hibbs et al., 1988; Marletta et al., 1988; Stuehr et al., 1989). A series of interesting studies led to the appreciation that NO plays some neurotransmission role in the central nervous system (CNS). NO formation was found to be associated with activation of cerebellar glutamate and kainate receptors, and the target sites for NO appeared to be the adjacent glial cells, presynaptic nerve terminals, and Purkinje fibers instead of the neuronal or granules cells that synthesize the NO (Garthwaite et al., 1988). These observations suggested that central excitatory neurotransmitters stimulate NO formation and that NO diffuses outward and communicates with nearby cells to modulate neurotransmission. NO was shown to mediate glutamate-linked enhancement of cyclic GMP formation in cerebellum (Garthwaite, 1991). These observations led to the isolation, purification, and characterization of NO synthase from cerebellum (Bredt and Snyder, 1989, 1990). The evidence is thus strong that NO functions as a CNS neurotransmitter.

NO is a small lipophilic and chemically unstable molecule that is well suited to carry out its many biological functions. Endothelium-derived NO and NO generated in cerebellar cells is synthesized from the basic amino acid L-arginine in a multistep oxidation reaction catalyzed by NO synthase, where NADPH, calcium, and calmodulin are required (Bredt and Snyder, 1989; Forstermann et al., 1990; Mayer et al., 1990a). NO readily permeates many biological membranes to elicit effects in target cells lying adjacent to the cells of origin in NO. NO is chemically unstable and undergoes spontaneous oxidation to NO_2^- with further oxidation to NO_3^- in the presence of superoxide anion and/or oxyhemoproteins.

II. EVIDENCE THAT EDRF IS NITRIC OXIDE OR A LABILE NITROSO PRECURSOR

On the basis of the inhibitory actions of several different types of inhibitors of arachidonic acid metabolism, EDRF was first proposed to be a lipoxygenase product of arachidonic acid metabolism (Furchgott et al., 1981). Subsequent studies suggested that EDRF might be a cytochrome P450 product of arachidonic acid metabolism (Singer and Peach, 1983). Another view was that EDRF was an unstable ketone, lactone, or aldehyde (Griffith et al., 1984), but these observations were never extended. Others reported that phospholipid metabolism and release of free fatty acid could be involved in EDRF formation (Forstermann et al., 1986). However, these investigators cautioned that the chemical probes employed were nonselective and the data were difficult to interpret unequivocally. Nevertheless, phospholipid metabolism is closely associated with EDRF formation, perhaps attributed to elevated intracellular calcium concentrations.

Based on the remarkable similarities between endothelium-dependent relaxation and NO-elicited relaxation in pulmonary blood vessels, and because EDRF was found to activate purified guanylate cyclase (Ignarro et al., 1986b), we proposed in 1986 that EDRF was NO or a related labile nitroso compound (see Ignarro et al., 1988c). The single major finding in our laboratory that prompted our hypothesis was the activation of cytosolic guanylate cyclase by EDRF (Ignarro et al., 1986b). This observation was consistent with the earlier observations that endothelium-dependent relaxation is associated with cyclic GMP formation in vascular smooth muscle cells and that guanylate cyclase inhibitors turn off both endothelium-dependent relaxation and cyclic GMP formation (Ignarro et al., 1984b). The subsequent finding that, like NO, EDRF from artery and vein causes heme-dependent activation of guanylate cyclase convinced us beyond any shadow of a doubt that EDRF must be NO or a labile nitroso precursor to NO (Ignarro et al., 1987b).

Subsequent studies in our laboratory employing the bioassay cascade technique in order to compare the pharmacological properties of arterial and venous EDRF (Fig. 1), authentic NO, and a variety of labile nitroso compounds revealed that EDRF was indistinguishable from authentic NO (Ignarro et al., 1987a,b; 1988a,b). The studies revealed not only the similarity between EDRF and NO but also that veins generate EDRF and that venous EDRF is indistinguishable from arterial EDRF, at least in the bovine pulmonary circulation. Chemical evidence was also provided that EDRF is identical with NO. Like authentic NO, EDRF reacted with hemoglobin to form the characteristic nitrosyl–heme adduct of hemoglobin, which displays a characteristic absorption spectrum in the Soret region (Ignarro et al., 1987b). Moreover, like authentic NO, EDRF reacted with sulfanilamide under acidic conditions to yield a diazo product that could be

FIGURE 1

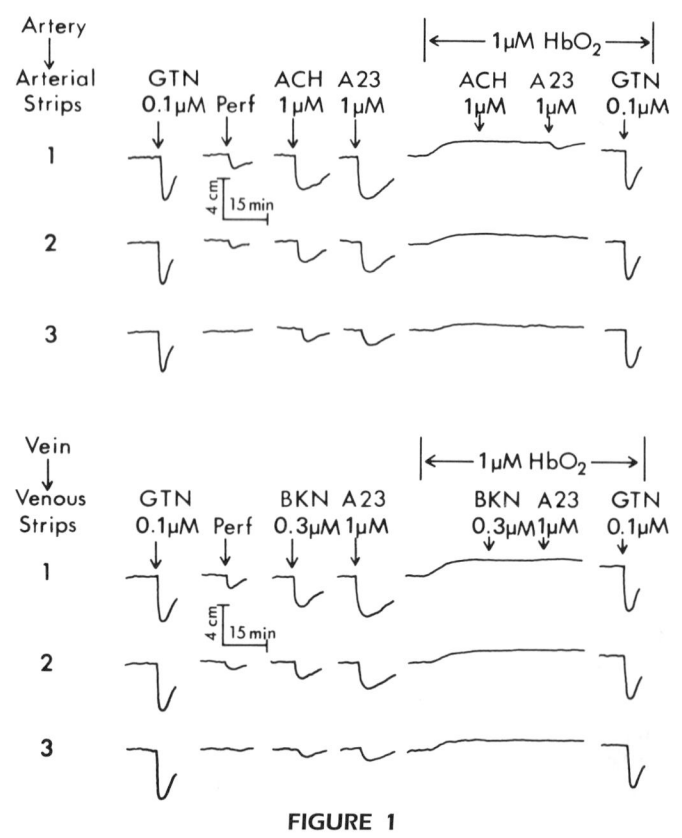

Bioassay cascade method illustrating the pharmacological similarity between EDRF and authentic NO. Endothelium-denuded helical strips of bovine pulmonary artery or vein were placed under optimal tension and precontracted by superfusion of 10 μM phenylephrine or 0.01 μM U46619 (a stable thromboxane analog) in Krebs bicarbonate solution at 37°C. The numbers 1, 2, and 3 signify the cascade arrangement of the strips. Glyceryl trinitrate (GTN) was superfused over the strips for 1 min in order to standardize the relaxation responses of each strip. Perf signifies perfusate from segments of artery or vein. Breaks in the tracing represent periods of tissue equilibration. Acetylcholine (ACH) and bradykinin (BKN) were perfused through the arterial and venous segments, respectively, for 3 min. Oxyhemoglobin (HbO$_2$) was superfused over the strips as indicated, and this was continued during perfusion with ACH or BKN and during superfusion with authentic NO and GTN.

monitored spectrophotometrically (Ignarro et al., 1987a). Another group employed the chemiluminescence procedure to demonstrate that EDRF is nitric oxide (Palmer et al., 1987). In the latter procedure, the EDRF collected from perfused vascular endothelial cells that had become readily oxidized to NO$_2^-$ was reduced to NO by refluxing in acidic iodide solution in an oxygen-free atmosphere, and the evolved NO was reacted with ozone to yield an activated form of NO_2 which emits photons that can be quantified by suitable photomultiplier detectors. Subsequent studies showed that, like authentic NO, EDRF caused relaxation of several different nonvascular smooth muscle preparations (Buga et al., 1989) and inhibited platelet adhesion and aggregation (Radomski et al., 1987). The above procedures or modifications of these procedures are now used routinely by many laboratories to detect and monitor the formation of EDRF or NO in diverse biological preparations (Ignarro, 1990a).

Although the experimental evidence is quite convincing that EDRF is NO, none of the techniques employed thus far have been able to distinguish NO from a very labile nitroso precursor to NO. That is, it is conceivable that EDRF is a labile nitroso compound that is spontaneously converted to free NO on release from its site of formation or on contact with biological membranes. One such candidate, S-nitrosocysteine (Myers et al., 1990), was demonstrated many years ago by our laboratory to possess pharmacological and biochemical properties that are indistinguishable from those of NO (Ignarro et al., 1980b, 1981). This chemically unstable nitroso compound is a potent vascular smooth muscle relaxant and vasodilator (Ignarro and Gruetter, 1980; Ignarro et al., 1981), inhibitor of platelet aggregation (Mellion et al., 1983), and activator of cytosolic guanylate cyclase (Ignarro et al., 1980a,b). S-Nitrosocysteine may also be an intermediate in the metabolism of organic nitrate esters to NO in vascular smooth muscle (Ignarro et al., 1981). Clear-cut experimental evidence that EDRF is anything but NO, however, has not been provided.

III. BIOSYNTHESIS AND METABOLISM OF ENDOTHELIUM-DERIVED NITRIC OXIDE

The discovery that EDRF is NO set the stage for the subsequent studies on the biosynthesis of NO (Fig. 2). Even prior to the finding that the vascular endothelium generates NO, Stuehr and Marletta showed that murine macrophages synthesize nitrite and nitrate in response to *Escherichia coli* lipopolysaccharide (Stuehr and Marletta, 1985). These observations provided an explanation for the earlier findings that nitrite and nitrate are synthesized in germfree rodents, thereby revealing the capacity of mammalian cells to generate nitrogen oxides (Green and Tannenbaum, 1981). These observations were extended to include macrophages activated by *Mycobacterium bovis* BCG infection, lymphokines, and gamma interferon (Stuehr and Marletta, 1987). Subsequent studies (Miwa et al., 1987) revealed that some reactive nitrosating species may have been formed as a precursor to nitrite and nitrate, although NO was not mentioned. Experiments by Hibbs and co-workers indicated that an L-arginine-dependent biochemical pathway was involved in the biosynthesis of nitrite and

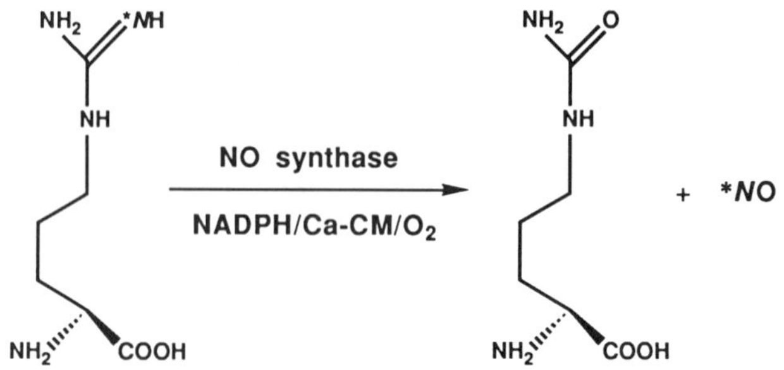

FIGURE 2

Schematic illustration of the conversion of L-arginine to NO plus L-citrulline by the enzyme NO synthase. Conversion requires the presence of NADPH, calcium (Ca), calmodulin (CM), and O_2. Calcium complexes with CM and the Ca–CM complex binds to the enzyme. The asterisk signifies the basic amino nitrogen atom that undergoes oxidation and cleavage to form NO. Both of the basic amino nitrogens are equivalent and either nitrogen can be incorporated into NO.

L-citrulline, and that this pathway was inhibited by N^G-methyl-L-arginine (Hibbs et al., 1987a,b). In an elegant series of experiments, L-arginine was shown to be obligatory for the formation of nitrite, nitrate, and N-nitrosylated amines by activated murine macrophages (Iyengar et al., 1987). The above observations with macrophages led to the discovery that vascular endothelial cells also utilize L-arginine in the biosynthesis of NO (Palmer et al., 1988a), a reaction that is inhibited by N^G-methyl-L-arginine (Palmer et al., 1988b). These observations led to a reexamination of the macrophage system and to the finding that NO was an intermediate in the formation of nitrite (Hibbs et al., 1988).

Studies on the isolation, purification, and characterization of NO synthase were preceded by studies showing that NO mediates glutamate-linked cyclic GMP formation in the cerebellum. Glutamate is the major excitatory neurotransmitter in the brain and elicits effects on ion channels, inositol metabolism, and cyclic GMP formation. The stimulation of cyclic GMP formation by glutamate is most prominent in the Purkinje cells of the cerebellum (Ferrendelli, 1978). Glutamate and related excitatory amino acids stimulate the conversion of L-arginine to NO plus L-citrulline, and N^G-methyl-L-arginine inhibits glutamate-elicited cyclic GMP formation (Bredt and Snyder, 1989). Subsequent studies focused on the isolation and purification of cytosolic NO synthase from rat cerebellum (Bredt and Snyder, 1990). Purified NO synthase was found to be dependent on NADPH, calcium, and calmodulin, and the enzyme appeared to be a monomer of 150 kDa with very low chemical stability. Shortly thereafter, the same group of investigators reported that cloned and expressed NO synthase structurally resembles cytochrome P450 reductase (Bredt et al., 1991). Cloning of a complementary DNA for cytosolic NO synthase from cerebellum revealed the presence of recognition sites for NADPH, FAD, flavin mononucleotide, and calmodulin, and phosphorylation reactions.

Another group of investigators purified cytosolic NO synthase from cultured N1E-115 neuroblastoma cells (Forstermann et al., 1990). NO synthase from this source was also dependent on NADPH, calcium, and calmodulin. Following the demonstration that a crude soluble endothelial cell fraction contained a calcium-dependent enzyme that catalyzed the conversion of L-arginine to a NO-like activator of guanylate cyclase (Mayer et al., 1989; Mulsch and Busse, 1991), NO synthase was isolated and purified from vascular endothelial cells (Forstermann et al., 1991a). Other studies showed that cultured vascular endothelial cells contained NO synthase that was primarily membrane-bound rather than cytosolic and required NADPH, calcium, and calmodulin for full activity (Forstermann et al., 1991a,b). Cytosolic NO synthase has also been purified from porcine cerebellum Mayer et al., 1990a,b; 1991) and rat cerebellum (Schmidt et al., 1991) and found to be dependent on NADPH, calcium, and calmodulin. The enzyme from porcine cerebellum was found to be dependent on tetrahydrobiopterin as well. NO synthase from rat cerebellum appears to have a molecular mass of 279 kDa and is a homodimer of 155 kDa (Schmidt et al., 1991). Further characterization of the soluble enzyme from porcine cerebellum revealed that NO synthase is a biopterin- and flavin-containing multifunctional oxidoreductase (Mayer et al., 1991). The enzyme is a nonheme iron-containing flavoprotein which acts as a calcium–calmodulin-dependent NADPH:oxygen oxidoreductase catalyzing the formation of hydrogen peroxide instead of NO or L-citrulline at low concentrations of L-arginine or tetrahydrobiopterin. Cytosolic NO synthase purified from activated rat macrophages was found to be a dimer (300 kDa) of similar or identical subunits (150 kDa), and was dependent on NADPH, tetrahydrobiopterin, and a thiol (Yui et al., 1991). Unlike the cytosolic enzymes from brain, neuroblastoma cells, and vascular endothelium, the NO synthase from rat macrophages does not require calcium or calmodulin for activity. Interestingly, the cytosolic form of NO synthase purified from activated rat peritoneal neutrophils was found to be calcium dependent but calmodulin independent (Hiki et al., 1991).

There are cytosolic and membrane-bound isoforms of NO synthase. Certain soluble and particulate isoforms are constitutive and other soluble isoforms are inducible. The constitutive enzyme is, by definition, present in the catalytically active form and needs only to be stimulated by an appropriate chemical species, following which there is immediate formation of NO plus L-citrulline. This form of NO synthase requires calcium, and for the most part calmodulin, for stimulation of enzymatic activity. It is likely that an increase in intracellular free calcium in the presence of calmodulin is the signal for stimulation of NO synthase, and therefore, the production of NO. This view is consistent with the general understanding that, in vascular tissue, all endothelium-dependent vasodilators that elicit their effects via EDRF or NO have an absolute requirement for calcium (Singer and Peach, 1982; Peach et al., 1985; Long and Stone, 1985), and such vasodilators cause calcium influx (Luckhoff et al., 1988). Contrariwise, the inducible form of NO synthase shows no appreciable enzymatic activity until at least several hours after an appropriate signal has been generated to promote the synthesis of new enzyme protein. Neither calcium nor calmodulin appears to be required for inducible NO synthase activity. The appropriate signal for induction of NO synthase is lipopolysaccharide, a variety of cytokines, and related chemical substances.

Studies have indicated that N^G-hydroxy-L-arginine is an intermediate in the biosynthesis of NO from L-arginine (Wallace et al., 1991; Stuehr et al., 1991). N^G-Hydroxy-L-arginine appears to be generated as an intermediate in the synthesis of NO through a NADPH-dependent hydroxylation of L-arginine. Consistent with this view is the observation that N^G-hydroxy-L-arginine causes vascular smooth muscle relaxation by mechanisms that are inhibited by the N^G-methyl-, N^G-nitro-, and N^G-amino-analogs of L-arginine (Wallace et al., 1991).

The discovery that N^G-substituted analogs of L-arginine inhibit NO formation from L-arginine (Palmer et al., 1988a) led to the wide use of such compounds as pharmacological probes to study the physiological relevance of the L-arginine–NO pathway. The first of such studies revealed that N^G-methyl-L-arginine, after i.v. injection, caused a sustained elevation of the systemic blood pressure that was reversed by administration of excess L-arginine but not D-arginine (Palmer et al., 1988b; Sakuma et al., 1988; Vargas et al., 1990, 1991). Subsequent studies showed that such analogs of L-arginine inhibit the basal formation of endothelium-derived NO and that this was associated with a decline in basal levels of cyclic GMP in vascular smooth muscle (Gold et al., 1990; Fukuto et al., 1990). Studies with such inhibitory L-arginine analogs are being focused on the role of endothelium-derived NO in modulating local blood flow. These N^G-substituted analogs of L-arginine should prove to be useful in elucidating the role of NO in other physiological and/or pathophysiological processes.

NO appears to be inactivated very rapidly under physiological conditions, possessing a half-life of 5 sec or less (Ignarro, 1990a). The inactivation process is most likely attributed to spontaneous and catalyzed oxidation to NO_2^- and NO_3^-. In the presence of oxygen alone, NO is converted to NO_2^-. In the presence of other oxidizing species, however, the NO_2^- formed is further oxidized to NO_3^-. For example, superoxide anion catalyzes the oxidation of NO to NO_3^-, via the intermediate formation of $ONOO^-$ (peroxynitrite anion) and subsequent rearrangement to NO_3^- in solution (Beckman et al., 1990). Oxyhemoglobin rapidly catalyzes the oxidation of NO to NO_3^-. Thus, in the presence of oxygen and superoxide anion and/or oxyhemoglobin and other oxidizing species, NO undergoes rapid oxidation to both NO_2^- and NO_3^-.

IV. UNIQUE CHEMICAL PROPERTIES FOR A UBIQUITOUS AND GENERAL PHYSIOLOGICAL MEDIATOR

The chemical and biological properties of NO are well suited for a local modulator of vascular smooth muscle, platelet, and neutrophil functions. The vascular endothelial cell origin of NO is ideal for the local and immediate delivery of such a labile small lipophilic molecule not only to the underlying smooth muscle but also to the vessel lumen for interaction with nearby blood elements and cells. The chemically labile nature of NO allows for truly local actions, as does the high capacity of NO to bind to and become sequestered by erythrocyte hemoglobin. The diffusible nature of NO endows this molecule with the unique capacity to engage in transcellular communication. In this manner, NO released locally from the vascular endothelium may communicate very quickly with adjacent cells containing cytosolic guanylate cyclase, and cause them to act in concert to attain a given physiological or pathophysiological response. It follows, therefore, that interference with basal generation of NO, which could result from local damage to the vascular endothelium, could cause increased platelet and neutrophil adhesion as well as increased intrinsic vascular smooth muscle tone and/or increased responsiveness to endogenous contractile agents.

The chemical and biological properties of NO are compatible with a physiological role in mediating shear force-induced or flow-dependent vasodilation (Buga et al., 1991). Moreover, the short half-life of NO renders the possibility feasible that NO could function as a unique neurotransmitter in that the NO would diffuse out of the neuron immediately after its synthesis. For example, electrically induced relaxation of corpus cavernosum smooth muscle in human and rabbit penis is mediated by the L-arginine-NO pathway (Ignarro et al., 1990b; Rajfer et al., 1992). The neuronal pathway responsible for corporal smooth muscle relaxation and penile erection in response to electrical stimulation is nonadrenergic–noncholinergic (NANC). The NANC neurotransmitter in the corpus cavernosum may be NO or, alternatively, an unknown substance that stimulates the innervated corporal smooth muscle to generate NO (Ignarro et al., 1990a).

NO is also synthesized and released by activated macrophages (Marletta et al., 1988; Hibbs et al., 1988; Stuehr et al., 1989) and neutrophils (Wright et al., 1989), and appears to be responsible for the cytotoxic functions of these cells. Cytotoxicity results from high local concentrations of NO, which are at least 100-fold higher than the concentrations of NO generated by vascular endothelial cells. High concentrations of NO can damage iron-containing and iron–sulfur-containing proteins and lead to disruption of cellular function. High levels of NO can cause a chemical reaction with superoxide anion to occur, whereby peroxynitrite anion and other cytotoxic radical species are generated (Beckman et al., 1990). This mechanism may explain the cytotoxic action of NO generated by activated macrophages (Mulligan et al., 1991). Hepatic Kupffer cells also synthesize NO, and the NO appears to diffuse out and into nearby hepatocytes to cause inhibition of hepatocyte protein synthesis (Billiar et al., 1989). Another function of endogenous NO may be as a central nervous system neurotransmitter, especially in the cerebellum (Garthwaite et al., 1988; Bredt and Snyder, 1989). This will be discussed below.

V. SIGNAL TRANSDUCTION MECHANISMS FOR TRANSCELLULAR COMMUNICATION

The most important hemoprotein interaction involving NO is the reaction between cytosolic guanylate cyclase and NO to yield the nitrosyl–heme–enzyme ternary complex, which represents the activated state of guanylate cyclase (Fig. 3). Interaction between enzyme–bound heme and NO to form NO–heme causes an immediate and marked increase in catalytic activity, resulting in a 50- to 200-fold increase in the velocity of conversion of GTP to cyclic GMP in the presence of excess magnesium. This interaction between NO and the heme prosthetic group of guanylate cyclase represents a novel and widespread transduction mechanism that links extracellular stimuli to the biosynthesis of the second messenger cyclic GMP in adjacent target cells. The importance of NO in signal transduction takes on new meaning and significance in view of the observations that not only vascular endothelial cells but also many other cell types can synthesize NO from L-arginine. NO generated in one cell may thereby recruit the cyclic GMP-mediated functions of adjacent cells, some of which may be different in type from the original NO-generating cell. In this manner, one cell type can signal different and complementary cellular responses within a localized environment.

The requirement of reduced iron (Fe^{2+}) in the form of heme for the activation of cytosolic guanylate cyclase by NO was first appreciated in 1978 in studies conducted with unpurified enzyme (Craven and DeRubertis, 1978). The activation of guanylate cyclase by NO in the presence of heme was accounted for by the reaction between NO and heme to form nitrosyl–heme (NO–heme), as preformed NO–heme complex markedly activated guanylate cyclase. These observations were later confirmed with guanylate cyclase purified from bovine lung (Ignarro et al., 1982a, 1984c, 1986a), rat liver (Ohlstein et al., 1982), and human platelets (Mellion et al., 1983). Soluble guanylate cyclase was found to be a hemoprotein containing 1 mol of heme per mole of holoenzyme (Gerzer et al., 1981; Wolin et al., 1982) and was markedly activated by NO and nitroso compounds in the absence of added heme. Heme can be readily detached from enzyme protein without causing denaturation, and this heme-deficient enzyme form cannot be activated by NO unless heme or a suitable hemoprotein is added back

FIGURE 3

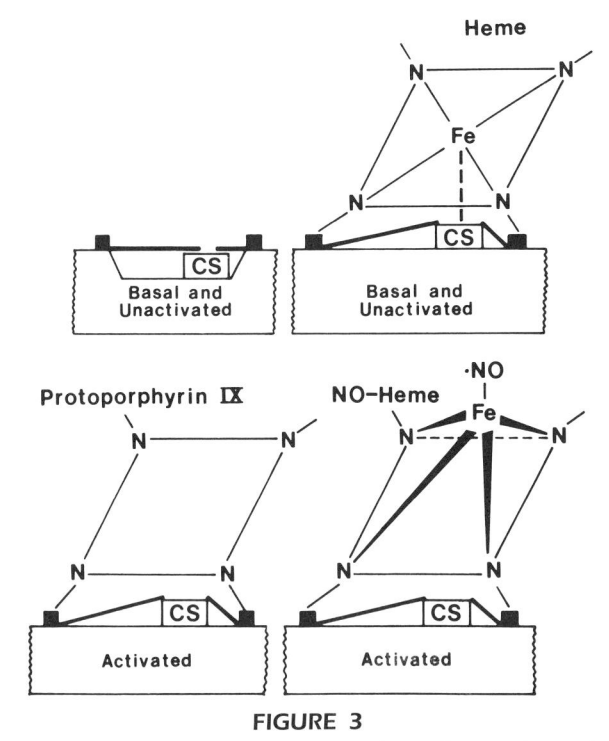

Schematic illustration of the mechanism of activation of cytosolic guanylate cyclase by NO and protoporphyrin IX. Heme-deficient guanylate cyclase and heme-containing guanylate cyclase are both inactivated but display significant basal catalytic activity (~1% of maximum). The catalytic site (CS) has only limited access to the substrate Mg^{2+}-GTP in the absence of heme as illustrated. Although bound heme may alter the structural configuration of the enzyme and allow the CS to surface, the coordinate bond or axial ligand between iron (Fe) and enzyme protein structurally limits the accessibility of the CS to substrate. In contrast, however, the absence of Fe from heme allows for the exposure of the CS to available substrate as long as the porphyrin ring is bound to the enzyme. Similarly, binding of NO to the Fe of heme causes protrusion of the heme iron up and out from the plane of the porphyrin ring configuration, thereby allowing exposure of the CS to available substrate. Thus, addition of either protoporphyrin IX or preformed NO–heme to heme-deficient enzyme or addition of NO to heme-containing enzyme causes enzyme activation, up to 100- to 200-fold over basal enzymatic activity.

to enzyme reaction mixtures. The heme-deficient enzyme can be readily reconstituted with heme, and this enzyme form behaves indistinguishably from the native heme-containing enzyme (Ignarro et al., 1984c, 1986a).

In a study designed to ascertain the characteristics of binding of heme to guanylate cyclase, we discovered that the iron of heme was not required for binding as protoporphyrin IX bound tightly to heme-deficient guanylate cyclase (Ignarro et al., 1982b). We observed, however, that protoporphyrin IX caused a marked activation of guanylate cyclase that was indistinguishable in kinetic mechanisms from that by which NO activated heme-containing guanylate cyclase or preformed NO–heme activated heme-deficient enzyme (Wolin et al., 1982). We postulated that the paramagnetic species NO alters the conformation of heme by pulling the iron away from the enzyme and out-of-plane relative to the planar porphyrin ring configuration (Ignarro et al., 1984d). This configurational change at the porphyrin binding site of guanylate cyclase stresses and breaks the proximal axial ligand bond thereby allowing the iron to detach from the enzyme protein. Thus, the plane of NO–heme that remains bound to guanylate cyclase superficially resembles protoporphyrin IX (heme without its iron). The binding of protoporphyrin IX to heme-deficient guanylate cyclase causes marked enzyme activation. Indeed, heme-containing guanylate cyclase is also markedly activated by protoporphyrin IX because the apparent binding affinity of enzyme for protoporphyrin IX is so high ($K_d = 1.4$ nM) that heme is displaced from the porphyrin binding site and replaced with protoporphyrin IX. The introduction of a divalent metal into the coordination complex of protoporphyrin IX abolishes the capacity for enzyme activation and can actually cause enzyme inhibition. Thus, heme (ferroprotoporphyrin IX) inhibits the basal catalytic activity of heme-deficient guanylate cyclase (Ignarro et al., 1984c, 1986a). Similarly, zinc- and manganese-protoporphyrin IX inhibit basal enzymatic activity and are potent competitive inhibitors of protoporphyrin IX-induced and NO–heme-induced enzyme activation (Ignarro et al., 1984a). Guanylate cyclase activation is characterized by increased affinities of enzyme for MgGTP substrate and excess uncomplexed Mg^{2+}, as well as an increased V_{max} (Wolin et al., 1982).

The "on" signal for guanylate cyclase activation by NO is represented by the binding of NO to heme and the accompanying detachment of the heme Fe^{2+} axial ligand from the enzyme protein. The "off" signal is represented by breakdown of the unstable NO–heme complex with liberation of NO and reestablishment of the heme Fe^{2+} axial ligand.

NO is a small lipophilic and chemically unstable molecule that is very well suited for what appears to be its probable biological role. NO readily permeates biological membranes as can be easily demonstrated experimentally by its release from cells and subsequent interaction with cytosolic guanylate cyclase located in other cells. Moreover, in a series of unpublished observations (L. J. Ignarro; 1981) we noted repeatedly that application of a small drop of aqueous NO, S-nitroso-L-cysteine, or even sodium nitroprusside solution under the tongue of anesthetized cats caused a discernible decrease in systolic blood pressure within 40 sec of application. These physiological properties of NO allow it to diffuse out of one cell and into one or more adjacent cells. This diffusion can and does occur rapidly, before the NO undergoes spontaneous oxidation to NO_2^- and higher oxides of nitrogen.

Studies conducted thus far indicate or suggest that NO synthesized in a given cell elicits its biological response, not in the cells of origin, but rather in nearby target cells. These observations are consistent with the physical properties of NO discussed earlier. Thus, the NO synthesized by vascular endothelial cells rapidly diffuses out of these cells of origin and into the underlying smooth muscle cells to cause relaxation and into nearby adhering platelets in the lumen of the blood vessel to inhibit platelet adhesion and aggregation. Endothelium-derived NO can therefore activate cytosolic guanylate cyclase present in diverse cell types located in very close proximity to its cell of origin. In this manner the NO generated by vascular endothelial cells communicates with nearby cells, calling on them to act in a complementary manner to increase local blood flow and diminish or prevent thrombosis. The very short half-life of NO (less than 5 sec), together with its capacity to bind hemoglobin, myoglobin, and other reduced hemoproteins, ensures that a highly localized action will prevail.
    Endothelium-derived NO may interact with other target cells in addition to vascular smooth muscle and platelets. For example, back in the mid 1970s we showed that human neutrophils and macrophages contained guanylate cyclase and that intracellular cyclic GMP levels could be elevated markedly in response to various immune reactants (Ignarro and George, 1974; Smith and Ignarro, 1975; Ignarro, 1977). The increase in cyclic GMP levels was closely associated with the initiation of lysosomal enzyme release and phagocytosis in the case of neutrophils. Thus, it is conceivable that NO-elicited cyclic GMP accumulation could turn on the cytotoxic functions of such cells. Also plausible is the possibility that endothelium-derived NO could trigger the pathophysiological functions of circulating neutrophils and tissue macrophages. In this manner the formation of NO by a single cell type can signal and thereby recruit the complementary actions of numerous cell types for the purpose of initiating a complete pathophysiological response. Such a response could represent a local inflammatory reaction to tissue injury.
    Because NO is synthesized by a variety of cell types including neuronal cells, as is discussed below, NO may play a widespread physiological and pathophysiological role in local transcellular communication by acting as a short-lived, potent activator of cytosolic guanylate cyclase. The biological significance of this rapid communication may be to signal the initiation of several localized complementary cellular responses, each contributing toward a more complete response. In the attainment of this end, NO would maintain local blood flow and prevent thrombosis in the coronary, cerebral, pulmonary, renal, and gastrointestinal mucosal circulatory beds. Other plausible functions of NO synthesized in critical organs such as the heart, brain, kidney, and lung could be to ensure patency of the associated extensive arteriolar networks in order to maintain flow to these organs. In this capacity NO may function as an autoregulatory molecule.

VI. NITRIC OXIDE AS A NEURONAL MESSENGER

The first study to suggest that NO is formed in the brain and that it functions as an intercellular messenger was published in 1988 by Garthwaite et al. They showed that activation of NMDA (N-methyl-D-aspartate) receptors on cerebellar cells by glutamate causes the release of a diffusible substrate with properties similar to those of EDRF or NO, including stimulation of cyclic GMP formation. Garthwaite et al. proposed that, in the CNS, NO may link activation of postsynaptic NMDA receptors to the modulation of function of neighboring postsynaptic terminals and glial cells. This was an intriguing observation as glutamate receptors (same as NMDA receptors) mediate most if not all excitatory synaptic neurotransmission in the brain. Moreover, this study stimulated interest in other cellular events associated with glutamate receptor activation such as excitotoxicity of certain neurons caused by excessive exposure to glutamate and long-term potentiation and memory, both of which are dependent on calcium, as is NMDA receptor-stimulated NO formation.

This observation led to the finding that a soluble enzyme from rat forebrain catalyzes the NADPH-dependent formation of NO and L-citrulline from L-arginine by calcium-dependent mechanisms (Knowles et al., 1989). A subsequent study revealed that glutamate and NMDA stimulate NO and L-citrulline formation in rat cerebellar slices and that this is associated with a concomitant stimulation of cyclic GMP formation (Bredt and Snyder, 1989), thus establishing a link between NMDA-receptor activation and cyclic GMP formation (Fig. 4). These investigators went on to purify and characterize NO synthase from cerebellum (Bredt and Snyder, 1990; Bredt et al., 1991) and to show the localization of NO synthase in the brain to discrete neuronal populations (Bredt et al., 1990).

NO is a unique CNS neurotransmitter and second messenger. One of the most striking properties of NO is its rapid diffusibility across biological membrane barriers, making unnecessary any active carriers or transporters. Because of its very short half-life, NO is not stored in secretory vesicles but is released from its cells of origin by simple diffusion. Moreover, because of its short half-life, NO is a unique second messenger in that it acts within a localized area in the CNS, thus allowing for the signal (NO) generated in the cells of origin to elicit effects in nearby target cells limited in distance only by the short half-life of 3–5 sec or less. Another highly conducive property to a neuromodulatory role of NO is the calcium-dependent mechanism by which an electrical current turns on NO biosynthesis. Calcium is the link between the extracellular stimulus (glutamate–NMDA receptor interaction) and intracellular generation of NO (calcium activates NO synthase).

It is not clear whether NO functions as a neurotransmitter per se or just as a neuromodulator in the CNS. A function of a neurotransmitter would require

FIGURE 4

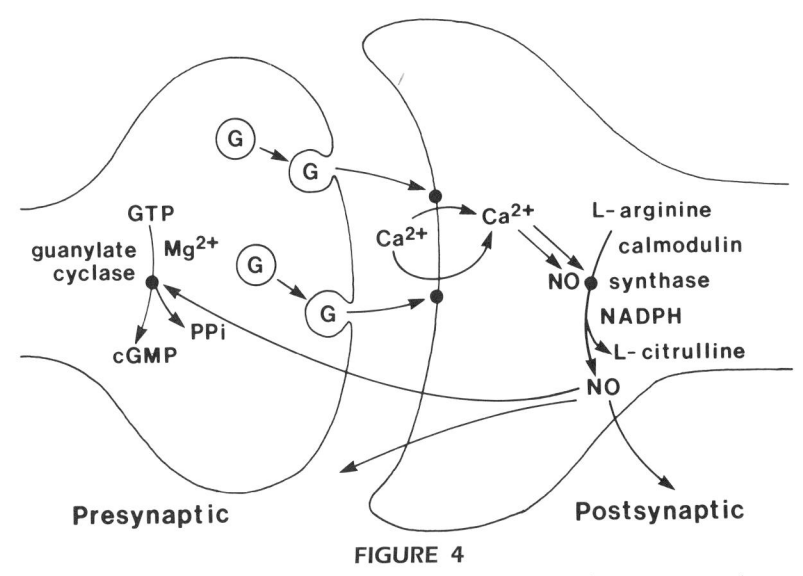

Schematic illustration of the interrelationships between glutamate and NO in synaptic function in the cerebellum. The presynaptic nerve terminal synthesizes, stores, and releases glutamate (G) as the neurotransmitter by exocytosis as illustrated. The glutamate diffuses across the synaptic cleft and interacts with postsynaptic NMDA receptors (●) that are coupled to calcium (Ca^{2+}) channels. Ca^{2+} influx occurs and the free intracellular Ca^{2+} complexes with calmodulin and activates NO synthase. NADPH is also required for conversion, and the products of the reaction are NO plus L-citrulline. NO diffuses out of the postsynaptic cell to interact with nearby target cells, one of which is the presynaptic neuron that released the glutamate in the first place. NO stimulates cytosolic guanylate cyclase and cyclic GMP (cGMP) formation presynaptically, but the consequence of this presynaptic modification is unknown.

that NO is released from the neuron by simple diffusion immediately after its biosynthesis. Alternatively, NO may react with specific thiols in neurons to form relatively stable S-nitrosothiols that can be stored in acidic vesicles and discharged by voltage-dependent mechanisms (Ignarro, 1990a). The S-nitrosothiol would rapidly liberate NO at neutral pH in the presence of tissue components.

Attention then turned to the mechanism of glutamate-induced neurotoxicity in the brain. Evidence was provided that NO mediates glutamate neurotoxicity in primary cortical cultures (Dawson *et al.*, 1991). NO synthase inhibitors and hemoglobin prevented NMDA- and glutamate-induced neurotoxicity; on the other hand, L-arginine reversed the effect of NO synthase inhibitors, and sodium nitroprusside (which decomposes to NO) caused neurotoxicity that paralleled cyclic GMP formation. In the cerebral cortex, NO synthase immunoreactivity was found to be confined to a discrete population of aspiny neurons comprising

1–2% of the total neuronal population (Bredt et al., 1990), and that these same neurons stained selectively for NADPH diaphorase (Snyder and Bredt, 1991). It is of interest that neurons containing NO synthase are resistant to degeneration in Huntington's chorea, Alzheimer's disease, and hypoxic–ischemic brain injury. These same neurons are resistant also to glutamate neurotoxicity (Koh et al., 1988; Koh and Choi, 1988).

The precise mechanism by which NO causes glutamase neurotoxicity is unknown. Calcium must be required because of the requirement for NMDA- and glutamate-induced NO formation in brain tissue (Garthwaite et al., 1988). Although both NMDA-receptor agonists and sodium nitroprusside induce specific neurotoxicity as well as cyclic GMP formation in brain tissue (Dawson et al., 1991), it is unlikely that cyclic GMP is the ultimate cause of the neurotoxicity. Instead, NO is most likely involved in producing target cell death. One possible mechanistic pathway is that locally synthesized NO and superoxide anion react with each other to yield peroxynitrite anion (Beckman et al., 1990), which can destroy cell membranes either directly via interaction with cellular thiols (Radi et al., 1991) or indirectly via decomposition to hydroxyl and other free radicals (Beckman et al., 1990).

The original studies showing a link between NMDA-receptor activation and NO formation were conducted on cerebellar tissue obtained from immature rats, and the question therefore arose as to whether such a relationship exists in the adult. One study designed to answer this important question showed that excitatory amino acid-receptor activation in adult cerebellum results in NO and cyclic GMP formation (Southam et al., 1991). However, NMDA becomes a less effective activator with age, which reflects a decline in endogenous glycine levels. On the other hand, quisqualate becomes a more effective activator of excitatory amino acid receptors with maturation. Why such changes in excitatory amino acid-receptor selectivity occur during maturation is unknown. NMDA also stimulates cyclic GMP formation in immature and adult hippocampal slices by mechanisms involving NO formation, suggesting that NMDA-receptor-elicited NO formation has significance outside the cerebellum (East and Garthwaite, 1991).

Although studies clearly indicate, as discussed above, that NMDA-receptor activation causes NO and cyclic GMP formation in certain target cells, it is still not at all clear what function NO has in the brain. It is likely that NO functions in the brain as an intercellular messenger to stimulate cyclic GMP biosynthesis in critical target cells, where cyclic GMP acts as the intracellular messenger to trigger the principal function of the target cell. In this manner, the physiological function of neuronal NO would be analogous to the mechanism by which endothelium-derived NO elicits its physiological effects on nearby vascular smooth muscle and circulating platelets and neutrophils. Moreover, neuronally released NO may trigger other local functions in nonneuronal cells such as vasodilation and inhibition of platelet and neutrophil adhesion in nearby arterioles.

As an unconventional second messenger molecule, NO may serve to relay information about postsynaptic NMDA-receptor activation to adjacent neuronal and glial cells, perhaps to trigger long-term changes in synaptic strength on which certain types of learning depend. NO may influence glial cells or astrocytes, by elevating cyclic GMP levels in these target cells, and thereby influence developmental plasticity of synaptic connections. What is lacking is a clear understanding of precisely what role cyclic GMP plays in mediating the neuromodulatory actions of NO.

A role for NO in NANC neurotransmission has been suggested by the observations that inhibitors of NO synthase block NANC neurotransmission (Gillespie et al., 1989; Ramagopal and Leighton, 1989; Bult et al., 1990; Gibson et al., 1990; Ignarro et al., 1990b). Moreover, NO synthase is concentrated in the myenteric plexus of nerves in both the cell bodies and fibers of the small intestine (Snyder and Bredt, 1991). The initial studies were conducted with the rodent anococcygeus and bovine retractor penis muscles (see Garthwaite, 1991). In the presence of adrenergic neuronal blocking agents and atropine, electrical nerve stimulation caused smooth muscle relaxation that could not be attributed to any known neurotransmitter. Hemoglobin was then found to be inhibitory and cyclic GMP appeared to be involved in the relaxation responses. Crude extracts of the inhibitor factor from the retractor penis were studied and found to possess similar pharmacological properties to those of EDRF. After the discovery that EDRF is NO, studies on NANC neurotransmission suggested that the neurotransmitter could be NO. Electrical stimulation of the NANC neuronal pathway in a variety of tissue preparations caused smooth muscle relaxation that appeared to be mediated by NO. Some of these preparations were the rat and mouse anococcygeus (Gibson et al., 1990; Hobbs and Gibson, 1990), bovine retractor penis (Martin and Gillespie, 1990), rabbit corpus cavernosum (Ignarro et al., 1990b; Rajfer et al., 1992), canine ileocolonic junction (Bult et al., 1990), canine mesenteric and cerebral arteries (Toda and Okamura, 1990a,b), guinea pig trachea (Tucker et al., 1990), and rat esophagus (Bieger and Triggle, 1990). Consistent with these observations was the subsequent finding that NO synthase is localized in nerve fibers that innervate the intestine, blood vessels, adrenal medulla, and retina (Bredt et al., 1990). Moreover, NO synthase has been identified in both cytosolic and particulate fractions prepared from rat anococcygeus muscle (Mitchell et al., 1991). Both forms of the enzyme were dependent on NADPH, tetrahydrobiopterin, and calcium, and were inhibited by N^G-nitro-L-arginine. Fractions prepared from rat vas deferens, which does not contain NANC neurons, were devoid of NO synthase activity.

In a series of studies conducted in our laboratory (Ignarro et al., 1990b; Rajfer et al., 1992), relaxation of corpus cavernosum smooth muscle from the penis of the rabbit and human caused by electrical stimulation of the NANC neuronal pathway was found to be inhibited by oxyhemoglobin, methylene blue, N^G-nitro-

L-arginine, and N^G-amino-L-arginine (Fig. 5). Corporal smooth muscle relaxation was associated with tissue accumulation of nitrite and cyclic GMP, both of which were abolished by NO synthase inhibitors. Authentic NO elicited similar rapidly developing but transient relaxation responses and tissue accumulation of cyclic GMP that were inhibited by oxyhemoglobin and methylene blue, and augmented by M&B 22948. These observations clearly indicated that NANC neurotransmission caused relaxation of the corpus cavernosum by mechanisms involving NO and cyclic GMP. Interestingly, whereas the NO-mediated relaxation response to added acetylcholine was found to be endothelium-dependent, relaxation in response to electrical stimulation was found to be endothelium-independent. These observations suggest that either NO is the NANC neuro-

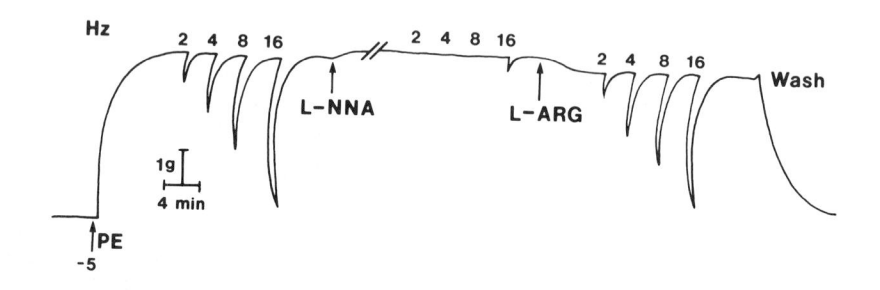

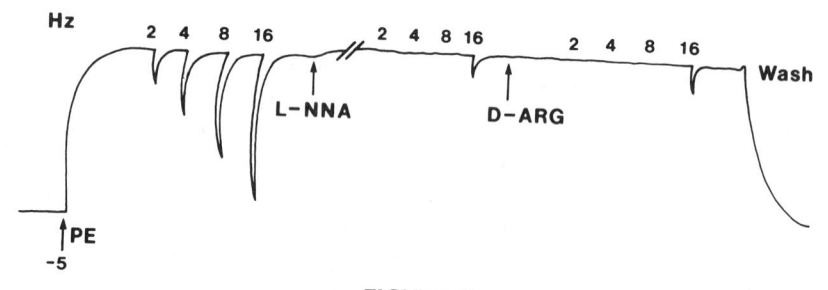

FIGURE 5

Tracing of relaxation of human corpus cavernosum smooth muscle in response to electrical stimulation of nonadrenergic–noncholinergic (NANC) neurons. Strips of corpus cavernosum from human penis were placed under optimal tension in Krebs bicarbonate solution at 37°C. Guanethidine (5 μM) and atropine (1 μM) were added to bathing media to block adrenergic and cholinergic function, respectively. Strips were precontracted by addition of 10 μM phenylephrine (PE). Electrical field stimulation (EFS) (square wave pulses of 0.2 msec duration and 10 V) was applied at various frequencies ranging from 2 to 16 Hz. EFS caused frequency-dependent relaxation as illustrated. N^G-Nitro-L-arginine (L-NNA; 0.1 mM; preincubated for 15 min) abolished relaxation and this was reversed by addition of 0.3 mM L-arginine (L-ARG) but not 0.3 mM D-arginine (D-ARG).

transmitter or that the NANC neurotransmitter is some other chemical substance that stimulates the underlying corporal smooth muscle to generate NO (Ignarro et al., 1990a).

VII. CONCLUSIONS

One of the most intriguing features of NO, considering the simplicity of the molecule, is that it elicits so many different regulatory functions and other effects. At relatively low concentrations, NO functions as a vasodilator, inhibitor of platelet adhesion and aggregation, inhibitor of neutrophil adhesion, a neuromodulator in the CNS, and possibly as an NANC neurotransmitter and neurotransmitter in other neuronal pathways as well. At relatively high concentrations, NO functions in a pathophysiological manner as a cytosolic product of cytokine-activated macrophages and possibly neutrophils and other phagocytic cells such as hepatic Kupffer cells. Under normal conditions, the NO generated by macrophages is cytotoxic against invading target cells but not host cells. The reason for this selective toxicity must be that NO reacts with only certain target cell proteins under the existing environmental conditions. In special situations such as reoxygenation of hypoxic tissues, the NO and superoxide anion generated react with each other to generate peroxynitrite anion, which is cytotoxic to host cells localized to the area involved (Matheis et al., 1992).

The chemical and biological properties of NO are well suited for all of its known physiological and pathophysiological functions. Its very small molecular size and lipophilic nature allow the rapid diffusion of NO through plasma membranes and enable NO to communicate with neighboring cells. The chemically unstable nature of NO allows for truly local actions, as does the high capacity of NO to bind to and become inactivated by erythrocyte hemoglobin and perhaps other hemoproteins. The small amounts of NO generated and released by vascular endothelial cells, brain tissue, neurons, and vascular smooth muscle in response to lipopolysaccharide and cytokines are inactivated before any appreciable cytotoxic effects can be elicited. The reactivity of NO with heme iron and iron–sulfur centers of proteins may account for the destructive action of NO on certain proteins. Relatively large amounts of locally generated NO are required to elicit such toxic reactions. One exception is the heme-containing soluble enzyme, guanylate cyclase, which is activated by NO. Another important chemical property of NO is its capacity to react with superoxide anion, not to serve as a free radical scavenger, but rather to generate the highly reactive intermediate, peroxynitrite anion, which reacts with tissue sulfhydryls (Radi et al., 1991) and decomposes to hydroxyl free radical and other radicals (Beckman et al., 1990).

Clearly, NO possesses many unique properties that endow this molecule with the capacity to exert its many varied biological functions. Other than its pathophysiological or cytotoxic actions, the physiological actions of endogenous NO are mediated by a common signal transduction mechanism involving cyclic GMP (Ignarro, 1990b). The endogenous receptor for NO is the heme moiety of cytosolic guanylate cyclase. NO interacts with guanylate cyclase-bound heme to form the nitrosyl–heme adduct, which is the activated form of the enzyme. Thus, NO generated from one cell diffuses out of the cell and into adjacent cells to interact with an intracellular soluble receptor rather than an extracellular membrane-bound receptor. This is the basis by which NO engages in cell-to-cell communication.

ACKNOWLEDGMENTS

This work was supported by research grants from the National Institutes of Health (HL-35014; HL-40922), the Laubisch Fund for Cardiovascular Research, and the Tobacco-Related Disease Research Program. The author is grateful to Diane Rome Peebles for preparing the illustrations.

REFERENCES

Beckman, J. S., Beckman, T. W., Chen, J., Marshall, P. A., and Freeman, B. A. (1990). Apparent hydroxyl radical production by peroxynitrite: Implications for endothelial injury from nitric oxide and superoxide. *Proc. Natl. Acad. Sci. U.S.A.* **87**, 1620–1624.

Bieger, W. S., and Triggle, C. R. (1990). NO: Possible role in TTX-sensitive field-stimulated relaxations of the rat oesophageal tunica muscularis mucosae. *Eur. J. Pharmacol.* **183**, 2419.

Billiar, T. R., Curran, R. D., Stuehr, D. J., West, M. A., Bentz, B. G., and Simmons, R. L. (1989). An L-arginine-dependent mechanism mediates Kupffer cell inhibition of hepatocyte protein synthesis *in vitro*. *J. Exp. Med.* **169**, 1467–1472.

Bredt, D. S., and Snyder, S. H. (1989). Nitric oxide mediates glutamate-linked enhancement of cGMP levels in the cerebellum. *Proc. Natl. Acad. Sci. U.S.A.* **86**, 9030–9033.

Bredt, D. S., and Snyder, S. H. (1990). Isolation of nitric oxide synthetase, a calmodulin-requiring enzyme. *Proc. Natl. Acad. Sci. U.S.A.* **87**, 682–685.

Bredt, D. S., Hwang, P. M., and Snyder, S. H. (1990). Localization of nitric oxide synthase indicating a neural role for nitric oxide. *Nature (London)* **347**, 768–770.

Bredt, D. S., Hwang, P. M., Glatt, C. E., Lowenstein, C., Reed, R. R., and Snyder, S. H. (1991). Cloned and expressed nitric oxide synthase structurally resembles cytochrome P-450 reductase. *Nature (London)* **351**, 714–718.

Buga, G. M., Gold, M. E., Wood, K. S., Chaudhuri, G., and Chaudhuri, G. (1989). Endothelium-derived nitric oxide relaxes nonvascular smooth muscle. *Eur. J. Pharmacol.* **161**, 61–72.

Buga, G. M., Gold, M. E., Fukuto, J. M., and Ignarro, L. J. (1991). Shear-stress-induced release of nitric oxide from endothelial cells grown on beads. *Hypertension (Dallas)* **17**, 187–193.

Bult, H., Boeckxstaens, G. E., Pelckmans, P. A., Jordaens, F. H., Van Maercke, Y. M., and Herman, A. G. (1990). Nitric oxide as an inhibitory non-adrenergic noncholinergic neurotransmitter. *Nature (London)* **345**, 346–347.

Craven, P. A., and DeRubertis, F. R. (1978). Restoration of the responsiveness of purified guanylate cyclase to nitrosoguanidine, nitric oxide, and related activators by heme and heme proteins: Evidence for the involvement of the paramagnetic nitrosyl–heme complex in enzyme activation. *J. Biol. Chem.* **253**, 8433–8443.

Dawson, V. L., Dawson, T. M., London, E. D., Bredt, D. S., and Snyder, S. H. (1991). Nitric oxide mediates glutamate neurotoxicity in primary cortical cultures. *Proc. Natl. Acad. Sci. U.S.A.* **88**, 6368–6371.

East, S. J., and Garthwaite, J. (1991). NMDA receptor activation in rat hippocampus induces cyclic GMP formation through the L-arginine–nitric oxide pathway. *Neurosci. Lett.* **123**, 17–19.

Ferrendelli, J. A. (1978). Distribution and regulation of cyclic GMP in the central nervous system. *Adv. Cyclic Nucleotide Res.* **9**, 453–464.

Forstermann, U., Goppelt-Strube, M., Frolich, J. C., and Busse, R. (1986). Inhibitors of acylcoenzyme A: Lysolecithin acyltransferase activate the production of endothelium-derived vascular relaxing factor. *J. Pharmacol. Exp. Ther.* **238**, 352–359.

Forstermann, U., Gorsky, L. D., Pollock, J. S., Ishii, K., Schmidt, H. H. H. W., Heller, M., and Murad, F. (1990). Hormone-induced biosynthesis of endothelium-derived relaxing factor/nitric oxide-like material in N1E-115 neuroblastoma cells requires calcium and calmodulin. *Mol. Pharmacol.* **38**, 7–13.

Forstermann, U., Pollock, J. S., Schmidt, H. H. H. W., Heller, M., and Murad, F. (1991a). Calmodulin-dependent endothelium-derived relaxing factor/nitric oxide synthase activity is present in the particulate and cytosolic fractions of bovine aortic endothelial cells. *Proc. Natl. Acad. Sci. U.S.A.* **88**, 1788–1792.

Forstermann, U., Schmidt, H. H. H. W., Pollock, J. S., Heller, M., and Murad, F. (1991b). Enzymes synthesizing guanylate cyclase-activating factors in endothelial cells, neuroblastoma cells, and rat brain. *J. Cardiovasc. Pharmacol.* **17**, S57–S64.

Fukuto, J. M., Wood, K. S., Byrns, R. E., and Ignarro, L. J. (1990). N^G-Amino-L-arginine: A new potent antagonist of L-arginine-mediated endothelium-dependent relaxation. *Biochem. Biophys. Res. Commun.* **168**, 458–465.

Furchgott, R. F., and Zawadzki, J. V. (1980). The obligatory role of endothelial cells in the relaxation of arterial smooth muscle by acetylcholine. *Nature (London)* **288**, 373–376.

Furchgott, R. F., Zawadzki, J. V., and Cherry, P. D. (1981). Role of endothelium in the vasodilator response to acetylcholine. *In* "Vasodilatation" (P. M. Vanhoutte and I. Leusen, eds.), pp. 49–66. Raven, New York.

Garthwaite, J. (1991). Glutamate, nitric oxide and cell–cell signalling in the nervous system. *Trends Neurosci.* **14**, 60–67.

Garthwaite, J., Charles, S. L., and Chess-Williams, R. (1988). Endothelium-derived relaxing factor release on activation of NMDA receptors suggests role as intercellular messenger in the brain. *Nature (London)* **336**, 385–388.

Gerzer, R., Bohme, E., Hofmann, F., and Schultz, G. (1981). Soluble guanylate cyclase purified from bovine lung contains heme and copper. *FEBS Lett.* **132,** 71–74.

Gibson, A., Mirzazadeh, S., Hoffs, A. J., and Moore, P. K. (1990). L-N^G-Monomethylarginine and L-N^G-nitroarginine inhibit non-adrenergic, noncholinergic relaxation of the mouse anococcygeus muscle. *Br. J. Pharmacol.* **99,** 602-606.

Gillespie, J. S., Liu, X., and Martin, W. (1989). The effects of L-arginine and N^G-monomethyl-L-arginine on the response of the rat anococcygeus muscle to NANC nerve stimulation. *Br. J. Pharmacol.* **98,** 1080–1082.

Gold, M. E., Wood, K. S., Byrns, R. E., Fukuto, J. M., and Ignarro, L. J. (1990). N^G-Methyl-L-arginine causes endothelium-dependent contraction and inhibition of cyclic GMP formation in artery and vein. *Proc. Natl. Acad. Sci. U.S.A.* **87,** 4430–4434.

Green, L. C., and Tannenbaum, S. E. (1981). Nitrate synthesis in the germfree and conventional rat. *Science* **212,** 56–58.

Griffith, T. M., Edwards, D. H., Lewis, M. J., Newby, A. C., and Henderson, A. H. (1984). The nature of endothelium-derived vascular relaxant factor. *Nature (London)* **308,** 645–647.

Gruetter, C. A., Barry, B. K., McNamara, D. B., Gruetter, D. Y., Kadowitz, P. J., and Ignarro, L. J. (1979). Relaxation of bovine coronary artery and activation of coronary arterial guanylate cyclase by nitric oxide, nitroprusside and a carcinogenic nitrosoamine. *J. Cyclic Nucleotide Protein Phosphorylation Res.* **5,** 211–224.

Hibbs, J. B., Taintor, R. R., and Vavrin, Z. (1987a). Macrophage cytotoxicity: Role for L-arginine deiminase and imino nitrogen oxidation to nitrite. *Science* **235,** 473–476.

Hibbs, J. B., Vavrin, Z., and Taintor, R. R. (1987b). L-Arginine is required for expression of the activated macrophage effector mechanism causing selective metabolic inhibition in target cells. *J. Immunol.* **138,** 550–565.

Hibbs, J. B., Taintor, R. R., Vavrin, Z., and Rachlin, E. M. (1988). Nitric oxide: A cytotoxic activated macrophage effector molecule. *Biochem. Biophys. Res. Commun.* **157,** 87–94.

Hiki, K., Yui, Y., Hattori, R., Eizawa, H., Kosuga, K., and Kawai, C. (1991). Three regulation mechanisms of nitric oxide synthase. *Eur. J. Pharmacol.* **206,** 163–164.

Hobbs, A. J., and Gibson, A. (1990). L-N^G-Nitroarginine and its methyl ester are potent inhibitors of non-adrenergic, non-cholinergic transmission in the rat anococcygeus. *Br. J. Pharmacol.* **100,** 747–752.

Ignarro, L. J. (1977). Regulation of PMN leukocyte, macrophage, and platelet function. In "Immunopharmacology" (J. W. Hadden, R. G. Coffey, and F. Spreafico, eds.), pp. 61–86. Plenum, New York.

Ignarro, L. J. (1990a). Biosynthesis and metabolism of endothelium-derived nitric oxide. *Annu. Rev. Pharmacol. Toxicol.* **30,** 535–560.

Ignarro, L. J. (1990b). Nitric oxide: A novel signal transduction mechanism for transcellular communication. *Hypertension (Dallas)* **16,** 477–483.

Ignarro, L. J., and George, W. J. (1974). Hormonal control of lysosomal enzyme release from human neutrophils: Elevation of cyclic nucleotides by autonomic neurohormones. *Proc Natl. Acad. Sci. U.S.A.* **71,** 2027–2031.

Ignarro, L. J., and Gruetter, C. A. (1980). Requirement of thiols for activation of coronary arterial guanylate cyclase by glyceryl trinitrate and sodium nitrite: Possible involvement of S-nitrosothiols. *Biochim. Biophys. Acta* **631**, 221–231.

Ignarro, L. J., Barry, B. K., Gruetter, D. Y., Edwards, J. C., Ohlstein, E. H., Gruetter, C. A., and Baricos, W. H. (1980a). Guanylate cyclase activation by nitroprusside and nitrosoguanidine is related to formation of S-nitrosothiol intermediates. *Biochem. Biophys. Res. Commun.* **94**, 93–100.

Ignarro, L. J., Edwards, J. C., Gruetter, D. Y., Barry, B. K., and Gruetter, C. A. (1980b). Possible involvement of S-nitrosothiols in the activation of guanylate cyclase by nitroso compounds. *FEBS Lett.* **110**, 275–278.

Ignarro, L. J., Lippton, H., Edwards, J. C., Baricos, W. H., Hyman, A. L., Kadowitz, P. J., and Gruetter, C. A. (1981). Mechanism of vascular smooth muscle relaxation by organic nitrates, nitrites, nitroprusside and nitric oxide: Evidence for the involvement of S-nitrosothiols as active intermediates. *J. Pharmacol. Exp. Ther.* **218**, 739–749.

Ignarro, L. J., Degnan, J. N., Baricos, W. H., Kadowitz, P. J., and Wolin, M. S. (1982a). Activation of purified guanylate cyclase by nitric oxide requires heme: Comparison of heme-deficient, heme-reconstituted and heme-containing forms of soluble enzyme from bovine lung. *Biochim. Biophys. Acta* **718**, 49–59.

Ignarro, L. J., Wood, K. S., and Wolin, M. S. (1982b). Activation of purified soluble guanylate cyclase by protoporphyrin IX. *Proc. Natl. Acad. Sci. U.S.A.* **79**, 2870–2873.

Ignarro, L. J., Ballot, B., and Wood, K. S. (1984a). Regulation of soluble guanylate cyclase activity by porphyrins and metalloporphyrins. *J. Biol. Chem.* **259**, 6201–6207.

Ignarro, L. J., Burke, T. M., Wood, K. S., Wolin, M. S., and Kadowitz, P. J. (1984b). Association between cyclic GMP accumulation and acetylcholine-elicited relaxation of bovine intrapulmonary artery. *J. Pharmacol. Exp. Ther.* **228**, 682–690.

Ignarro, L. J., Wood, K. S., Ballot, B., and Wolin, M. S. (1984c). Guanylate cyclase from bovine lung: Evidence that enzyme activation by phenylhydrazine is mediated by iron–phenyl hemoprotein complexes. *J. Biol. Chem.* **259**, 5923–5931.

Ignarro, L. J., Wood, K. S., and Wolin, M. S. (1984d). Regulation of purified soluble guanylate cyclase by porphyrins and metalloporphyrins: A unifying concept. *Adv. Cyclic Nucleotide Protein Phosphorylation Res.* **17**, 267–274.

Ignarro, L. J., Adams, J. B., Horwitz, P. M., and Wood, K. S. (1986a). Activation of soluble guanylate cyclase by NO-hemoproteins involves NO-heme exchange: Comparison of heme-containing and heme-deficient enzyme forms. *J. Biol. Chem.* **261**, 4997–5002.

Ignarro, L. J., Harbison, R. G., Wood, K. S., and Kadowitz, P. J. (1986b). Activation of purified soluble guanylate cyclase by endothelium-derived relaxing factor from intrapulmonary artery and vein: Stimulation by acetylcholine, bradykinin and arachidonic acid. *J. Pharmacol. Exp. Ther.* **237**, 893–900.

Ignarro, L. J., Buga, G. M., Wood, K. S., Byrns, R. E., and Chaudhuri, G. (1987a). Endothelium-derived relaxing factor produced and released from artery and vein is nitric oxide. *Proc. Natl. Acad. Sci. U.S.A.* **84**, 9265–9269.

Ignarro, L. J., Byrns, R. E., Buga, G. M., and Wood, K. S. (1987b). Endothelium-derived relaxing factor from pulmonary artery and vein possesses pharmacological and chemical properties that are identical to those for nitric oxide radical. *Circ. Res.* **61**, 866–879.

Ignarro, L. J., Buga, G. M., Byrns, R. E., Wood, K. S., and Chaudhuri, G. (1988a). Endothelium-derived relaxing factor (EDRF) and nitric oxide (NO) possess identical properties as relaxants of bovine arterial and venous smooth muscle. *J. Pharmacol. Exp. Ther.* **246,** 218–226.

Ignarro, L. J., Byrns, R. E., Buga, G. M., Wood, K. S., and Chaudhuri, G. (1988b). Pharmacological evidence that endothelium-derived relaxing factor is nitric oxide: Use of pyrogallol and superoxide dismutase to study endothelium-dependent and nitric oxide-elicited vascular smooth muscle relaxation. *J. Pharmacol. Exp. Ther.* **244,** 181–189.

Ignarro, L. J., Byrns, R. E., and Wood, K. S. (1988c). Biochemical and pharmacological properties of endothelium-derived relaxing factor and its similarity to nitric oxide radical. In "Mechanisms of Vasodilatation" (P. M. Vanhoutte, ed.), pp. 427–435. Raven, New York.

Ignarro, L. J., Bush, P. A., Buga, G. M., and Rajfer, J. (1990a). Neurotransmitter identity in doubt. *Nature (London)* **347,** 131–132.

Ignarro, L. J., Bush, P. A., Buga, G. M., Wood, K. S., Fukuto, J. M., and Rajfer, J. (1990b). Nitric oxide and cyclic GMP formation upon electrical field stimulation cause relaxation of corpus cavernosum smooth muscle. *Biochem. Biophys. Res. Commun.* **170,** 843–850.

Iyengar, R., Stuehr, D. J., and Marletta, M. A. (1987). Macrophage synthesis of nitrite, nitrate, and N-nitrosamines: Precursors and role of the respiratory burst. *Proc. Natl. Acad. Sci. U.S.A.* **84,** 6369–6373.

Knowles, R. G., Palacios, M., Palmer, R. M. J., and Moncada, S. (1989). Formation of nitric oxide from L-arginine in the central nervous system: A transducer mechanism for stimulation of the soluble guanylate cyclase. *Proc. Natl. Acad. Sci. U.S.A.* **86,** 5159–5162.

Koh, J. Y., and Choi, D. W. (1988). Vulnerability of cultured cortical neurons to damage by endotoxins: Differential susceptibility of neurons containing NADPH-diaphorase. *J. Neurosci.* **8,** 2153–2163.

Koh, J. Y., Peters, S., and Choi, D. W. (1988). Neurons containing NADPH-diaphorase are selectively resistant to quinolinate toxicity. *Science* **234,** 73–76.

Long, C. J., and Stone, T. W. (1985). The release of endothelium-derived relaxing factor is calcium-dependent. *Blood Vessels* **22,** 205–208.

Luckhoff, A., Pohl, U., Mulsch, A., and Busse, R. (1988). Differential role of extra- and intracellular calcium in the release of EDRF and prostacyclin from cultured endothelial cells. *Br. J. Pharmacol.* **95,** 189–196.

Marletta, M. A., Yoon, P. S., Iyengar, R., Leaf, C. D., and Wishnok, J. S. (1988). Macrophage oxidation of L-arginine to nitrite and nitrate: Nitric oxide is an intermediate. *Biochemistry* **27,** 8706–8711.

Martin, W., and Gillespie, J. S. (1990). In "Novel Peripheral Transmitters" (C. Bell, ed.), pp. 65–79. Pergamon, New York.

Matheis, G., Sherman, M. P., Buckberg, G. D., Haybron, D. M., Young, H. H., and Ignarro, L. J. (1992). Role of the L-arginine–nitric oxide pathway in myocardial reoxygenation injury. *Am. J. Physiol.* **262,** H616–H620.

Mayer, B., Schmidt, K., Humbert, P., and Bohme, E. (1989). Biosynthesis of endothelium-derived relaxing factor: A cytosolic enzyme in porcine aortic endothelial cells Ca^{2+}-

dependently converts L-arginine into an activator of soluble guanylate cyclase. *Biochem. Biophys. Res. Commun.* **164**, 678–685.

Mayer, B., John, M., and Bohme, E. (1990a). Purification of a Ca^{2+}/calmodulin-dependent nitric oxide synthase from porcine cerebellum. *FEBS Lett.* **277**, 215–219.

Mayer, B., John, M., and Bohme, E. (1990b). Partial purification of a Ca^{2+}/calmodulin-dependent endothelium-derived relaxing factor from porcine cerebellum. *J. Cardiovasc. Pharmacol.* **17**, S46–S51.

Mayer, B., John, M., Heinzel, B., Werner, E. R., Wachter, H., Schultz, G., and Bohme, E. (1991). Brain nitric oxide synthase is a biopterin- and flavin-containing multifunctional oxido-reductase. *FEBS Lett.* **288**, 187–191.

Mellion, B. T., Ignarro, L. J., Ohlstein, E. H., Pontecorvo, E. G., Hyman, A. L., and Kadowitz, P. J. (1981). Evidence for the inhibitory role of guanosine 3′,5′-monophosphate in ADP-induced human platelet aggregation in the presence of nitric oxide and related vasodilators. *Blood* **57**, 946–955.

Mellion, B. T., Ignarro, L. J., Myers, C. B., Ohlstein, E. H., Ballot, B. A., Hyman, A. L., and Kadowitz, P. J. (1983). Inhibition of human platelet aggregation by S-nitrosothiols. Heme-dependent activation of soluble guanylate cyclase and stimulation of cyclic GMP accumulation. *Mol. Pharmacol.* **23**, 653–664.

Mitchell, J. A., Sheng, H., Forstermann, U., and Murad, F. (1991). Characterization of nitric oxide synthases in nonadrenergic noncholinergic nerve containing tissue from the rat anococcygeus muscle. *Br. J. Pharmacol.* **104**, 289–291.

Miwa, M., Stuehr, D. J., Marletta, M. A., Wishnok, J. S., and Tannenbaum, S. R. (1987). Nitrosation of amines by stimulated macrophages. *Carcinogenesis (London)* **8**, 955–958.

Mulligan, M. S., Hevel, J. M., Marletta, M. A., and Ward, P. A. (1991). Tissue injury caused by deposition of immune complexes is L-arginine dependent. *Proc. Natl. Acad. Sci. U.S.A.* **88**, 6338–6342.

Mulsch, A., and Busse, R. (1991). Nitric oxide synthase in native and cultured endothelial cells: Calcium/calmodulin and tetrahydrobiopterin are cofactors. *J. Cardiovasc. Pharmacol.* **17**, S52–S56.

Myers, P. R., Minor, R. L., Jr., Guerra, R., Jr., Bates, J. N., and Harrison, D. G. (1990). Vasorelaxant properties of the endothelium-derived relaxing factor more closely resemble S-nitrosocysteine than nitric oxide. *Nature (London)* **345**, 161–163.

Ohlstein, E. H., Wood, K. S., and Ignarro, L. J. (1982). Purification and properties of heme-deficient hepatic soluble guanylate cyclase: Effects of heme and other factors on enzyme activation by NO, NO–heme, and protoporphyrin IX. *Arch. Biochem. Biophys.* **218**, 187–198.

Palmer, R. M. J., Ferrige, A. G., and Moncada, S. (1987). Nitric oxide release accounts for the biological activity of endothelium-derived relaxing factor. *Nature (London)* **327**, 524–526.

Palmer, R. M. J., Ashton, D. S., and Moncada, S. (1988a). Vascular endothelial cells synthesize nitric oxide from L-arginine. *Nature (London)* **333**, 664-666.

Palmer, R. M. J., Rees, D. D., Ashton, D. S., and Moncada, S. (1988b). L-Arginine is the physiological precursor for the formation of nitric oxide in endothelium-dependent relaxation. *Biochem. Biophys. Res. Commun.* **153**, 1251–1256.

Peach, M. J., Loeb, A. L., Singer, H. A., and Saye, J. A. (1985). Endothelium-derived relaxing factor. *Hypertension (Dallas)* **7,** I-94–I-100.

Radi, R., Beckman, J. S., Bush, K. M., and Freeman, B. A. (1991). Peroxynitrite oxidation of sulfhydryls: The cytotoxic potential of superoxide and nitric oxide. *J. Biol. Chem.* **266,** 4244–4250.

Radomski, M. W., Palmer, R. M. J., and Moncada, S. (1987). Comparative pharmacology of endothelium-derived relaxing factor, nitric oxide and prostacyclin in platelets. *Br. J. Pharmacol.* **92,** 181–187.

Rajfer, J., Aronson, W. J., Bush, P. A., Dorey, F. J., and Ignarro, L. J. (1992). Nitric oxide as a mediator of relaxation of the corpus cavernosum elicited by nonadrenergic, noncholinergic transmission. *N. Engl. J. Med.* **326,** 90–94.

Ramagopal, M. W., and Leighton, H. J. (1989). Effects of N^G-monomethyl-L-arginine on field stimulation-induced decreases in cytosolic Ca^{2+} levels and relaxation in the rat anococcygeus muscle. *Eur. J. Pharmacol.* **174,** 297–299.

Sakuma, I., Stuehr, D. J., Gross, S. S., Nathan, C. F., and Levi, R. (1988). Identification of arginine as a precursor of endothelium-derived relaxing factor. *Proc. Natl. Acad. Sci. U.S.A.* **85,** 8664–8667.

Schmidt, H. H. H. W., Pollock, J. S., Nakane, M., Gorsky, L. D., Forstermann, U., and Murad, F. (1991). Purification of a soluble isoform of guanylyl cyclase-activating-factor synthase. *Proc. Natl. Acad. Sci. U.S.A.* **88,** 365–369.

Singer, H. A., and Peach, M. J. (1982). Calcium- and endothelium-mediated vascular smooth muscle relaxation in rabbit aorta. *Hypertension (Dallas)* **4,** II-19–II-25.

Singer, H. A., and Peach, M. J. (1983). Endothelium-dependent relaxation of rabbit aorta. I. Relaxation stimulated by arachidonic acid. *J. Pharmacol. Exp. Ther.* **226,** 790–795.

Smith, R. J., and Ignarro, L. J. (1975). Bioregulation of lysosomal enzyme secretion from human neutrophils: Roles of cyclic GMP and calcium in stimulus–secretion coupling. *Proc. Natl. Acad. Sci. U.S.A.* **72,** 108–112.

Snyder, S. H., and Bredt, D. S. (1991). Nitric oxide as a neuronal messenger. *Trends Pharmacol. Sci.* **12,** 125–128.

Southam, E., East, S. J., and Garthwaite, J. (1991). Excitatory amino acid receptors coupled to the nitric oxide/cyclic GMP pathway in rat cerebellum during development. *J. Neurochem.* **56,** 2072–2081.

Stuehr, D. J., and Marletta (1985). Mammalian nitrate biosynthesis: Mouse macrophages produce nitrite and nitrate in response to *Escherichia coli* lipopolysaccharide. *Proc. Natl. Acad. Sci. U.S.A.* **82,** 7738–7742.

Stuehr, D. J., and Marletta (1987). Induction of nitrite/nitrate synthesis in murine macrophages by BCG infection, lymphokines, or interferon-gamma. *J. Immunol.* **139,** 518–525.

Stuehr, D. J., Gross, S. S., Sakuma, I., Levi, R., and Nathan, C. F. (1989). Activated murine macrophages secrete a metabolite of arginine with the bioactivity of EDRF and the chemical reactivity of nitric oxide. *J. Exp. Med.* **169,** 1011–1020.

Stuehr, D. J., Kwon, N. S., Nathan, C. F., Griffith, O. W., Feldman, P. L., and Wiseman, J. (1991). N^ω-Hydroxy-L-arginine is an intermediate in the biosynthesis of nitric oxide from L-arginine. *J. Biol. Chem.* **266,** 6259–6263.

Toda, N., and Okamura, T. (1990a). Modification of L-N^G-monomethyl arginine (L-NMMA) of the response to nerve stimulation in isolated dog mesenteric and cerebral arteries. *Jpn. J. Pharmacol.* **52,** 170–173.

Toda, N., and Okamura, T. (1990b). Possible role of nitric oxide in transmitting information from vasodilator nerves to cerebroarterial muscle. *Biochem. Biophys. Res. Commun.* **170,** 308–313.

Tucker, J. F., Brave, S. R., Charalambous, L., Hobbs, A. J., and Gibson, A. (1990). L-N^G-nitro-arginine inhibits nonadrenergic, noncholinergic relaxations of guinea pig isolated tracheal smooth muscle. *Br. J. Pharmacol.* **100,** 663–664.

Vargas, H. M., Ignarro, L. J., and Chaudhuri, G. (1990). Physiological release of nitric oxide is dependent on the level of vascular tone. *Eur. J. Pharmacol.* **190,** 393–397.

Vargas, H. M., Cuevas, J. M., Ignarro, L. J., and Chaudhuri, G. (1991). Comparison of the inhibitory potencies of N^G-methyl, N^G-nitro-, and N^G-amino-L-arginine on EDRF function in the rat: Evidence for continuous basal EDRF release. *J. Pharmacol. Exp. Ther.* **257,** 1208–1215.

Wallace, G. C., Gulati, P., and Fukuto, J. M. (1991). N^w-Hydroxy-L-arginine: A novel arginine analog capable of causing vasorelaxation in bovine intrapulmonary artery. *Biochem. Biophys. Res. Commun.* **176,** 528–534.

Wolin, M. S., Wood, K. S., and Ignarro, L. J. (1982). Guanylate cyclase from bovine lung: A kinetic analysis of the regulation of the purified soluble enzyme by protoporphyrin IX, heme, and nitrosyl–heme. *J. Biol. Chem.* **257,** 13312–13320.

Wright, C. D., Mulsch, A., Busse, R., Osswald, H. (1989). Generation of nitric oxide by human neutrophils. *Biochim. Biophys. Acta* **160,** 813–819.

Yui, Y., Hattori, R., Kosuga, K., Eizawa, H., Hiki, K., and Kawai, C. (1991). Purification of nitric oxide synthase from rat macrophages. *J. Biol. Chem.* **266,** 12544–12547.

4

The Intracellular Reactions of Nitric Oxide in the Immune System and Its Enzymatic Synthesis

Jack Lancaster, Jr.
Departments of Physiology and Medicine
Louisiana State University Medical Center
New Orleans, Louisiana 70112

Dennis J. Stuehr
The Cleveland Clinic
Immunology Section NN-1
Cleveland, Ohio 44195

I. INTRODUCTION

In this chapter, we will survey the molecular and biochemical basis of the actions of nitric oxide (NO) in the immune system and the enzymatic synthesis of NO. In line with the overall theme of this book, our attention is focused on the more molecular aspects of NO, in particular, those systems where information is available regarding the biochemical interactions of NO with its intracellular targets (primarily iron-containing proteins) in the immune system and the enzymology of its synthesis. Other chapters in this volume deal with more specific aspects of NO in the immune system, including the liver (see Chapter 6), allograft rejection (see Chapter 7), and type I diabetes (see Chapter 5). For more comprehensive descriptions of the multiple biological roles of NO in the immune system the reader is referred to several excellent reviews (Moncada et al., 1991; Green et al., 1991; Nathan and Hibbs, 1991; James, 1991; Kolb and Kolb-Bachofen, 1992; Corbett and McDaniel, 1992; Nathan, 1992; Moncada and Higgs, 1993).

II. INTRACELLULAR REACTIONS OF NITRIC OXIDE IN THE IMMUNE SYSTEM

A. Mammalian Synthesis of Inorganic Nitrogen Oxides

The study of the biosynthesis of inorganic nitrogen oxides has a long and distinguished history, principally in the field of microbial and plant nitrogen metabolism (Zumft, 1993). Relatively recently, it has been found that NO, in addition to other species such as nitrite, nitrate, and N_2O, is a true intermediate in the biological nitrogen cycle, as described in detail in Chapter 9.

Mammalian inorganic nitrogen oxide synthesis has only been discovered relatively recently, although an early report utilizing nutritional balance studies claimed such synthesis in humans (Mitchell et al., 1916). In a series of careful studies by Tannenbaum and colleagues, it was definitively demonstrated that rats and humans excrete more nitrate in the urine than is ingested (Green et al., 1981a,b). In 1982, the Tannenbaum group (Wagner and Tannenbaum, 1982, Wagner et al., 1983) and Hegesh and Shiloah (1982) independently reported data that had far-reaching consequences for one physiological role of nitrogen oxide synthesis: Adults or infants with diarrhea excreted much more urinary nitrate than uninfected individuals. In addition, the Tannenbaum group demonstrated that injection of rats with a component of the cell wall of gram-negative bacteria, lipopolysaccharide (LPS), induces the formation of $^{15}NO_3^-$ from $^{15}NH_3$ (Wagner and Tannenbaum, 1982). Subsequent work from several groups demonstrated the formation of NO_3^- from NH_3 induced by oxygen radicals (Saul and Archer, 1984; Dull and Hotchkiss, 1984; Nagano and Fridovich, 1985), suggesting at that time that the NO_3^- may be formed as a result of production of such radicals during immune activation. The relevance of this mechanism to the mammalian formation of nitrogen oxides is not known.

In 1985, Stuehr and Marletta demonstrated that the macrophage is the major source of NO_2^- and NO_3^- in response to LPS, at least in the mouse. Subsequently, this group showed that addition of LPS and the immune cytokine interferon-γ (cytokines are discussed in more detail below) to macrophages results in N-nitrosation of morpholine; the N-nitrosamine was not formed by addition of NO_2^- and morpholine to the macrophages, and the highest levels of N-nitrosamine occurred many hours prior to the peak NO_2^- formation (Miwa et al., 1987). Thus, treated macrophages are stimulated to produce a reactive precursor to NO_2^- and NO_3^-, which is capable of N-nitrosamine formation.

What is the immunological significance and the mechanism of this stimulation of nitrogen oxide formation in response to bacterial products? The answers to these questions originate in an extensive body of medical and immunological literature prior to 1987.

B. Immunological Studies prior to 1987

For over 100 years it has been known that under certain circumstances injection of bacterial products or pathogenic organisms can induce dramatic cross-resistance to other pathogens and even regression of tumors. The initial observations were made with human patients, where it was shown that injection with pathogenic bacterial extracts (which induces an intense immune activation response) into patients with incurable cancer resulted in a dramatic necrosis (death) of tumors in many (but not all) cases (Coley, 1893). This type of immune response is called nonspecific, because unlike antibody-dependent responses the activated immune defense mechanism is not strictly specific for the stimulus that induces it. That is, in experimental animals injection of extracts from one pathogenic organism can under some conditions induce cross-resistance to another, unrelated pathogen or, as described above, even rejection of host tissue that has become cancerous (Abbas, 1991). Interest in this phenomenon as a potential therapy in human patients was discontinued early in this century, because of its obvious toxicity and also because of the simultaneous advent of radiotherapy (Nauts and McLaren, 1990).

In 1975, Carswell *et al.* discovered that this bacterial-induced tumor killing activity could be transferred from one animal to another. When the serum of a mouse that is pretreated with LPS (also called endotoxin) is injected into another animal that has not been treated with endotoxin but harbors a transplantable skin tumor, the tumor undergoes necrosis. In addition, Carswell *et al.* also showed that addition of this substance (which they dubbed tumor necrosis factor, or TNF) to cultures of transformed cells results in death of the cells (Helson *et al.*, 1975). It is now known that TNF is a member of a class of proteins called cytokines, which are "immune hormones," that is, proteins that are secreted by certain immune cells during stimulation. These proteins are extraordinarily important in the immune response, by acting as the messengers between different cells to relay information (Abbas, 1991).

There are two "arms" of the immune system. Humoral immunity is the formation of antibodies against foreign antigens, which act as "flags" to trigger specific destructive mechanisms against invading pathogen, cancerous cell, or material recognized as non-self. Cell-mediated (also called nonspecific or natural) immunity is the collection of destructive mechanisms that recognizes foreign or abnormal material without the immediate participation of antibodies in the destruction, although specific mechanisms often act as initial triggering events. Although cytokines play a critical role in orchestrating both responses, a great deal of what we know about cell-mediated immunity has come from studies on the functions of these proteins.

Studies in the 1960s and early 1970s demonstrated that the macrophage plays a central role in nonspecific killing of tumor cells (Mackaness, 1964; Evans and Alexander, 1970; Alexander and Evans, 1971; Hibbs *et al.*, 1971; Keller and

Jones, 1971), and it is now known that these cytotoxic activated macrophages (CAMs) can be formed *in vitro* from macrophages by a combination of two signals, usually interferon-γ and lipopolysaccharide (a major component of the cell wall of gram-negative bacteria).

When exposed in culture dishes to CAMs, tumor cells respond in either of two phenotypic ways, relatively immediate death (lysis) or cytostasis (i.e., the cells no longer proliferate but do not lyse). Use of tumor cell targets of the cytostatic response has allowed investigation of the specific biochemical lesions induced in the cell by CAM exposure. In 1980 it was found that tumor cell DNA synthesis is inhibited at quite early times after exposure to CAMs, as early as 2 hr (Krahenbuhl, 1980). In 1980 Granger, et al. found evidence that CAMs induce a specific pattern of aerobic metabolic dysfunction in tumor cell targets. It was found that although uninjured (by CAMs) tumor cells proliferate in the absence of glucose in the medium, it was required in order for the injured cells to survive the injury. Although unable to proliferate, the injured cells did not die (lyse) if glucose was present. Studies demonstrated that this requirement was due to an increased demand for glycolytic substrate, because aerobic metabolism (complete oxidation to CO_2) was greatly inhibited. By using specific inhibitors and electron donors, in 1982 Granger and Lehninger demonstrated that there was a specific lesion in the mitochondrial electron transfer chain, localized to the complexes responsible for the transfer of electrons from NADH (Complex I) and succinate (Complex II) to coenzyme Q, from which electrons proceed through a series of common carriers to ultimately reduce O_2 to water (Complexes III and IV), which was relatively unaffected. They suggested that the mechanism of inhibition may involve oxidative injury to the mitochondrial membrane.

It is noteworthy that except for the Rieske center in Complex III, Complexes I and II are home to all the iron–sulfur clusters in the mitochondrial electron transfer chain and consequently most of the iron-containing carriers in the entire sequence. Hibbs subsequently showed that CAM-injured cells lose a substantial portion of their total intracellular iron (Hibbs et al., 1984) [later studies specifically identified loss of mitochondrial iron (Wharton et al., 1988)] and Drapier and Hibbs (1986) showed that the activity of another iron–sulfur-containing enzyme, aconitase, is also lost. In early 1987 Hibbs reported that the cytostatic actions of CAMs requires the presence of only one component in culture medium, L-arginine (Hibbs et al., 1987b). Thus, the stage was set for the discovery of a unique reactive species that targets intracellular iron, produced by CAMs.

C. Mammalian Synthesis of Nitric Oxide

In 1987, Hibbs published a landmark paper in the nitric oxide field (Hibbs et al., 1987a), by demonstrating that the L-arginine that is required for CAM cytostatic activity is metabolized to L-citrulline (similar to L-arginine but less one nitrogen atom from the guanidinium group) and nitrite plus nitrate. In addition, an analogue of L-arginine that is modified at the guanidinium group (N^G-monomethyl-L-arginine, NMMA) prevents both CAM cytostasis and formation of citrulline and nitrite plus nitrate. This paper was the first demonstration of nitrogen oxide formation from L-arginine (with its accompanying cytodamaging effects) and of the utilization of L-arginine analogues as selective inhibitors.

As described in more detail in Chapter 3, at the same time as these studies with CAMs, progress in delineating the mechanism of endothelium-dependent vascular relaxation was also providing evidence for mammalian nitrogen oxide synthesis. In particular, Furchgott (1988) and Ignarro et al. (1988) proposed at a conference in 1986 that the endothelium-derived relaxing factor (EDRF), first described by Furchgott and Zawadzki in 1980, is nitric oxide. Ignarro et al. (1987b) and Palmer et al. (1987) showed that EDRF and NO are indistinguishable in terms of inactivation by hemoproteins and superoxide anion. Iyengar et al. (1987) demonstrated that the nitrogen in nitrite and nitrate and also in the nitrosamine functionality produced by CAMs derives from the guanidino group of L-arginine. Shortly thereafter, chemical detection of the production of nitric oxide (as opposed to nitrite or nitrate) was made. Ignarro et al. (1987a) measured formation of the nitric oxide complex of hemoglobin exposed to stimulated endothelial cells, and Palmer et al. (1988a) detected endothelial NO production using mass spectrometry and also demonstrated that the NO originates from the guanidino group of L-arginine and that relaxation is inhibited by NMMA (Palmer et al., 1988b). Kelm et al. (1988) subsequently showed that the concentrations of NO formed by stimulated endothelial cells are quantitatively sufficient to explain EDRF-mediated relaxation. Nitric oxide formation from the guanidino group of L-arginine by CAMs and CAM cytosol was demonstrated by Marletta et al. (1988) using mass spectrometry, and Stuehr et al. (1989) showed that CAMs produce EDRF which is inhibited by NMMA and enhanced by L-arginine. Finally, Garthwaite et al. (1988) demonstrated that NO is produced in cerebellar cells on receptor stimulation.

D. Iron Enzyme Dysfunction Resulting from Nitric Oxide Synthesis

In their paper in 1987, Hibbs et al. (1987a) proposed that the characteristic pattern of metabolic dysfunction inflicted by CAMs is due to iron loss from aconitase and other iron–sulfur-containing enzymes resulting from nitrite or oxygenated nitrogen intermediates in the pathway of nitrite and nitrate synthesis. Much data have since been published to support this proposal, although as described below the chemical details of this process are still not clear.

As early as 1976 it was demonstrated that NO treatment of iron–sulfur-containing enzymes of the mitochondrial electron transfer chain results in liberation of iron from the clusters (Salerno et al., 1976). Hibbs et al. (1988) and Stuehr and Nathan (1989) demonstrated that treatment of tumor target cells with NO results in the same pattern of inhibition in mitochondrial electron transfer (sites confined to Complexes I and II) that is induced by CAMs (Granger and Lehninger, 1982). NO treatment also induces loss of iron and inhibition of proliferation (Hibbs et al., 1988), similar to CAMs (Hibbs et al., 1984). Additionally, as with CAM treatment (Granger et al., 1980) cell survival after NO treatment is preserved in the presence, but not in the absence, of glucose (Hibbs et al., 1988). NO treatment also results in loss of aconitase activity (Hibbs et al., 1988; Drapier et al., 1993) as does CAM treatment (Drapier and Hibbs, 1986). Stuehr et al. (1989) showed that CAMs produce a substance from L-arginine which results in irreversible destruction of the iron–sulfur cluster of ferredoxin added to the medium (as does NO or NO_2) and this destruction correlates directly with the acquisition of CAM cytostatic activity. In a related study, Stuehr and Nathan (1989) demonstrated that CAM cytostasis is prevented by ferromyoglobin as well as superoxide, which react with and inactivate NO. Taken together, these results support the proposal (Hibbs et al., 1987a) that the cytostatic/cytotoxic properties of CAMs under some conditions is due to L-arginine-dependent nitrogen oxide production, which affects the cellular targets [including the CAM itself (Drapier and Hibbs, 1988)] by the destruction of iron-containing enzyme function. Stronger (although still not conclusive) evidence comes from the direct observation of the formation of thiol–metal–nitrosyl complexes, as described below.

E. Electron Paramagnetic Resonance Observation of Dinitrosylirondithiol Complexes

In the case of certain oxidation states of particular metals (most notably Fe^{2+}) binding of NO results in an unpaired electron that is partially localized on the metal and consequently possesses magnetic properties characteristic for the atomic identity of the metal and also for the other ligands bound to the metal in addition to NO. Thus, by examining the magnetic properties of biological samples with electron paramagnetic resonance (EPR) spectroscopy, direct observation and identification of metal–nitrosyl complexes can be made, as described in more detail in Chapter 2. Primarily because of its abundance, the most commonly observed metal–nitrosyl is Fe–NO, and the two major forms of cellular iron (heme and nonheme) give distinctive and different EPR signals (Henry et al. 1991, 1993).

In 1989, Vanin et al. detected the LPS-induced formation *in vivo* of EPR signals in the liver of mice treated with the iron chelator diethylthiocarbamate (DETC) (Kubrina et al., 1989). This signal results from the nitrosylation of iron bound to the DETC with a stoichiometry of one NO bound to one iron. In 1990 two groups independently reported the NMMA-inhibitable EPR detection of nonheme iron–nitrosyl EPR signals in NO-producing CAMs *in vitro* (Lancaster and Hibbs, 1990; Pellat et al., 1990). That the nitrosyl group in the complexes was derived from the guanidino group of L-arginine was shown by demonstrating differences in the signal between cells incubated with L-arginine labeled with either ^{14}N or ^{15}N in the guanidino group (Lancaster and Hibbs, 1990). Perhaps more importantly, these complexes were also observed in tumor cell targets of CAMs, and approximately 50% of the total intracellular signal was found in the mitochondrial subcellular fraction (Drapier et al., 1991). In 1991, the Vanin group demonstrated that the appearance of the EPR signal of DETC–Fe–NO complexes *in vivo* in mice is decreased by injection of NMMA, showing that the NO originates from L-arginine (Vanin et al., 1991). Since that time, EPR signals from nonheme iron–nitrosyl complexes have been observed in several cell types producing NO *in vitro* in addition to macrophages, including rat pancreatic islet β cells (Corbett et al., 1991), hepatocytes (Stadler et al., 1993), and vascular smooth muscle cells (Geng et al., 1994). In addition, EPR signals from both nonheme and heme iron–nitrosyl complexes have been detected, in isolated cells (Geng et al., 1994) as well as in tissue samples undergoing rejection (Lancaster et al. 1992; Langrehr et al., 1992; Bastian et al., 1994) and during inflammation (Chamulitrat et al., 1994) and also intense signals attributable to hemoglobin–NO complexes in blood during sepsis (Westenberger et al., 1990; Wang et al., 1991).

This was not the first time that nonheme iron–nitrosyl signals had been observed in cells or tissues. Beginning in the 1960s, numerous studies described the appearance of two novel types of EPR signals present in cancerous tissue (Brennan et al., 1966; Emanuel et al., 1969; Vanin et al., 1970; Maruyama et al., 1971; Nagata et al., 1973) and in the livers of carcinogen-treated rats (Vithayathil et al., 1965; Woolum and Commoner, 1970; Chiang et al., 1972). Based on the characteristic features of these spectra it was speculated that the signals arise from an unpaired electron on a nitrogen atom complexed with a metal, most probably iron. This conclusion was strengthened by the results of a series of EPR experiments involving exposure to NO or NO-generating systems of small inorganic complexes (Mcdonald et al., 1965; Vanin, 1967; Woolum et al., 1968; Woolum and Commoner, 1970; Jezowska-Trezebiatowska and Jezierski, 1973; Basosi et al., 1975; Salerno et al., 1976) and of biological samples (Azhipa et al., 1965; Vanin, 1967; Woolum et al., 1968; Kon, 1968; Woolum and Commoner, 1970; Chiang et al., 1972; Dervartanian et al., 1973; Vanin et al., 1975), as well as studies of animals fed with nitrite or nitrate plus or minus iron (Chiang et al., 1972; Vanin et al., 1975, 1989; Vanin and Varich, 1980, 1981; Varich, 1980; Varich and Vanin, 1983; Foster and Hutchison, 1974; Varich et al., 1987) and also nitrite-treated bacteria (Reddy et al., 1983). In the latter case, it was found that appearance of the nonheme iron–NO signal correlated with the disappearance of the EPR signal from native iron–sulfur centers, consistent with the attack and destruction of these clusters by NO. It is now known that the two types of EPR signals correspond to complexes of NO with heme iron and with nonheme iron. However, the significance of the formation of these signals in carcinogenesis has not been systematically investigated further, due to the lack at that time of a mechanism for the mammalian formation of nitrogen oxides and also to the demonstration that similar signals can be formed as a result of nonspecific necrosis of tissues (Vanin et al., 1970; Maruyama et al., 1971).

F. Are the "g = 2.04" Electron Paramagnetic Resonance Signals Direct Observation of Nonheme Iron–Sulfur Nitrosylation?

The EPR observation of the formation of nonheme iron–nitrosyl complexes in cells producing NO provided strong evidence for the Hibbs proposed mechanism for the effector mechanism of NO synthesis. This signal (which exhibits a principal spectral feature at $g = 2.04$) is characteristic for complexes which contain one iron complexed with two thiols and two NO molecules, thus designated dinitrosylirondithiol complexes (DNIC). Thus, it may seem plausible to suggest that these signals originate from complexation of NO to the iron–thiol in the iron–sulfur cluster of target proteins. Indeed, this signal (as well as the signal from heme–NO complexes) has been observed in a variety of conditions of inflammation or infection and in many *in vitro* conditions of cellular NO synthesis (Lancaster, 1992; Henry et al., 1991). However, there is evidence that the appearance of this signal, while characteristic for endogenous NO synthesis, is not necessarily a direct observation of the iron–thiol groups on the target enzymes which have been transformed into dinitrosyl complexes. In essence, the critical question is whether the nonheme iron and thiol which constitute the DNIC complexes originate from the enzymatic targets of NO production (i.e., mitochondrial nonheme iron–sulfur centers, aconitase) or other cellular forms of nonheme iron or thiol. In addition, as is true for all chemical/biochemical effects of NO, it is important to realize that under biological conditions (in the presence of O_2, which is a cosubstrate for the enzyme, as described below) there will be several different nitrogen oxide species formed as a result of the action of nitric oxide synthase (NOS) on L-arginine, as described in more detail in Chapter 1. Probably with the only exception of nitrate, any of these species could be reactive enough to inactivate nonheme iron centers in proteins, although no studies have appeared examining the mechanistic aspects of the reaction of nonheme iron centers with NO in the presence of O_2. We will address each of these questions in turn below.

Is it NO (alone) which reacts directly with the iron–sulfur clusters of target enzymes, thus explaining the metabolic dysfunction observed in CAM-mediated injury? Under anaerobic conditions, NO can react with isolated iron–sulfur proteins to inhibit activity (Meyer, 1981; Michalski and Nicholas, 1987; Liang and Burris, 1988; Hyman and Arp, 1988; Payne et al., 1990; Hyman and Arp, 1991; Hyman et al., 1992) and generate DNIC EPR signals (Hyman et al., 1992; Dervartanian et al., 1973; Salerno et al., 1976; Petroulas and Diner, 1990; Drapier et al., 1991, 1992). Bleaching of the brown color of ferredoxin has been used to assay for cellular

NO formation, although this reaction occurred under aerobic conditions and was also observed with NO_2 (Stuehr and Nathan, 1989). It is important to point out, however, that individual iron–sulfur clusters do not exhibit uniform sensitivity to NO reaction, and inhibition of enzymatic activity can occur in either the absence or presence of the appearance of the $g = 2.04$ DNIC signal (Hyman and Arp, 1988; Hyman et al., 1992). It is thus unclear whether the loss of iron enzyme function is due to NO or oxidation products.

Does the $g = 2.04$ DNIC signal in cells arise from the attack of NO (or another nitrogen oxide) on the iron–sulfur clusters of vulnerable target enzymes, destroying their function? Thiol ligation to the iron–nitrosyl moiety could theoretically arise from the cysteine residues which had participated in the native cluster prior to NO exposure. However, Drapier et al. (1992) showed that a majority of the signal is present in the cytosolic fraction of tumor cells exposed to CAMs. The cytosolic signals were apparently heterogeneous in that the signal was distributed among fractions of different molecular weight. This raises the possibility that either (or both) NO directly reacts with cytosolic, nonmitochondrial iron–thiol species or that the iron–dinitrosyl moiety, formed initially from iron–sulfur clusters, exchanges from mitochondrial to cytosolic thiols. Indeed, previous work with small model complexes have shown that the iron–dinitrosyl moiety undergoes facile exchange with different thiol ligands (Butler et al., 1985, 1988), thus raising the possibility that, once formed, the iron–dinitrosyl formed from the reaction (and destruction) of nonheme iron–sulfur complexes can exchange onto other cellular thiol-containing species.

$$(RS^-)_2Fe(NO)_2 + 2\ R'S^- \longrightarrow (R'S^-)_2Fe(NO)_2 + 2\ RS^-$$

DNIC formation may thus be a dynamic process, that is, the iron–dinitrosyl moiety may transfer between different thiol donors.

The native iron–sulfur clusters of many enzymes can be directly observed by EPR spectroscopy (Beinert, 1978). In several cases, appearance of the DNIC signal from nitrogen oxide-treated enzyme (Salerno et al., 1976; Dervartanian et al., 1973; Hyman et al., 1992) or cells (Reddy et al., 1983; Payne et al., 1990) results in the disappearance of these native signals, indicating conversion of the iron from the native cluster coordination to DNIC complexes. However, loss of EPR signal from native clusters does not necessarily correlate with appearance of the DNIC signal (Hyman et al., 1992).

Vanin et al. (1992) have performed a careful EPR study of the treatment of cells from a cultured macrophage line with NO. A 5-min treatment of these cells with low (~20 μM) NO results in the appearance of DNIC signals, but without concomitant decrease in the intensities of the signals from mitochondrial iron–sulfur clusters. These results indicate that under at least some conditions DNIC formation occurs with iron which is not part of the mitochondrial electron transfer chain, as suggested by Drapier et al., (1992) and Bergonia and co-workers

(Stadler et al., 1993) based on subcellular localization of this signal (Drapier et al., 1992). Similar to results with whole bacterial cells (Reddy et al., 1983; Michalski and Nicholas, 1987) and mitochondrial succinate dehydrogenase (Salerno et al., 1976), production of large amounts of NO by reduction of nitrite results in appearance of a large DNIC signal and also loss of the native mitochondrial cluster signals in these cells. Finally, Vanin et al. (1993) demonstrated the release of DNIC from macrophages into the medium, providing a possible explanation for the release of intracellular iron first observed by the Hibbs group (Hibbs, et al., 1984).

In summary, these results demonstrate that intracellular iron is a major target for cellular damage resulting from NO synthesis, and that EPR spectroscopy is a valuable tool in assessing this damage. However, it cannot be assumed that observation of these iron–nitrosyl complexes is direct observation of the damage inflicted by NO, although it may well be an accurate predictor of the extent of injury. In addition, the nitrogen oxide species which is most responsible for iron-enzyme dysfunction in the aerobic cellular environment is not known.

III. ENZYMOLOGY OF NITRIC OXIDE SYNTHESIS

Over the past several years, an increasing portion of research has been directed toward understanding the biochemistry of NO synthesis in mammals. NO synthase (NOS) activities were first reported in cytokine activated macrophages (Stuehr and Marletta, 1985) in endothelium (Palmer et al., 1987; Ignarro et al., 1987a), and in brain cerebellum (Garthwaite et al., 1988; Bredt and Snyder, 1990). The isoforms purified from macrophages and brain (Bredt and Snyder, 1990; Bredt et al., 1991; Stuehr et al., 1991a; Hevel et al., 1991; Schmidt et al., 1991; Mayer et al., 1990) have fueled the majority of biochemical studies to date. As summarized below, a fascinating picture has begun to emerge regarding the structural and catalytic properties of this unique family of enzymes.

A. Reaction Catalyzed by Nitric Oxide Synthases

All NOS isoforms appear to catalyze a stepwise oxidation of the amino acid L-arginine to form NO and L-citrulline as products (Fig. 1). The reaction formally represents a five-electron oxidation of the guanidino nitrogen, involves an initial N-hydroxylation at that position, and requires exogenous NADPH and O_2. The NADPH stoichiometry of approximately 1.5 NADPH consumed per NO formed that was determined independently for macrophage and brain NOS isoforms 'NO (Stuehr et al., 1991b; Klatt et al., 1993) is somewhat unusual, given that monooxygenase reactions typically utilize even numbers of electrons from NADPH. The implications of these results are discussed later in the chapter. The NOS reaction is biologically novel in that L-arginine was previously known only to be hydroxylated at its delta carbon or nitrogen by microbes or plants (Sulser and Sager, 1976; Widmer and Keller-Schierlein, 1974). It also differs considerably from microbial pathways that generate NO via reduction of nitrite or oxidation of ammonia (see Chapter 9). Stable isotope and stoichiometry studies have identified the source of the nitrogen and oxygen atoms in the products (Iyengar et al., 1987; Kwon et al., 1990; Leone et al., 1991) (Fig. 1), which has helped narrow down the number of possible mechanisms for the enzyme.

B. Expression of Nitric Oxide Synthase

In general, the NOS isoforms can be grouped into two classes as determined by their expression in cells. Certain NOS isoforms, termed constitutive, appear to be expressed at a fairly constant level in their host cells. However, these isoforms are inactive in their native state and require that the Ca^{2+}-binding protein calmodulin associate with them in order to generate NO (Schmidt et al., 1991; Busse and Mulsch, 1990; Bredt and Snyder, 1994). Thus NO production by constitutive NOS isoforms is often linked to Ca^{2+}-mediated signal transduction cascades that involve soluble guanylate cyclase, which is a hemeprotein that is activated by NO (Arnold et al., 1977).

Other NOS isoforms, termed inducible, are typically not present in cells until the cell is exposed to immunostimulants of bacterial or host origin (Nathan, 1992). Expression of these NOS isoforms is controlled primarily at the level of transcription and translation (Xie et al., 1992; Lorsbach et al., 1993; Deng et al., 1993). Calmodulin has been discovered to associate with inducible macrophages

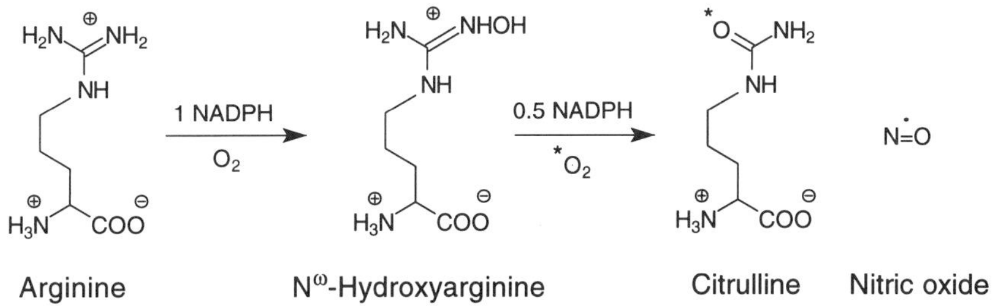

FIGURE 1

Stoichiometry and general mechanism of NO synthesis. Conversion of L-arginine to NO and citrulline is carried out in steps by NOS and represents a five-electron oxidation of an L-arginine guanidino nitrogen. The first step (a two-electron oxidation) is a hydroxylation reaction that forms N^{ω}-hydroxy-L-arginine as an enzyme-bound intermediate. The second step (an overall three-electron oxidation) involves electron removal, oxygen insertion, and C–N bond scission to form citrulline and the free radical NO from N-hydroxy-L-arginine. The biological electron donor NADPH is required for both steps in the reaction. A total of 1.5 NADPH are oxidized per molecule NO formed, with 1 NADPH being used in the initial hydroxylation reaction. The oxygen atoms that are incorporated into NO and citrulline derive from distinct molecules of O_2.

NOS as a tightly bound prosthetic group (Cho et al., 1992), possibly explaining why this isoform is active as isolated and does not require addition of exogenous Ca^{2+} or calmodulin. Related inducible NOS purified from hepatocytes (Iida et al., 1992; Geller et al., 1993) and smooth muscle (Gross and Levi, 1992) also appear to function primarily independent of exogenous calmodulin. This suggests that the inducible NOS exhibit a spectrum of relatively high binding affinities toward calmodulin, dependent on the exact sequence of their individual calmodulin binding domains.

C. Composition and Properties of Nitric Oxide Synthase

NOS isoforms have now been purified from several sources, some of their properties are listed in Table 1.

The rat brain and macrophage NOS isoforms are soluble, dimeric enzymes, each comprised of two identical subunits (Stuehr et al., 1991a; Schmidt et al., 1991). Denatured molecular weights for the various NOS isoforms range from 130 to 150 kDa. The endothelium NOS is of unknown quaternary composition, and localizes primarily in the membrane fraction (Pollock et al., 1991), due to posttranslational myristoylation near the N-terminus (Lamas et al., 1992). Phosphorylation consensus motifs are present on the brain and endothelial isoforms, with phosphorylation possibly down regulating NOS activity or preventing localization onto the membrane (Bredt et al., 1992; Michel et al., 1993).

Although the reported specific activities for the NOS isoforms have varied widely, this may be due to differences in assay conditions (e.g., failure to add cofactors whose requirement was not known at the time) or differential stabilities of the preparations. When brain and macrophage isoforms are assayed in the

TABLE 1
Two-Column Purification of Dimeric iNos and nNOS

Fraction	Protein (mg)	Total activity[a]	Specific activity[b]	Yield (%)	Purification factor
iNOS					
Lysate sup.[c]	1620	5022	3.1	100	1
2',5'-ADP	11	2090	190	42	61
MonoQ[d]	2.1	1380	650	27	210
nNOS					
Lysate sup.[c]	2656	6640	2.5	100	1
2',5'-ADP	20	4300	215	65	86
MonoQ	4.1	2710	661	41	264

[a] Nanomoles of nitrite/min (lysate supernatants) or nmol NO/min (2',5'-ADP and MonoQ).
[b] Nanomoles of nitrite (or NO for the column concentrates) per min per mg.
[c] Supernatant of 100,000g centrifugation of cell lysate.
[d] This preparation also yielded 1.3 mg of purified monomeric iNOS subunits.

presence of all required cofactors a 37°C, similar specific activities of 0.6–1.3 nmol NO/min per mg have been obtained (Stuehr and Ikeda-Saito, 1992; and Marletta, 1992; McMillan et al., 1992), corresponding to turnover numbers of 2.0–3.2 molecules of NO per second per subunit. These are respectable in light of the multistep process that generates NO from L-arginine. It should also be noted that differences in the isoform specific activities probably do not account for the vast range in NO production reported for different cell types and tissues, which may instead reflect differences in enzyme copy numbers per cell, phosphorylation level, or dependence of a given isoform on transient increases in intracellular Ca^{2+}, redox factors (tetrahydrobiopterin), or L-arginine (Bredt et al., 1992; Michel et al., 1993; Werner-Felmayer et al., 1990; Morris, and Billar, 1994; Gross and Levi, 1992).

The macrophage and brain NOS both contain four distinct prosthetic groups (Fig. 2): Flavin adenine dinucleotide (FAD), flavin mononucleotide (FMN), tetrahydrobiopterin (H4biopterin), and iron protoporphorin IX (heme). Macrophage or neuronal NOS contains an average of 1 heme, 1 FAD, 1 FMN, and variable amounts (0.1–1) of H4biopterin bound per subunit (Stuehr et al., 1991a; Hevel et al., 1991; Schmidt et al., 1991, 1992; Mayer et al., 1990; Stuehr and Ikeda-Saito, 1992; White and Marletta, 1992; McMillan et al., 1992; Hevel and Marletta, 1992). Although these four prosthetic groups are utilized throughout nature to catalyze a vast number of oxidative, reductive, and electron transfer reactions, their coincident presence in a single protein is unprecedented. NOS thus presents a unique opportunity to study their function and interactions within a single enzyme.

The cloning and sequencing of several NOS isoforms (Bredt et al., 1991; Xie et al., 1992; Geller et al., 1993; Lamas et al., 1992; Nakane et al., 1993) has revealed important insights into how some of the prosthetic groups are positioned and how they might function within the enzyme. First, all NOS appear to be comprised of a reductase domain and an oxygenase domain, each representing roughly one-half of the protein molecule (Fig. 3). A binding sequence for the Ca^{2+}-binding protein calmodulin is invariably located between the NOS reductase and oxygenase domains.

All NOS reductase domains contain binding sites for NADPH, FAD, and FMN. Because these molecules either provide electrons (NADPH) or participate in electron transfer reactions (FAD and FMN), the domain is thought to be the site of electron entry into NOS. Its amino acid sequence is strikingly similar to sequences present within three other FAD- and FMN-containing proteins found in nature (cytochrome P450) reductase (Fig. 3), the α-subunit of sulfite reductase, and cytochrome $P450_{BM3}$) (Porter, 1991). This suggests that the NOS flavins, as in these related proteins, function to store electrons derived from NADPH and transfer them to a catalytic center located within the oxygenase domain of NOS.

Although the NOS oxygenase domain does not contain clear consensus sequences that identify heme, H4biopterin, of L-arginine binding domains, it is

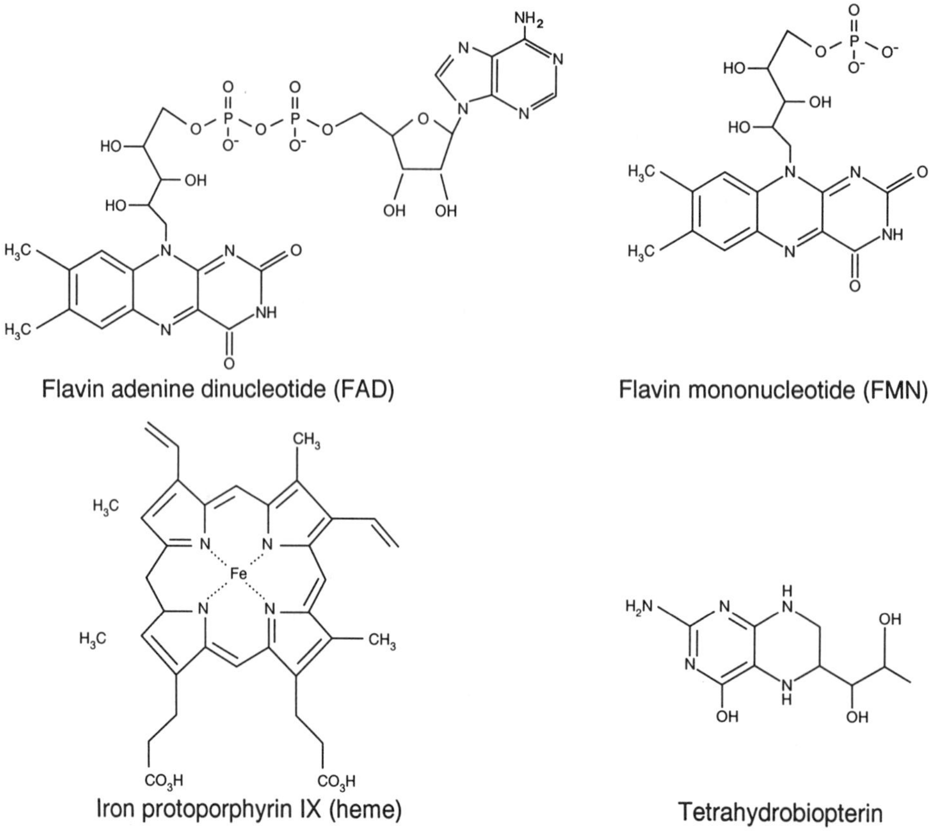

Flavin adenine dinucleotide (FAD)

Flavin mononucleotide (FMN)

Iron protoporphyrin IX (heme)

Tetrahydrobiopterin

FIGURE 2

Prosthetic groups contained within NOS. Nitric oxide synthases are isolated containing approximately one molecule each of heme, FAD, and FMN per subunit, and also contain variable quantities of tetrahydrobiopterin (0.1 to 1 molecule per subunit).

presumed to be the region where the binding sites for these molecules are located. In fact, a 320-amino-acid region that is highly conserved is present within all NOS oxygenase domains. Modeling studies based on a partial sequence similarity between NOS and the cytochromes P450 have identified a cysteine residue located within this shaded region which may serve to ligate heme iron (McMillan et al., 1992; Renaud et al., 1993).

Overall sequences, similarity between the constitutive endothelium and neuronal NOS isoforms is approximately 50%, and both are roughly 50% identical to the inducible macrophage NOS (Xie et al., 1992; Lamas et al., 1992). Studies have located the genes for human neuronal, endothelial, and inducible NOS on

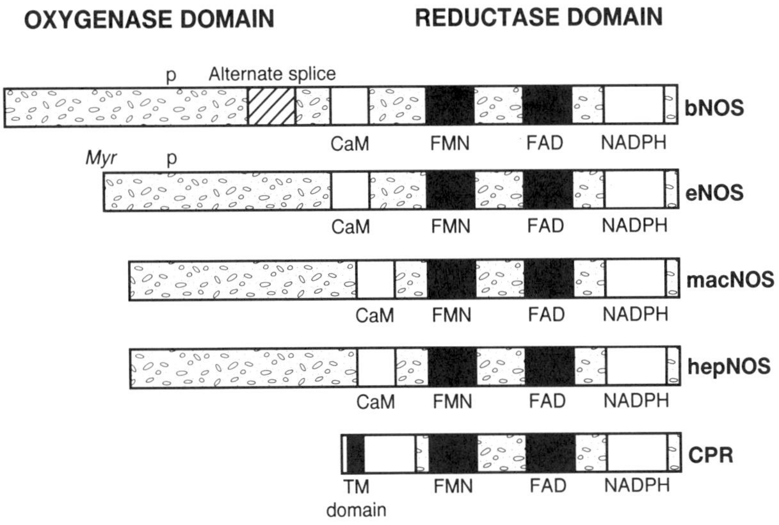

OXYGENASE DOMAIN   REDUCTASE DOMAIN

FIGURE 3
Diagram displaying the structural relationship between the NOS isoforms and NADPH–cytochrome P450 reductase (CPR). The cloned NOS isoforms from brain (bNOS), endothelium (eNOS), macrophage (macNOS), and hepatocytes (hepNOS) are all comprised of an oxygenase and reductase domain, with a binding domain for calmodulin between them. The reductase domain contains binding sites for NADPH, FAD, and FMN, and is homologous with the dual flavin enzyme NADPH–cytochrome P450 reductase. The oxygenase domain of bNOS and eNOS contain phosphorylation sites (P). Also noted are a myristoylation site near the N terminus of eNOS, a hydrophobic transmembrane anchoring domain in CPR.

chromosomes 12, 7, and 17, respectively. The variability in NOS sequences has conferred immunologic specificity only in some cases (polyclonal antibodies raised against inducible macrophage NOS have uniformly failed to cross-react with constitutive NOS isoforms from endothelium or brain, whereas antibodies raised against brain NOS sometimes recognize endothelial NOS). Inducible NOS cloned from various cells (macrophages, hepatocytes, chondrocytes, and smooth muscle) are highly homologous (roughly 80–90%), perhaps reflecting their localization on a single chromosome.

D. Role of Heme in Nitric Oxide Synthesis

The hemes of NOS are envisioned to play a central role in catalyzing NO synthesis. A variety of biophysical measurements suggest that the macrophage and neuronal NOS heme environment is somewhat similar to that within the cytochromes P450 (Stuehr and Ikeda-Saito, 1992; White and Marletta, 1992; Wang et al., 1993). The heme iron in its resting state is ferric, predominantly high spin, and five-coordinate, and is bound to the protein through coord-

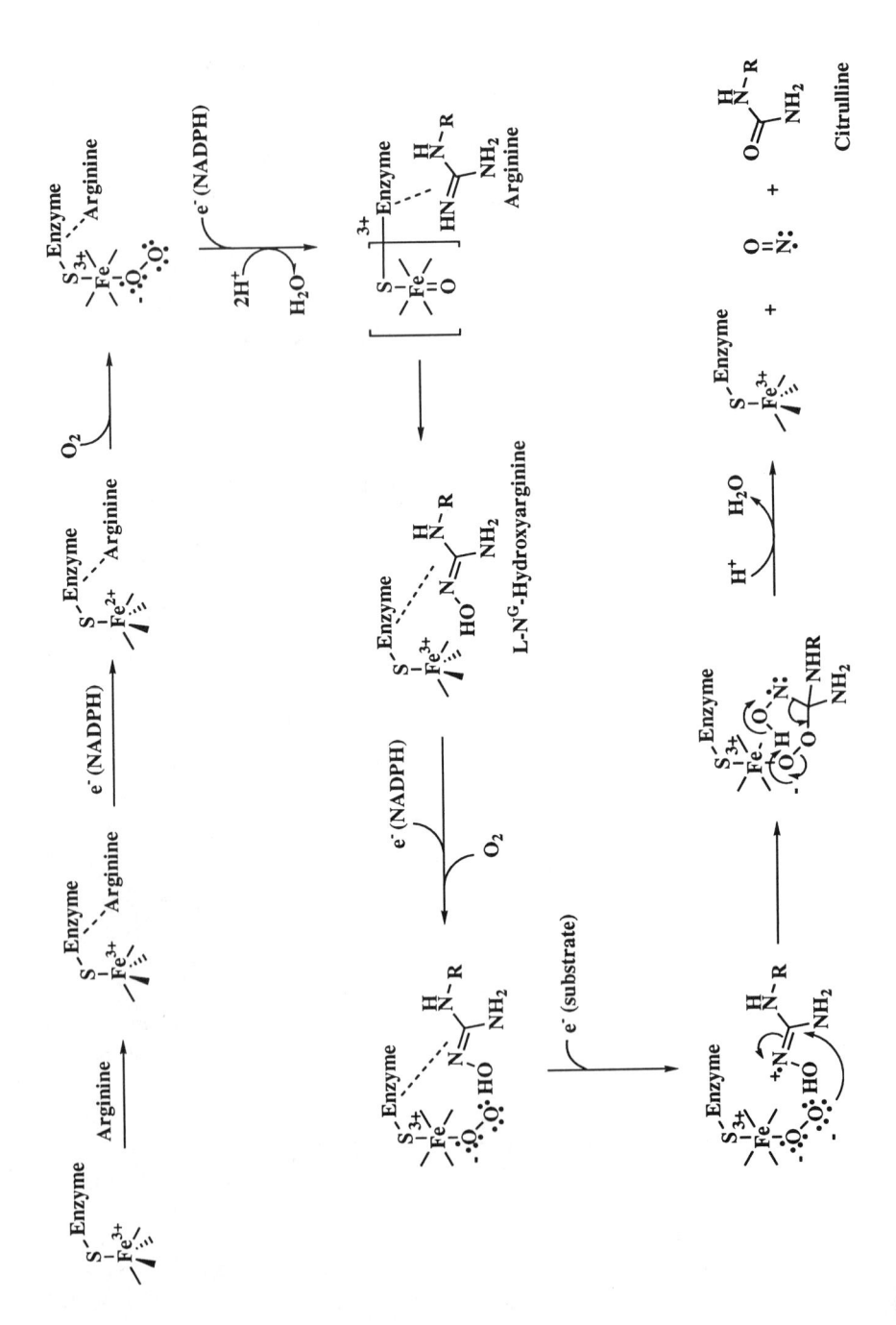

ination to a thiolate provided by an enzymatic cysteine residue. The sixth ligand position of the heme iron is available to bind cyanide in the ferric form, and can bind CO when reduced to its ferrous form. This suggests that the NOS heme, as in the cytochromes P450, is likely to be the site where binding and activation of O_2 occurs, with subsequent oxidative transformation of L-arginine to NO. Figure 4 shows a proposed heme-based mechanism for NO synthesis that is consistent with all the results to date.

One intriguing aspect of the proposed mechanisms is that NOS utilizes a distinct heme-based oxidant in either step of the reaction. An electrophilic perferryl species is postulated to hydroxylate a terminal guanidino nitrogen of L-arginine and form N-hydroxy-L-arginine. In the second step, a nucleophilic iron peroxide species is shown to attack the quanidino carbon of N-hydroxy-L-arginine cation radical, generating a tetrahedral intermediate that breaks down to form NO and L-citrulline. Both heme-based oxidants are thought to form during the normal course of oxygen activation by cytochrome P450s, with the iron–peroxide species being generated prior to the perferryl species (Fig. 5). The electrophilic perferryl oxidant has been implicated in the vast majority of cytochrome P450 reactions, including heteroatom oxidation (Ortiz de Montellano, 1986; Guengerich, 1991; Clement et al., 1991), while the nucleophilic iron peroxide has been implicated in few, notably demethylation reactions during steroidogenesis (Vaz et al., 1991).

Conversion of the iron peroxide to the perferryl species, with accompanying formation of water, is envisioned to proceed relatively quickly. Thus, the ability of the iron peroxide to participate in the reaction may depend to some extent on whether the bound substrate is correctly positioned and is activated toward

---

FIGURE 4

Proposed mechanism of NO synthesis involving heme-based catalysis. The NOS reaction can be divided into two steps. The first is a two-electron oxidation that forms N-hydroxy-l-arginine, and likely proceeds as a typical mixed-function monooxygenase reaction. L-Arginine binds near the ferric heme, and an electron derived from NADPH is transferred to the heme, enabling oxygen to bind. Transfer of a second NADPH-derived electron to the heme leads to scission of the oxygen–oxygen bond, with one atom of oxygen being released as water and the other remaining bound to form an oxo iron oxidant. This electron-deficient oxygen is inserted into a terminal guanidino N–H bond of L-arginine, reforming ferric heme and generating N-hydroxy-L-arginine as an enzyme-bound intermediate. The second step of the reaction involves a three-electron oxidation of nitrogen, incorporation of a second oxygen atom, and C–N bond scission. To start, an NADPH-derived electron is again transferred to the ferric heme, and a second molecule of dioxygen binds. At this point, the ferrous oxy heme may act as an oxidant, removing an electron from bound N-hydroxy-L-arginine, and generating a peroxide–iron species in conjunction with the N-hydroxy-L-arginine cation radical. The peroxide iron species carries out nucleophilic attack at the guanidino carbon, generating a tetrahedral intermediate that rearranges to yield citrulline, NO, water, and ferric heme.

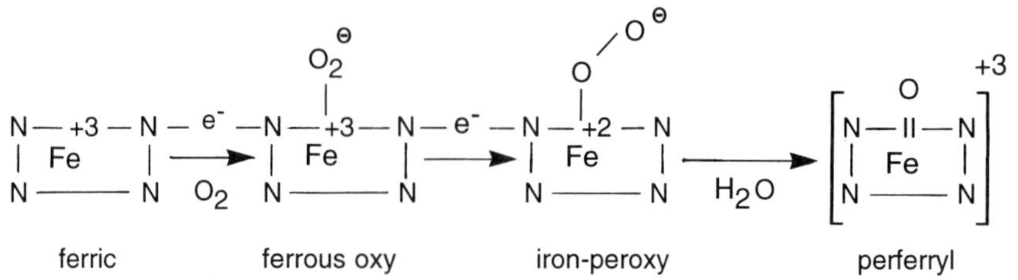

FIGURE 5

Proposed model for oxygen activation in cytochrome P450. Reduction by a single electron enables the ferrous iron to bind oxygen. Addition of a second electron generates an iron peroxide heme, which then cleaves to form water and an electrophilic perferryl species.

nucleophilic attack. Evidence suggests that these criteria may be fulfilled in the case of N-hydroxy-L-arginine. The guanidino carbon of N-hydroxyarginine reacts readily with aqueous nucleophiles such as hydroxide ion or ammonia, resulting in a loss of hydroxylamine and generation of L-citrulline or L-arginine, respectively (D. J. Stuehr, unpublished results). In addition, chemical model studies by Fukuto and colleagues (Fukuto et al., 1993) have shown that organic peracids can transform N-hydroxyguanidines to their corresponding ureas in good yield, with oxygen transfer to the urea occurring from the nucleophilic peracid (Fig. 6). This result is consistent with both the proposed mechanism and the $^{18}O_2$ isotope incorporation studies carried out with NOS (Kwon et al., 1990; Leone

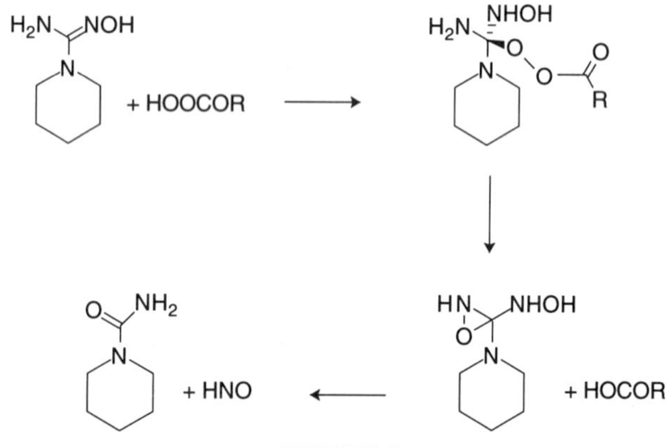

FIGURE 6

Chemical mechanism for reaction of a peracid with a N-hydroxyguanidine to generate the urea product and nitroxyl. This reaction is a chemical model for the second phase of the proposed NOS reaction, which involves nucleophilic attack by an activated oxygen species, C–N bond scission, and incorporation of molecular oxygen.

et al., 1991). Ongoing work with heme-based iron–peroxide model compounds should determine if nucleophiles more similar to those in the proposed mechanism can carry out similar chemistry.

A second important tenet of the proposal is that the heme iron obtains an electron from the bound intermediate N-hydroxy-L-arginine to generate the iron peroxide oxidant in the second step. Electron donation from N-hydroxy-L-arginine was originally proposed by Stuehr and Griffith (1992) to account for their stoichiometry studies which showed that only 0.5 mol of NADPH are required to generate 1 NO from N-hydroxy-L-arginine (Stuehr *et al.*, 1991b). Because conversion of a ferric to an iron–peroxide heme requires two electrons, obtaining an electron from bound N-hydroxy-L-arginine would allow NOS to generate the nucleophilic peroxide while obtaining only one electron from NADPH (i.e., 0.5 NADPH), and thus satisfy the stoichiometry studies. This would also enable NOS to carry out a three-electron oxidation of the N-hydroxylated nitrogen to generate NO, an odd-electron species, as a product. Although the identity of the NOS-based oxidant which may remove an electron from N-hydroxy-L-arginine is unknown, the heme iron in its ferric state appears to be incapable (Pufahl and Marletta, 1993). However, previous work has shown that ferrous oxy hemoproteins can carry out one-electron oxidations of N-hydroxylated compounds such as hydroxylamine and hydroxyurea (Stolze and Nohl, 1990a,b), and thus it is shown as the electron acceptor in the proposed mechanism. A curious result of this electron donation mechanism is that it would generate the N-hydroxy-L-arginine cation radical within the enzyme just prior to nucleophilic reaction with the iron peroxide species. Thus far, the formation and chemistry of N-hydroxyguanidine cation radicals has not been extensively studied, although evidence indicates that N-hydroxyguanidines react rapidly with one-electron oxidants such as ferricyanide, and further oxidize to form their corresponding cyanamides (Fig. 7) (Fukuto *et al.*, 1992, 1993). Finally, it should be noted that electron removal from N-hydroxy-L-arginine is only one of three possible points for electron donation back to the enzyme (Stuehr and Griffith, 1992; Fukuto *et al.*, 1993).

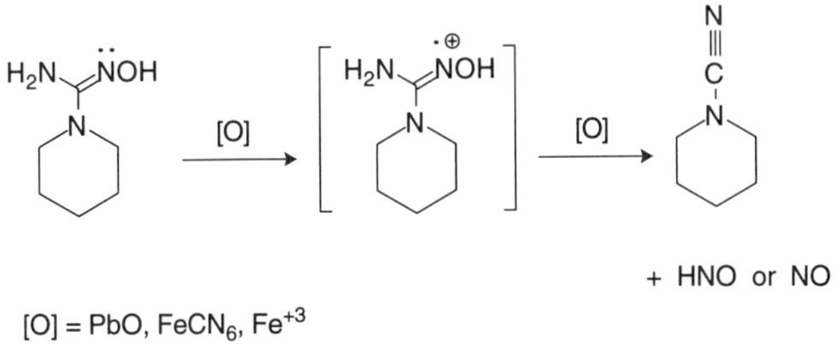

[O] = PbO, FeCN$_6$, Fe^{+3}

FIGURE 7
Reaction of N-hydroxyguanidine with various one-electron oxidants.

E. Roles for Nitric Oxide Synthase Flavins in Electron Transfer

Nitric oxide synthases represent only the fourth group of enzymes known to contain both FAD and FMN. The three related dual flavin enzymes, NADPH–cytochrome P450 reductase, sulfite oxidase α subunit, and cytochrome P450$_{BM3}$, have been studied in some detail (Otvos et al., 1986; Iyanagi et al., 1981; Vermilion et al., 1981; Ostrowski et al., 1989; Narhi and Fulco, 1987; Kurzban and Strobel, 1986; Peterson and Boddupalli, 1992). Each enzyme utilizes its flavins to store NADPH-derived electrons and transfer them from the FMN group to a heme located either within the protein or in an adjacent protein. This suggests that the NOS flavins may also function to transfer electrons to the heme iron in the following manner:

$$\text{NADPH} \longrightarrow \text{FAD} \longrightarrow \text{FMN} \longrightarrow \text{heme}$$

Flavins are well-equipped to function in electron transfer. Each can exist in three oxidation states: zero, one-, or two-electron reduced. Thus, the flavin pair in NOS can in theory achieve five levels of reduction (Fig. 8). Which levels are reached during catalysis depend in part on the reduction potentials of the flavins relative to those of NADPH and the heme iron. Proteins usually alter their flavin potentials relative to free flavins in solution (Ghisla and Massey, 1989). For example, flavin reduction potentials for the related enzyme NADPH–cytochrome P450 reductase at pH 7.0 range from -110 mV for addition of the first electron to -365 mV for addition of the fourth; this enables the flavins to readily accept a total of three electrons from the physiological donor NADPH ($E° = -0.320$ mV) under normal turnover condition (Iyanagi et al., 1981; Vermilion et al., 1981). Also, addition of the second and third electrons occur at a relatively similar potential (-270 and -290 mV, respectively), enabling the three-electron reduced enzyme to donate two single electrons in discreet steps at almost equal potential to its heme protein acceptors (Vermilion et al., 1981). The flavins cycle only between their one- and three-electron reduced states during catalysis, and the reductase maintains an air-stable semiquinone flavin radical in the resting state, due to a kinetic barrier regarding electron transfer from the one-electron reduced form of the enzyme (Otvos et al., 1986; Iyanagi et al., 1981; Vermilion et al., 1981). Interestingly, a similar kinetic barrier is apparently absent in the related dual-flavin enzyme cytochrome P450$_{BM3}$ (Peterson and Boddupalli, 1992). Thus, this enzyme does not maintain a flavin semiquinone radical (Peterson and Boddupalli, 1992).

How the NOS isoforms compare to these related dual flavin enzymes is a matter of ongoing investigation. Characterization of the neuronal NOS revealed that it normally exists in its one-electron reduced form and maintains an air-stable, flavin semiquinone radical (Stuehr and Ikeda-Saito, 1992), as seen for NADPH-cytochrome P450 reductase. It is unknown which flavin in NOS contains the odd electron, although precedent argues that it probably resides on

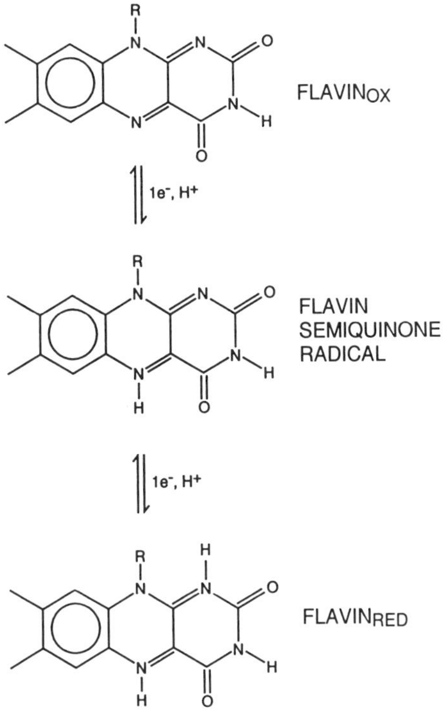

	Number of electrons
FAD FMN	0
FAD FMNH•	1
FAD FMNH$_2$	2
FADH• FMNH$_2$	3
FADH$_2$ FMNH$_2$	4

FIGURE 8
Flavin redox states in a dual flavin enzyme. *(Left)* Single-electron reduction of the isoalloxazine ring generates the semiquinone radical, while reduction by two electrons generates the fully reduced species. *(Right)* Five possible oxidation levels of a dual flavin enzyme, where the FMN reduction potential is held at a more positive value relative to FAD. The flavins can theoretically accept a maximum of four electrons obtained from two NADPH. However, in NADPH–cytochrome P450, reductase, full reduction of the flavins is not normally reached when NADPH serves as the reductant.

the terminal flavin with regard to electron transfer. The spectral characteristics of the NOS flavin radical suggest that it is positioned close to another paramagnetic atom within the enzyme, the ferric heme iron being a likely candidate. Close proximity between the terminal flavin and heme iron would be consistent with a role for the flavins in mediating electron transfer to the heme.

F. Nitric Oxide Synthase

Nitric oxide synthase is an unusual enzyme in that substrates with even numbers of electrons (L-arginine, O_2, NADPH) are transformed into an odd-electron product (NO) in addition to even-electron products (L-citrulline, NADP, water). In fact, the overall conversion of L-arginine to NO represents a five-electron oxidation of a guanidino nitrogen. A key issue in resolving the reaction mechanism is to understand how the enzyme can catalyze successive rounds of odd-electron oxidation. Although speculative, one means that is consistent with the available data utilizes the dual flavin nature of NOS. As noted above, NOS can accept and store up to four electrons in its two flavins, and several schemes can be drawn in which the flavins are used to store an electron between catalytic cycles. The overall effect would be to turn two five-electron oxidations into a single ten-electron (even-numbered) oxidation, requiring three NADPH to make two NO (Fig. 9).

G. Role for Calmodulin in Controlling Electron Transfer in Nitric Oxide Synthase

Calmodulin binds to and activates neuronal NOS (Bredt and Snyder, 1990; Schmidt et al., 1991), and also functions as a tightly bound prosthetic group to keep macrophage NOS in its active state (Cho et al., 1992). Work with the neuronal NOS has uncovered the basis for its calmodulin activation: Calmodulin binding triggers electrons to transfer onto the NOS heme iron (Abu-Soud and Stuehr, 1993) (Fig. 10). Because this transfer is associated with initiation of

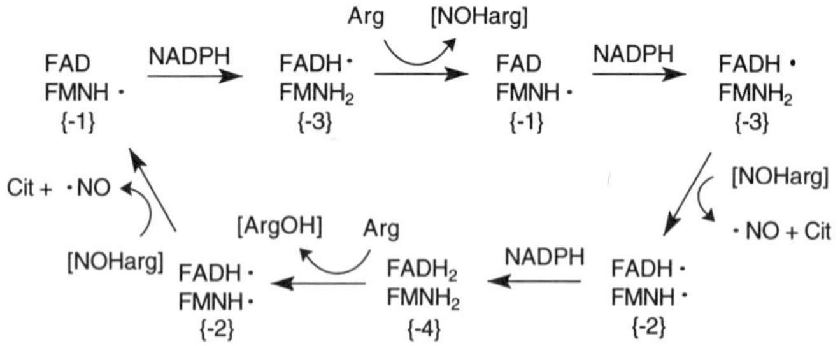

FIGURE 9

Possible role for flavins in promoting NOS odd-electron chemistry. Hypothetical flavin utilization during NO synthesis. The NOS flavins start in the one-electron reduced state (−1, upper left), and are reduced to the −3 state by the first NADPH. Formation of the N^ω-hydroxy-L-arginine [ArgOH] uses the two electrons obtained from NADPH and returns the flavins to the −1 state. The flavins are then reduced again by a second NADPH to the −3 state. However, conversion of [ArgOH] to NO only requires one electron derived from NADPH, this leaves the flavins in the two-electron reduced state −2 to start the second round of catalysis. Two electrons from a third NADPH reduce the flavins fully −4. Alternatively, one electron could be transferred from the flavins onto the ferric iron prior to their reduction by NADPH, leaving the flavins in their −3 reduced state (not shown). In either case, NOS would use two of the electrons to form the second [ArgNOH], returning the flavins to the −2 state. The residual "stored" electron on the flavins is then used in forming the second NO, returning the flavins to their original oxidation state −1.

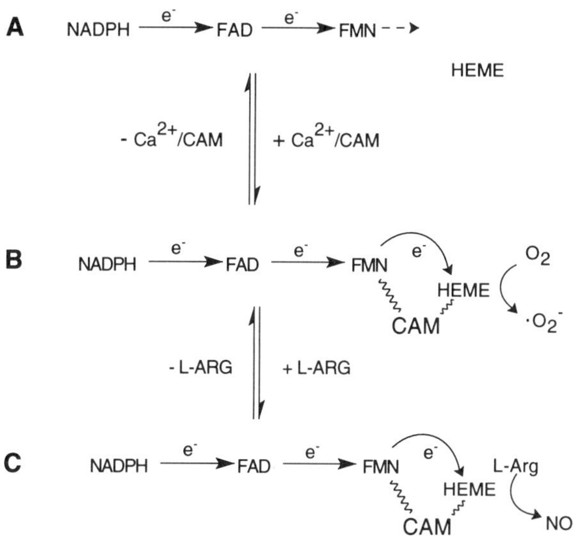

4 Nitric Oxide in the Immune System 161

A NADPH $\xrightarrow{e^-}$ FAD $\xrightarrow{e^-}$ FMN - - ▸ HEME

$-Ca^{2+}/CAM$ ⇅ $+Ca^{2+}/CAM$

B NADPH $\xrightarrow{e^-}$ FAD $\xrightarrow{e^-}$ FMN $\xrightarrow{e^-}$ HEME O_2 → $\cdot O_2^-$
CAM

$-L\text{-}ARG$ ⇅ $+L\text{-}ARG$

C NADPH $\xrightarrow{e^-}$ FAD $\xrightarrow{e^-}$ FMN $\xrightarrow{e^-}$ HEME L-Arg → NO
CAM

FIGURE 10
Role for calmodulin in control of heme reduction in NOS. (A) In the absence of bound calmodulin, electrons derived from NADPH can load into the flavins but cannot be transferred onto the heme iron. (B) On calmodulin binding, electrons transfer from the flavins onto the heme. In the absence of bound L-arginine, heme reduction generates superoxide (C), whereas in the presence of L-arginine, heme reduction can lead to NO synthesis.

NADPH oxidation and NO synthesis by the enzyme, it supports a role for reduction of the heme iron in catalysis, and may explain why NOS functions only as an NADPH-dependent reductase in the absence of bound calmodulin (Klatt et al., 1993). The mechanism of calmodulin gating is envisioned to involve a conformational change between the reductase and oxygenase domains of NOS, such that an electron transfer between the terminal flavin and heme iron becomes possible. Calmodulin may also have a distinct role within the NOS reductase domain, in that its binding dramatically increases reductase activity of the enzyme toward cytochrome c (Klatt et al., 1993; Heinzel et al., 1992). However, it is clear that several other NOS functions occur independent of calmodulin, including the binding of L-arginine and NADPH, and transfer of NADPH-derived electrons into the flavins (Abu-Soud and Stuehr, 1993).

H. Role for Tetrahydrobiopterin

Studies with the macrophage NOS were the first to show that NO synthesis was partially dependent on added H4biopterin (Kwon et al., 1989; Stuehr et al., 1990), which is a redox active cofactor utilized by the aromatic amino acid hydroxylases (Nichol et al., 1985). The requirement for H4biopterin has since been expanded to include all NOS isoforms studied to date. Mayer et al. (1990)

were the first to discover that NOS contain variable amounts of tightly bound H4biopterin. This finding was unusual and unexpected, in that H4biopterin acts as a freely diffusible cofactor in all other enzyme systems. This makes NOS the first proteins known to contain stably bound H4biopterin.

Tetrahydrobiopterin has been postulated to have both catalytic and stabilizing roles in NO (Stuehr et al., 1990; Giovanelli et al., 1991). Within the aromatic amino acid hydroxylases, H4biopterin functions catalytically by forming a hydroxylating species within the enzyme active site in conjunction with dioxygen and nonheme iron (Nichol et al., 1985; Lazarus et al., 1981) (Fig. 11). The structure of the hydroxylating agent is postulated to approach a 4a-peroxypterin. After participating in the hydroxylation, the oxidized form of the cofactor dis-

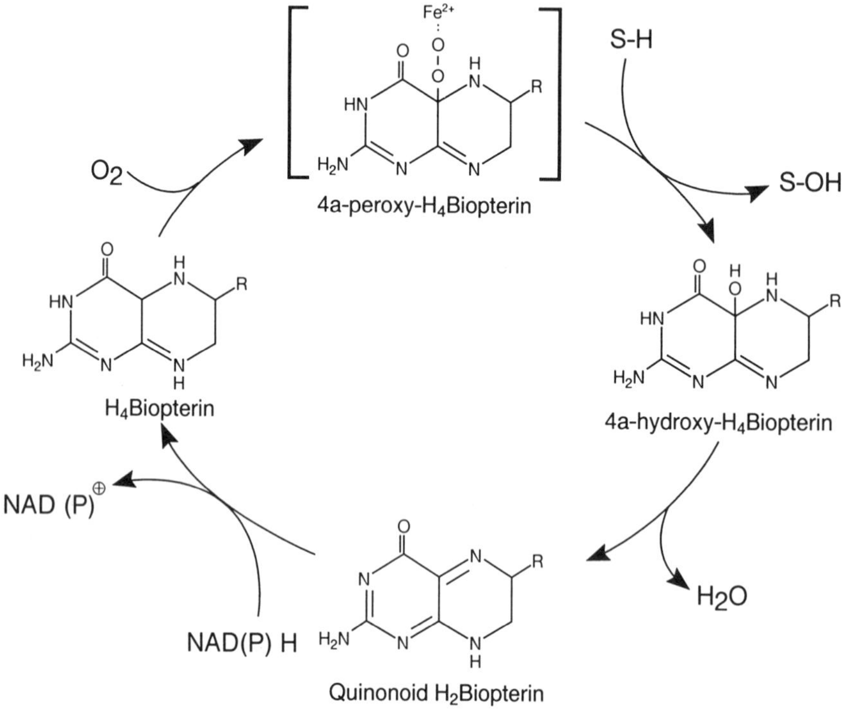

FIGURE 11

Model for participation in H4biopterin in hydroxylation reactions. Tetrahydrobiopterin [R = CH₃CH(OH)CH(OH)—] binds to phenylalanine hydroxylase and forms a 4a-peroxy hydroxylating species at the active site in conjunction with bound O_2 and nonheme iron. The cofactor participates in a single hydroxylation, being oxidized in the process to the quinonoid dihydro form, and leaves the enzyme active site. Subsequent reduction by exogenous enzymes like dihydropteridine reductase regenerates fully reduced H4biopterin, which may bind and participate in a second round of catalysis. Thus far, an identical system does not appear to operate in NOS, which maintains tightly bound H4biopterin and may not oxidize its bound H4biopterin during each round of catalysis.

sociates from the enzyme to make way for a new round of catalysis. In NOS, a 4a-peroxy H4biopterin could conceivably substitute for the peroxy iron heme in carrying out nucleophilic attack on N-hydroxy-L-arginine (Stuehr and Griffith, 1992a), as was depicted in Fig. 4. However, current evidence suggests that this is unlikely. For example, catalytic, and not stoichiometric amounts of H4biopterin are needed to drive NO synthesis. Although this could be explained if NOS could retain its bound dihydrobiopterin and rereduce its H4biopterin after most catalytic cycles, the search for NOS dihydropteridine reductase-like activity has proven negative thus far (Giovanelli et al., 1991). It is unknown if the bound H4biopterin is redox active within the enzyme. NOS that is partially depleted of bound H4biopterin displays a curious phenotype, it continues to oxidize NADPH, but partially loses its ability to couple the electron flow to NO synthesis (Heinzel et al., 1992). This may help explain an earlier report that neuronal or macrophage NOS gradually lost their ability to generate NO when undergoing catalysis in the absence of added H4biopterin (Kwon et al., 1989; Giovanelli et al., 1991). H4biopterin-deficient NOS transfers a significant percentage of electrons obtained from NADPH onto oxygen to form superoxide (Heinzel et al., 1992). Thus, bound H4biopterin appears at least to have a role in coupling NADPH oxidation to NO synthesis in NOS. While a definitive answer is still forthcoming, work with the macrophage NOS shows that H4biopterin is required for enzyme subunits to associate into an active dimeric enzyme (see below). Thus, a structural role does appear to exist for H4biopterin regarding enzyme assembly.

I. Dimeric Structure and Enzyme Function

Both the macrophage and brain NOS isoforms are reported to be homodimers (Stuehr et al., 1991a; Hevel et al., 1991; Schmidt et al., 1991). How the enzyme dimeric structure relates to its function is a matter of current interest. The macrophage NOS can be isolated as a mixture of its monomeric and dimeric forms (Baek et al., 1993). The monomeric form of macrophage NOS contains bound FAD, FMN, and calmodulin, but does not contain bound heme or H4biopterin. The dissociated NOS subunits are incapable of NO synthesis, and only function as NADPH-dependent reductases toward acceptors such as cytochrome c or ferricyanide (Baek et al., 1993). Thus, the data support a role for the heme and H4biopterin groups in catalysis, and suggest that the subunit intrinsically functions as a reductase.

Macrophage subunits can be reassociated into their active dimeric form in a reaction that requires the coincident presence of H4biopterin, L-arginine, and stoichiometric amounts of heme (Baek et al., 1993). The reaction proceeds with incorporation of H4biopterin and heme into the dimer and generates an enzyme that is catalytically and spectrally identical to the native dimeric NOS (Fig. 12). These findings have established a role for heme, L-arginine, and H4biopterin in NOS assembly *in vitro*. How the substrate and prosthetic groups interact to induce

FIGURE 12

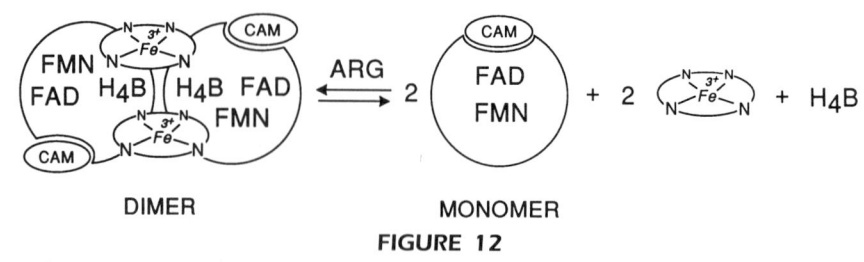

Role for heme, H4biopterin, and L-arginine in subunit assembly of macrophage NOS. Assembly of NOS from its isolated subunits *in vitro* requires that subunits be coincubated with heme, L-arginine, and H4biopterin. Tetrahydrobiopterin and heme become bound within the dimer during its assembly and enable NOS to catalyze NO synthesis. Dissociation of dimeric NOS leads to loss of its bound heme and H4biopterin, and it generates subunits which are inactive regarding NO synthesis.

subunit dimerization is unknown, but the data suggest that their effects may be cooperative.

Whether L-arginine, H4biopterin, or heme availability limits the assembly of NOS subunits in cells is an open question. A repurification of authentic dimeric macrophage NOS showed that it did not disassemble into monomers during the purification procedure (Baek *et al.* 1993), suggesting that the dimermonomer mixture normally observed following purification of macrophage NOS results from something limiting dimer formation during biosynthesis of the enzyme in the intact cells. Clearly, further work is required to determine if and under what conditions dimerization is limited, and whether dimerization can be controlled *in vivo*.

REFERENCES

Abbas, A. K. (1991). "Cellular and Molecular Immunology." Saunders, Philadelphia, Pennsylvania.

Abu-Soud, H. M., and Stuehr, D. J. (1993). Nitric oxide synthases reveal a role for calmodulin in controlling electron transfer. *Proc. Natl. Acad. Sci. U.S.A.* **90,** 10769–10772.

Alexander, P., and Evans, R. (1971). Endotoxin and double stranded RNA render macrophages cytotoxic. *Nature (London) New Biol.* **232,** 76–78.

Arnold, W. P., Mittal, C. K., Katsuki, S., and Murad, F. (1977). Nitric oxide activates guanylate cyclase and increases guanosine $3':5'$-cyclic monophosphate levels in various tissue preparations. *Proc. Natl. Acad. Sci. U.S.A.* **74,** 3203–3207.

Azhipa, Y. I., Kayushin, L. P., and Nikishkin, E. I. (1965). Title unavailable. *Biophysics (USSR)* **10,** 167.

Baek, K. J., Thiel, B. A., Lucas, S., and Stuehr, D. J. (1993). Macrophage nitric oxide synthase subunits. Purification, characterization, and role of prosthetic groups and substrate in regulating their association into a dimeric enzyme. *J. Biol. Chem.* **268,** 21120–21129.

Basosi, R., Gaggelli, E., Tiezzi, E., and Valensin, G. (1975). Nitrosyliron complexes with mercapto-purines and -pyrimidines studied by nuclear magnetic and electron spin resonance spectroscopy. *J. Chem. Soc. Perkin Trans.* **2,** 423–428.

Bastian, N. R., Xu, S., Shao, X. L., Shelby, J., Granger, D. L., and Hibbs, J. B., Jr. (1994). N omega-monomethyl-L-arginine inhibits nitric oxide production in murine cardiac allografts but does not affect graft rejection. *Biochim. Biophys. Acta* **1226,** 225–231.

Beinert, H. (1978). EPR spectroscopy of components of the mitochondrial electron-transfer system. *In* "Methods in Enzymology" (S. Fleischer and L. Packer, eds.), Vol. 54, pp. 133–150. Academic Press, New York.

Bredt, D. S., and Snyder, S. H. (1990). Isolation of nitric oxide synthetase, a calmodulin-requiring enzyme. *Proc. Natl. Acad. Sci. U.S.A.* **87,** 682–685.

Bredt, D. S., and Snyder, S. H. (1994). Nitric oxide: A physiologic messenger molecule. *Annu. Rev. Biochem.* **63,** 175–195.

Bredt, D. S., Hwang, P. M., Glatt, C. E., Lowenstein, C., Reed, R. R., and Snyder, S. H. (1991). Cloned and expressed nitric oxide synthase structurally resembles cytochrome P-450 reductase. *Nature (London)* **351,** 714–718.

Bredt, D. S., Ferris, C. D., and Snyder, S. H. (1992). Nitric oxide synthase regulatory sites. Phosphorylation by cyclic AMP-dependent protein kinase, protein kinase C, and calcium/calmodulin protein kinase; identification of flavin and calmodulin binding sites. *J. Biol. Chem.* **267,** 10976–10981.

Brennan, M. J., Cole, T., and Singley, J. A. (1966). A unique hyperfine ESR spectrum in mouse neoplasms analyzed by computer simulation. *Proc. Soc. Exp. Biol. Med.* **123,** 715–718.

Busse, R., and Mulsch, A. (1990). Calcium-dependent nitric oxide synthesis in endothelial cytosol is mediated by calmodulin. *FEBS Lett.* **265,** 133–136.

Butler, A. R., Glidewell, C., Hyde, A. R., and Walton, J. C. (1985). Formation of paramagnetic mononuclear iron nitrosyl complexes from diamagnetic di- and tetranuclear iron–sulphur nitrosyls: Characterization by EPR spectroscopy and study of thiolate and nitrosyl ligand exchange reactions. *Polyhedron* **4,** 797–809.

Butler, A. R., Glidewell, C., and Li, M.-H. (1988). Nitrosyl complexes of iron–sulfur clusters. *Adv. Inorg. Chem.* **32,** 335–393.

Carswell, E. A., Old, L. J., Kassel, R. L., Green, S., Fiore, N., and Williamson, B. (1975). An endotoxin-induced serum factor that causes necrosis of tumors. *Proc. Natl. Acad. Sci. U.S.A.* **72,** 3666–3670.

Chamulitrat, W., Jordan, S. J., and Mason, R. P. (1994). Nitric oxide production during endotoxic shock in carbon tetrachloride-treated rats. *Mol. Pharmacol.* **46,** 391–397.

Chiang, R. W., Woolum, J. C., and Commoner, B. (1972). Further study on the properties of the rat liver protein involved in a paramagnetic complex in the livers of carcinogen-treated rats. *Biochim. Biophys. Acta* **257,** 452–460.

Cho, H. J., Xie, Q. W., Calaycay, J., Mumford, R. A., Swiderek, K. M., Lee, T. D., and Nathan, C. (1992). Calmodulin is a subunit of nitric oxide synthase from macrophages. *J. Exp. Med.* **176,** 599–604.

Clement, B., Immel, N., Pfunder, H., Schmitt, F., and Zimmerman, M. (1991). New aspects of the microsomal N-hydroxylation of benzamidines. *In* "N-Oxidation of Drugs:

Biochemistry, Pharmacology, and Toxicology" (P. Hlavica and L. A. Dumani, eds.), pp. 185–205. Chapman & Hall, London.

Coley, W. B. (1893). The treatment of malignant tumors by repeated inoculations of erysipelas; with a report of ten original cases. *Am. J. Med. Sci.* **105**, 487–490.

Corbett, J. A., and McDaniel, M. L. (1992). Does nitric oxide mediate autoimmune destruction of beta-cells? Possible therapeutic interventions in IDDM. *Diabetes* **41**, 897–903.

Corbett, J. A., Lancaster, J. R., Jr., Sweetland, M. A., and McDaniel, M. L. (1991). Interleukin-1 beta-induced formation of EPR-detectable iron–nitrosyl complexes in islets of Langerhans. Role of nitric oxide in interleukin-1 beta-induced inhibition of insulin secretion. *J. Biol. Chem.* **266**, 21351–21354.

Deng, W., Thiel, B., Tannenbaum, C. S., Hamilton, T. A., and Stuehr, D. J. (1993). Synergistic cooperation between T cell lymphokines for induction of the nitric oxide synthase gene in murine peritoneal macrophages. *J. Immunol.* **151**, 322–329.

Dervartanian, D. V., Albracht, S. P. J., Berden, J. A., Van Gelder, B. F., and Slater, E. C. (1973). The EPR spectrum of isolated complex III. *Biochim. Biophys. Acta* **292**, 496–501.

Drapier, J. C., and Hibbs, J. B., Jr. (1986). Murine cytotoxic activated macrophages inhibit aconitase in tumor cells. Inhibition involves the iron–sulfur prosthetic group and is reversible. *J. Clin. Invest.* **78**, 790–797.

Drapier, J. C., and Hibbs, J. B., Jr. (1988). Differentiation of murine macrophages to express nonspecific cytotoxicity for tumor cells result in L-arginine-dependent inhibition of mitochondrial iron–sulfur enzymes in the macrophage effector cells. *J. Immunol.* **140**, 2829–2838.

Drapier, J. C., Pellat, C., and Henry, Y. (1991). Generation of EPR-detectable nitrosyl–iron complexes in tumor target cells cocultured with activated macrophages. *J. Biol. Chem.* **266**, 10162–10167.

Drapier, J. C., Pellat, C., and Henry, Y. (1992). Characterization of the nitrosyl–iron complexes generated in tumour cells after co-culture with activated macrophages. In "The Biology of Nitric Oxide. 2. Enzymology, Biochemistry and Immunology." (S. Moncada, M. A. Marletta, J. B. Hibbs, Jr., and E. A. Higgs, eds.), pp. 72–76, Portland Press, London.

Drapier, J. C., Hirling, H., Wietzerbin, J., Kaldy, P., and Kuhn, L. C. (1993). Biosynthesis of nitric oxide activates iron regulatory factor in macrophages. *EMBO J.* **12**, 3643–3649.

Dull, B. J., and Hotchkiss, J. H. (1984). Activated oxygen and mammalian nitrate biosynthesis. *Carcinogenesis (London)*, **5**, 1161–1164.

Emanuel, N. M., Saprin, A. N., Shabalkin, V. A., Kozlova, L. E., and Kruglijakova, K. E. (1969). Detection and investigation of a new type of ESR signal characteristic of some tumour tissues. *Nature (London)* **222**, 165–167.

Evans, R., and Alexander, P. (1970). Cooperation of immune lymphoid cells with macrophages in tumour immunity. *Nature (London)* **228**, 620–622.

Foster, M. A., and Hutchison, J. M. (1974). The origin of an E.S.R. signal at g equals 2.03 from normal rabbit liver and the effects of nitrites upon it. *Phys. Med. Biol.* **19**, 289–302.

Fukuto, J. M., Wallace, G. C., Hszieh, R., and Chaudhuri, G. (1992). Chemical oxidation of N-hydroxyguanidine compounds. Release of nitric oxide, nitroxyl and possible relationship to the mechanism of biological nitric oxide generation. *Biochem. Pharmacol.* **43,** 607–613.

Fukuto, J. M., Stuehr, D. J., Feldman, P. L., Bova, M. P., and Wong, P. (1993). Peracid oxidation of an N-hydroxyguanidine compound: A chemical model for the oxidation of N-omega-hydroxyl-L-arginine by nitric oxide synthase. *J. Med. Chem.* **36,** 2666–2670.

Furchgott, R. F. (1988). Studies on relaxation of rabbit aorta by sodium nitrite: The basis for the proposal that the acid-activatable factor from bovine retractor penis is inorganic nitrite and the endothelium-derived relaxing factor is nitric oxide. *In* "Vasodilatation: Vascular Smooth Muscle, Peptides, Autonomic Nerves, and Endothelium" (P. M. Vanhoutte, ed.), pp. 401–414. Raven, New York.

Furchgott, R. F., and Zawadzki, J. V. (1980). The obligatory role of endothelial cells in the relaxation of arterial smooth muscle by acetylcholine. *Nature (London)* **288,** 373–376.

Garthwaite, J., Charles, S. L., and Chess-Williams, R. (1988). Endothelium-derived relaxing factor release on activation of NMDA receptors suggests role as intercellular messenger in the brain. *Nature (London)* **336,** 385–388.

Geller, D. A., Lowenstein, C. J., Shapiro, R. A., Nussler, A. K., Di Silvio, M., Wang, S. C., Nakayama, D. K., Simmons, R. L., Snyder, S. H., and Billiar, T. R. (1993). Molecular cloning and expression of inducible nitric oxide synthase from human hepatocytes. *Proc. Natl. Acad. Sci. U.S.A.* **90,** 3491–3495.

Geng, Y. J., Petersson, A. S., Wennmalm, A., and Hansson, G. K. (1994). Cytokine-induced expression of nitric oxide synthase results in nitrosylation of heme and nonheme iron proteins in vascular smooth muscle cells. *Exp. Cell Res.* **214,** 418–428.

Ghisla, S., and Massey, V. (1989). Mechanisms of flavoprotein-catalyzed reactions. *Eur. J. Biochem.* **181,** 1–17.

Giovanelli, J., Campos, K. L., and Kaufman, S. (1991). Tetrahydrobiopterin, a cofactor for rat cerebellar nitric oxide synthase, does not function as a reactant in the oxygenation of arginine. *Proc. Natl. Acad. Sci. U.S.A.* **88,** 7091–7095.

Granger, D. L., and Lehninger, A. L. (1982). Sites of inhibition of mitochondrial electron transport in macrophage-injured neoplastic cells. *J. Cell Biol.* **95,** 527–535.

Granger, D. L., Taintor, R. R., Cook, J. L., and Hibbs, J. B., Jr. (1980). Injury of neoplastic cells by murine macrophages leads to inhibition of mitochondrial respiration. *J. Clin. Invest.* **65,** 357–370.

Green, L. C., Ruiz de Luzuriaga, K., Wagner, D. A., Rand, W., Istfan, N., Young, V. R., and Tannenbaum, S. R. (1981a). Nitrate biosynthesis in man. *Proc. Natl. Acad. Sci. U.S.A.* **78,** 7764–7768.

Green, L. C., Tannenbaum, S. R., and Goldman, P. (1981b). Nitrate synthesis in the germfree and conventional rat. *Science* **212,** 56–58.

Green, S. J., Nacy, C. A., and Meltzer, M. S. (1991). Cytokine-induced synthesis of nitrogen oxides in macrophages: A protective host response to *Leishmania* and other intracellular pathogens. *J. Leukocyte Biol.* **50,** 93–103.

Gross, S. S., and Levi, R. (1992). Tetrahydrobiopterin synthesis. An absolute requirement for cytokine-induced nitric oxide generation by vascular smooth muscle. *J. Biol. Chem.* **267,** 25722–25729.

Guengerich, F. P. (1991). Reactions and significance of cytochrome P-450 enzymes. *J. Biol. Chem.* **266,** 10019–10022.

Hegesh, E., and Shiloah, J. (1982). Blood nitrates and infantile methemoglobinemia. *Clin. Chim. Acta* **125,** 107–115.

Heinzel, B., John, M., Klatt, P., Bohme, E., and Mayer, B. (1992). Ca^{2+}/calmodulin-dependent formation of hydrogen peroxide by brain nitric oxide synthase. *Biochem. J.* **281,** 627–630.

Helson, L., Green, S., Carswell, E., and Old, L. J. (1975). Effect of tumour necrosis factor on cultured human melanoma cells. *Nature (London)* **258,** 731–732.

Henry, Y., Ducrocq, C., Drapier, J. C., Servent, D., Pellat, C., and Guissani, A. (1991). Nitric oxide, a biological effector. Electron paramagnetic resonance detection of nitrosyl-iron-protein complexes in whole cells. *Eur. Biophys. J.* **20,** 1–15.

Henry, Y., Lepoivre, M., Drapier, J. C., Ducrocq, C., Boucher, J. L., and Guissani, A. (1993). EPR characterization of molecular targets for NO in mammalian cells and organelles. *FASEB J.* **7,** 1124–1134.

Hevel, J. M., and Marletta, M. A. (1992). Macrophage nitric oxide synthase: Relationship between enzyme-bound tetrahydrobiopterin and synthase activity. *Biochemistry* **31,** 7160–7165.

Hevel, J. M., White, K. A., and Marletta, M. A. (1991). Purification of the inducible murine macrophage nitric oxide synthase. Identification as a flavoprotein. *J. Biol. Chem.* **266,** 22789–22791.

Hibbs, J. B., Jr., Lambert, L. H., and Remington, J. S. (1971). Resistance to murine tumors conferred by chronic infection with intracellular protozoa, *Toxoplasma gondii* and *Besnoitia jellisoni. J. Infect. Dis.* **124,** 587–592.

Hibbs, J. B., Jr., Taintor, R. R., and Vavrin, Z. (1984). Iron depletion: Possible cause of tumor cell cytotoxicity induced by activated macrophages. *Biochem. Biophys. Res. Commun.* **123,** 716–723.

Hibbs, J. B., Jr., Taintor, R. R., and Vavrin, Z. (1987a). Macrophage cytotoxicity: Role for L-arginine deiminase and imino nitrogen oxidation to nitrite. *Science* **235,** 473–476.

Hibbs, J. B., Jr., Vavrin, Z., and Taintor, R. R. (1987b). L-Arginine is required for expression of the activated macrophage effector mechanism causing selective metabolic inhibition in target cells. *J. Immunol.* **138,** 550–565.

Hibbs, J. B., Jr., Taintor, R. R., Vavrin, Z., and Rachlin, E. M. (1988). Nitric oxide: A cytotoxic activated macrophage effector molecule. *Biochem. Biophys. Res. Commun.* **157,** 87–94.

Hyman, M. R., and Arp, D. J. (1988). Reversible and irreversible effects of nitric oxide on the soluble hydrogenase from *Alcaligenes eutrophus* H16. *Biochem. J.* **254,** 469–475.

Hyman, M. R., and Arp, D. J. (1991). Kinetic analysis of the interaction of nitric oxide with the membrane-associated, nickel and iron-sulfur-containing hydrogenase from *Azotobacter vinelandii. Biochim. Biophys. Acta* **1076,** 165–172.

Hyman, M. R., Seefeldt, L. C., Morgan, T. V., Arp, D. J., and Mortenson, L. E. (1992). Kinetic and spectroscopic analysis of the inactivating effects of nitric oxide on the individual components of *Azotobacter vinelandii* nitrogenase. *Biochemistry* **31**, 2947–2955.

Ignarro, L. J., Buga, G. M., Wood, K. S., Byrns, R. E., and Chaudhuri, G. (1987a). Endothelium-derived relaxing factor produced and released from artery and vein is nitric oxide. *Proc. Natl. Acad. Sci. U.S.A.* **84**, 9265–9269.

Ignarro, L. J., Byrns, R. E., Buga, G. M., and Wood, K. S. (1987b). Endothelium-derived relaxing factor from pulmonary artery and vein possesses pharmacologic and chemical properties identical to those of nitric oxide radical. *Circ. Res.* **61**, 866–879.

Ignarro, L. J., Byrns, R. E., and Wood, K. S. (1988). Biochemical and pharmacological properties of endothelium-derived relaxing factor and its similarity to nitric oxide radical. *In* "Vasodilatation: Vascular Smooth Muscle, Peptides, Autonomic Nerves, and Endothelium" (P. M. Vanhoutte, ed.), pp. 427–435. Raven, New York.

Iida, S., Ohshima, H., Oguchi, S., Hata, T., Suzuki, H., Kawasaki, H., and Esumi, H. (1992). Identification of inducible calmodulin-dependent nitric oxide synthase in the liver of rats. *J. Biol. Chem.* **267**, 25385–25388.

Iyanagi, T., Makino, R., and Anan, F. K. (1981). Studies on the microsomal mixed-function oxidase system: Mechanism of action of hepatic NADPH-cytochrome P-450 reductase. *Biochemistry* **20**, 1722–1730.

Iyengar, R., Stuehr, D. J., and Marletta, M. A. (1987). Macrophage synthesis of nitrite, nitrate, and N-nitrosamines: Precursors and role of the respiratory burst. *Proc. Natl. Acad. Sci. U.S.A.* **84**, 6369–6373.

James, S. L. (1991). The effector function of nirogen oxides in host defense against parasites. *Exp. Parasitol.* **73**, 223–226.

Jezowska-Trezebiatowska, B., and Jezierski, A. (1973). Electron spin resonance spectroscopy of iron nitrosyl complexes with organic ligands. *J. Mol. Struct.* **19**, 635–640.

Keller, R., and Jones, V. E. (1971). Role of activated macrophages and antibody in inhibition of tumour growth in rats. *Lancet* **2**, 847–849.

Kelm, M., Feelisch, M., Spahr, R., Piper, H. M., Noack, E., and Schrader, J. (1988). Quantitative and kinetic characterization of nitric oxide and EDRF released from cultured endothelial cells. *Biochem. Biophys. Res. Commun.* **154**, 236–244.

Klatt, P., Schmidt, K., Uray, G., and Mayer, B. (1993). Multiple catalytic functions of brain nitric oxide synthase. Biochemical characterization, cofactor-requirement, and the role of N omega-hydroxy-L-arginine as an intermediate. *J. Biol. Chem.* **268**, 14781–14787.

Kolb, H., and Kolb-Bachofen, V. (1992). Nitric oxide: A pathogenetic factor in autoimmunity. *Immunol. Today* **13**, 157–160.

Kon, H. (1968). Paramagnetic resonance study of nitric oxide hemoglobin. *J. Biol. Chem.* **243**, 4350–4357.

Krahenbuhl, J. L. (1980). Effects of activated macrophages on tumor target cells in discrete phases of the cell cycle. *Cancer Res.* **40**, 4622–4627.

Kubrina, L. N., Mordvintsev, P. I., and Vanin, A. F. (1989). Nitric oxide production in animal tissues on inflammation. *Bull. Exp. Biol. Med.* **1**, 31–34.

Kurzban, G. P., and Strobel, H. W. (1986). Preparation and characterization of FAD-dependent NADPH-cytochrome P-450 reductase. *J. Biol. Chem.* **261**, 7824–7830.

Kwon, N. S., Nathan, C. F., and Stuehr, D. J. (1989). Reduced biopterin as a cofactor in the generation of nitrogen oxides by murine macrophages. *J. Biol. Chem.* **264**, 20496–20501.

Kwon, N. S., Nathan, C. F., Gilker, C., Griffith, O. W., Matthews, D. E., and Stuehr, D. J. (1990). L-Citrulline production from L-arginine by macrophage nitric oxide synthase. The ureido oxygen derives from dioxygen. *J. Biol. Chem.* **265,** 13442–13445.

Lamas, S., Marsden, P. A., Li, G. K., Tempst, P., and Michel, T. (1992). Endothelial nitric oxide synthase: Molecular cloning and characterization of a distinct constitutive enzyme isoform. *Proc. Natl. Acad. Sci. U.S.A.* **89,** 6348–6352.

Lancaster, J. R., Jr. (1992). Nitric oxide in cells. *Am. Sci.* **80,** 248–259.

Lancaster, J. R., Jr., and Hibbs, J. B., Jr. (1990). EPR demonstration of iron–nitrosyl complex formation by cytotoxic activated macrophages. *Proc. Natl. Acad. Sci. U.S.A.* **87,** 1223–1227.

Lancaster, J. R., Jr., Langrehr, J. M., Bergonia, H. A., Murase, N., Simmons, R. L., and Hoffman, R. A. (1992). EPR detection of heme and nonheme iron-containing protein nitrosylation by nitric oxide during rejection of rat heart allograft. *J. Biol. Chem.* **267,** 10994–10998.

Langrehr, J. M., Muller, A. R., Bergonia, H. A., Jacob, T. D., Lee, T. K., Schraut, W. H., Lancaster, J. R., Jr., Hoffman, R. A., and Simmons, R. L. (1992). Detection of nitric oxide by electron paramagnetic resonance spectroscopy during rejection and graft-versus-host disease after small-bowel transplantation in the rat. *Surgery* **112,** 395–401; discussion.

Lazarus, R. A., Dietrich, R. F., Wallick, D. E., and Benkovic, S. J. (1981). On the mechanism of action of phenylalanine hydroxylase. *Biochemistry* **20,** 6834–6841.

Leone, A. M., Palmer, R. M., Knowles, R. G., Francis, P. L., Ashton, D. S., and Moncada, S. (1991). Constitutive and inducible nitric oxide synthases incorporate molecular oxygen into both nitric oxide and citruline. *J. Biol. Chem.* **266,** 23790–23795.

Liang, J., and Burris, R. H. (1988). Inhibition of nitrogenase by NO. *Indian J. Biochem. Biophys.* **25,** 636–641.

Lorsbach, R. B., Murphy, W. J., Lowenstein, C. J., Snyder, S. H., and Russell, S. W. (1993). Expression of the nitric oxide synthase gene in mouse macrophages activated for tumor cell killing. Molecular basis for the synergy between interferon-gamma and lipopolysaccharide. *J. Biol. Chem.* **268,** 1908–1913.

Mcdonald, C. C., Phillips, W. D., and Mower, H. F. (1965). An electron spin resonance study of some complexes of iron, nitric oxide, and anionic ligands. *J. Am. Chem. Soc.* **87,** 3319–3326.

Mackaness, G. B. (1964). The immunological basis of acquired cellular resistance. *J. Exp. Med.* **120,** 105–120.

McMillan, K., Bredt, D. S., Hirsch, D. J., Snyder, S. H., Clark, J. E., and Masters, B. S. (1992). Cloned, expressed rat cerebellar nitric oxide synthase contains stoichiometric amounts of heme, which binds carbon monoxide. *Proc. Natl. Acad. Sci. U.S.A.* **89,** 11141–11145.

Marletta, M. A., Yoon, P. S., Iyengar, R., Leaf, C. D., and Wishnok, J. S. (1988). Macrophage oxidation of L-arginine to nitrite and nitrate: Nitric oxide is an intermediate. *Biochemistry* **27,** 8706–8711.

Maruyama, T., Kataoka, N., Nagase, S., Nakada, H., Sato, H., and Sasaki, H. (1971). Identification of three-line electron spin resonance signal and its relationship to ascites tumors. *Cancer Res.* **31,** 179–184.

Mayer, B., John, M., and Bohme, E. (1990). Purification of a Ca^{2+}/calmodulin-dependent nitric oxide synthase from procine cerebellum. Cofactor-role of tetrahydrobiopterin. *FEBS Lett.* **277,** 215–219.

Meyer, J. (1981). Comparison of carbon monoxide, nitric oxide, and nitrite as inhibitors of the nitrogenase from *Clostridium pasteurianum*. *Arch. Biochem. Biophys.* **210,** 246–256.

Michalski, W. P., and Nicholas, D. J. D. (1987). Inhibition of nitrogenase by nitrite and nitric oxide in *Rhodopseudomonas sphaeroides* f. sp. *denitrificans*. *Arch. Microbiol.* **147,** 304–308.

Michel, T., Li, G. K., and Busconi, L. (1993). Phosphorylation and subcellular translocation of endothelial nitric oxide synthase. *Proc. Natl. Acad. Sci. U.S.A.* **90,** 6252–6256.

Mitchell, H. H., Shonle, H. A., and Grindley, H. S. (1916). The origin of nitrates in the urine. *J. Biol. Chem.* **24,** 461–490.

Miwa, M., Stuehr, D. J., Marletta, M. A., Wishnok, J. S., and Tannenbaum, S. R. (1987). Nitrosation of amines by stimulated macrophages. *Carcinogenesis (London)* **8,** 955–958.

Moncada, S. and Higgs, A. (1993). The L-arginine–nitric oxide pathway. (Review). *N. Engl. J. Med.* **329,** 2002–2012.

Moncada, S., Palmer, R. M., and Higgs, E. A. (1991). Nitric oxide: Physiology, pathophysiology, and pharmacology. *Pharmacol. Rev.* **43,** 109–142.

Morris, S. M., Jr., and Billiar, T. R. (1994). New insights into the regulation of inducible nitric oxide synthesis. *Am. J. Physiol.* **266,** E829–E839.

Nagano, T., and Fridovich, I. (1985). The co-oxidation of ammonia to nitrite during the aerobic xanthine oxidase reaction. *Arch. Biochem. Biophys.* **241,** 596–601.

Nagata, C., Ioki, Y., Kodama, M., Tagashira, Y., and Nakadate, M. (1973). Free radical induced in rat liver by a chemical carcinogen, *N*-methyl-*N'*-nitro-*N*-nitrosoguanidine. *Ann. N.Y. Acad. Sci.* **222,** 1031–1047.

Nakane, M., Schmidt, H. H., Pollock, J. S., Forstermann, U., and Murad, F. (1993). Cloned human brain nitric oxide synthase is highly expressed in skeletal muscle. *FEBS Lett.* **316,** 175–180.

Narhi, L. O., and Fulco, A. J. (1987). Identification and characterization of two functional domains in cytochrome P-450BM-3, a catalytically self-sufficient monooxygenase induced by barbiturates in Bacillus megaterium. *J. Biol. Chem.* **262,** 6683–6690.

Nathan, C. (1992). Nitric oxide as a secretory product of mammalian cells. *FASEB J.* **6,** 3051–3064.

Nathan, C. F., and Hibbs, J. B., Jr. (1991). Role of nitric oxide synthesis in macrophage antimicrobial activity. *Curr. Opin. Immunol.* **3,** 65–70.

Nauts, H. C., and McLaren, J. R. (1990). Coley toxins—The first century. *Adv. Exp. Med. Biol.* **267,** 483–500.

Nichol, C. A., Smith, G. K., and Duch, D. S. (1985). Biosynthesis and Metabolism of Tetrahydrobiopterin and Molybdopterin. *Annu. Rev. Biochem.* **54,** 729–764.

Ortiz de Montellano, P. R. (1986). Oxygen activation and transfer. *In* "Cytochrome P-450" (P. R. Ortiz de Montellano, ed.), pp. 217–271. Plenum, New York.

Ostrowski, J., Barber, M. J., Rueger, D. C., Miller, B. E., Siegel, L. M., and Kredich, N. M. (1989). Characterization of the flavoprotein moieties of NADPH–sulfite reductase from *Salmonella typhimurium* and *Escherichia coli*. Physicochemical and catalytic properties, amino acid sequence deduced from DNA sequence of *cysJ*, and comparison with NADPH–cytochrome P-450 reductase. *J. Biol. Chem.* **264,** 15796–15808.

Otvos, J. D., Krum, D. P., and Masters, B. S. (1986). Localization of the free radical on the flavin mononucleotide of the air-stable semiquinone state of NADPH–cytochrome P-450 reductase using 31P NMR spectroscopy. *Biochemistry* **25,** 7220–7228.

Palmer, R. M., Ferrige, A. G., and Moncada, S. (1987). Nitric oxide release accounts for the biological activity of endothelium-derived relaxing factor. *Nature (London)* **327,** 524–526.

Palmer, R. M. J., Ashton, D. S., and Moncada, S. (1988a). Vascular endothelial cells synthesize nitric oxide from L-arginine. *Nature (London)* **333,** 664–666.

Palmer, R. M. J., Rees, D. D., Ashton, D. S., and Moncada, S. (1988b). L-Arginine is the physiological precursor for the formation of nitric oxide in endothelium-dependent relaxation. *Biochem. Biophys. Res. Commun.* **153,** 1251–1256.

Payne, M. J., Woods, L. F., Gibbs, P., and Cammack, R. (1990). Electron paramagnetic resonance spectroscopic investigation of the inhibition of the phosphoroclastic system of *Clostridium sporogenes* by nitrite. *J. Gen. Microbiol.* **136,** 2067–2076.

Pellat, C., Henry, Y., and Drapier, J. C. (1990). IFN-gamma-activated macrophages: Detection by electron paramagnetic resonance of complexes between L-arginine-derived nitric oxide and nonheme iron proteins. *Biochem. Biophys. Res. Commun.* **166,** 119–125.

Peterson, J. A., and Boddupalli, S. S. (1992). P450BM-3: Reduction by NADPH and sodium dithionite. *Arch. Biochem. Biophys.* **294,** 654–661.

Petroulas, V., and Diner, B. A. (1990). Formation by NO of nitrosyl adducts of redox components of the photosystem II reaction center. I. NO binds to the acceptor-side non-heme iron. *Biochim. Biophys. Acta* **1015,** 131–140.

Pollock, J. S., Forstermann, U., Mitchell, J. A., Warner, T. D., Schmidt, H. H., Nakane, M., and Murad, F. (1991). Purification and characterization of particulate endothelium-derived relaxing factor synthase from cultured and native bovine aortic endothelial cells. *Proc. Natl. Acad. Sci. U.S.A.* **88,** 10480–10484.

Porter, T. D. (1991). An unusual yet strongly conserved flavoprotein reductase in bacteria and mammals. *Trends Biochem. Sci.* **16,** 154–158.

Pufahl, R. A., and Marletta, M. A. (1993). Oxidation of N^G-Hydroxy-L-arginine by nitric oxide synthase: Evidence for the involvement of the heme in catalysis. *Biochem. Biophys. Res. Commun.* **193,** 963–970.

Reddy, D., Lancaster, J. R., Jr., and Cornforth, D. P. (1983). Nitrite inhibition of *Clostridium botulinum*: Electron spin resonance detection of iron–nitric oxide complexes. *Science* **221,** 769–770.

Renaud, J. P., Boucher, J. L., Vadon, S., Delaforge, M., and Mansuy, D. (1993). Particular ability of liver P450s3A to catalyze the oxidation of N omega-hydroxyarginine to citrulline and nitrogen oxides and occurrence in no synthases of a sequence very similar to the heme-binding sequence in P450s. *Biochem. Biophys. Res. Commun.* **192,** 53–60.

Salerno, J. C., Ohnishi, T., Lim, J., and King, T. E. (1976). Tetranuclear and binuclear iron–sulfur clusters in succinate dehydrogenase: A method of iron quantitation by formation of paramagnetic complexes. *Biochem. Biophys. Res. Commun.* **73,** 833– 839.

Saul, R. L., and Archer, M. C. (1984). Oxidation of ammonia and hydroxylamine to nitrate in the rat and *in vitro*. *Carcinogenesis (London)* **5,** 77–81.

Schmidt, H. H., Pollock, J. S., Nakane, M., Gorsky, L. D., Forstermann, U., and Murad, F. (1991). Purification of a soluble isoform of guanylyl cyclase-activating-factor synthase. *Proc. Natl. Acad. Sci. U.S.A.* **88,** 365–369.

Schmidt, H. H., Smith, R. M., Nakane, M., and Murad, F. (1992). Ca^{2+}/calmodulin-dependent NO synthase type I: A biopteroflavoprotein with Ca^{2+}/calmodulin-independent diaphorase and reductase activities. *Biochemistry* **31,** 3243–3249.

Stadler, J., Bergonia, H. A., Di Silvio, M., Sweetland, M. A., Billiar, T. R., Simmons, R. L., and Lancaster, J. R., Jr. (1993). Nonheme iron–nitrosyl complex formation in rat hepatocytes: Detection by electron paramagnetic resonance spectroscopy. *Arch. Biochem. Biophys.* **302,** 4–11.

Stolze, K., and Nohl, H. (1990a). Free radical intermediates in the oxidation of N-methylhydroxylamine and N,N-dimethylhydroxylamine by oxyhemoglobin. *Free Radical Res. Commun.* **8,** 123–131.

Stolze, K., and Nohl, H. (1990b). EPR studies on the oxidation of hydroxyurea to paramagnetic compounds by oxyhemoglobin. *Biochem. Pharmacol.* **40,** 799–802.

Stuehr, D. J., and Griffith, O. W. (1992). Mammalian nitric oxide synthases. *Adv. Enz.* **65,** 287–346.

Stuehr, D. J., and Ikeda-Saito, M. (1992). Spectral characterization of brain and macrophage nitric oxide synthases. Cytochrome P-450-like heme proteins that contain a flavin semiquinone radical. *J. Biol. Chem.* **267,** 20547–20550.

Stuehr, D. J., and Marletta, M. A. (1985). Mammalian nitrate biosynthesis: Mouse macrophages produce nitrite and nitrate in response to *Escherichia coli* lipopolysaccharide. *Proc. Natl. Acad. Sci. U.S.A.* **82,** 7738–7742.

Stuehr, D. J., and Nathan, C. F. (1989). Nitric oxide. A macrophage product responsible for cytostasis and respiratory inhibition in tumor target cells. *J. Exp. Med.* **169,** 1543–1555.

Stuehr, D. J., Gross, S. S., Sakuma, I., Levi, R., and Nathan, C. F. (1989). Activated murine macrophages secrete a metabolite of arginine with the bioactivity of endothelium-derived relaxing factor and the chemical reactivity of nitric oxide. *J. Exp. Med.* **169,** 1011–1020.

Stuehr, D. J., Kwon, N. S., and Nathan, C. F. (1990). FAD and GSH participate in macrophage synthesis of nitric oxide. *Biochem. Biophys. Res. Commun.* **168,** 558–565.

Stuehr, D. J., Cho, H. J., Kwon, N. S., Weise, M. F., and Nathan, C. F. (1991a). Purification and characterization of the cytokine-induced macrophage nitric oxide synthase: An FAD- and FMN-containing flavoprotein. *Proc. Natl. Acad. Sci. U.S.A.* **88,** 7773–7777.

Stuehr, D. J., Kwon, N. S., Nathan, C. F., Griffith, O. W., Feldman, P. L., and Wiseman, J. (1991b). N omega-hydroxy-L-arginine is an intermediate in the biosynthesis of nitric oxide from L-arginine. *J. Biol. Chem.* **266,** 6259–6263.

Sulser, H., and Sager, F. (1976). Identification of uncommon amino acids in the lentil seed *(Lens culinaris Med.)*. *Experientia* **32,** 422–423.
Vanin, A. F. (1967). Identification of divalent iron complexes with cysteine in biological systems by the EPR method. *Biochemistry (USSR)* **32,** 228–232.
Vanin, A. F., and Varich, V. Y. (1980). Formation of nitrosyl complexes of non-haeme iron (complexes 2.03) in animal tissues *in vivo. Biophysics (USSR)* **24,** 686–690.
Vanin, A. F., and Varich, V. Y. (1981). Nitrosyl non-heme iron complexes in animal tissues. *Stud. Biophys.* **86,** 177–185.
Vanin, A. F., Vakhina, L. V., and Chetverikov, A. G. (1970). Nature of the EPR signals of a new type found in cancer tissues. *Biophysics USSR (Engl. trans)* **15,** 1082–1089.
Vanin, A. F., Osipov, A. N., Kubrina, L. N., Burbaev, D. S., and Nalbandyan, R. M. (1975). On the origin of paramagnetic centers with g = 2.03 in animal tissues and microorganisms. *Stud. Biophys.* **49,** 13–25.
Vanin, A. F., Kubrina, L. N., Kurbanov, I. S., Mordvintsev, P. I., Khrapova, N. V., Galagan, M. E., and Matkhanov, E. I. (1989). Iron as an inducer of nitric oxide formation in the animal body. *Biokhimiia* **54,** 1974–1979.
Vanin, A. F., Kubrina, L. N., Malenkova, I. V., and Mordintsev, P. I. (1991). L-Arginine—an endogenous source of nitric oxide in animal tissues *in vivo.* (Russian). *Biokhimiia* **56,** 935–939.
Vanin, A. F., Men'shikov, G. B., Moroz, I. A., Mordvintcev, P. I., Serezhenkov, V. A., and Burbaev, D. S. (1992). The source of non-heme iron that binds nitric oxide in cultivated macrophages. *Biochim. Biophys. Acta* **1135,** 275–279.
Vanin, A. F., Mordvintcev, P. I., Hauschildt, S., and Mulsch, A. (1993). The relationship between L-arginine-dependent nitric oxide synthesis, nitrite release and dinitrosyl–iron complex formation by activated macrophages. *Biochim. Biophys. Acta* **1177,** 37–42.
Varich, V. Y. (1980). Formation of nitrosyl complexes of non-haeme iron in the blood of animals *in vivo. Biophys. (USSR)* **24,** 1146–1148.
Varich, V. Y., and Vanin, A. F. (1983). Mechanism of formation of nitrosyl complexes of non-haeme iron in animal tissues *in vivo. Biophys. (USSR)* **28,** 1125–1131.
Varich, V. Y., Vanin, A. F., and Ovsyannikova, L. M. (1987). Detection of endogenous nitrogen oxide in mouse liver by electron paramagnetic resonance. *Biophys. (USSR)* **32,** 1158–1160.
Vaz, A. D. N., Roberts, E. S., and Coon, M. J. (1991). Olefin formation in the oxidative deformylation of aldehydes by cytochrome P-450. Mechanistic implications for catalysis by oxygen-derived peroxide. *J. Am. Chem. Soc.* **113,** 5886–5887.
Vermillion, J. L., Ballou, D. P., Massey, V., and Coon, M. J. (1981). Separate roles for FMN and FAD in catalysis by liver microsomal NADPH–cytochrome P-450 reductase. *J. Biol. Chem.* **256,** 266–277.
Vithayathil, A. J., Ternberg, J. L., and Commoner, B. (1965). Changes in electron spin resonance signals of rat liver during chemical carcinogenesis. *Nature (London)* **207,** 1246–1249.
Wagner, D., and Tannenbaum, S. R. (1982). Enhancement of nitrate biosynthesis by *Escherichia coli* lipopolysaccharide. *In* Nitrosamines and Human Cancer (P. N. Magee, ed.), Cold Spring Harbor Laboratory, Cold Spring Harbor, New York.

Wagner, D. A., Young, V. R., and Tannenbaum, S. R. (1983). Mammalian nitrate biosynthesis: Incorporation of $^{15}NH_3$ into nitrate is enhanced by endotoxin treatment. *Proc. Natl. Acad. Sci. U.S.A.* 4518–4521.

Wang, J., Stuehr, D. J., Ikeda-Saito, M., and Rousseau, D. L. (1993). Heme coordination and structure of the catalytic site in nitric oxide synthase. *J. Biol. Chem.* **268,** 22255–22258.

Wang, Q. Z., Jacobs, J., DeLeo, J., Kruszyna, H., Kruszyna, R., Smith, R., and Wilcox, D. (1991). Nitric oxide hemoglobin in mice and rats in endotoxic shock. *Life Sci.* **49,** PL55–PL60.

Werner-Felmayer, G., Werner, E. R., Fuchs, D., Hausen, A., Reibnegger, G., and Wachter, H. (1990). Tetrahydrobiopterin-dependent formation of nitrite and nitrate in murine fibroblasts. *J. Exp. Med.* **172,** 1599–1607.

Westenberger, U., Thanner, S., Ruf, H. H., Gersonde, K., Sutter, G., and Trentz, O. (1990). Formation of free radicals and nitric oxide derivative of hemoglobin in rats during shock syndrome. *Free Radical Res. Commun.* **11,** 167–178.

Wharton, M., Granger, D. L., and Durack, D. T. (1988). Mitochondrial iron loss from leukemia cells injured by macrophages. A possible mechanism for electron transport chain defects. *J. Immunol.* **141,** 1311–1317.

White, K. A., and Marletta, M. A. (1992). Nitric oxide synthase is a cytochrome P-450 type hemoprotein. *Biochemistry* **31,** 6627–6631.

Widmer, J., and Keller-Schierlein, W. (1974). Metabolites of microorganisms. 130. Synthesis of delta-N-hydroxy-L-arginine. *Helv. Chim. Acta* **57,** 657–664.

Woolum, J. C., and Commoner, B. (1970). Isolation and identification of a paramagnetic complex from the livers of carcingen-treated rats. *Biochim. Biophys. Acta* **201,** 131–140.

Woolum, J. C., Tiezzi, E., and Commoner, B. (1968). Electron spin resonance of iron–nitric oxide complexes with amino acids, peptides and proteins. *Biochim. Biophys. Acta* **160,** 311–320.

Xie, Q. W., Cho, H. J., Calaycay, J., Mumford, R. A., Swiderek, K. M., Lee, T. D., Ding, A., Troso, T., and Nathan, C. (1992). Cloning and characterization of inducible nitric oxide synthase from mouse macrophages. *Science* **256,** 225–228.

Zumft, W. G. (1993). The biological role of nitric oxide in bacteria. *Arch. Microbiol.* **160,** 253–264.

5
The Role of Nitric Oxide in Autoimmune Diabetes

John A. Corbett[1] and Michael L. McDaniel
Department of Pathology
Washington University School of Medicine
St. Louis, Missouri 63110

I. INTRODUCTION

Over 5% of the population of western nations is afflicted with diabetes. The most prevalent form of diabetes, non-insulin-dependent diabetes mellitus (NIDDM, or type II), is commonly associated with obesity and hypertension, and is believed to be the consequence of altered insulin action or insulin secretion (for review see Defronzo, 1988; Defronzo and Ferrannini, 1991). Insulin-dependent diabetes mellitus (IDDM, or type I diabetes) accounts for approximately 10% of all cases of diabetes. IDDM is characterized by specific destruction of insulin secreting β-cells found in islets of Langerhans. Destruction of 80–90% of islet β-cells causes insulin deficiency and the inability to regulate blood glucose levels.

β-Cell destruction during the development of IDDM is the result of autoimmune processes. Considerable evidence supports the autoimmune nature of IDDM. This evidence includes (1) observation of lymphocyte infiltration into pancreatic islets in biopsy specimens taken from patients in early stages of IDDM; (2) the identification of islet-cell autoantibodies from patients with IDDM;

[1] Address correspondence to Dr. Corbett at Department of Biochemistry and Molecular Biology, Saint Louis University School of Medicine, St. Louis, Missouri 63104.

(3) IDDM is associated with other diseases thought to involve autoimmune components; (4) activated cell-mediated immunity toward pancreatic islet cells and cell antigens from IDDM patients; and (5) the correlation of IDDM with HLA haplotypes associated with other autoimmune diseases (Maclaren, 1981; Lernmark et al., 1991; Andreani et al., 1991; Rossini et al., 1991).

The mechanisms of autoimmune β-cell destruction are unknown. Evidence has implicated the involvement of helper and cytotoxic T lymphocytes, macrophages, cytokines released from infiltrating lymphocytes, and free radicals. Much attention has focused on the involvement of nitrogen oxide free radicals. In this chapter we review the mechanism(s) associated with β-cell destruction focusing on the role of cytokines. Particular attention will be drawn to cytokine-induced nitric oxide production by islet β-cells, because this cell type is selectively destroyed during the development of IDDM. Also the mechanisms by which T-cells and macrophages contribute to β-cell destruction will be examined. We will conclude this chapter by examining the role of cytokines in the production of nitric oxide by human islets, and the effects of nitric oxide on human islet function as this directly relates to the progression of insulin-dependent diabetes mellitus.

A. Autoimmune Nature of Insulin-Dependent Diabetes Mellitus

The first evidence implicating IDDM as an autoimmune disease was provided by the morphological studies of Gepts (1965). In these studies the presence of an insulitis or inflammatory reaction in and around islets was observed in children who perished during the onset of IDDM. The cellular infiltration (insulitis), comprised of lymphocytic cells, was later shown to contain both CD8[+] and CD4[+] T cells, macrophages, and monocytes (Bottazzo et al., 1985; Sibley et al., 1985). Animal models of type I diabetes, the nonobese spontaneous diabetic mouse (NOD) and the BioBreeding rat (BB), have been used to more fully characterize the cellular content of this lymphocyte infiltration. In these animal models, macrophages and monocytes appear to dominate the early islet infiltrate, and at later times T cells appear in the infiltrate (Kolb et al., 1986; Walker et al., 1988; Lee et al., 1988; Cooke, 1990).

Macrophages and T cells are believed to participate in the process of β-cell destruction. Treatment of NOD mice or BB rats with silica particles completely prevents insulitis and diabetes in these animal models (Oschilewski et al., 1985; Charlton et al., 1988). Silica particles are selectively toxic to macrophages (Levy and Wheelock, 1975; Brosnan et al., 1981), suggesting that macrophages have a permissive role in the development of insulitis and diabetes in both animal models. Evidence for the involvement of T cells has come from the observations that diabetes can be transferred to nondiabetic BB rats or NOD mice by splenic lymphocytes from newly diagnosed diabetic animals (Like et al., 1985; Wicker et al., 1986). Also, neonatal thymectomy and depletion of T cells by monoclonal antibody therapy prevents spontaneous diabetes in both the NOD mouse and the BB rat (Shizuru et al., 1988; Sempe et al., 1991; Maki et al., 1992; Like et al., 1982). Based on this evidence, and many other studies reviewed previously (Rossini et al., 1991; Crisa et al., 1992), both T cells and macrophages appear to participate in the development of diabetes.

The cellular homing of lymphocytes and macrophages into the islet implicates the presence of β-cell specific autoantigens. Several autoantibodies of β cells or β-cell determinants, as well as specific proteins, have been described (Sigurdsson and Baekkeskov, 1990). These include islet-cell antibodies (Marner et al., 1985), islet-cell surface antibodies against normal β cells (Lernmark et al., 1978), cytotoxic islet-cell antibodies (Dobersen et al., 1980), antibodies directed against the insulin secretory granule proteins carboxypeptidase H (Castano et al., 1991) and insulin (Roep et al., 1990), and autoantibodies directed against two forms of glutamic acid decarboxylase (GAD), 65 and 67 kDa (Baekkeskov et al., 1990; Kaufman et al., 1992). Antibodies to these proposed autoantigens have been identified from patients several years prior to the onset of diabetes, and have proven to be useful predictive tools in determining the onset of IDDM.

Two groups (Kaufman et al., 1993; Tisch et al., 1993) have shown that helper T-cell responses to GAD occur in a temporal fashion that coincides with the development of insulitis during the onset of diabetes in the NOD mouse. Also, interthymic injection of GAD65 prevents both T-cell proliferative response to GAD and the development of diabetes in the NOD mouse. These studies indicate that an immune response to GAD may play an important early role in the development of autoimmune diabetes.

We have examined β-cell surface expression of potential autoantigens and have shown surface colocalization of the autoantigens insulin (Kaplan et al., 1983) and carboxypeptidase H (Aguilar-Diosdado et al., 1994), and that cell-surface expression of both insulin and carboxypeptidase H is stimulated by glucose (Aguilar-Diosdado et al., 1994). However, in these studies we were unable to demonstrate GAD65 expression on the surface of the β cell. These findings suggest that an immune response to GAD may require β-cell destruction for the release of this proposed autoantigen. The potential mechanisms of cell destruction are unknown; however, we have obtained evidence which suggests that nitric oxide may play an early role in β-cell destruction (Corbett et al., 1993d; see Section X,B).

The details of this complex interplay between macrophages, T cells, and autoantigens during the development of diabetes are incompletely understood. Macrophages which comprise the early islet infiltrate potentially function as both the antigen presenting cell and a contributor to β-cell destruction. Antigen presentation by macrophages may signal T-cell infiltration into the islet, followed by T-cell mediated β-cell death. The release of cytokines and lymphokines may also contribute to β-cell destruction either by direct interactions with receptors on the β cell, or by the activation of T cells and macrophages. The presence of β-cell specific autoantigens are not essential for insulitis and β-cell destruction. Islets from MHC compatible non-BB rats treated to prevent allograft rejection and then transplanted under the kidney capsule of nondiabetic BB rats develop insulitis similar to that observed in islets of the host (Weringer and Like, 1985). These data indicate that a β-cell antigen is not required for insulitis, suggesting that environmental factors may also induce autoimmunity.

B. Proposed Models of Insulin-Dependent Diabetes Mellitus

Many hypotheses have been proposed to explain the processes of β-cell destruction during the onset of IDDM. Nerup and Lernmark (1981) suggested that susceptible persons have an abnormal reaction to exogenous stimuli (β-cell cytotropic virus or β-cell cytotoxic chemicals). This reaction leads to the destruction of β cells either directly by endogenous agents, or indirectly by autoimmune mechanisms. β-cell destruction would require helper T-cell proliferation, the induction of cytotoxic T cells, and autoantibodies. This hypothesis suggests that the primary event is the result of an etiologic factor, and the resulting autoimmune reaction is responsible for the progression of β-cell destruction.

More recently Mandrup-Poulsen et al. (1990) have proposed that cytokines released during islet infiltration may participate in the development of diabetes. In this model an initial triggering event occurs in an islet under immunosurveillance (i.e., in the presence of macrophages and helper T lymphocytes). The triggering event induces the immune response of antigen processing, presentation, and the release of cytokines (possibly regulated by the HLA region of immune-response genes). These cytokines [interferon (IFN) and interleukin 1 (IL-1)] activate macrophages and natural killer cells to release soluble mediators such as IL-1 and tumor necrosis factor (TNF). The cytokine IL-1 is proposed to directly impair β-cell function as a result of local increases in IL-1 concentrations during islet infiltration. The process continues until all islet β cells are destroyed. β-cell destruction does not continue automatically in other islets but is proposed to require a second triggering event before continuation of the cycle. The frequency of further triggering events are proposed to be exponentially more likely to occur because the specific T-cell clone involved in β-cell destruction would be expanded with each islet destroyed. Although evidence is lacking for either of these models, they have provided working hypotheses for the study of the immunological and biochemical processes which appear to mediate β-cell destruction during the onset of diabetes.

II. EFFECTS OF CYTOKINES ON β-CELL FUNCTION

A. Initial Demonstration of the Effects Interleukin 1 on Islet Function

In 1985 Nerup and co-workers demonstrated that treatment of isolated rat and human islets with conditioned media derived from activated mononuclear cells resulted in the inhibition of glucose-stimulated insulin secretion (Mandrup-Poulsen et al., 1985). The cytokine IL-1β was found to be the active component of the conditioned media (Bendtzen et al., 1986). Mandrup-Poulsen et al. (1986) further showed that continuous exposure of islets to IL-1 is toxic.

The pioneering work of Nerup and co-workers provided the initial evidence suggesting that local fluctuations of cytokine concentrations may influence insulin secretion by islets. The effects of IL-1β on insulin secretion have been confirmed and extended to show that IL-1β-induced inhibition of glucose stimulated insulin secretion by isolated islets is both time- and concentration-dependent as shown in Fig. 1 (Spinas et al., 1986; Comens et al., 1987; McDaniel et al., 1988a; Palmer et al., 1989). Treatment of islets for short periods of time with high concentrations of IL-1β (1–10 U/ml) or for longer periods of time with

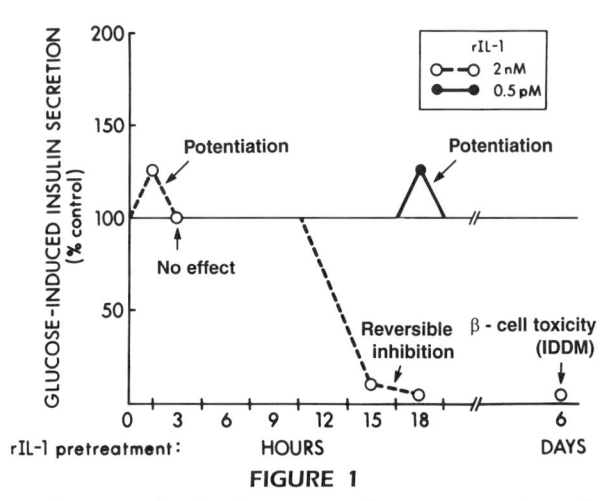

FIGURE 1

Effects of IL-1 on glucose-stimulated insulin secretion: Concentration and time dependence. The ability of IL-1 to modulate insulin secretion by isolated rat islets is illustrated. Short exposures of islets to high concentrations of IL-1 [2 nM as shown here, or 5 U/ml (5 pM) as used in other studies], or prolonged exposure to low concentrations of IL-1 (0.5 pM or 0.5 U/ml) result in a stimulation of glucose-induced insulin secretion. Incubation of islets for 18 hr with 2 nM (or 5 U/ml) IL-1 results in a potent and reversible inhibition of glucose stimulated insulin secretion. Islet destruction is observed when islets are treated for 4 to 6 days with 2 nM (or 5 U/ml) IL-1. Reproduced with permission from *Diabetes* (McDaniel et al., 1988), by the American Diabetes Association, Inc.

low concentrations of IL-1β (0.1 U/ml) potentiate glucose-stimulated insulin secretion. Incubation of islets with IL-1β at higher concentrations (1–10 U/ml) results in a time-dependent inhibition of glucose-stimulated insulin secretion with the initial inhibitory effects observed between 5 and 8 hr after the addition of IL-1. Following 18 hr of continuous exposure to IL-1β, nearly complete inhibition of glucose stimulated insulin secretion is observed. Treatment of islets for 4–6 days with IL-1 results in islet destruction. These results provided the initial evidence that IL-1 is a potent modulator of islet function and viability.

B. Effects of Interleukin 1 on mRNA Transcription and Protein Translation

The time- and concentration-dependent inhibitory and stimulatory effects of IL-1 on insulin secretion by islets are paralleled by similar changes in preproinsulin mRNA levels. Treatment of rat islets with concentrations of IL-1 that inhibit insulin secretion reduce preproinsulin mRNA content. Conditions of IL-1 treatment which stimulate insulin secretion increase preproinsulin mRNA content of islets (Spinas et al., 1987; Eizirik, et al., 1990). Islet-cell replication is also inhibited by exposure to IL-1. The importance of these findings are unknown, although increases in preproinsulin mRNA levels have been suggested to participate in IL-1-induced stimulation of glucose-induced insulin secretion (Spinas et al., 1987). The reduction of preproinsulin mRNA levels by IL-1 treatment on the other hand does not appear to participate in IL-1-induced inhibition of glucose-stimulated insulin secretion (see below).

In contrast to the inhibitory effects of IL-1 on preproinsulin mRNA content, IL-1-induced inhibition of glucose-stimulated insulin secretion requires mRNA transcription and new protein synthesis. Hughes et al. (1990a) were the first to demonstrate that both the mRNA transcriptional inhibitor actinomycin D and the protein synthesis inhibitor cycloheximide, completely prevent the inhibitory effects of IL-1 on glucose-stimulated insulin secretion by isolated rat islets. It was observed that a 1-hr pulse of islets with IL-1 followed by removal of this cytokine results in a time-dependent inhibition of glucose-stimulated insulin secretion that mimics the effects of continuous exposure of islets to IL-1 (Fig. 2). Using this experimental design the inhibitory effects of IL-1 following a 1-hr pulse could be prevented by actinomycin D only when the transcriptional inhibitor was present during the 1-hr exposure. Actinomycin D did not prevent the inhibitory effects of IL-1 when it was added during the interval following the 1-hr pulse. Cycloheximide also prevents IL-1-induced inhibition of insulin secretion when present during the interval following the 1-hr exposure (Fig. 3). Based on evidence obtained from this pulse model, the effects of IL-1 can be separated into an early mRNA transcriptional phase followed by a later protein translational phase.

The effects of IL-1 in this pulse model have been confirmed and extended to show that actinomycin D prevents IL-1-induced potentiation of glucose-stimu-

5 Role of Nitric Oxide in Autoimmune Diabetes

FIGURE 2

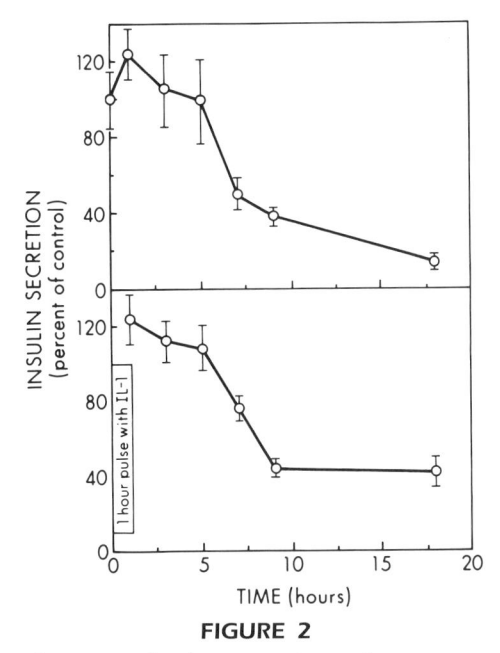

Effects of IL-1β on insulin secretion by islets, comparing continuous exposure with the effects of a 1-hr pulse of IL-1β. *(Top)* Islets were continuously exposed to 5 U/ml IL-1β for the indicated times, and insulin secretion was examined as described previously (Hughes et al., 1990a). *(Bottom)* Islets were pulsed with 5 U/ml IL-1β for 1 hr at which time the cytokine was removed by washing and insulin secretion was examined following culture in the absence of IL-1β for the indicated times. This graph demonstrates that a 1-hr exposure of islets to IL-1 induces an inhibition of glucose-stimulated insulin secretion in a time-dependent fashion that mimics the effects of continuous exposure of islets to IL-1β. Reproduced with permission from *J. Clin. Invest.* (Hughes et al., 1990), by the American Society for Clinical Investigation.

lated insulin secretion (following short exposures of islets to 1 U/ml IL-1), and that repeated daily 1-hr exposures of islets to IL-1 for 4 days results in a complete inhibition of glucose-stimulated insulin secretion and a 50% reduction in islet insulin content (Eizirik et al., 1992b). The loss of 50% of islet insulin content is believed to represent a loss of β cells. The inhibitory effect of IL-1 following short 1-hr exposure periods for 4 days indicates that local changes in cytokine concentrations can modulate insulin secretion and, if repeated frequently, can result in islet-cell destruction.

The demonstration that protein synthesis is required for IL-1-induced inhibition of insulin secretion stimulated a number of studies directed at examining proteins whose synthesis is modulated by IL-1. The synthesis of a number of islet proteins has been shown to be stimulated by IL-1 (Hughes et al., 1990a; Sandler et al., 1991; Helqvist et al., 1989; Welsh et al., 1991b). Treatment of islets for

FIGURE 3

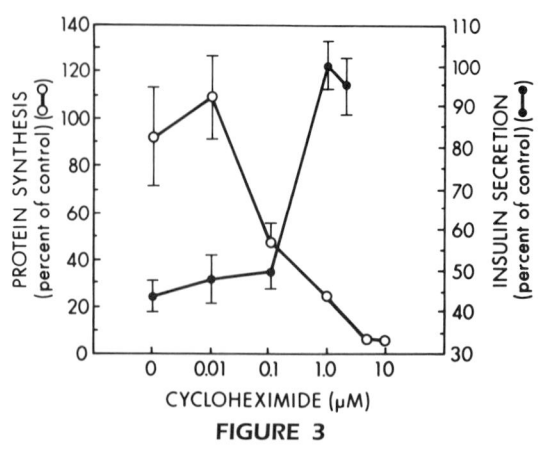

Effects of cycloheximide on IL-1-induced inhibition of glucose-stimulated insulin secretion and islet protein synthesis. Isolated islets were pulsed for 1 hr with 5 U/ml IL-1β, washed to remove IL-1, and then incubated for 8 additional hr in the presence of [^{35}S] methionine and the indicated concentrations of cycloheximide. Following this treatment, insulin secretion and protein synthesis were determined as described previously (Hughes et al., 1990a). The graph demonstrates that IL-1-induced inhibition of glucose-stimulated insulin secretion is prevented by the inhibition of new protein synthesis, indicating that IL-1 induces the expression of a protein which may mediate the inhibitory effects of IL-1 on glucose-stimulated insulin secretion. Reproduced with permission from *J. Clin. Invest.* (Hughes et al., 1990), by the American Society for Clinical Investigation.

24 hr with IL-1 induces the expression of the heat shock protein, hsp 70 (Eizirik et al., 1990; Helqvist et al., 1991; Welsh et al., 1991b). It has been suggested that hsp 70 may attenuate IL-1β-induced islet damage, possibly by refolding damaged proteins similar to its function during heat stress (Pelham, 1989). IL-1 also stimulates the synthesis of a 64-kDa protein (Hughes et al., 1990a) that has been speculated to be either the proposed autoantigen GAD, or a related hsp protein (Jones et al., 1990). Because the molecular weight and isoelectric point of the IL-1-induced 64-kDa protein are similar to GAD (Baekkeskov et al., 1989) GAD6 monoclonal antibodies were used to determine if the IL-1-induced protein is GAD. Even though GAD6 antibodies did not immunoprecipitate the IL-1-induced 64-kDa protein, preliminary evidence suggests that IL-1 may induce the expression of GAD in addition to this 64-kDa protein as determined by immunoprecipitation of GAD from islets treated for 8 hr with IL-1 (J. A. Corbett and M. L. McDaniel, unpublished observation). Although IL-1 modulates the expression of a number of proteins in islets, only more recently has a protein which appears to mediate the inhibitory effects of IL-1 on insulin secretion been identified. IL-1 appears to stimulate the expression of the inducible isoform of nitric oxide synthase (see below) resulting in the overproduction of nitric oxide and

consequent β-cell dysfunction and destruction (Corbett and McDaniel, 1992; Rabinovitch, 1993).

C. Effects of Interleukin 1 on Islet Mitochondrial Function

Islets are uniquely sensitive to perturbations in oxidative metabolism because glucose-stimulated insulin secretion absolutely requires the oxidation of glucose to generate ATP. It is believed that increases in ATP facilitate β-cell depolarization by regulating ATP sensitive K$^+$ channel activity. Impairment of oxidative metabolism severely attenuates the ability of islets to secrete insulin following a glucose challenge because it prevents β-cell depolarization (McDaniel et al., 1988b; Misler et al., 1992). Current evidence indicates that IL-1-induced inhibition of insulin secretion by islets is due to an impairment in the mitochondrial oxidative capacity of islets. Pretreatment of islets for 18 hr with IL-1 results in a potent inhibition of the oxidation of uniformly labeled [^{14}C]-D-glucose to ^{14}CO$_2$, and inhibition of oxygen uptake by islets (Sandler et al., 1987, 1989). These inhibitory effects are believed to be mitochondrial in origin as treatment of islets with IL-1 has little effect on glucose utilization via glycolysis (Sandler et al., 1989), or glucose transport (Corbett et al., 1992a). IL-1 also has no effect on the activities of islet glycolytic enzymes hexokinase, glucokinase, or glyceraldehyde 3-phosphate dehydrogenase (Eizirik et al., 1989; Eizirik, 1988). Importantly, the inhibitory effects of IL-1β on glucose oxidation mimic the time-dependent inhibitory effects of IL-1β on glucose-stimulated insulin secretion (Sandler et al., 1991). Both cycloheximide and actinomycin D also prevent IL-1-induced inhibition of glucose oxidation (Eizirik et al., 1990; Corbett et al., 1992a). These studies have identified an impairment in mitochondrial function as one mechanisms by which treatment of islets with IL-1 results in a potent inhibition of glucose-stimulated insulin secretion.

III. WHAT MEDIATES INTERLEUKIN 1-INDUCED ISLET DYSFUNCTION?

A. Oxygen Radicals

It has been speculated that oxygen free radicals may mediate both the inhibitory effects of IL-1 on glucose-stimulated insulin secretion by islets and islet β-cell destruction associated with diabetes (Mandrup-Poulsen et al., 1990; Oberley, 1988; Rabinovitch, 1993). This idea was originally based on the ability of alloxan to induce diabetes in animals. Alloxan appears to destroy β cells by producing highly reactive oxygen free radicals, hydrogen peroxide, superoxide, and hydroxyl radicals during redox cycling between alloxan and dialuric acid (reduced form of alloxan) in an iron-dependent mechanism (Wilson et al., 1984; Heikkila et al., 1976). Based on the sensitivity of β cells to oxygen free radicals (see below), IL-1-induced islet damage has been proposed to be mediated by either cytokine-induced intracellular production of oxygen radicals by the β-cell or oxygen radical release by islet infiltrating macrophages (Mandrup-Poulsen et al., 1990). In both cases the β cell may be more susceptible to oxygen radical damage because of a lower radical scavenging potential relative to other islet cells (Asayama et al., 1986; Malaisse et al., 1982). Scavengers of oxygen free radicals have been shown to attenuate the development of diabetes following alloxan treatment of rats and mice (Mandrup-Poulsen et al., 1990; Wilson et al., 1984), and destruction of murine islets following allograft transplantation (Mendola et al., 1989). Nicotinamide has been shown to delay the onset of diabetes in children predisposed to insulin-dependent diabetes based on the presence of islet-cell antibodies (Elliott and Chase, 1991). The free radical scavenger Probucol and an antioxidant analogue of Probucol, MDL 29,311, attenuate the development of diabetes in the NOD mouse (Uehara et al.,1991; Heineke et al., 1993). Oxygen radical scavengers also partially prevent both cytokine-induced inhibition of glucose-stimulated insulin secretion by isolated islets, and cytokine-mediated islet-cell lysis (Sumoski et al., 1989; Rabinovitch et al., 1992; Kroncke et al., 1991a; Buscema et al., 1992). These studies implicate oxygen free radicals in the development of islet dysfunction and destruction, however complete prevention of oxygen free radical damage to islets has yet to be demonstrated.

Although oxygen radicals are destructive to islet cells, the inability of nicotinamide, Probucol, and other free radical scavengers to completely prevent cytokine mediated islet destruction suggests that other cytotoxic mechanisms may be involved in cytokine-induced islet-cell lysis. The possible interactions of superoxide with nitric oxide resulting in the generation of peroxynitrite and hydroxyl radicals may contribute to islet-cell lysis. The chemistry of these free radical interactions, and potential biological roles of these toxic radicals are reviewed in this book (see Chapter 1).

B. Nitric Oxide

Southern et al. (1990) were the first to implicate nitric oxide as an effector molecule that may mediate the inhibitory effects of IL-1 on islet function. They showed that pretreatment of islets with IL-1 for 48 hr inhibits glucose-stimulated insulin secretion and that coincubation of islets with the nitric oxide synthase inhibitor, N^G-nitro-L-arginine methylester (NAME), attenuates this inhibition. In the same study IL-1 was shown to induce the formation of nitrite (an oxidative metabolite of nitric oxide), an effect also blocked by NAME. In two independent and concurrent studies, our laboratory (Corbett et al., 1991a) and Welsh et al. (1991a) demonstrated that the nitric oxide synthase inhibitor, N^G-monomethyl-L-arginine (NMMA), completely prevents IL-1-induced inhibition of insulin secretion and nitrite formation by islets (Fig. 4). IL-1 was also shown to stimulate nitrite formation and cGMP accumulation in a time-dependent fashion that mimicked IL-1-induced inhibition of glucose-stimulated insulin secretion by is-

FIGURE 4

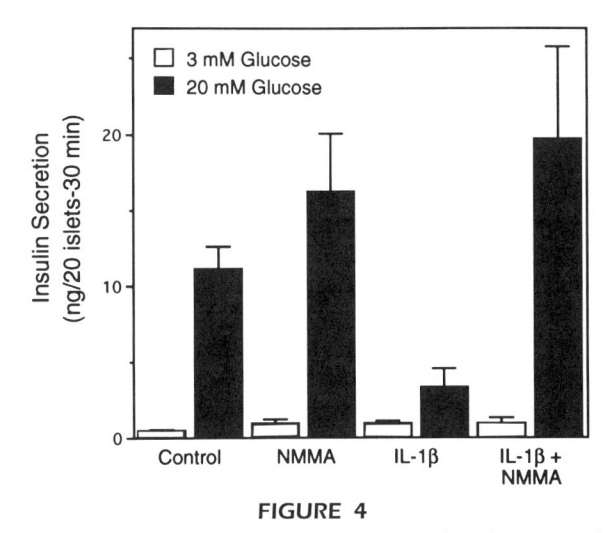

Effects of NMMA on IL-1-induced inhibition of glucose-stimulated insulin secretion by isolated islets. Islets were incubated for 18 hr in the presence of 5 U/ml IL-1β, 0.5 mM NMMA, or IL-1β and NMMA. Insulin secretion was examined as described previously (Corbett et al., 1991a). Results demonstrate that inhibition of nitric oxide synthase activity completely prevents the inhibitory effects of IL-1β on glucose-stimulated insulin secretion by islets. Reproduced with permission from J. Biol. Chem. (Corbett et al., 1991a), from the American Society for Biochemistry and Molecular Biology.

lets (Welsh et al., 1991a; Corbett et al., 1992a). Using EPR (electron paramagnetic resonance) spectroscopy in collaboration with Dr. Jack Lancaster and Michael Sweetland at Utah State University, we confirmed that IL-1β induces the formation of nitric oxide by islets. We showed that IL-1β induces the formation of an axial $g = 2.04$ iron-dithio-dinitrosyl EPR signal in islets identical to that previously described for model iron–nitrosyl complexes (Corbett et al., 1991a, 1992a). NMMA prevented IL-1-induced iron–nitrosyl complex formation, indicating the requirement for nitric oxide synthase activity (Fig. 5). These studies provided the initial evidence implicating nitric oxide as an effector molecule that may mediate cytokine-induced islet dysfunction and provided the first demonstration of an agent (NMMA) that completely prevents cytokine-induced inhibition of glucose-stimulated insulin secretion.

IV. MECHANISM OF NITRIC OXIDE-MEDIATED ISLET DYSFUNCTION AND DESTRUCTION

How does nitric oxide mediate β-cell dysfunction and destruction? The observation of an IL-1β-induced iron–dithio–dinitrosyl complex in islets suggests

FIGURE 5

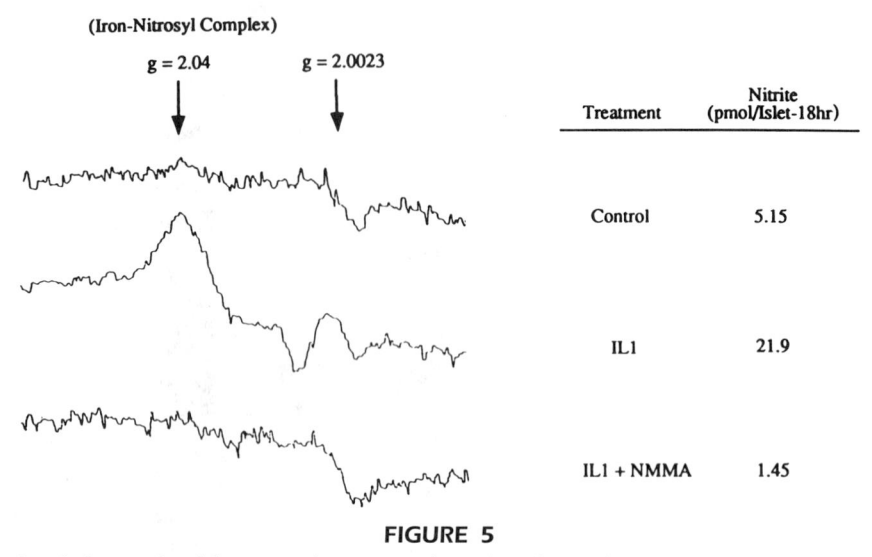

Interleukin 1-induced formation of iron–nitrosyl complexes by rat islets. Rat islets were incubated for 18 hr in the presence or absence of 5 U/ml IL-1β, 0.5 mM NMMA, or IL-1β and NMMA. The islets were isolated and the formation of nitric oxide was examined by EPR spectroscopy as described previously (Corbett et al., 1991a). IL-1 induces the formation of a $g = 2.04$ feature that is characteristic of the formation of iron–nitrosyl complexes, and NMMA prevents the formation of this axial $g = 2.04$ iron–nitrosyl feature. Also shown is the simultaneous formation of nitrite by the same islets used for EPR spectroscopy. Reproduced with permission from *J. Biol. Chem.* (Corbett et al., 1991a), from the American Society for Biochemistry and Molecular Biology.

Treatment	Nitrite (pmol/Islet-18hr)
Control	5.15
IL1	21.9
IL1 + NMMA	1.45

that nitric oxide may mediate the destruction of iron–sulfur proteins. Evidence for this mechanism was initially obtained by Reddy et al. (1982) who showed the formation of EPR detectable iron–nitrosyl complexes following treatment of *Clostridium botulinum* with sodium nitrite. Nitrite treatment of meat is used to prevent botulism (see Chapter 8). The appearance of this iron–nitrosyl feature occurred with the simultaneous loss of an EPR detectable $g = 1.94$ feature from iron–sulfur proteins, providing evidence for nitric oxide-mediated destruction of iron–sulfur centers of iron-containing proteins (Reddy et al., 1982). The inhibitory effects of nitric oxide on iron-containing proteins have been demonstrated for a number of cell types. Cytokine and endotoxin-induced nitric oxide production by rat macrophages potently inhibits mitochondrial aconitase activity and also inhibits electron transport at complexes I and II of the electron transport chain (Granger and Lehninger, 1982; Drapier and Hibbs, 1986; Hibbs et al., 1990). Aconitase activity is partially restored by the addition of ferrous sulfate and the sulfhydryl donors cysteine and thiocyanate (Drapier and Hibbs, 1986), further suggesting that nitric oxide mediates iron–sulfur protein destruction.

Treatment of isolated hepatocytes with authentic nitric oxide inhibits the electron transport chain at complexes I and II, and mitochondrial aconitase activity (Stadler et al., 1991).

The inhibitory actions of IL-1 on glucose metabolism by islets are believed to result from the inhibition of mitochondrial function, specifically the citric acid cycle enzyme, aconitase and possibly the electron transport chain at complexes I and II. IL-1 has no effect on glucose uptake (Corbett et al., 1992a), glycolysis, or the oxidation of acetyl-CoA (Sandler et al., 1991). IL-1 also has no effect on the mitochondrial oxidation of glutamine, which enters the citric acid cycle at α-ketoglutarate, or on the activities of citrate synthase, isocitrate dehydrogenase, or α-ketoglutarate (Sandler et al., 1991). Treament of islets with IL-1, however, results in an inhibition of oxygen uptake (Sandler et al., 1987) and reduced cellular levels of ATP (Sandler et al., 1990; Corbett et al., 1992a).

Dimmeler et al. (1993) have shown that IL-1-induced nitric oxide production stimulates auto-ADP-ribosylation of glyceraldehyde-3-phosphate dehydrogenase in RINm5F cells. ADP-ribosylation appears to inhibit glyceraldehyde-3-phosphate dehydrogenase activity, and this inhibition is attenuated by NMMA. Also, nitric oxide donor compounds induce both auto-ADP-ribosylation and the inhibition of this enzymatic activity. These findings indicate that nitric oxide may also inhibit glycolysis as well as mitochondrial function, suggesting that mitochondrial dysfunction (see below) may not be the only site of nitric oxide-induced inhibition of oxidative metabolism by islets following IL-1 treatment.

A. Mitochondrial Function

The effects of nitric oxide on islet oxidative metabolism were initially demonstrated by Sandler and co-workers (Welsh et al., 1991a), who showed that IL-1β treatment of rat islets results in a 60% inhibition of islet aconitase activity, and an 80% inhibition of glucose oxidation. Both of these effects are prevented by NMMA. These studies have been extended to show similar results for aconitase activity of mouse islets treated with IL-1 (Welsh and Sandler, 1992). We have demonstrated that IL-1β-induced nitric oxide formation inhibits mitochondrial aconitase activity of rat islets by 80%, and also inhibits mitochondrial aconitase activity of rat insulinoma RINm5F cells by approximately 50% (Corbett et al., 1992b). The inhibition of islet mitochondrial aconitase activity was shown to occur simultaneously with the formation of nitrite (Fig. 6). We and others have also shown that IL-1-induced inhibition of mitochondrial glucose oxidation is prevented by both NMMA (Corbett et al., 1992b; Welsh et al., 1991a) and cycloheximide (Corbett et al., 1992b). Inhibition of mitochondrial aconitase and the demonstration of IL-1-induced iron–nitrosyl complex formation suggest that mitochondrial iron–sulfur protein destruction (specifically aconitase, and possibly electron transport chain complexes I and II) partic-

FIGURE 6

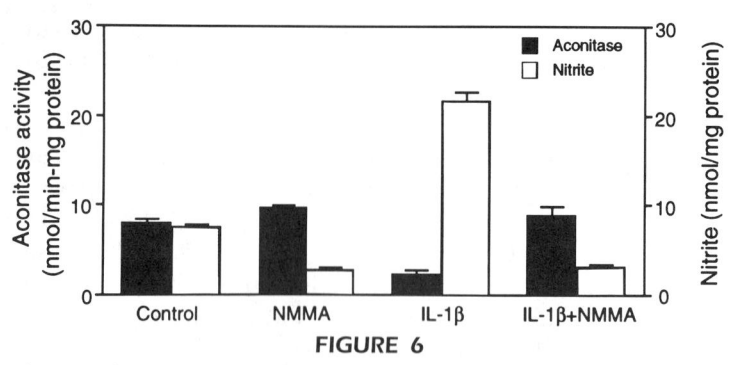

Effects of IL-1β and NMMA on islet-cell mitochondrial aconitase activity. Islets were treated for 18 hr with 5 U/ml IL-1, 0.5 mM NMMA, or IL-1β and NMMA. The islets were isolated and aconitase activity and the simultaneous release of nitrite were determined as described previously (Corbett et al., 1992b). Treatment of islets with IL-1β results in an 80% inhibition of mitochondrial aconitase activity, which is completely prevented by NMMA. IL-1 also stimulates a twofold increase in the level of nitrite released by islets, and NMMA prevents this nitrite formation. Reproduced with permission from J. Clin. Invest. (Corbett et al., 1992b), by the American Society for Clinical Investigation.

ipates in the inhibitory effects of IL-1 on glucose-stimulated insulin secretion by islets.

B. Cyclic GMP Is Not Involved in Interleukin 1-Induced Inhibition of Insulin Secretion by Islets

Nitric oxide is known to bind to the heme moiety of soluble guanylate cyclase. This interaction activates guanylate cyclase and stimulates the accumulation of cGMP (see Chapter 4). Although IL-1 induces the accumulation of cGMP by islets and β cells, cGMP does not appear to participate in the inhibitory effects of IL-1 on insulin secretion (Corbett et al., 1992a). As shown in Fig. 7, under conditions in which cycloheximide (1 μM) inhibits 90% of IL-1-induced protein synthesis (Fig. 3) and prevents the inhibitory effects of IL-1 on glucose-stimulated insulin secretion (Fig. 7a), cycloheximide has no effect on IL-1-induced cGMP accumulation by islets (Fig. 7b). Cycloheximide at a concentration of 10 μM prevents both IL-1-induced cGMP accumulation and the inhibitory effects on glucose-stimulated insulin secretion. The inability of cycloheximide to prevent IL-1-induced cGMP accumulation (at a concentration of 1 μM) is believed to be due to the expression of low levels of inducible nitric oxide synthase (Fig. 3). Treatment of islets with exogenous cGMP or nonhydrolyzable cGMP analogs also fails to inhibit insulin secretion, further indicating that cGMP does not mediate the inhibitory effects of IL-1 on glucose-stimulated insulin secretion by islets (Laychock, 1983; Verspohl et al., 1988).

FIGURE 7

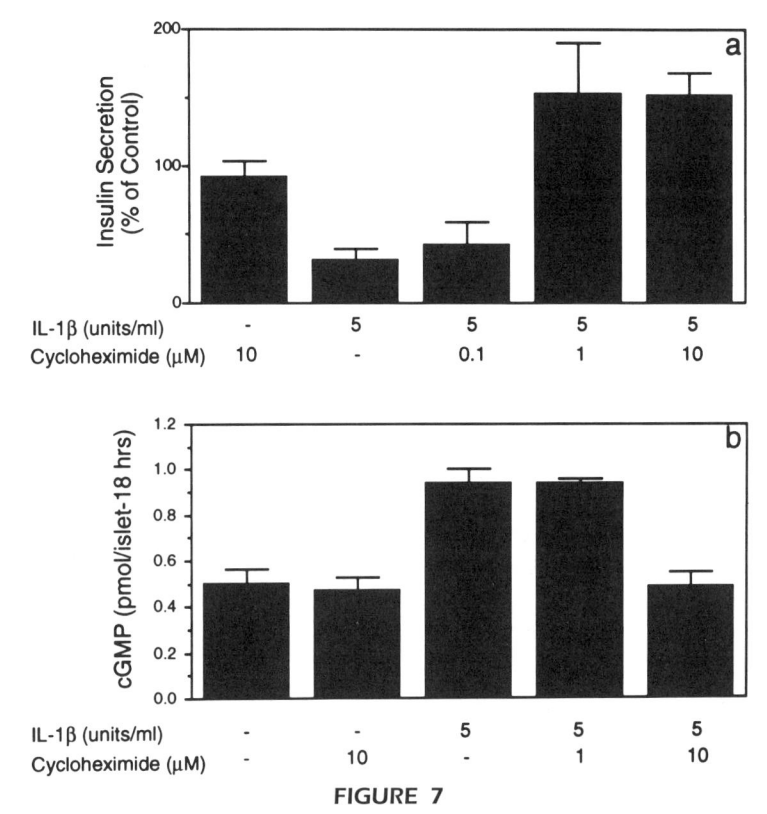

Effects of cycloheximide on IL-1-induced inhibition of insulin secretion and cGMP accumulation by rat islets. Islets were treated for 18 hr with the indicated concentrations of cycloheximide and IL-1β, and then insulin secretion and cGMP accumulation were determined as described previously (Corbett et al., 1992a). Cycloheximide prevents IL-1-induced inhibition of glucose-stimulated insulin secretion (a) and cGMP accumulation (b) in a concentration-dependent fashion. Complete recovery of insulin secretion is observed at a cycloheximide concentration of 1 μM, while it requires 10 μM cycloheximide to prevent IL-1-induced cGMP formation. These results demonstrate that cGMP does not appear to mediate the inhibitory effects of IL-1 on insulin secretion since complete recovery of secretion occurs at a concentration of cycloheximide (1 μM) which does not prevent IL-1-induced cGMP accumulation by islets. Reproduced with permission from *Biochem. J.* (Corbett et al., 1992a), by the Biochemical Society and Portland Press.

V. CELLULAR SOURCE OF INTERLEUKIN 1-INDUCED NITRIC OXIDE FORMATION BY ISLETS OF LANGERHANS

The cellular sources of IL-1-induced nitric oxide production by islets have stimulated considerable controversy (Kolb and Kolb-Bachofen, 1992a; Mandrup-

Poulsen et al., 1993). Islets are a heterogeneous population of approximately 2000 endocrine and nonendocrine cells. The endocrine cell composition includes approximately 60–70% insulin secreting β cells, 15–20% glucagon secreting α cells, and small populations of somatostatin secreting δ cells and pancreatic polypeptide secreting cells (Bonner-Weir, 1989). Islets also contain low levels of nonendocrine cells including macrophages, endothelial cells, fibroblasts, and dendritic cells (Bonner-Weir, 1989; Setum et al., 1991). Macrophages in an activated state produce high levels of nitric oxide (Hibbs et al., 1990), and both endothelial and dendritic cells contain the inducible isoform of nitric oxide synthase (Nathan, 1992). Although macrophages comprise approximately 0.5% of the islet-cell population (~10 macrophages/islet, Setum et al., 1991), Kolb and co-workers have proposed that islet macrophages are the source of IL-1-induced nitric oxide production (Kolb and Kolb-Bachofen, 1992b; Bergmann et al., 1992). Since the β cell is selectively destroyed during IDDM, it is interesting to implicate the β cell as a source of IL-1-induced nitric oxide production.

A. Pancreatic Islet β Cell

To address this controversy we have examined the production of nitric oxide by primary rat β cells purified by fluorescence-activated cell sorting (FACS) (Pipeleers et al., 1985, Wang and McDaniel, 1990). This purification procedure produces two populations of cells, one population that contains greater than 95% β cells, and the other that contains about 80% α cells. The α-cell population also contains some nonendocrine islet cells and approximately 5% β cells. Treatment of FACS purified β cells with IL-1 for 18 hr results in an inhibition of glucose-stimulated insulin secretion which is prevented by NMMA as shown in Fig. 8. The inhibitory effects of IL-1 on insulin secretion by purified β cells are similar to the effects produced by IL-1 on insulin secretion from isolated intact islets (Fig. 4), indicating that nitric oxide may mediate the inhibitory effects of IL-1 on purified β cells. The endocrine β cell appears to be a source of nitric oxide as evidenced by the ability of NMMA to prevent IL-1-induced cGMP accumulation by purified β cells. Treatment of purified α cells with IL-1 has no effect on cGMP levels, indicating that α cells are not a source of IL-1-induced nitric oxide production (Fig. 9; Corbett et al., 1992b).

IL-1 also appears to inhibit insulin secretion by modulating the mitochondrial oxidative metabolism of purified β cells. Treatment of purified β cells for 18 hr with IL-1 results in nearly complete inhibition of the oxidation of [^{14}C]-D-glucose to $^{14}CO_2$ (Fig. 10). This inhibition is completely prevented by NMMA (Corbett et al., 1992b). Treatment of purified α cells with IL-1 has no effect on glucose oxidation, suggesting that the effects of IL-1 are specific to the endocrine β cell (Fig. 10). These findings are further supported by the observation that IL-1 induces iron–nitrosyl complex formation and the accumulation of cGMP in the

5 Role of Nitric Oxide in Autoimmune Diabetes

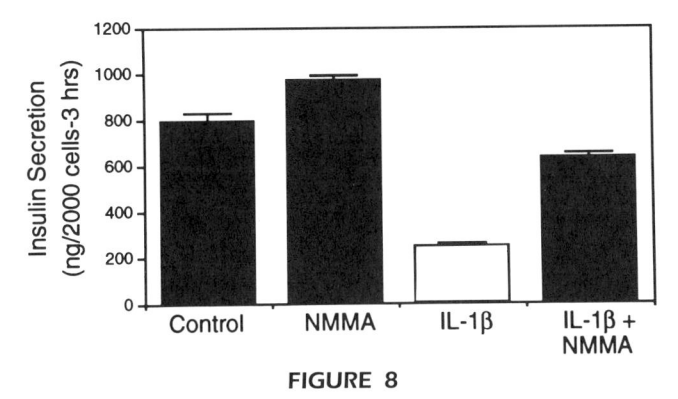

FIGURE 8

Effects of IL-1β on glucose-stimulated insulin secretion by β cells purified by FACS. Islets were isolated, dispersed into single cells, and β cells were then purified by FACS. Purified β cells were incubated for 18 hr with 5 U/ml IL-1β, 0.5 mM NMMA, or IL-1β and NMMA and then insulin secretion was examined as described previously (Corbett et al., 1992b). The IL-1 pretreatment results in a potent inhibition of glucose-stimulated insulin secretion which is significantly attenuated by treatment of purified β cells with NMMA in addition to IL-1. Reproduced with permission from J. Clin. Invest. (Corbett et al., 1992b), by the American Society for Clinical Investigation.

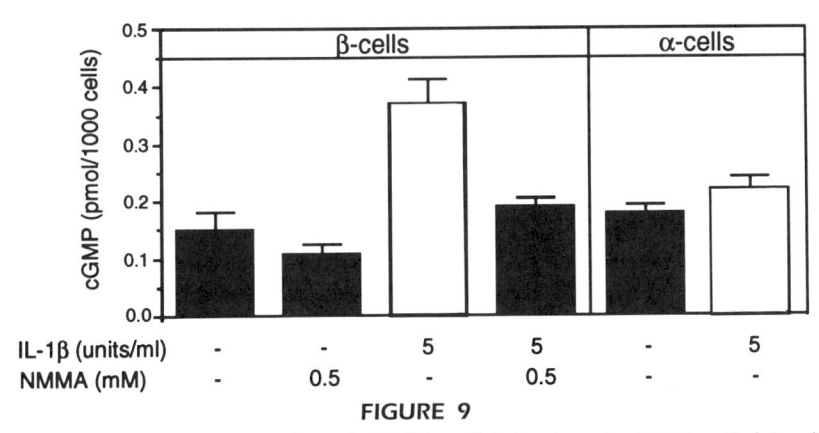

FIGURE 9

Effects of IL-1β on cGMP accumulation by FACS-purified β and α cells. FACS-purified β and α cells were obtained from the same FACS. The cells (5 × 10^5) were incubated for 18 hr with 5 U/ml IL-1β and 0.5 NMMA as indicated, and then incubated for an additional 3 hr in the presence of 1 mM isobutylmethylxanthine (phosphodiesterase inhibitor). The cells were isolated by centrifugation and the cellular content of cGMP was determined by radioimmunoassay. Results demonstrate that IL-1 induces the accumulation of cGMP by purified β cells which is prevented by NMMA, and that IL-1 has no effect on cGMP levels of purified α cells. The slight stimulation of cGMP levels by IL-1 in α cells is consistent with the level of β cell contamination of the purified α cell preparation (~10%). Reproduced with permission from the J. Clin. Invest. (Corbett et al., 1992b), by the American Society for Clinical Investigation.

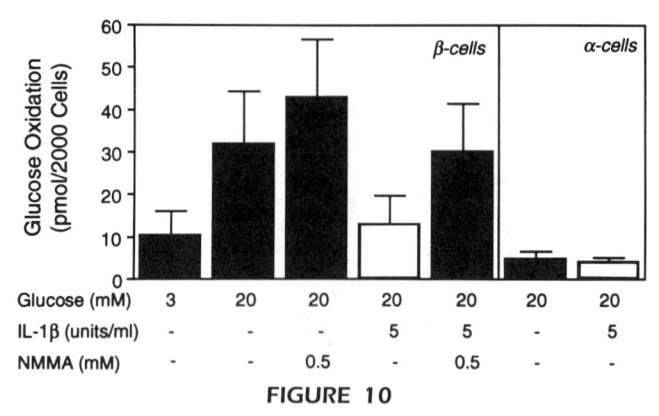

FIGURE 10

Effects of IL-1β and NMMA on glucose oxidation by FACS-purified β and α cells. FACS-purified β and α cells (150,000 cells/condition) were pretreated for 18 hr with 5 U/ml IL-1β or 0.5 mM NMMA as indicated, the cells were then isolated and the oxidation of [^{14}C]-D-glucose to $^{14}CO_2$ was determined as described previously (Corbett et al., 1992c). Results show that IL-1 completely prevents glucose oxidation by purified β cells, and has no effect on the ability of purified α cells to oxidize glucose. NMMA also prevents the inhibitory effects of IL-1 on glucose oxidation by purified β cells. Reproduced with permission from the J. Clin. Invest. (Corbett et al., 1992b), by the American Society for Clinical Investigation.

rat insulinoma cell line RINm5F, which represents a homogenous population of β cells (Corbett et al., 1991b, 1992b).

IL-1-Induced expression of the inducible isoform of nitric oxide synthase (iNOS) by purified β cells has been confirmed. We have immunoprecipitated a 130-kDa protein from both FACS purified β cells and RINm5F cells treated for 18 hr with IL-1 using polyclonal antiserum specific for the C-terminal 27 amino acids of mouse macrophage iNOS (Corbett et al., 1994; Corbett and McDaniel, 1995). This protein was absent in untreated control β cells, or β cells treated with IL-1 in the presence of the transcriptional inhibitor, actinomycin D. We were unable to immunoprecipitate iNOS from purified α-cell populations treated with or without IL-1 (Corbett et al., 1994). Furthermore, IL-1 has been shown to stimulate expression of iNOS mRNA in the hamster insulinoma cell line (HIT) and the rat insulinoma cell line RINm5F, as determined by Northern blot analysis using a cDNA probe derived from mouse macrophage iNOS (Eizirik et al., 1992a, 1993; Corbett et al., 1994). These results strongly suggest that β cells, which are selectively destroyed during IDDM, are a source of IL-1β-induced nitric oxide production resulting from IL-1-induced expression of the inducible isoform of nitric oxide synthase.

B. Resident Islet Macrophage

In contrast to these findings, Kolb and co-workers have suggested that resident macrophage and endothelial cells are the sole sources of IL-1-induced nitric oxide formation by islets (Kolb and Kolb-Bachofen 1992a, b, 1993). They find that isolated islets produce nitric oxide in response to IL-1, but dispersed islet cells allowed to reaggregate (which are believed to be devoid of endothelial cells and macrophages) do not generate nitric oxide in response to IL-1 (Bergmann et al., 1992). These authors conclude that IL-1β-induces the formation of nitric oxide by the nonendocrine cells of the islet and that the major source of nitric oxide is resident islet macrophages. The notion that IL-1 induces nitric oxide production by rat macrophages is not consistent with the findings of Hibbs and co-workers, who have shown that macrophage production of nitric oxide requires two signals [TNF, IFN, or lipopolysaccharide (LPS)] and that IL-1 alone or in combination with TNF or IFN does not stimulate nitric oxide production by murine macrophages (Amber et al., 1991; Drapier et al., 1988; Hibbs et al., 1990). In addition, we have been unable to demonstrate IL-1 stimulated nitric oxide production from rat peritoneal macrophages, or from the murine macrophage cell line RAW 264.7 (J. A. Corbett and M. L. McDaniel, unpublished observation). Based on these findings, we believe that IL-1 induces the formation of nitric oxide by the endocrine β cell and not resident islet macrophages. It is possible, however, that other nonendocrine islet cell types produce nitric oxide in response to IL-1, and this production may also participate in IL-1-induced islet destruction.

VI. SIGNALING MECHANISM OF INTERLEUKIN 1-INDUCED EXPRESSION OF NITRIC OXIDE SYNTHASE

Limited information is known concerning the signaling mechanism induced by IL-1 following receptor binding, and how these signals stimulate the expression of iNOS. Indirect evidence supports the presence of IL-1 receptors on islet cells. Binding studies have revealed the presence of high and low affinity IL-1 binding sites on HIT cells (Hammonds et al., 1990). Also, the interleukin 1 receptor antagonist protein (IRAP) prevents the inhibitory effects of IL-1 on glucose-stimulated insulin secretion by islets and RINm5F cells (Eizirik et al., 1990, 1991, 1992b). We have shown that IRAP prevents the inhibitory effects of IL-1 on glucose-stimulated insulin secretion, and IL-1-induced expression of iNOS by islets and FACS purified β cells (Corbett and McDaniel, 1995). These studies provide support for the presence of IL-1 receptors on the islet β cell. The lack of an inhibitory effect of IL-1 on glucose oxidation and cGMP formation (Corbett et al., 1992b), and the inability of IL-1 to stimulate iNOS expression by purified islet α cells (Corbett et al., 1994) indicates that the α cell may not express IL-1 receptors.

Following receptor binding, IL-1 induces the expression of the transcriptional regulator c-fos in islets and FACS purified β cells, while IL-1 does not induce c-fos expression in purified populations of α cells (Hughes et al., 1990b).

In addition we have shown that IL-1 also induces the expression of c-*jun* in both islets and RINm5F cells, and have obtained preliminary evidence for nuclear factor RB (NF-κB) activation (J. A. Corbett and M. L. McDaniel, unpublished observation). These three transcriptional regulators alone or in combination are believed to participate in IL-1-induced expression of nitric oxide synthase by the islet β cell. Importantly, Nathan and co-workers have shown the presence of NF-κB response elements upstream of the mouse macrophage iNOS gene (Xie et al., 1993).

The mechanism by which IL-1 binding induces the expression or activation of these transcriptional regulators is unknown. It is speculated that IL-1-induced expression of c-*fos* (and possibly other transcriptional regulators) may in turn regulate the expression of nitric oxide synthase. Studies have examined the potential role of G proteins (Eizirik et al., 1991), cAMP formation (Sandler et al., 1990), and phospholipid products (Hughes et al., 1989) as potential signaling second messengers following IL-1 receptor binding. However, none of these potential signaling systems appear to be involved in IL-1-induced inhibition of insulin secretion, or expression of nitric oxide synthase by islets. We have demonstrated that the tyrosine kinase inhibitors, genistein and herbimycin A, prevent IL-1-induced inhibition of glucose-stimulated insulin secretion and nitric oxide formation (as determined by nitrite formation) by isolated islets and RINm5F cells (Corbett et al., 1993a, 1994). These inhibitors also prevent IL-1-induced expression of nitric oxide synthase by isolated islets and FACS purified β cells as determined by immunoprecititation (Corbett et al., 1994). These results suggest that β-cell IL-1 receptors may be linked to a tyrosine kinase, and that IL-1 binding may stimulate the activation of this kinase and induction of intracellular signaling processes which result in the transcription of nitric oxide synthase. The role of c-*fos* or other transcriptional regulators in the induction of iNOS expression is unknown, but they are induced by IL-1 in islets and β cells. Determination of the upstream regulatory elements of β-cell nitric oxide synthase should provide invaluable insights to decipher the molecular regulatory mechanisms of nitric oxide synthase expression by the β cell.

VII. CONSTITUTIVE NITRIC OXIDE SYNTHASE AND INSULIN SECRETION

As reviewed in Chapter 3 of this book, nitric oxide is the product of the enzymatic oxidation of one of the guanidino nitrogen groups of L-arginine to the free radical nitric oxide and L-citrulline (Marletta et al., 1988). At present there appears to be at least three distinct isoforms of nitric oxide synthase that are primarily differentiated at the level of gene expression (Schmidt, 1992; Nathan,

1992). The constitutive isoform is a Ca^{2+}- and calmodulin-dependent enzyme that produces low levels of nitric oxide in response to physical- or receptor-mediated stimuli (Moncada et al., 1991). The expression of the other isoform of nitric oxide synthase is induced by cytokines and by endotoxin and this isoform produces much higher levels of nitric oxide following induction (Hibbs et al., 1990). This chapter has emphasized islet nitric oxide production by inducible nitric oxide synthase, but islets also contain the constitutive isoform of nitric oxide synthase. Laychock et al. (1991) initially showed that both glucose and L-arginine stimulate cGMP formation by rat islets and RINm5F cells, and that these effects are prevented by NMMA. It was also shown that inhibitors of soluble guanylate cyclase reduce insulin release from both islets and RINm5F cells. Schmidt et al. (1992) have extended these findings to show that the constitutive isoform of nitric oxide synthase is present in islets as determined by immunohistochemical staining and Western blot analysis using whole pancreatic tissue and the hamster insulinoma cell line HIT, respectively. The β cell appears to be the islet cellular source of constitutive nitric oxide synthase as demonstrated by immunohistochemical colocalization of constitutive nitric oxide synthase with insulin (Corbett et al., 1993c).

The role of the constitutive isoform of nitric oxide synthase during insulin secretion is controversial (Vincent, 1992). Schmidt et al. (1992) reported that NMMA and NAME inhibit glucose-stimulated insulin secretion by HIT cells, suggesting that nitric oxide may mediate glucose-stimulated insulin release by islets. In contrast a second report has shown that NMMA and NAME stimulate insulin secretion by isolated islets suggesting that the production of nitric oxide by the constitutive isoform of nitric oxide synthase does not participate in glucose-stimulated insulin secretion (Jansson and Sandler, 1991). Also both D- and L-arginine stimulate insulin secretion by islets, but D-arginine is not a substrate for nitric oxide synthase (Jansson and Sandler, 1991; Panagiotidis et al., 1992). Arginine is believed to stimulate insulin secretion by direct depolarization of the β cell due to its charge and not as a consequence of its metabolism (Misler et al., 1992). Jones et al. (1992) and Green et al. (1993) have concluded that nitric oxide generation by the constitutive isoform of nitric oxide synthase is not required for insulin secretion, but may participate in long-term glucose-dependent increases in cGMP content of islets. This conclusion is consistent with our findings that NMMA slightly stimulates both glucose-induced insulin secretion and mitochondrial aconitase activity of isolated islets following prolonged incubation of islets with the nitric oxide synthase inhibitor (Corbett et al., 1991a, 1992b). Nitric oxide production by the constitutive isoform of nitric oxide synthase in islets may function by downregulating oxidative metabolism of islets. This would be consistent with NMMA stimulation of both mitochondrial aconitase activity and insulin secretion by islets and suggests a novel role for constitutive nitric oxide synthase in the regulation of islet-cell metabolic function.

VIII. EFFECTS OF ACTIVATED MACROPHAGES ON ISLET FUNCTION

Although macrophages do not appear to participate in IL-1 induced nitric oxide mediated islet dysfunction, they kill isolated islet cells following activation *in vivo*. Kroncke et al. (1991b) have demonstrated that peritoneal macrophages isolated from Wistar rats 4 days after the injection of *Corynebacterium parvum* produce high levels of nitric oxide. When these activated macrophages are cocultured with islet cells at an effector to target ratio of 2:1 islet-cell lysis is observed (Kroncke et al., 1991b). NMMA prevents macrophage-mediated cell lysis, but antisera to either IL-1 or TNF are unable to prevent cell lysis. These results demonstrate that macrophage production of nitric oxide is cytotoxic to islet cells and does not require the presence of the cytokines TNF and IL-1.

The production of oxygen free radicals by macrophages has also been proposed to be cytotoxic to islet cells. Nicotinamide (a free radical scavenger and poly ADP-ribose synthase inhibitor) prevents activated macrophage mediated islet-cell lysis (Kroncke et al., 1991a). Nicotinamide also prevents damage to islet-cell mitochondria by oxygen free radicals. However, the interpretations of these experiments have been complicated by the demonstration that nicotinamide also inhibits the induction of iNOS stimulated by TNF in L929 cells, and IL-1-induced expression of iNOS in RINm5F cells (Hauschildt et al., 1992; Cetkovic-Cvrlje et al., 1993). It is also possible that interaction of both nitric oxide and oxygen radicals in the vicinity of the islet produces substances which are more toxic to β cells than nitric oxide or hydroxyl radicals alone, such as peroxynitrite (Beckman et al., 1990). More work is needed to determine the chemistry of nitric oxide and oxygen radical interactions, but the ability of oxygen radicals and nitric oxide to induce β-cell damage suggests that the β cell is very susceptible to free radical damage.

IX. PROPOSED MECHANISM BY WHICH INTERLEUKIN 1 INDUCES β-CELL DYSFUNCTION AND DESTRUCTION

We have proposed a mechanism by which IL-1 exerts its deleterious effects on islet function and viability (Fig. 11; Corbett et al., 1992). In this proposed mechanism, IL-1 is released by macrophages during the initial stages of islet infiltration. IL-1 binds to a specific IL-1 receptors on the β cell activating a tyrosine kinase. Tyrosine kinase phosphorylation stimulates second messengers to induce the expression of c-*fos*, c-*jun*, the activation of NF-κB, and possibly other early transcriptional regulators. These early-immediate transcriptional response elements may activate or stimulate the expression of inducible nitric oxide

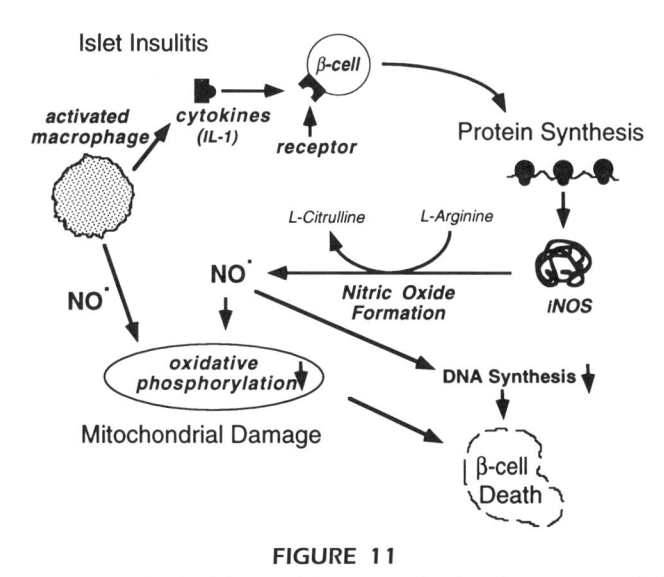

FIGURE 11
Proposed model of IL-1-induced inhibition of glucose-stimulated insulin secretion and islet destruction.

synthase. Nitric oxide is then overproduced by the oxidation of L-arginine to L-citrulline via the L-arginine-dependent nitric oxide synthase pathway. Nitric oxide interacts with iron–sulfur centers of enzymes resulting in the destruction of these clusters and the inhibition of enzymatic activity. This has been demonstrated for islet aconitase, and the inhibitory effects of nitric oxide on the electron transport chain complexes I and II have been demonstrated for other cell types (Hibbs et al., 1990). Nitric oxide may also inhibit ribonucleotide reductase and thereby prevent DNA synthesis (Hibbs et al., 1990). IL-1 has been shown to inhibit islet DNA synthesis (Hansen et al., 1988). The effects of NMMA (or other inhibitors of nitric oxide synthase) on this inhibition, however, have not been investigated. The inhibition of mitochondrial function and DNA synthesis may ultimately result in β-cell destruction.

IDDM is a disease in which β cells of islets are selectively destroyed. How does a nonspecific effector molecule (nitric oxide) mediate the selective destruction of the β cell? We propose that the specificity of cytokine interactions with the β cell, and the intrinsic sensitivity of the β cell to oxidative damage relative to other endocrine cells may impart this selective destruction. In rats the effects of IL-1 appear to be specific to the β cell. IL-1 induces iNOS expression and the overproduction of nitric oxide by β cells, but does not induce expression of iNOS or nitric oxide formation in other islet endocrine or nonendocrine cells of the islet (Corbett et al., 1992b). IL-1-induced nitric oxide production by β cells results in inhibition of insulin secretion due to an impairment of mitochondrial function. Production of toxic radicals (nitrogen and oxygen) by other cellular sources have also been shown to kill islet cells in a nonspecific fashion (Kroncke et al., 1991a, b).

Macrophage involvement in this process is also required. Macrophages are one of the earliest cells to infiltrate into the islet in animal models of IDDM. IL-1 released from these infiltrating macrophages may trigger the induction of nitric oxide by the β cell causing β-cell dysfunction and destruction. A subpopulation of β cells may be initially destroyed and release autoantigens. Macrophages present in the islet then process and present antigens, which trigger an autoimmune reaction that results in T-cell-mediated islet destruction. In this model, nitric oxide may mediate the initial destruction of β cells and may also be involved in the secondary wave of destruction followintg antigen presentation.

X. ANIMAL MODELS OF INSULIN-DEPENDENT DIABETES MELLITUS

A. Low Dose Streptozotocin

In vitro studies have provided valuable evidence implicating nitric oxide as an effector molecule which may participate in β-cell destruction. *In vivo* studies also support a role for nitric oxide-mediated β-cell destruction. Liew and co-workers initially demonstrated that NMMA prevents diabetes in the low dose streptozotocin (STZ) model of IDDM (Lukic et al., 1991). In this model, male mice injected for 5 days with STZ become hyperglycemic approximately 5 days after the last injection. STZ produces insulitis characterized by infiltrating lymphocytes and macrophages and β-cell necrosis. Using this model, Lukic et al. (1991) demonstrated that five daily low dose injections of STZ (40–45 mg/kg body weight i.p.) followed by five daily injections with L-NMMA (5 mg/kg body weight) significantly reduced hyperglycemia in these animals. Injections with the inactive enantiomer D-NMMA did not prevent hyperglycemia induced by STZ injections. These results have been confirmed by Kolb et al. (1991), suggesting that nitric oxide may participate in β-cell destruction in this animal model of IDDM.

Nitric oxide may be the active moiety of STZ that induces diabetes in this animal model. STZ contains a nitroso moiety and may release nitric oxide by a process analogous to the nitric oxide donor compounds SIN-1 and nitroprusside. Turk et al. (1993) have shown that incubation of rat islets with STZ at concentrations that impair insulin secretion results in the generation of nitrite and the accumulation of cGMP. STZ also inhibits mitochondrial aconitase activity of islets to a degree similar to that achieved by IL-1. These findings provide the first evidence that STZ impairs islet function by liberating nitric oxide.

B. Nonobese Diabetic Mouse

In collaboration with Drs. Emil Unanue and Paul Lacy at Washington University School of Medicine we have examined the role of nitric oxide in the development of diabetes in the NOD mouse. We have used an adoptive transfer protocol (see Section I, A, Wicker et al., 1986) in which splenocytes from diabetic female NOD mice are injected into the tail vein of lightly irradiated nondiabetic male NOD mice. This procedure induces diabetes in the male recipient mice approximately 10 days following transfer. Using this adoptive transfer model, we have examined the production of nitrite and the effects of the transfer process on insulin secretion by islets isolated at days 3, 6, 9, and 13 after splenocyte transfer (Corbett et al., 1993d). Normal insulin secretion was observed from islets isolated 3 days after adoptive transfer of splenocytes. However, at days 6, 9, and 13, a 40–60% inhibition of glucose-stimulated insulin secretion was observed. Importantly, islets isolated from recipient male mice produced nitric oxide at days 6, 9, and 13 following transfer, but at day 3 following transfer isolated islets did not produce nitrite. At day 13 after adoptive transfer a 50% reduction in the number of islets isolated was observed, suggesting that 50% of the islets had been destroyed by the transfer procedure. Also, lymphocyte infiltration into NOD mouse islets was first observed at day 6 following transfer, the same time in which islets began to produce nitric oxide, and inhibition of insulin secretion was initially observed. These results indicate that NOD mouse islets produce nitric oxide in a time-dependent manner that mimics the time-dependent lymphocyte infiltration and inhibition of insulin secretion from NOD mouse islets following transfer of diabetes. Treatment of recipient male NOD mice with two daily injections (6 mg/mouse/day) of aminoguanidine, a selective inhibitor of the inducible isoform of nitric oxide synthase (Corbett et al., 1991b; Misko, 1993) delayed the development of diabetes in this model by 7–10 days (Corbett et al., 1993d), suggesting that nitric oxide plays an early role in diabetes in this animal model. Also, Kleemann et al. (1993) have shown the presence of iNOS mRNA in the pancreas of prediabetic BB rats. These results implicate the involvement of nitric oxide in the development of diabetes in both the NOD mouse and BB rat models of autoimmune diabetes, but the cells which produce nitric oxide, and the effects of *in vivo* inhibition of nitric oxide production need to be more fully characterized.

XI. NITRIC OXIDE AND ISLET INFLAMMATION

β-cell destruction during the development of IDDM is characterized by an inflammatory reaction which occurs in and around islets. Products of cyclooxygenase (COX), a heme-containing enzyme that catalyzes the first reaction in the biosynthetic pathway responsible for the production of the inflammatory prostaglandins, prostacyclin and thromoxane, are believed to participate in inflammatory processes (Needleman et al., 1986). Nitric oxide is known to interact with heme-containing enzyme (see Chapter 4). Binding of nitric oxide to the heme group of soluble guanylate cyclase activates this enzyme and stimulates the accumulation of cGMP. In 1972 Yonetani et al. first demonstrated that nitric oxide binds to the heme moiety of COX. This interaction was used to determine the electromagnetic properties and heme coordination of COX (Karthein et al., 1987). The effects of nitric oxide on the activity of COX have proven to be paradoxical. Nitric oxide appears to both stimulate prostaglandin E_2 (PGE_2) production by murine macrophage RAW 264.7 cells, fetal fibroblasts (Salvemini et al., 1993), and norepinephrine-stimulated hypothalamic fragments (Rettori et al., 1992, 1993), and to inhibit COX products produced by resident liver macrophages (Stadler et al., 1993).

Similar to nitric oxide synthase, constitutive and inducible isoforms of COX have been identified. We have examined the effects of nitric oxide on both constitutive COX and IL-1-induced COX activity (Corbett et al., 1993e). Treatment of isolated rat islets with IL-1 results in a similar time-dependent accumulation of both PGE_2 and nitrite. NMMA completely prevents IL-1-induced nitrite formation and attenuates IL-1-induced PGE_2 production by rat islets, but has no inhibitory effects on the expression of either inducible nitric oxide synthase or COX. The addition of exogenous arachidonic acid to islets either untreated or treated with IL-1 stimulates a greater than 10-fold increase in the production of PGE_2 by islets, and NMMA significantly attenuates this PGE_2 production. Also the nitric oxide donor compound SIN-1 stimulates a greater than 10-fold increase in the production of PGE_2 by islets either untreated or cultured with IL-1, and this effect is attenuated by hemoglobin (J. A. Corbett and M. L. McDaniel, unpublished observation). These findings indicate that nitric oxide may directly activate both constitutive COX and inducible COX, and implicate a role for nitric oxide in both β-cell destruction and the inflammatory reaction that occurs during the development of autoimmune diabetes.

XII. NITRIC OXIDE PRODUCTION BY HUMAN ISLETS

Although nitric oxide production has been extensively examined in rodents, the generation of nitric oxide by human tissue has been difficult. Nussler et al. (1992) have shown that human hepatocytes produce nitric oxide in response to IL-1, TNF, IFN, and LPS. Nitric oxide, produced by human monocytes treated with LPS, may play an important role in liver disease associated with alcoholism. Monocytes isolated from alcoholics with liver disease produce approximately the same level of nitric oxide as monocytes from control subjects simulated with LPS

(Hunt and Goldin, 1992). Schmidt and Murad (1991) have purified the constitutive isoform of nitric oxide synthase from human brain. *In vivo* evidence for the production of nitric oxide by human inducible nitric oxide synthase has come from studies by Hibbs *et al.* (1992), who demonstrated that IL-2 therapy increases serum nitrate levels 10-fold and urinary nitrate excretions 6.5- to 9-fold. Using L-[*guanidino*-^{15}N]arginine as a tracer, IL-2-induced nitrate production was determined to be derived from the terminal guanidino nitrogen of L-arginine. Similar results have also been obtained by Ochoa *et al.* (1991, 1992) who showed increased nitrogen oxide levels in patients after trauma, during sepsis, and after tumor immunotherapy. Although these studies have demonstrated the formation of nitric oxide by the human inducible isoform of nitric oxide synthase, few studies have shown that cytokines induce the *in vitro* production of nitric oxide by human macrophages (Denis, 1991; Munoz-Fernandez *et al.*, 1992).

If β-cell production of nitric oxide participates in IDDM, human islets must produce nitric oxide in response to cytokines. We have shown that a combination of cytokines (IL-1, IFN, and TNF) induce the formation of nitric oxide by isolated human islets (Corbett *et al.*, 1993b). The formation of nitric oxide has been demonstrated by cytokine-induced cGMP accumulation, nitrite formation, and EPR-detectable iron–nitrosyl complex formation (Fig. 12), all of which were prevented by NMMA. The cytokine combination of IFN and IL-1 are required for nitrite production, while TNF potentiates IL-1 and IFN-induced nitrite formation by human islets. The cytokine combination of IL-1, TNF, and IFN also influences the physiological function of insulin secretion by human islets. Low concentrations of this cytokine combination slightly stimulate insulin secretion, while high concentrations inhibit insulin secretion, similar to the concentration-dependent effects of IL-1 on rat islet function. NMMA partially prevents the inhibitory effects of this cytokine combination on insulin secretion from human islets, suggesting that nitric oxide may participate in β-cell dysfunction associated with IDDM.

We believe that the β cell is a source of nitric oxide production by human islets because: (1) IL-1 and IFN-induced nitric oxide production by human macrophages has not been clearly demonstrated; (2) the cytokine combination of IL-1, IFN, and TNF induces the formation of nitric oxide by human islets either freshly isolated or cultured for 7 days at 25°C (a procedure which removes ~80–90% of nonendocrine cells from the islet) and also by islets cryoperserved; and (3) NADPH–diaphorase staining reveals that approximately 60–70% of human islet cells treated with cytokines stain for NADPH–diaphorase (J. A. Corbett and M. L. McDaniel, unpublished data). This staining procedure has been shown to colocalize with nitric oxide synthase in a number of cells including rat islets (Corbett *et al.*, 1993c), and nitric oxide synthase has been demonstrated to contain NADPH–diaphorase enzymatic activity (Dawson *et al.*, 1991; Hope *et*

FIGURE 12

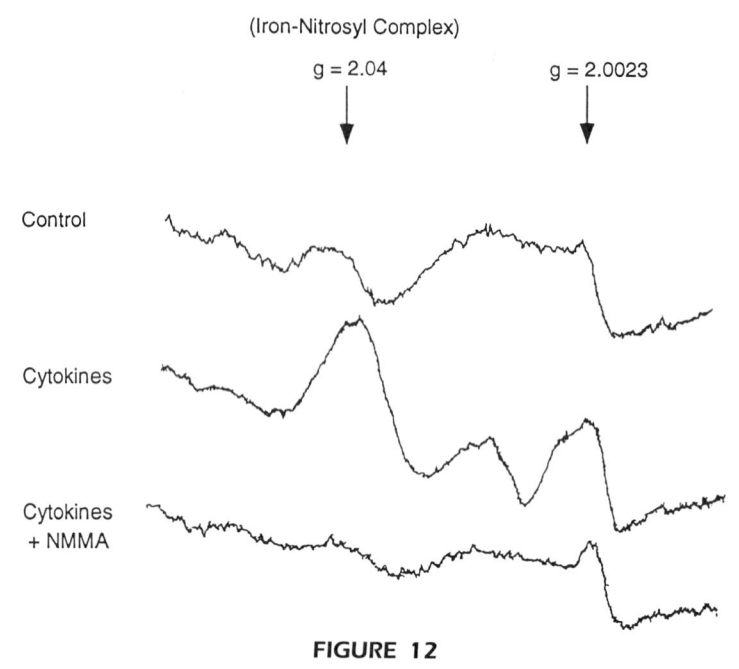

Effects of cytokines on the formation of nitric oxide by human islets as determined by EPR spectroscopy. Human islets were treated for 18 hr with 75 U/ml IL-1β, 3.5 nM TNF-α, and 750 U/ml IFN-γ, the islets were then isolated, and EPR spectroscopy was performed as described previously (Corbett et al., 1993b). Cytokine induced nitric oxide formation is demonstrated by the generation of an EPR detectable $g = 2.04$ iron–nitrosyl complex which is prevented by 0.5 mM NMMA. Reproduced with permission from *Proc. Natl. Acad. Sci. U.S.A.* (Corbett et al., 1993b).

al., 1991). Based on these results, we suggest that cytokine-induced nitric oxide formation by human islet β cells may participate in the onset of IDDM in a manner similar to that characterized by *in vitro* studies using rodent islets.

XIII. INTERVENTIONS

Nitric oxide has multiple functions depending on the amount of nitric oxide generated, the enzymatic source of nitric oxide (inducible versus constitutive), and the site at which nitric oxide is generated. If nitric oxide mediates or participates in IDDM, then inhibition of nitric oxide generation by the inducible isoform of nitric oxide synthase should delay or prevent the onset of IDDM. Many inhibitors have been developed to block nitric oxide synthase activity, however only more recently has an inhibitor been identified that is selective for the inducible isoform. In collaboration with Ronald Tilton and Joseph Williamson at Washington University School of Medicine, we have demonstrated that aminoguanidine is a potent and selective inhibitor of the inducible isoform of nitric oxide synthase, while having minimal effects on the regulation of blood pressure by constitutive nitric oxide synthase (Corbett et al., 1991b). These findings have been extended by Misko et al. (1993) who demonstrated that aminoguanidine is 20- to 40-fold more effective at inhibiting the enzymatic activity of inducible nitric oxide synthase purified from a macrophage cell line (RAW 264.7), than constitutive nitric oxide synthase purified from rat brain. Aminoguanidine is an analogue of L-arginine in that it contains two equivalent guanidino nitrogen groups (see structures, Fig. 13). It also contains a hydrazine moiety which is believed to confer its selectivity for the inducible isoform of nitric oxide synthase. Replacement of the hydrazine moiety with a methyl group (methyl guanidine) results in complete loss of selectivity for inducible nitric oxide synthase (Tilton et al., 1993). Methylguanidine is also 100-fold less effective than aminoguanidine at inhibiting both constitutive and inducible nitric oxide synthase enzymatic activities. We have evaluated the effects of aminoguanidine on the development of diabetes in the NOD mouse and have shown that aminoguanidine delays the onset of diabetes in this model by 7–10 days (Corbett et al., 1993d).

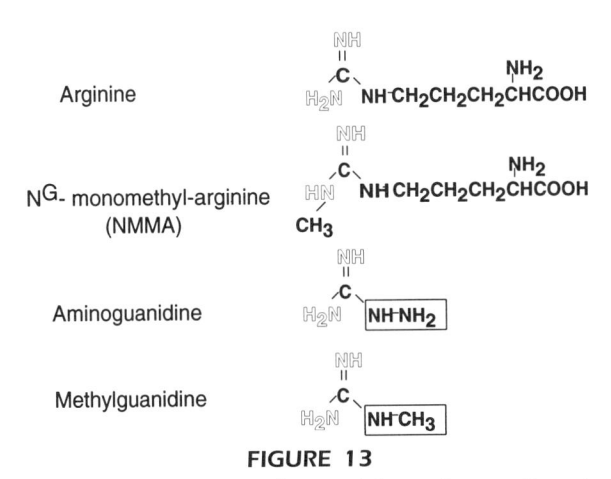

FIGURE 13

Structures of arginine and analogs of arginine that are inhibitors of nitric oxide synthase. Highlighted in the box is the hydrazine moiety of aminoguanidine that appears to confer selectivity of this inhibitor for the inducible isoform of nitric oxide synthase.

We have also evaluated the effectiveness of the corticosteroid, dexamethasone, to prevent IL-1-induced inhibition of glucose-stimulated insulin secretion and nitric oxide production by isolated islets. Dexamethasone, which prevents the expression of the inducible isoform of nitric oxide synthase (Radomski et al., 1990; Rosa et al., 1990), blocks both IL-1-induced inhibition of glucose-stimulated insulin secretion and nitrite formation by rat islets (Corbett et al., 1993d; Eizirik et al., 1993). Dexamethasone also attenuates IL-1-induced islet destruction, while inhibitors of nitric oxide synthase (NMMA, and aminoguanidine) completely prevent islet destruction (Corbett et al., 1993c). These studies suggest that IL-1-induced islet destruction may be mediated by nitric oxide, and that a combination of corticosteroid treatment with inhibitors specific for inducible nitric oxide synthase may represent a new therapeutic strategy for preventing the overproduction of nitric oxide *in vivo*.

XIV. CONCLUSIONS

Elucidation of the autoimmune-mediated mechanisms responsible for β-cell destruction associated with IDDM should allow for the development of therapeutic agents which may prevent or delay the onset of diabetes. *In vitro* studies have provided valuable evidence implicating the over production of nitric oxide as one mechanism that may mediate β-cell destruction during the development of autoimmune diabetes. Although limited, animal studies have further implicated the involvement of nitric oxide in autoimmune diabetes. The demonstration that human islets produce nitric oxide in response to cytokines provides important additional support for a role of this toxic free radical in IDDM. Also the ability of nitric oxide to directly activate islet COX suggests that nitric oxide may play a dual role in both islet inflammation and β-cell destruction. Even though this evidence strongly supports a role for nitric oxide in the process of autoimmune destruction of the pancreatic islet β cell, the exact function of nitric oxide in IDDM has yet to be established. IDDM is a complex disease in which macrophages, cytotoxic and helper T cells, autoantigens, and possibly viral components all appear to participate. Continued efforts directed at both the biochemical and immunological determination of the cytotoxic mechanisms of β-cell destruction should facilitate further advances to uncover the role of nitric oxide in the process of autoimmune destruction of pancreatic β cells in IDDM.

ACKNOWLEDGMENTS

We thank Joan Fink and Connie Marshall for assistance in the preparation of this manuscript, and David Scharp for providing human islets. We also thank John Turk and

Guim Kwon for critical evaluation of the manuscript. This work was supported by National Institutes of Health Grant DK-06181 and T32 DK0-07296.

REFERENCES

Aguilar-Diosdado, M., Parkinson, D., Corbett, J. A., Kwon, G., Marshall, C. A., Gingerich, R. L., Santiago, J. V., and McDaniel, M. L. (1994). Potential autoantigens in IDDM: Expression of carboxypeptidase-H and insulin but not glutamate decarboxylase on the β-cell surface. *Diabetes* **43**, 418–425.

Amber, I. J., Hibbs, J. B., Jr., Parker, C. J., Johnson, B. B., Taintor, R. R., and Vavrin, Z. (1991). Activated macrophage conditioned medium: Identification of the soluble factors inducing cytotoxicity and the L-arginine dependent effector mechanism. *J. Leukocyte Biol.* **49**, 610–620.

Andreani, D., Di Mario, U., and Pozzilli, P. (1991). Prediction, prevention, and early intervention in insulin-dependent diabetes. *Diabetes/Metab. Rev.* **7**, 61–77.

Asayama, K., Kooy, N. W., and Burr, I. M. (1986). Effect of vitamin E deficiency and selenium deficiency on insulin secretory reserve and free radical scavenging systems in islets: Decrease of islet manganosuperoxide dismutase. *J. Lab. Clin. Med.* **107**, 459–464.

Baekkeskov, S., Warnock, G., Christie, M., Rajotte, R. V., Larsen, P. M., and Fey, S. (1989). Revelation of specificity of 64K autoantibodies in IDDM serums by high-resolution 2-D gel electrophoresis. *Diabetes* **38**, 1133–1141.

Baekkeskov, S., Ranstoot, H.-J., Christgau, S., Reetz, A., Solimena, M., Cascalho, M., Folli, F., Richter-Olesen, H., and De Camilli, P. (1990). Identification of the 64K autoantigen in insulin-dependent diabetes as the GABA-synthesizing enzyme glutamic acid decarboxylase. *Nature (London)* **347**, 151–156.

Beckman, J. S., Beckman, T. W., Chen, J., Marshall, P. A., and Freeman, B. A. (1990). Apparent hydroxyl radical production by peroxynitrite: Implications for endothelial injury from nitric oxide and superoxide. *Proc. Natl. Acad. Sci. U.S.A.* **87**, 1620–1624.

Bendtzen, K., Mandrup-Poulsen, T., Nerup, J., Nielsen, J. H., Dinarello, C. A., and Svenson, M. (1986). Cytotoxicity of human pI 7 interleukin-1 for pancreatic islets of Langerhans. *Science* **232**, 1545–1547.

Bergmann, L., Kroncke, K.-D., Suschek, C., Kolb, H., and Kolb-Bachofern, V. (1992). Cytotoxic action of IL-1β against pancreatic islets is mediated via nitric oxide formation and is inhibited by N^G-monomethyl-L-arginine. *FEBS Lett.* **299**, 103–106.

Bonner-Weir, S. (1989). Pancreatic islets: Morphology, organization and physiological implication. *In* "Insulin Secretion" (B. Drazin, S. Melmed, and D. LeRoith, eds.), pp. 1–11. Alan R. Liss, New York.

Bottazzo, G. F., Dean, B. M., McNally, J. M., MacKay, E. H., Swift, P. G. F., and Gamble, D. R. (1985). In situ characterization of autoimmune phenomena and expression of HLA molecules in the pancreas in diabetic insulitis. *N. Engl. J. Med.* **313**, 353–360.

Brosnan, C. F., Bronstein, M. B., and Bloom, B. R. (1981). The effects of macrophage depletion on the clinical and pathological expression of experimental allergic encephalomyelitis. *J. Immunol.* **126**, 614–620.

Buscema, M., Vinci, C., Gatta, C., Rabuazzo, M. A., Vignen, R., and Purrello, F. (1992). Nicotinamide partially reverses the interleukin 1β inhibition of glucose-induced insulin release in pancreatic islets. *Metabolism* **41**, 296–300.

Castano, L., Russo, E., Zhou, L., Lipes, M. A., and Eisengarth, G. S. (1991). Identification and cloning of a granule autoantigen (carboxypeptidase-H) associated with type I diabetes. *J. Clin. Endocrinol. Metab.* **73**, 1197–1201.

Cetkovic-Cvrlje, M., Sandler, S., and Eizirik, D. L. (1993). Nicotinamide and dexamethasone inhibit interleukin-1-induced nitric oxide production by RINm5F cells without decreasing messenger ribonucleic acid expression for nitric oxide synthase. *Endocrinology (Baltimore)* **133**, 1739–1743.

Charlton, B., Bacelj, A., and Mandel, T. E. (1988). Administration of silica particles or anti-Lyt2 antibody prevents β-cell destruction in NOD mice given cyclophosphamide. *Diabetes* **37**, 930–935.

Comens, P. G., Wolf, B. A., Unanue, E. R., Lacy, P. E., and McDaniel, M. L. (1987). Interleukin 1 is a potent modulator of insulin secretion from isolated rat islets of Langerhans. *Diabetes* **36**, 963–970.

Cooke, A. (1990). An overview on possible mechanisms of destruction of the insulin-producing beta cell. *Curr. Top. Microbiol. Immunol.* **164**, 125–142.

Corbett, J. A., and McDaniel, M. L. (1992). Does nitric oxide mediate autoimmune destruction of β-cells? Possible therapeutic interventions in IDDM. *Diabetes* **41**, 897–903.

Corbett, J. A., Lancaster, J. R., Jr., Sweetland, M. A., and McDaniel, M. L. (1991a). Interleukin-1β-induced formation of EPR-detectable iron–nitrosyl complexes in islets of Langerhans. *J. Biol. Chem.* **266**, 21351–21354.

Corbett, J. A., Tilton, R. G., Chang, K., Hasan, K. S., Ido, Y., Wang, J. L., Sweetland, M. A., Lancaster, J. R., Jr., Williamson, J. R., and McDaniel, M. L. (1991b). Aminoguanidine, a novel inhibitor of nitric oxide formation, prevents diabetic vascular dysfunction. *Diabetes* **41**, 552–556.

Corbett, J. A., Wang, J. L., Hughes, J. H., Wolf, B. A., Sweetland, M. A., Lancaster, J. R., Jr., and McDaniel, M. L. (1992a). Nitric oxide and cyclic GMP formation induced by interleukin 1β in islets of Langerhans. *Biochem. J.* **287**, 229–235.

Corbett, J. A., Wang, J. L., Sweetland, M. A., Lancaster, J. R., Jr., and McDaniel, M. L. (1992b). IL-1β induces the formation of nitric oxide by β cells purified from rodent islets of Langerhans. *J. Clin. Invest.* **90**, 2384–2391.

Corbett, J. A., Sweetland, M. A., Lancaster, J. R., Jr., and McDaniel, M. L. (1993a). A 1 hour pulse with IL-1b induces the formation of nitric oxide and inhibits insulin secretion by rat islets of Langerhans: Evidence for a tyrosine kinase signaling mechanism. *FASEB J.* **7**, 369–374.

Corbett, J. A., Sweetland, M. A., Wang, J. L., Lancaster, J. R., Jr., and McDaniel, M. L. (1993b). Nitric oxide mediates cytokine-induced inhibition of insulin secretion by human islets of Langerhans. *Proc. Natl. Acad. Sci. U.S.A.* **90**, 1731–1735.

Corbett, J. A., Wang, J. L., Misko, T. P., Zhao, W., Hickey, W. F., and McDaniel, M. L. (1993c). Nitric oxide mediates IL-1β-induced islet dysfunction and destruction: Prevention by dexamethasone. *Autoimmunity* **15**, 145–153.

Corbett, J. A., Mikhael, A., Shimizu, J., Frederick, K., Misko, T. P., McDaniel, M. L., Kanagawa, O., and Unanue, E. R. (1993d). Nitric oxide production in islets from nonobese diabetic mice: Aminoguanidine-sensitive and -resistant stages in the immunological diabetic process. *Proc. Natl. Acad. Sci. U.S.A.* **90,** 8992–8995.

Corbett, J. A., Kwon, G., Turk, J., and McDaniel, M. L. (1993e). IL-1β induces the coexpression of both nitric oxide synthase and cyclooxygenase by islets of Langerhans: Activation of cyclooxygenase by nitric oxide. *Biochemistry* **32,** 13767–13770.

Corbett, J. A., Kwon, G., Misko, T. P., Rodi, C. P., and McDaniel, M. L. (1994). Tyrosine kinase involvement in IL-1β-induced expression of iNOS by β-cells purified from islets of Langerhans. *Am. J. Physiol.* **267,** C48–54.

Corbett, J. A., and McDaniel, M. L. (1995). Intraislet release of interleukin 1 inhibits β-cell function by inducing β-cell expression of inducible nitric oxide synthase. *J. Exp. Med.* **181,** 559–568.

Crisa, L., Mordes, J. P., and Rossini, A. A. (1992). Autoimmune diabetes mellitus in the BB rat. *Diabetes/Metab. Rev.* **8,** 9–37.

Dawson, T. M., Bredt, D. S., Fotuhi, M., Hwang, P. M., and Snyder, S. H. (1991). Nitric oxide synthase and neuronal NADPH diaphorase are identical in brain and peripheral tissues. *Proc. Natl. Acad. Sci. U.S.A.* **88,** 7797–7801.

Defronzo, R. A. (1988). The triumvirate: β-cell, muscle, liver. *Diabetes* **37,** 667–686.

Defronzo, R. A., and Ferrannini, E. (1991). Insulin resistance: A multifaceted syndrome responsible for NIDDM, obesity, hypertension, dyslipidemia, and atherosclerotic cardiovascular disease. *Diabetes Care* **14,** 173–194.

Denis, M. (1991). Tumor necrosis factor and granulocyte macrophage-colony stimulating factor stimulate human macrophages to restrict growth of virulent *Mycobacterium avium* and to kill avirulent *M. avium*: Killing effector mechanism depends on the generation of reactive nitrogen intermediates. *J. Leukocyte Biol.* **49,** 380–387.

Dimmeler, S., Ankarcrona, M., Nicotera, P., and Brune, B. (1993). Exogenous nitric oxide (NO) generation or IL-1β-induced intracellular NO production stimulates inhibitory auto-ATP-ribosylation of glyceraldehyde-3-phosphate dehydrogenase in RINm5F cells. *J. Immunol.* **150,** 2964–2971.

Dobersen, M. J., Scharff, J. E., Ginsberg-Fellner, F., Notkins, A. L. (1980). Cytotoxic autoantibodies to beta cells in the serum of patients with insulin-dependent diabetes mellitus. *N. Engl. J. Med.* **303,** 1493–1498.

Drapier, J.-C., and Hibbs, J. B., Jr. (1986). Murine cytotoxic activated macrophages inhibit aconitase in tumor cells. *J. Clin. Invest.* **78,** 790–797.

Drapier, J.-C., Wietzerbin, J., and Hibbs, J. B., Jr. (1988). Interferon-γ and tumor necrosis factor induce the L-arginine-dependent cytotoxic effector mechanism in murine macrophages. *Eur. J. Immunol.* **18,** 1587–1592.

Eizirik, D. L. (1988). Interleukin-1 induced impairment in pancreatic islet oxidative metabolism of glucose is potentiated by tumor necrosis factor. *Acta Endocrinol.* **119,** 321–325.

Eizirik, D. L., Sandler, S., Hallberg, A., Bendtzen, K., Sener, A., and Malaisse, W. J. (1989). Differential sensitivity to β-cell secretagogues in rat pancreatic islets exposed to human interleukin-1β. *Endocrinology (Baltimore)* **125,** 752–759.

Eizirik, D. L., Welsh, M., Strandell, E., Welsh, N., and Sandler, S. (1990). Interleukin-1β depletes insulin messenger ribonucleic acid and increases the heat shock protein hsp 70 in mouse pancreatic islets without impairing the glucose metabolism. *Endocrinology (Baltimore)* **127,** 2290–2297.

Eizirik, D. L., Tracey, D. E., Bendtzen, K., and Sandler, S. (1991). An interleukin-1 receptor antagonist protein protects insulin-producing beta cells against suppressive effects of interleukin-1β. *Diabetologia* **34,** 445–448.

Eizirik, D. L., Cagliero, E., Bjorklund, A., and Welsh, N. (1992a). Interleukin-1β induces the expression of an isoform of nitric oxide synthase in insulin-producing cells, which is similar to that observed in activated macrophages. *FEBS Lett.* **308,** 249–252.

Eizirik, D. L., Tracey, D. E., Bendtzen, K., and Sandler, S. (1992b). Role of receptor binding and gene transcription for both the stimulatory and inhibitory effects of interleukin-1 in pancreatic β cells. *Autoimmunity* **12,** 127–133.

Eizirik, D. L., Cagliero, E., Bjorklund, A., and Welsh, N. (1993). Interleukin-1-induced expression of nitric oxide synthase in insulin-producing cells is preceded by c-fos induction and depends on gene transcription and protein synthesis. *FEBS Lett.* **317,** 62–66.

Elliott, R. B., and Chase, H. P. (1991). Prevention or delay of type 1 (insulin-dependent) diabetes mellitus in children using nicotinamide. *Diabetologia* **34,** 362–365.

Gepts, W. (1965). Pathologic anatomy of the pancrease in juvenile diabetes mellitus. *Diabetes* **14,** 619–633.

Granger, D. L., and Lehninger, A. L. (1982). Sites of inhibition of mitochondrial electron transport in macrophage-injured neoplastic cells. *J. Cell Biol.* **95,** 527–535.

Green, I. C., Delaney, C. A., Cunningham, J. M., Karmirism, V., and Southern, C. (1993). Interleukin-1β effects on cyclic GMP and cyclic AMP in cultured rat islets of Langerhans—Arginine dependence and relationship to insulin secretion. *Diabetologia* **36,** 9–16.

Hammonds, P., Beggs, M., Beresford, G., Espinal, J., Clarke, J., and Mertz, R. J. (1990). Insulin-secreting β-cells possess specific receptors for interleukin-1β. *FEBS Lett.* **261,** 97–100.

Hansen, B. S., Nielsen, J. H., Linde, S., Spinas, G. A., Welinder, B. S., Mandrup-Poulsen, T., and Nerup, J. (1988). Effects of interleukin-1 on the biosynthesis of proinsulin and insulin in isolated rat pancreatic islets. *Biomed. Biochim. Acta* **47,** 305–309.

Hauschildt, S., Scheipers, P., Bessler, W. G., and Mulsch, A. (1992). Induction of nitric oxide synthase in L929 cells by tumour-necrosis factor α is prevented by inhibitors of poly(ADP-ribose) polymerase. *Biochem. J.* **288,** 255–260.

Heikkila, R. E., Winston, B., and Cohen, G. (1976). Alloxan-induced diabetes. Evidence for hydroxyl radical as a cytotoxic intermediate. *Biochem. Pharmacol.* **25,** 1085–1092.

Heineke, E. W., Johnson, M. B., Dillberger, J. E., and Robinson, K. M. (1993). Antioxidant MDL 29,311 prevents diabetes in nonobese diabetic and multiple low-dose STZ-injected mice. *Diabetes* **42,** 1721–1730.

Helqvist, S., Hansen, B. S., Johannesen, J., Andersen, H. U., Nielsen, J. H., and Nerup, J. (1989). Interleukins 1 induces new protein formation in isolated rat islets of Langerhans. *Acta Endocrinol.* **121,** 136–140.

Helqvist, S., Polla, B. S., Johannesen, J., and Nerup, J. (1991). Heat shock protein induction in rat pancreatic islets by human recombinant interleukin 1β. *Diabetologia* **34,** 150–156.

Hibbs, J. B., Jr., Taintor, R. R., Vavrin, Z., Granger, D. L., Drapier, J.-C., Amber, I. J., and Lancaster, J. R., Jr. (1990). Synthesis of nitric oxide from a terminal guanidino nitrogen atom of L-arginine: A molecular mechanism regulating cellular proliferation that targets intracellular iron. In "Nitric Oxide from L-arginine: A Bioregulatory System" (S. Moncada and E. A. Higgs, eds.), pp. 189–223. Elsevier, New York.

Hibbs, J. B., Westenfelder, C., Taintor, R., Vavrin, Z., Kablitz, C., Baranowski, R. L., Ward, J. H., Menlove, R. L., McMurry M. P., Kushner, J. P., and Samlowski, W. E. (1992). Evidence for cytokine-inducible nitric oxide synthesis from L-arginine in patients receiving interleukin-2 therapy. *J. Clin. Invest.* **89,** 867–877.

Hope, B. T., Michael, G. J., Knigge, K. M., and Vincent, S. R. (1991). Neuronal NADPH diaphorase is a nitric oxide synthase. *Proc. Natl. Acad. Sci. U.S.A.* **88,** 2811–2814.

Hughes, J. H., Easom, R. A., Wolf, B. A., Turk, J., and McDaniel, M. L. (1989). Interleukin 1-induced prostaglandin E_2 accumulation by isolated pancreatic islets. *Diabetes* **38,** 1251–1257.

Hughes, J. H., Colca, J. R., Easom, R. A., Turk, J., and McDaniel, M. L. (1990a). Interleukin 1 inhibits insulin secretion from isolated rat pancreatic islets by a process that requires gene transcription and mRNA translation. *J. Clin. Invest.* **86,** 856–863.

Hughes, J. H., Watson, M. A., Easom, R. A., Turk, J., and McDaniel, M. L. (1990b). Interleukin-1 induces rapid and transient expression of the c-*fos* proto-oncogene in isolated pancreatic islets and in purified β cells. *FEBS Lett.* **266,** 33–36.

Hunt, N. C. A., and Goldin, R. D. (1992). Nitric oxide production by monocytes in alcoholic liver disease. *J. Hepatol.* **14,** 146–150.

Jansson, L., and Sandler, S. (1991). The nitric oxide synthase II inhibitor N^G-nitro-L-arginine stimulates pancreatic islet insulin release *in vitro*, but not in the perfused pancrease. *Endocrinology (Baltimore)* **128,** 3081–3085.

Jones, D. B., Hunter, N. R., and Duff, G. W. (1990). Heat-shock protein 65 as a β-cell antigen of insulin-dependent diabetes. *Lancet* **336,** 583–585.

Jones, P. M., Persaud, S. J., Bjaaland, T., Pearson, J. D., and Howell, S. L. (1992). Nitric oxide is not involved in the initiation of insulin secretion from rat islets of Langerhans. *Diabetologia* **35,** 1020–1027.

Kaplan, D. R., Colca, J. R., and McDaniel, M. L. (1983). Insulin as a surface marker on isolated cells from rat pancreatic islets. *J. Cell Biol.* **97,** 433–437.

Karthein, R., Nastainczyk, W., and Ruf, H. H. (1987). EPR study of ferric native prostaglandin G synthase and its ferrous NO derivative. *Eur. J. Biochem.* **166,** 173–180.

Kaufman, D. L., Erlander, M. G., Clare-Salzler, M., Atkinson, M. A., Maclaren, N. K., and Tobin, A. J. (1992). Autoimmunity to two forms of glutamate decarboxylase in insulin-dependent diabetes mellitus. *J. Clin. Invest.* **89,** 283–292.

Kaufman, D. L., Clare-Salzler, M., Tian, J., Forsthuber, T., Ting, G. S. P., Sercarz, E. E., Tobin, A. J., and Lehmann, P. V. (1993). Spontaneous loss of T-cell tolerance to glutamic acid decarboxylase in murine insulin-dependent diabetes. *Nature (London)* **366,** 69–72.

Kleemann, R., Rothe, H., Kolb-Bachofen, V., Xie, Q. W., Nathan, C., Martin, S., and Kolb, H. (1993). Transcription and translation of inducible nitric oxide synthase in the pancreas of prediabetic BB rats. *FEBS Lett.* **328,** 9–12.

Kolb, H., and Kolb-Bachofen, V. (1992a). Type 1 (insulin-dependent) diabetes mellitus and nitric oxide. *Diabetologia* **35,** 796–797.

Kolb, H., and Kolb-Bachofen, V. (1992b). Nitric oxide: A pathogenetic factor in autoimmunity. *Immunol. Today* **13,** 157–160.

Kolb, H., and Kolb-Bachofen, V. (1993). Type 1 (insulin-dependent) diabetes mellitus and nitric oxide. *Diabetologia* **35,** 796–797.

Kolb, H., Kantwert, G., Treichal, U., Kurner, T., Keisel, U., Hoppe, T., and Kolb-Bachofen, V. (1986). Natural history of insulitis in BB rats. *Pancreas* **1,** 370.

Kolb, H., Kiesel, U., Kroncke, K-D., Kolb-Bachofen, V. (1991). Suppression of low dose streptozotocin induced diabetes in mice by administration of nitric oxide synthase inhibitor. *Life Sci* **49,** PL213–PL217.

Kroncke, K.-D., Funda, J., Berschick, B., Kolb, H., and Kolb-Bachofen, V. (1991a). Macrophage cytocoxicity towards isolated rat islet cells: Neither lysis nor its protection by nicotinamide are beta-cell specific. *Diabetologia* **34,** 232–238.

Kroncke, K.-D., Kolb-Bachofen, V., Berschick, B., Burkart, V., and Kolb, H. (1991b). Activated macrophages kill pancreatic syngeneic islet cells via arginine-dependent nitric oxide generation. *Biochem. Biophys. Res. Commun.* **175,** 752–758.

Laychock, S. G. (1983). Mediation of insulin release by cGMP and cAMP in a starved animal model. *Mol. Cell. Endocrinol.* **32,** 157–170.

Laychock, S. G., Modica, M. E., and Cavanaugh, C. T. (1991). L-arginine stimulates cyclic guanosine 3′,5′-monophosphate formation in rat islets of Langerhans and RINm5F insulinoma cells: Evidence for L-arginine: Nitric oxide synthase. *Endocrinology (Baltimore)* **129,** 3043–3052.

Lee, K., Amano, K., and Yoon, J.-W. (1988). Evidence for initial involvement of macrophage in development of insulitis in NOD mice. *Diabetes* **37,** 989–991.

Lernmark, A., Freedman, Z. R., Hofmann, C., Rubenstein, A. H., Steiner, D. F., Jackson, R. L., Winter, R. J., and Traisman, H. S. (1978). Islet-cell-surface antibodies in juvenile diabetes mellitus. *N. Engl. J. Med.* **299,** 375–380.

Lernmark, A., Barmeier, H., Dube, S., Hagopian, W., Karlsen, A., and Wassmuth, R. (1991). Autoimmunity of diabetes. *Endocrinol. Metab. Clin. North Am.* **20,** 589–617.

Levy, M. H., and Wheelock, E. F. (1975). Effects of intravenous silica on immune and nonimmune functions of the murine host. *J. Immunol.* **115,** 41–48.

Like, A. A., Kislauskis, E., Williams, R. M., and Rossini, A. A. (1982). Neonatal thymectomy prevents spontaneous diabetes mellitus in the BB/W rat. *Science* **216,** 644–646.

Like, A. A., Weringer, E. J., Holdash, A., McGill, P., Atkinson, D., and Rossini, A. A. (1985). Adoptive transfer of autoimmune diabetes mellitus in Biobreeding/Worcester (BB/W) inbred and hybrid rats. *J. Immunol.* **134,** 1583–1587.

Lukic, M. L., Stosic-Grujicic, S., Ostojic, N., Chan, W. L., and Liew, F. Y. (1991). Inhibition of nitric oxide generation affects the induction of diabetes by streptozotocin in mice. *Biochem. Biophys. Res. Commun.* **178,** 913–920.

McDaniel, M. L., Hughes, J. H., Wolf, B. A., Easom, R. E., and Turk, J. (1988a). Descriptive and mechanistic considerations of interleukin 1 and insulin secretion. *Diabetes* **37,** 1311–1315.

McDaniel, M. L., Wolf, B. A., and Turk, J. (1988b). Signal transduction and insulin secretion: Role of intracellular Ca^{2+}, inositoltrisphosphate, and arachidonic acid. *Prog. Clin. Biol. Res.* **265,** 99–116.

Maclaren, N. K. (1981). Autoimmunity and diabetes. In "The Islets of Langerhans" (S. J. Cooperstein and D. Watkins, eds.), pp. 453–466. Academic Press, New York.

Maki, T., Ichikawa, T., Blanco, R., and Porter, J. (1992). Long-term abrogation of autoimmune diabetes in nonobese diabetic mice by immunotherapy with anti-lymphocyte serum. *Proc. Natl. Acad. Sci. U.S.A.* **89,** 3434–3438.

Malaisse, W. J., Malaisse-Lagae, F., Sener, A., and Pipeleers, D. G. (1982). Determinants of the selective toxicity of alloxan to the pancreatic B cell. *Proc. Natl. Acad. Sci. U.S.A.* **79,** 927–930.

Mandrup-Poulsen, T., Bendtzen, K., Nielsen, H., Bendixen, G., and Nerup, J. (1985). Cytokines cause functional and structural damage to isolated islets of Langerhans. *Allergy* **40,** 424–429.

Mandrup-Poulsen, T., Bendtzen, K., Nerup, J., Egeberg, J., and Nielsen, H. (1986). Mechanisms of pancreatic islet cell destruction. *Allergy* **41,** 250–259.

Mandrup-Poulsen, T., Helqvist, S., Wogensen, L. D., Mølvig, J., Pociot, F., Johannesen, J., and Nerup, J. (1990). Cytokines and free radicals as effector molecules in the destruction of pancreatic beta cells. *Curr. Top. Microbiol. Immunol.* **164,** 169–193.

Mandrup-Poulsen, T., Corbett, J. A., McDaniel, M. L., and Nerup, J. (1993). What are the types and cellular sources of free radicals in the pathogenesis of type 1 (insulin-dependent) diabetes mellitus? *Diabetologia* **36,** 470–471.

Marletta, M. A., Yoon, P. S., Lyengar, R., Leaf, C. D., and Wishnok, J. S. (1988). Macrophage oxidation of L-arginine to nitrite and nitrate: Nitric oxide is an intermediate. *Biochemistry* **27,** 8706–8711.

Marner, B., Knutson, C., Lernmark, A., Nerup, J., and the Hagedorn Study Group (1985). Immunological investigations: Islet cell antibodies. *In* "Methods in Diabetes Research" (J. L. Larner and S. L. Pohl, eds.), Vol. 1, pp. 181–194. Wiley, New York.

Mendola, J., Wright, J. R., Jr., and Lacy, P. (1989). Oxygen free-radical scavengers and immune destruction of murine islets in allograft rejection and multiple low-dose streptozocin-induced insulitis. *Diabetes* **38,** 379–385.

Misko, T. P., Moore, W. M., Kasten, T. P., Nickols, G. A., Corbett, J. A., Tilton, R. G., McDaniel, M. L., Williamson, J. R., and Currie, M. G. (1993). Selective inhibition of the inducible isoform of nitric oxide synthase by aminoguanidine. *Eur. J. Pharmacol.* **233,** 119–125.

Misler, S., Barnett, D. W., Gillis, K. D., and Pressel, D. M. (1992). Electrophysiology of stimulus-secretion coupling in human β cells. *Diabetes* **41,** 1221–1228.

Moncada, S., Palmer, R. M. J., and Higgs, E. A. (1991). Nitric oxide: Physiology, pathophysiology, and pharmacology. *Pharmacol. Rev.* **43,** 109–142.

Munoz-Fernandez, M. A., Fernandez, M. A., and Fresno, M. (1992). Activation of human macrophages for the killing of intracellular *Trypanosoma cruzi* by TNF-alpha and IFN-gamma through a nitric oxide-dependent mechanism. *Immunol. Lett.* **33,** 35–40.

Nathan, C. (1992). Nitric oxide as a secretory product of mammalian cells. *FASEB J.* **6,** 3051–3064.

Needleman, P., Turk, J., Jakschik, B. A., Morrison, A. R., and Lefkowith, J. B. (1986). Arachidonic acid metabolism. *Annu. Rev. Biochem.* **55,** 69–102.

Nerup, J., and Lernmark, A. (1981). Autoimmunity in insulin-dependent diabetes mellitus. *Am. J. Med.* **70,** 135–141.

Nussler, A. K., Di Silvio, M., Billiar, T. R., Hoffman, R. A., Geller, D. A., Selby, R., Madariage, J., and Simmons, R. L. (1992). Stimulation of the nitric oxide synthase pathway in human hepatocytes by cytokines and endotoxin. *J. Exp. Med.* **176**, 261–264.

Oberley, L. W. (1988). Free radicals and diabetes. *Free Radical Biol. Med.* **5**, 113–124.

Ochoa, J. B., Udekwu, A. O., Billiar, T. R., Curran, R. D., Cerra, F. B., Simmons, R. L., and Peitzman, A. B. (1991). Nitrogen oxide levels in patients after trauma and during sepsis. *Ann. Surg.* **214**, 621–626.

Ochoa, J. B., Curti, B., Peitzman, A. B., Simmons, R. L., Billiar, T. R., Hoffman, R., Rault, R., Longo, D. L., Urba, W. J., and Ochoa, A. C. (1992). Increased circulation nitrogen oxides after human tumor immunotherapy: Correlation with toxic hemodynamic changes. *J. Natl. Cancer Inst.* **84**, 864–867.

Oschilewski, U., Kiesel, U., and Kolb, H. (1985). Administration of silica prevents diabetes in BB rats. *Diabetes* **34**, 197–199.

Palmer, J. P., Helqvist, S., Spinas, G. A., Molvig, J., Mandrup-Poulsen, T., Andersen, H. U., and Nerup, J. (1989). Interactions of β-cell activity and IL-1 concentration and exposure time in isolated rat islets of Langerhans. *Diabetes* **38**, 1211–1216.

Panagiotidis, G., Alm, P., and Lundquist, I. (1992). Inhibition of islet nitric oxide synthase increases arginine-induced insulin release. *Eur. J. Pharmacol.* **229**, 277–278.

Pelham, H. R. B. (1989). Heat shock and the sorting of luminal ER proteins. *EMBO J.* **8**, 3171–3176.

Pipeleers, D. G., Int Veld, P. A., Van De Winkel, M., Maes, E., Schuit, F. C., and Gepts, W. (1985). A new *in vitro* model for the study of pancreatic A and B cells. *Endocrinology (Baltimore)* **117**, 806–816.

Rabinovitch, A., Suarez, W. L., Thomas, P. D., Strynadka, K., and Simpson, I. (1992). Cytotoxic effects of cytokines on rat islets: Evidence for involvement of free radicals and lipid peroxidation. *Diabetologia* **33**, 409–413.

Rabinovitch, A. (1993). Roles of cytokines in IDDM pathogenesis and islet β-cell destruction. *Diabetes Rev.* **1**, 215–240.

Radomski, M. W., Plamer, R. M. J., and Moncada, S. (1990). Glucocorticoids inhibit the exsspression of an inducible, but not the constitutive isoform of nitric oxide synthase in vascular endothelial cells. *Proc. Natl. Acad. Sci. U.S.A.* **87**, 10043–10047.

Reddy, D., Lancaster, J. R., Jr., and Cornforth, D. P. (1982). Nitrite inhibition of *Clostridium botulinum*: Electron spin resonance detection of iron–nitric oxide complexes. *Science* **221**, 769–770.

Rettori, V., Gimeno, M., Lyson, K., and McCann, S. M. (1992). Nitric oxide mediates norepinephrine-induced prostaglandin E_2 release from the hypothalamus. *Proc. Natl. Acad. Sci. U.S.A.* **89**, 11543–11546.

Rettori, V., Belova, N., Dees, W. L., Nyberg, C. L., Gimeno, M., and McCann, S. M. (1993). Role of nitric oxide in the control of luteinizing hormone–releasing hormone release *in vivo* and *in vitro*. *Proc. Natl. Acad. Sci. U.S.A.* **90**, 10131–10134.

Roep, B. O., Arden, S. D., de Vries, R. R. P., and Hutton, J. C. (1990). T-cell clones from type I diabetes patient respond to insulin secretory granule proteins. *Nature (London)* **345**, 632–634.

Rosa, M. D., Radomski, M., Camuccio, R., and Moncada, S. (1990). Glucocorticoids inhibit the induction of nitric oxide synthase in macrophages. *Biochem. Biophys. Res. Commun.* **172**, 1246–1252.

Rossini, A. A., Handler, E. S., Greiner, D. L., and Mordes, J. P. (1991). Insulin dependent diabetes mellitus hypothesis of autoimmunity. *Autoimmunity* **8**, 221–235.

Salvemini, D., Misko, T. P., Masferrer, J. L., Seibert, K., Currie, M. G., and Needleman, P. (1993). Nitric oxide activates cyclooxygenase enzymes. *Proc. Natl. Acad. Sci. U.S.A.* **90**, 7240–7244.

Sandler, S., Andersson, R., and Hellerstrom, C. (1987). Inhibitory effects of interleukin 1 on insulin secretion, insulin biosynthesis, and oxidative metabolism of isolated rat pancreatic islets. *Endocrinology (Baltimore)* **121**, 1424–1431.

Sandler, S., Bendtzen, K., Hakan Borg, L. A., Eizirik, D. L., Strandell, W., and Welsh, N. (1989). Studies on the mechanisms causing inhibition of insulin secretion in rat pancreatic islets exposed to human interleukin-1β indicate a perturbation in the mitochondrial function. *Endocrinology (Baltimore)* **124**, 1492–1501.

Sandler, S., Bendtzen, K., Eizirik, D. L., Strandell, E., Welsh, M., and Welsh, N. (1990). Metabolism and β-cell function of rat pancreatic islets exposed to human interleukin-1β in the presence of a high glucose concentration. *Immunol. Lett.* **26**, 245–252.

Sandler, S., Eizirik, D. L., Svensson, C., Strandell, E., Welsh, M., and Welsh, N. (1991). Biochemical and molecular actions of interleukin-1 on pancreatic β-cells. *Autoimmunity* **10**, 241–253.

Schmidt, H. H. H. W. (1992). NO·, CO and ·OH endogenous soluble guanylyl cyclase-activating factors. *FEBS Lett.* **307**, 102–107.

Schmidt, H. H. H. W., and Murad, F. (1991). Purification and characterization of a human NO synthase. *Biochem. Biophys. Res. Commun.* **181**, 1372–1377.

Schmidt, H. H. H. W., Warner, T. D., Ishii, K., Sheng, H., and Murad, F. (1992). Insulin secretion from pancreatic β-cells caused by L-arginine-derived nitrogen oxides. *Science* **255**, 721–723.

Sempe, P., Bedossa, P., Richard, M.-F., Villa, M.-C., Bach, J.-F., and Boitard, C. (1991). Anti-a/b T cell receptor monoclonal antibody provides an efficient therapy for autoimmune diabetes in nonobese diabetic (NOD) mice. *Eur. J. Immunol* **21**, 1163–1169.

Setum, C. M., Serie, J. R., and Hegre, O. D. (1991). Confocal microscopic analysis of the nonendocrine cellular component of isolated adult rat islets of Langerhans. *Transplantation* **51**, 1131–1133.

Shizuru, J. A., Taylor-Edwards, C., Banks, B. A., Gregory, A. K., and Fathman, C. G. (1988). Immunotherapy of the nonobese diabetic mouse: Treatment with an antibody to T-helper lymphocytes. *Science* **240**, 659–662.

Sibley, R. K., Sutherland, D. E., Goetz, F., and Michael, A. F. (1985). Recurrent diabetes mellitus in the pancrease iso- and allograft. A light and electron microscopic and immunohistochemical analysis of four cases. *Lab. Invest.* **53**, 132–144.

Sigurdsson, E., and Baekkeskov, S. (1990). The 64-kDa beta cell membrane autoantigen and other target molecules of humoral autoimmunity in insulin-dependent diabetes mellitus. *Curr. Top. Microbiol. Immunol.* **164**, 143–168.

Southern, C., Schulster, D., and Green, I. C. (1990). Inhibition of insulin secretion by interleukin-1β and tumor necrosis factor-α via an L-arginine-dependent nitric oxide generation mechanism. *FEBS Lett.* **276**, 42–44.

Spinas, G. A., Mandrup-Poulsen, T., Molvig, J., Baek, L., Bendtzen, K., Dinarello, C. A., and Nerup, J. (1986). Low concentrations of interleukin-1 stimulate and high concentrations inhibit insulin release from isolated rat islets of Langerhans. *Acta Endocrinol.* **113,** 551–586.

Spinas, G. A., Hansen, B. S., Linde, S., Kastern, W., Molvig, J., Mandrup-Poulsen, T., Dinarello, C. A., Nielsen, J. H., and Nerup, J. (1987). Interleukin 1 dose-dependently affects the biosynthesis of (pro)insulin in isolated rat islets of Langerhans. *Diabetologia* **30,** 474–480.

Stadler, J., Billiar, T. B., Curran, R. D., Stuehr, D. J., Ochoa, J. B., and Simmons, R. L. (1991). Effect of exogenous and endogenous nitric oxide on mitochondrial respiration of rat hepatocytes. *Am. J. Physiol.* **260,** C910–C916.

Stadler, J., Harbrecht, B. G., Di Silvio, M., Curran, R. D., Jordan, M. L., Simmons, R. L., and Billiar, T. R. (1993). Endogenous nitric oxide inhibits the synthesis of cyclooxygenase products and interleukin-6 by rat Kupffer cells. *J. Leukocyte Biol.* **53,** 165–172.

Sumoski, W., Baquerizo, H., and Rabinovitch, A. (1989). Oxygen free radical scavengers protect rat islet cell from damage by cytokines. *Diabetologia* **32,** 792–796.

Tilton, R. G., Chang, K., Hasan, K. S., Smith, S., Petrash, J. M., Misko, T. P., Moore, W. M., Currie, M. G., Corbett, J. A., McDaniel, M. L., and Williamson, J. R. (1993). Prevention of diabetic vascular dysfunction by guanidines. *Diabetes* **42,** 221–232.

Tisch, R., Yang, X. D., Singer, S. M., Liblau, R. S., Fugger, L., and McDevitt, H. O. (1993). Immune response to glutamic acid decarboxylase correlates with insulitis in non-obese diabetic mice. *Nature (London)* **366,** 72–75.

Turk, J., Corbett, J. A., Ramanadham, S., Bohrer, A., and McDaniel, M. L. (1993). Biochemical evidence for nitric oxide formation from streptozotocin in isolated pancreatic islets. *Biochem. Biophys. Res. Commun.* **197,** 1458–1464.

Uehara, Y., Shimizu, H., Sato, N., Shimomura, Y., Mori, M., and Kobayashi, I. (1991). Probucol partially prevents development of diabetes in NOD mice. *Diabetes Res.* **17,** 131–134.

Verspohl, E. J., Kuhn, M., and Ammon, H. P. T. (1988). RINm5F (Rat insulinoma) cells possess receptors for atrial natriuretic peptide (ANP) and a functioning cGMP system. *Horm. Metab. Res.* **20,** 770–771.

Vincent, S. R. (1992). Nitric oxide and arginine-evoked insulin secretion. *Science* **258,** 1376–1377.

Wang, J. L., and McDaniel, M. L. (1990). Secretagogue-induced oscillations of cytoplasmic Ca^{2+} in single β and α cells obtained from pancreatic islets by fluorescence-activated cell sorting. *Biochem. Biophys. Res. Commun.* **166,** 813–818.

Walker, R., Bone, A. J., Cooke, A., and Baird, J. D. (1988). Distinct macrophage subpopulations in pancreas of prediabetic BB/E rats. *Diabetes* **37,** 1301–1304.

Welsh, N., and Sandler, S. (1992). Interleukin-1b induces nitric oxide production and inhibits the activity of aconitase without decreasing glucose oxidation rates in isolated mouse pancreatic islets. *Biochem. Biophys. Res. Commun.* **182,** 333–340.

Welsh, N., Eizirik, D. L., Bendtzen, K., and Sandler, S. (1991a). Interleukin-1β-induced nitric oxide production in isolated rat pancreatic islets requires gene transcription and may lead to inhibition of Krebs cycle enzyme aconitase. *Endocrinology (Baltimore)* **129,** 3167–3173.

Welsh, N., Welsh, M., Lindquist, S., Eizirik, D. L., Bendtzen, K., and Sandler, S. (1991b). Interleukin 1β increases the biosynthesis of the heat shock protein hsp 70 and selectively decreases the biosynthesis of five proteins in rat pancreatic islets. *Autoimmunity* **9,** 33–40.

Weringer, E. J., and Like, A. A. (1985). Immune attack on pancreatic islet transplants in the spontaneously diabetic BioBreeding/Worcester (BB/W) rat is not MHC restricted. *J. Immunol.* **134,** 2383–2386.

Wicker, L. S., Miller, B. J., and Mullen, Y. (1986). Transfer of autoimmune diabetes mellitus with splenocytes from nonobese diabetic (NOD) mice. *Diabetes* **35,** 855–860.

Wilson, G. L., Patton, N. J., McCord, J. M., Mullins, D. W., and Mossman, B. T. (1984). Mechanisms of streptozotocin- and alloxan-induced damage in rat β cells. *Diabetologia* **27,** 587–591.

Xie, Q. W., Whisnant, R., and Nathan, C. (1993). Promoter of the mouse gene encoding calcium-independent nitric oxide synthase confers inducibility by interferon γ and bacterial lipopolysaccharide. *J. Exp. Med.* **177,** 1779–1784.

Yonetani, T., Yamamoto, H., Erman, J. E., Leigh, J. S., and Reed, G. H. (1972). Electromagnetic properties of hemoproteins. *J. Biol. Chem.* **247,** 2447–2455.

6

A Role for Nitric Oxide in Liver Inflammation and Infection

Mauricio Di Silvio, Andreas K. Nussler, David A. Geller, and Timothy R. Billiar

Department of Surgery
University of Pittsburgh
Pittsburgh, Pennsylvania 15261

I. INTRODUCTION

It is now known that nitric oxide (·NO) is synthesized by a number of cell types ranging in diversity from neurons to hepatocytes (Lancaster, 1992; Moncada et al., 1991; Bredt and Snyder, 1992). It is also established that the discovery of ·NO biosynthesis represents a novel mammalian biochemical pathway in that this enzymatic reaction was not known to exist in mammals prior to 1987. Although the details of the enzymatic reaction have not been fully elucidated, certain aspects of the biochemistry are well understood. In this reaction, one of the chemically equivalent guanido nitrogens is acted on by a ·NO synthase (NOS) and converted to ·NO. In this process, N^ω-hydroxyarginine is formed as an intermediate (Stuehr et al., 1991a) while ·NO and citrulline are the products of this reaction, and molecular oxygen is incorporated into both ·NO and citrulline (Kwon et al., 1990; Leone et al., 1991). ·NO has a very short life span (timed in seconds) in oxygenated, aqueous solutions and is further oxidized to nitrite (NO_2^-) and nitrate (NO_3^-), which are the stable and inactive end products of ·NO formation. Critical to the study of ·NO is the use of N-substituted

L-arginine analogues, such as N^ω-monomethyl-L-arginine (Hibbs et al., 1987b), which acts as an extremely effective and competitive inhibitor of ·NO formation.

We, like many other research groups, became interested in ·NO when we found that ·NO production provided key answers to unanswered questions in our research. The early research which led to our understanding of the L-arginine–·NO pathway and its application to our experimental systems can be summarized by the following four observations reported in the 1980s. First, Tannenbaum and co-workers (Green et al., 1981; Wagner et al., 1983) reported conclusive evidence in 1981 that mammals synthesized nitrogen oxides by demonstrating that rodents excreted more NO_3^- than they ingested. They went on to show that sterile irritants increased nitrogen oxide biosynthesis in animals. The second key observation was reported when Stuehr and Marletta (1985) showed for the first time that a specific mammalian cell type, macrophages, had the capacity to synthesize NO_2^- and NO_3^- if exposed to bacterial endotoxin. Hibbs et al. (1987a,b) then reported two seminal observations in 1987 when they reported that L-arginine was the substrate for macrophage NO_2^- and NO_3^- biosynthesis and that N-guanido substituted derivatives of L-arginine served as potent competitive inhibitors of macrophage nitrogen oxide biosynthesis. Palmer et al. (1987, 1988) provide the final piece of the puzzle when, in two separate reports, they showed that endothelial cells produced ·NO and then that L-arginine was the substrate for ·NO, providing evidence that ·NO was the biologically active intermediate of the pathway. Parallel studies by Ignarro and co-workers (1987) confirmed that ·NO was endothelium-derived relaxing factor (EDRF), and linked ·NO production with cGMP increases in vascular smooth muscle. These final observations also identified ·NO as EDRF. Hibbs et al. (1988) and Stuehr et al. (1989a) subsequently identified ·NO as an intermediate in macrophage NO_2^- and NO_3^- biosynthesis and provided evidence that ·NO is an important macrophage cytotoxic effector molecule. This series of four independent observations spanning nearly a decade not only provided critical answers to long-standing questions in the area of macrophage cytotoxicity and endothelium regulation of vascular tone, they opened an entirely new field of biomedical research, the biology of ·NO.

Before discussing some of the observations concerning ·NO in our laboratory, it is important to review a few of the basic principles of ·NO production and mechanism of action. More detail on these topics can be found in other chapters in this book. First, ·NO production can occur by one of two categories of NOS enzymes. Constitutive enzymes are expressed in various cell types, including endothelial cells, neurons, and neutrophils. Production of ·NO by this type of enzyme is typically calcium/calmodulin-dependent and occurs immediately. This type of production has also been referred to as low output. Cells such as macrophages, hepatocytes, and smooth muscle cells express an inducible ·NO synthase. The induction of this enzyme typically results from exposure of the cells to immunologic stimuli such as bacterial endotoxin and/or cytokines. The onset of production is delayed but then is much greater than constitutive ·NO formation and continues over long periods. Although the activity of this inducible NOS is not calcium sensitive, there is recent evidence that a calmodulin molecule is bound to the NOS and causes the enzyme to remain in a fully activated state (Cho et al., 1992). Therefore, inducible ·NO production can be viewed as high output production. Endothelial cells appear unique in that they are capable of expressing both types of ·NO production (Kilbourne and Belloni, 1990).

·NO has only a few described actions and these can be divided into enzyme activation and enzyme inactivation. In most instances, the target proteins (enzymes) have a prosthetic iron group for which ·NO has a strong avidity. The best studied and characterized action of ·NO is the activation of soluble guanylate cyclase. This results in an increase in cellular cGMP levels. It is through this action that ·NO acts as a signaling molecule involved in the relaxation of vascular smooth muscle (Ignarro et al., 1987), inhibition of platelet adherence (Dimmeler et al., 1992) and aggregation (Mellion et al., 1981), neutrophil chemotaxis (Kaplan et al., 1989), and signal transduction in the central (Bredt et al., 1990) and peripheral (Bredt and Snyder, 1992; Burnett et al., 1992) nervous systems. ·NO may also directly activate ADP-ribosyltransferase and thereby act as a messenger molecule by a second pathway (Brune and Lapetina, 1989). ·NO also has well-characterized actions as an enzyme inhibitor. Aconitase of the Krebs cycle and NADPH:ubiquinone oxidoreductase and succinate:ubiquinone oxidoreductase of the electron transport chain are inhibited by ·NO, especially when ·NO is produced in large amounts (Drapier and Hibbs, 1988; Stuehr and Nathan, 1989). Macrophage tumor and microbial toxicity may be mediated by this mechanism in some instances. The demonstration of the inhibition of tumor cell ribonucleotide reductase (Kwon et al., 1991; Lapoivre et al., 1990), an enzyme essential for DNA synthesis, by ·NO may provide an explanation for the antiproliferative action of ·NO. Evidence also suggests that the activation of ADP-ribosyltransferase with subsequent ribosylation of glyceraldehyde-phosphate dehydrogenase (GAP-DH) and simultaneous nitrosylation of enzyme sulfhydryl groups may represent another mechanism whereby ·NO may inhibit enzyme activity (Molina et al., 1992).

II. NITRIC OXIDE AND THE LIVER IN SEPSIS AND INFLAMMATION

Our work on ·NO in the liver stems from an interest in the mechanism leading to the altered liver function in sepsis. We hypothesized that in the liver in sepsis, the Kupffer cells (KC) are the cells sensitive to circulating septic stimuli such as bacteria and their products [i.e., endotoxin or lipopolysaccharide (LPS)], and that once stimulated, the KC, like other macrophages, would secrete substances that regulate or interfere with the function of the adjacent heptocytes (HC). For example, cytokines such as interleukin-6 (IL-6) or interleukin-1 (IL-1) released by KC could induce acute phase reactant synthesis by HC while cytotoxic molecules (i.e., oxygen radicals) may damage HC, and contribute to the hepatocellular dysfunction often seen in sepsis and organ failure in critically ill patients.

To study the importance of KC–HC cell interaction in sepsis, we developed a rat KC and HC coculture model. An entire range of KC to HC ratios were tested (0.5:1 to 10:1). LPS was used as the relevant septic stimulus. As anticipated, the addition of even small amounts of LPS resulted in induction of acute phase reactant synthesis (Kispert et al., 1990). Of greater interest was a profound decrease in total protein synthesis seen when higher concentrations of LPS (0.1 µg/ml or greater) was added to KC–HC cocultures with KC:HC ratios greater than 2:1 (Figure 1). The decrease was first detected at 6–8 hr and became maximal at 24 hr.

Lipopolysaccharide had no effect on the protein synthesis ([^{3}H]leucine incorporation) of HC cultured alone. A number of studies concluded that this decrease in protein synthesis was not due to a single cytokine, eicosanoid, or oxygen radical (West et al., 1989). Also, the activity could not be transferred in conditioned supernatant for KC stimulated with LPS alone. However, when

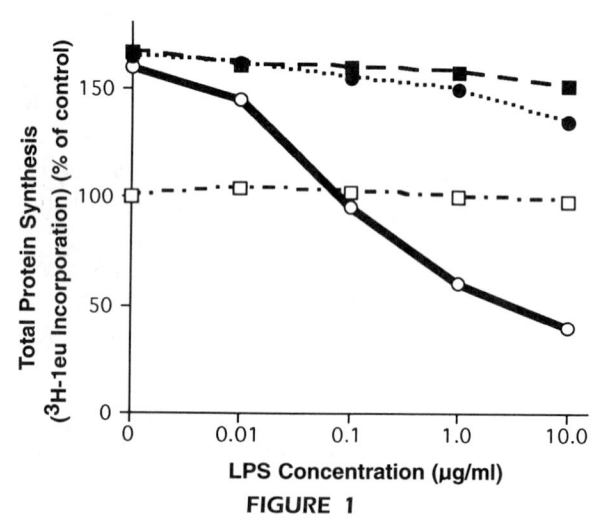

FIGURE 1

Rat HC cultured alone or with KC (5:1 KC:HC ratio) were exposed to increasing concentrations of LPS. After 20 hr, total protein synthesis was measured. HC alone (□) were cultured in the presence of 0.5 mM L-arginine while cocultures were cultured in the presence of L-arginine (○), in the absence of L-arginine (●), or in the presence of L-arginine with 0.2 mM N^G-monomethyl-L-arginine (■).

L-arginine was excluded from the culture medium, the decrease in protein synthesis was not seen (Billiar et al., 1989a). In fact, L-arginine was the only amino acid needed for the decrease to occur. That the inhibition was due to the metabolism of L-arginine to ·NO came with studies demonstrating simultaneous L-arginine-dependent increases in NO_2^- + NO_3^- and citrulline. In addition, the competitive inhibitor of L-arginine metabolism, N^G-monomethyl-L-arginine (NMA), blocked both the decrease in protein synthesis and NO_2^- + NO_3^- and citrulline production. The NMA effect was overcome by excess arginine. The addition of authentic ·NO or ·NO-releasing compounds such as sodium nitroprusside or S-nitrosyl-N-acetylpenicillamine added to cultured HC reproduced the inhibition in protein synthesis while NO_2^- + NO_3^- had no effect (Billiar et al., 1989b; Curran et al., 1991). Therefore, the LPS inhibition of HC protein synthesis in KC–HC coculture fit all the criteria for involvement of the L-arginine-dependent production of ·NO, that is, L-arginine and only L-arginine was needed, the production of the end products of the pathway corresponded with the effect, NMA competitively reversed the effect, and the biologic effect could be reproduced by the addition of ·NO or ·NO donors. It should also be pointed out that the decrease in protein synthesis seen either with the addition of ·NO or endogenous ·NO production was associated with almost no HC death. The HC damage that could be detected did require L-arginine but also high KC to HC ratios (8:1) (Billiar et al., 1989c).

An important question remained unanswered from these initial studies. What cell in the KC and HC cocultures was the source of the ·NO? At the time of these initial studies, it was already known that macrophages produce ·NO, therefore it was not unexpected to find that KC also produce ·NO in response to LPS. However, KC ·NO production only accounted for a fraction of all of the ·NO released by KC–HC coculture (Figure 2).

A series of supernatant transfer studies uncovered the fact that HC also produce ·NO, and in relatively large quantities (Curran et al., 1989). A supernatant generated from KC 8 hr after exposure to LPS and interferon-gamma (IFNγ) stimulated HC ·NO biosynthesis and caused a simultaneous decrease in protein synthesis. As seen in the coculture model, a 6- to 8-hr delay in measurable NO_2^- + NO_3^- synthesis was seen in the HC exposed to conditioned KC supernatant. However, unlike KC or KC + HC where equimolar concentrations of citrulline were released into the culture supernatant, no increase in citrulline release paralleled the NO_2^- + NO_3^- synthesis by stimulated HC. It is most likely that the citrulline enters the HC urea cycle and is not secreted.

A specific combination of cytokines plus LPS were found to mimic the KC supernatant. A mixture of IL-1, tumor necrosis factor (TNF), IFNγ, and LPS acted synergistically to induce HC ·NO synthesis from L-arginine in an identical manner to KC supernatant (Curran et al., 1990). Binary or tertiary combinations of these signals also induced ·NO biosynthesis but to lesser degrees. This finding

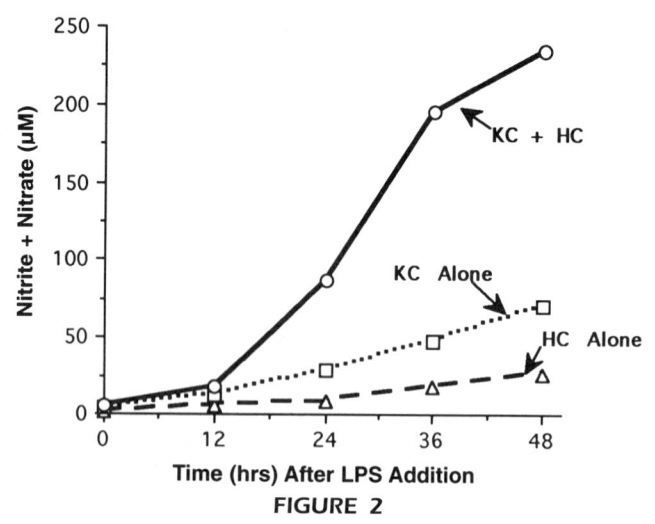

FIGURE 2

NO_2^- and NO_3^- levels in cultures of HC, KC, or cocultures (KC + HC) are shown following the addition of 10 μg/ml LPS.

was significant because it was the first description of inducible ·NO synthesis by a parenchymal and nonimmune cell type. It also represents the first demonstration of novel enzyme biosynthesis by HC in response to cytokines, and the first known response by HC to inflammatory stimuli that is regulated and promoted by four signals acting synergistically. Most recently, we have demonstrated inducible ·NO biosynthesis by cultured human HC exposed to these same four signals (Nussler et al., 1992a). Human macrophages fail to respond as rodent cells to similar signals. However, HC respond in time, quantities, and signals in a near-identical manner to cells obtained from rats. Until this finding, the presence of an inducible ·NO synthase (iNOS) in human cells remained in question. This finding permitted us to isolate and sequence a cDNA for the iNOS clone from human HC, representing the first iNOS cloned from human tissue. The approximately 4.4-kb clone shares about 80% nucleic acid sequence similarity with the murine macrophage cDNA, which was used to isolate the human NOS clone (Nussler et al., 1992b).

The availability of the molecular probe for inducible NOS has permitted a more detailed examination of the regulation of iNOS expression in HC (Geller et al., 1993a). Using Northern blot analysis, we found that message is present for iNOS 2 hr following exposure to cytokines plus LPS. The mRNA levels peak at 6–8 hr. As single agents, TNF, IL-1, and IFNγ all induce gene expression while LPS has no effect. Strong synergy can be seen between TNF and IL-1, as well as TNF and IFNγ, but the greatest increase in mRNA levels occurs with the combination of all four signals. A similar response has been observed in cultured human liver cells exposed to all four signals with the only major difference being a much sharper reduction in mRNA levels by 24 hr in human cells. Studies into the *in vivo* induction of iNOS mRNA in liver cells has revealed a strong and chronic induction following *Corynebacterium parvuym* injection and small but significant induction 6 hr following LPS injection.

III. BIOCHEMISTRY OF HEPATOCYTE NITRIC OXIDE SYNTHESIS

Models of *in vivo* induction of HC ·NO synthesis were pursued in order to obtain adequate quantities of materials to study HC ·NO synthase regulation. The logical stimulation for *in vivo* induction was LPS along with IFNγ, but minimal ·NO synthase activity could be detected in HC cytosol obtained from rats injected with LPS (Billiar et al., 1990a) or LPS + IFNγ (Billiar et al., 1990b). In striking contrast, but in good correlation with the mRNA levels, HC isolated from rats previously injected intravenously with killed *C. parvum* (28 mg/kg) produced large amounts of $NO_2^- + NO_3^-$ spontaneously when placed in culture and contained high amounts of cytosolic ·NO synthase activity. This expression of ·NO synthase activity was first detected 3 days after *C. parvum* injection and persisted until day 8 or 9.

When subjected to the purification scheme described for the macrophage ·NO synthase (Stuehr et al., 1991b), a near-identical band on polyacrylamide gel electrophoresis was identified and corresponded to the samples containing large amounts of ·NO synthase activity. As with the macrophage, two closely spaced bands, 125,000–130,000 in molecular weight, were isolated (Figure 3).

Studies on partially purified preparations of HC ·NO synthase indicated a similar dependency on tetrahydrobiopterin (BH_4), reduced glutathione (GSH), FAD, and NADPH as described for the macrophage (Kwon et al., 1989; Stuehr et al., 1989b, Tayeh and Marletta, 1989) (see Table 1). Similar data have been obtained using cytosol from human HC stimulated with cytokines + LPS in culture.

We have further investigated the dependency of HC ·NO synthesis on cellular GSH levels and BH_4 availability. Inhibition of GSH synthesis using the γ-glutamylcysteine synthetase inhibitor buthionine sulfoximine, which blocks *de novo* GSH synthesis, markedly reduced GSH levels in cultured HC, approximately 5% of control, but resulted in only a 40–50% reduction in NO_2^- biosynthesis in response to cytokines and LPS. More effective was the inhibition of GSH reductase with 1,3-bis(chloroethyl)-1-nitrosurea, which prevents the recycling of GSSH back to GSH. GSH levels also fell in these cells, but a marked decrease in NO_2^- formation was seen, suggesting that GSH recycling was an important aspect of ·NO formation. Similar results had been reported for con-

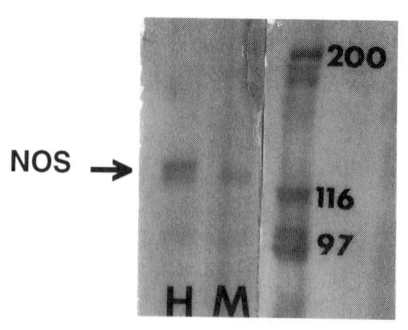

FIGURE 3

Analysis by SDS–polyacrylamide gel electrophoresis of purified NOS from isolated HC from rats injected with killed *Corynebacterium parvum* (H), and from the murine macrophage cell line RAW 264.7 (M) (courtesy of D. Stuehr, The Cleveland Clinic, Cleveland, OH) which was exposed to LPS and IFNγ. Crude cytosols were separated using ion exchange and affinity 2′5′-ADP-Sepharose chromatography. Last step by gel filtration is equivalent to separation by molecular weight.

stitutive ·NO production by endothelial cells (Murphy et al., 1991); however, cell damage was associated with a reduction in cell GSH levels, suggesting that nonspecific cell injury contributed to the unpaired ·NO synthesis. No evidence for injury paralleled the decreased ·NO production seen with reduced GSH in HC.

Perhaps even more important to inducible ·NO production than GSH levels is BH_4 availability. BH_4 levels are dependent on both *de novo* synthesis and recycling pathways. *De novo* synthesis occurs from GTP where the first step is

TABLE 1

Hepatocyte Mitochondrial Enzyme Activity after Exposure to Exogenous Nitric Oxide[a] or 18 hr after Induction of Endogenous Nitric Oxide Synthesis Using Cytokines plus Lipopolysaccharide[b]

Enzyme	Exogenous ·NO	Endogenous ·NO (−)NMA[c]	(+)NMA
Aconitase	21	71 (17)	88
Complex I	43	77 (4)	81
Complex II	63	79 (6)	85
Complex III	108	ND[d]	ND
Complex IV	98	86 (5)	91

[a] Exogenous ·NO was provided as a 20% saturated solution for 5 min.
[b] Activities are given as the percentage of activities of control HC.
[c] Values in parentheses show degree of decrease due to ·NO [(−)NMA − (+)NMA].
[d] ND, Not determined.

catalyzed by the enzyme GTP cyclohydrolase-I (Hatakeyama et al., 1989), which is the rate-limiting enzyme for BH_4 formation. Recycling of dihydrobiopterin (BH_2, oxidized form) back to the reduced cofactor BH_4 occurs via the action of either dihydrofolate reductase or dihydropteridine reductase. In HC we have found that inhibition of GTP cyclohydrolase (i.e., de novo synthesis) has a pronounced effect on ·NO synthesis in response to cytokine + LPS. In cells such as fibroblasts, stimuli for ·NO production also stimulate GTP cyclohydrolase activity (Werner-Felmayer et al., 1990). HC constitutively express moderate levels of mRNA for GTP cyclohydrolase, most likely to provide BH_4 for other HC enzymes which utilize this cofactor. When HC are stimulated to produce ·NO either in vivo (C. parvum) or in vitro (cytokines + LPS), a marked increase in GTP cyclohydrolase mRNA levels have been measured by Northern blot analysis. Following in vivo induction of ·NO synthesis, inhibitors of BH_2–BH_4 recycling have little effect on HC ·NO production, suggesting that ongoing de novo BH_4 production is more essential for supporting inducible HC ·NO production.

IV. STUDIES ON *IN VITRO* ACTIONS OF NITRIC OXIDE IN LIVER CELLS

In our original assay system, we detected the presence of ·NO in liver cells when we attempted to explain the decrease in HC protein synthesis seen in our endotoxin-treated cocultures. We have since performed a number of studies to determine the effects of ·NO on liver cell function in vitro. Perhaps the most striking effect is the profound decrease in total protein synthesis. This was seen in KC–HC coculture (Billiar et al., 1989a), as well as HC cultured alone, exposed to conditioned KC supernatant (Curran et al., 1989) or cytokines + LPS (Curran et al., 1990). Like ·NO biosynthesis, the decrease in protein synthesis was blocked by NMA and could be reproduced by exposure to authentic ·NO gas as ·NO-releasing compounds, including sodium nitroprusside or S-nitro-N-acetylpenicillamine (Curran et al., 1991). In fact, profound decreases in [^{3}H]leucine or [^{35}S]methionine were seen in cultured HC exposed to ·NO-saturated solutions for as little as 2 min. The decrease persisted for up to 12–16 hr following the removal of the ·NO-containing solution. Simultaneous decreases were measured in the synthesis of the two HC proteins, albumin and fibrinogen. No decreases in HC mRNA levels for albumin were seen at time points where albumin synthesis was depressed by ·NO, suggesting that ·NO decreased total protein synthesis by a translational or posttranslational mechanism (Curran et al., 1991).

Two actions of ·NO had been well-described in other cell types. Induction of cGMP synthesis by ·NO produced by the constitutive enzyme in the brain (Bredt et al., 1990) and endothelial cells (Ignarro et al., 1987) indicated that small amounts of ·NO acted as an intra- or intercellular messenger molecule. The larger amounts of ·NO generated by the inducible enzyme in macrophages were found to have cytostatic properties by interfering with the function of three enzymes containing Fe–sulfur (Drapier and Hibbs, 1988; Stuehr and Nathan, 1989), namely aconitase of the Krebs cycle, and complex I and complex II of the electron transport chain. The effects of exogenous and endogenous ·NO were tested on both cGMP synthesis and mitochondrial respiration in cultured HC.

HC are known to contain a soluble guanylate cyclase which is responsive to ·NO-releasing substances (Wood and Ignarro, 1987). Therefore, it seemed likely that ·NO generated by HC would result in increases in HC cGMP levels. We first confirmed that exogenous ·NO would increase HC cGMP levels in culture (West et al., 1989). Second, we showed that HC stimulated to produce ·NO by exposure to TNF, IL-1, IFNγ, and LPS, in combination, greatly enhanced cGMP biosynthesis but that most of the cGMP appeared in the culture supernatant and not in the cells. Addition of the nonspecific phosphodiesterase inhibitor, iso-methyl-butyl-xanthine, increased cGMP levels while the specific cGMP phosphodiesterase inhibitor had no effect (Billiar et al., 1992). Single cytokines or LPS alone had minimal effects on cGMP release when compared to the four signals together (Figure 4).

Hepatocytes isolated from rats injected with C. parvum also had increased ·NO cGMP levels while cGMP levels were approximately 30% higher in these animals. These data suggested that inducible ·NO synthesis in HC may act as

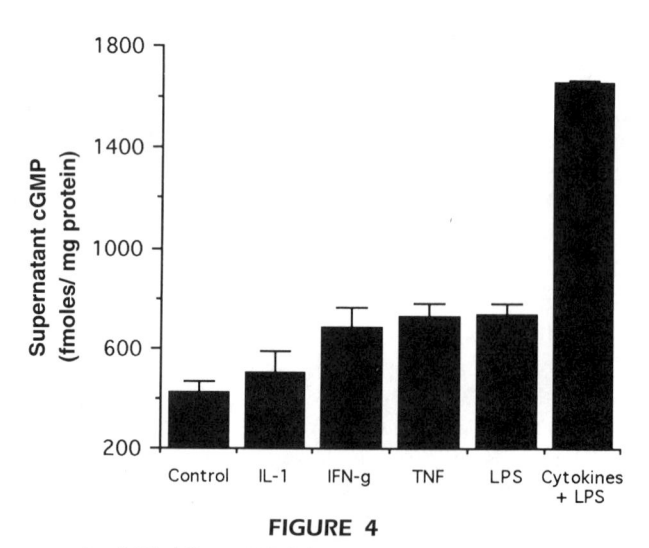

FIGURE 4

Rat HC were exposed to 5 U/ml IL-1, 100 U/ml IFNγ, 500 U/ml TNF, 10 μg/ml LPS, or a combination of all four agents. Supernatant cGMP levels were determined 18 hr later.

an intracellular messenger resulting in chronic increases in cGMP synthesis and release by HC. The action of cGMP in HC is poorly understood. We do know from experiments where 8-bromo-cGMP was added directly to HC that cGMP does not appear to account for the ·NO-induced decrease in protein synthesis (West et al., 1989).

Inhibition of certain mitochondrial enzymes with impairment of oxidative metabolism is an important feature of the cytostatic action of macrophage ·NO on tumor cells (Drapier and Hibbs, 1988; Stuehr and Nathan, 1989). In addition, macrophages stimulated to produce ·NO *in vitro* have been shown to impair their own respiration through the production of ·NO. The hypothesis that HC may also be subject to such an inhibition was tested in both *in vitro* and *in vivo* models of ·NO production in rat HC. However, prior to these studies, we first exposed HC to 20% ·NO-saturated solutions to determine if HC mitochondrial aconitase, Complex I, and Complex II were sensitive to ·NO inhibition. In these experiments, HC in solution were exposed to 20% ·NO-saturated solutions (using deoxygenated buffer) for 5 min (Stadler et al., 1991a). The HC were then permeated and substrate-dependent O_2 consumption was determined using a Clark-type electrode. As seen with tumor targets, a significant degree of inhibition was seen in the activity of mitochondrial aconitase, but not cytosolic aconitase, complex I, and complex II (Table 1). No decrease was seen in the activity of complex III or IV, indicating the selective nature of the ·NO effect. In addition, no change in activity of cytosolic aconitase was seen, suggesting that ·NO may target certain subcellular sites. HC were then induced to produce ·NO by exposure to cytokines and LPS, then permeated, and mitochondrial enzyme activity determined. In contrast to the exogenous ·NO, endogenously produced ·NO resulted in only minor (~10–15%) decreases in mitochondrial aconitase activity, as indicated by a reversal by NMA (Table 1). Subsequent studies have shown that the majority of the suppression seen with cytokine addition is due to TNF via a poorly characterized mechanism (Stadler et al., 1992). Virtually no inhibition of mitochondrial activity was seen in HC isolated from rats induced to synthesize ·NO *in vivo* by C. parvum injection (Stadler et al., 1991b). Taken together, these data suggest that HC possess protective mechanisms to prevent or reverse the action of ·NO on mitochondrial respiration.

Our studies with mitochondrial respiration also ruled out the hypothesis that ·NO-induced impairment of mitochondrial function accounted for the decrease in protein synthesis. The reasons for this conclusion include: first, a minimal decrease in aconitase activity and essentially no decrease in complex I or complex II activity was seen in HC exposed to KC supernatant or cytokines + LPS, despite a marked reduction in protein synthesis; and second, exposure of HC to ·NO resulted in a profound and prolonged decrease in HC protein synthesis (up to 18 hr), but a very short-lived decrease in mitochondrial respiration (90 min)

(Figure 5). We have postulated that the decrease in protein synthesis occurs via an independent mechanism and that this action of ·NO may be important in its cytostatic properties or in the regulation of certain cellular functions.

Further studies on the role of ·NO in regulating cellular functions have been performed in KC stimulated to produce ·NO. KC represent the largest population of fixed macrophages in the body, and the primary site for clearance of circulating endotoxin. As with other macrophages, it is not surprising that KC produce ·NO when stimulated by LPS and that this ·NO production is further upregulated by IFNγ. A somewhat unique property of KC is the fact that they do not appear to produce ·NO in response to IFNγ alone (Stadler et al., 1993) as do other macrophages. We have investigated whether ·NO has regulatory actions on the secretory activity of KC. In the presence of NMA, which blocks ·NO synthesis, KC produce 3–5 times more prostaglandin E_2 (PGE_2) in response to LPS and IFNγ, an effect reversed by excess L-arginine. Using an assay which directly measures cyclooxygenase activity, we were able to determine that ·NO synthesis was associated with a marked reduction in the activity of this enzyme essential to PGE_2 formation. Cytokine synthesis may also be influenced by ·NO synthesis. Greater IL-6, but not TNF nor IL-1, activity was detected in the supernatants of KC exposed to IFNγ + LPS + NMA than in cells not incubated with NMA. The data suggest that other intracellular targets for ·NO (i.e., cyclooxygenase) may exist and that ·NO may selectively influence or regulate all enzyme activity.

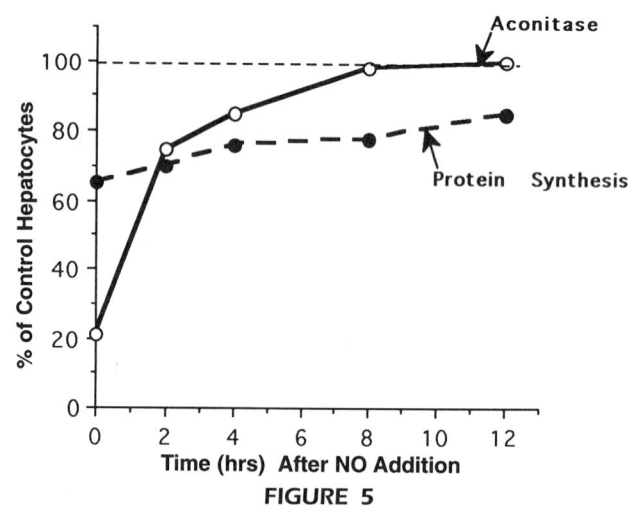

FIGURE 5

Hepatocyte total protein synthesis and aconitase activity following exposure to a 50% saturated ·NO solution under anaerobic conditions for 5 min.

V. STUDIES ON *IN VIVO* ACTIONS OF NITRIC OXIDE IN THE LIVER

Turning again to our original hypothesis, we had postulated that KC released substances that altered HC function in sepsis. We had suggested that KC may even interfere with HC function and believed that the depressed protein synthesis could represent an example of KC interference of normal HC function. In discovering ·NO synthesis in our cocultures, as well as KC and HC cultured alone, we believed that we might have identified an important effector mechanism for altering liver function in inflammatory or septic states, and sought models of *in vivo* ·NO synthesis to study the role of ·NO *in vivo*. Injections of killed C. *parvum*, a known stimulus for hepatic inflammation and endotoxin hypersensitivity, resulted in massive and prolonged induction of NOS in HC. This induction was first detected at day 3 and persisted beyond day 8 (Geller *et al.*, 1993b) as measured by the detection of ·NO synthase activity in the cytosol of isolated HC and by measuring mRNA levels for inducible ·NO synthase. The *in vivo* signals which induce this production of ·NO in HC have not been completely elucidated but we do know that antibodies to TNF partially prevent the induction of ·NO synthesis in C. *parvum*-treated animals. In addition, it is not currently whether this induction of ·NO synthesis occurs predominantly in a specific zone of the liver lobule.

This model of known HC ·NO synthesis and hypersensitivity to endotoxin was then used to examine the role of ·NO in the liver under inflammatory conditions (Billiar *et al.*, 1990c). Mice were injected with killed C. *parvum* (1 mg/mouse), and 7 days later they received 1 μg *Escherichia coli* LPS intravenously. Then 5 hr later, plasma was obtained for circulating NO_2^- + NO_3^- levels and for liver enzymes [ornithine carbamoyltransferase (OCT), aspartate aminotransferase (AST), lactate dehydrogenase (LDH)] to biochemically assess the degree of liver damage. The injection of C.*parvum* alone resulted in a nearly 10-fold increase in circulating NO_2^- + NO_3^- levels and mild hepatic injury. Injection of NMA (250 mg/kg, IV), 5 hr prior to obtaining the blood, had minimal influences on nitrogen oxide levels or liver damage. *Corynebacterium parvum*-treated mice exhibited a sudden and dramatic increase in plasma NO_2^- + NO_3^- levels on LPS injection, which was effectively prevented by NMA treatment. As expected, LPS injection also resulted in liver necrosis. NMA treatment resulted in a two- to threefold increase in the degree of liver damage. The NMA effect was partially prevented by simultaneous L-arginine administration. Lower doses of NMA were less effective at inhibiting the NO_2^- + NO_3^- increases and caused less hepatic damage. Histologic evaluation of the livers from these animals revealed that the LPS-injected mice which also received NMA had increased liver necrosis in the form of microinfarcts and platelet thrombi within the small vessels. Similar, although less severe, results have been obtained in our laboratory using LPS without C. paravum pretreatment (Harbrecht et al., 1992a). L-Arginine injection had no effect in the C. parvum + LPS model, but did increase NO_2^- + NO_3^- levels late when injected with LPS alone (Billiar et al., 1990d).

We have investigated further the mechanism by which ·NO inhibitor increases end organ (liver) damage in endotoxemia. Two hypotheses have been pursued. It has been proposed that ·NO can combine with superoxide (O_2^-) to form NO_3^-, a nontoxic end product. Others have suggested that peroxynitrite with subsequent formation of hydroxyl radical may result from the ·NO + O_2^- interaction (Beckman et al., 1990). Because it is known that O_2^- plays a role in the liver damage in the C. parvum + LPS model (Arthur et al., 1985), we first looked into the role of O_2^- in the enhancement injury found with NMA treatment. Administration of superoxide dismutase (SOD) with NMA in C. parvum + LPS treated mice abrogated the NMA-associated liver injury, suggesting that O_2^- did contribute to the damage (Harbrecht et al., 1992b). Deferoxamine had a similar effect, supporting a role for hydroxyl radicals. The O_2^--induced injury could result from a loss of the O_2^- neutralizing effect of ·NO. Alternatively, the inhibition of ·NO could result in vasoconstriction, sludging, and clot formation with subsequent ischemia and enhanced O_2^- production. Heparin treatment also partially reversed the NMA-induced injury, supporting the latter hypothesis.

To further investigate the role of disturbances in blood flow resulting from platelet adhesion and aggregation, studies were performed using aspirin (ASA) to inhibit prostacyclin formation. Administration of ASA 3 hr prior to LPS injection inhibits prostacyclin synthesis. When ASA was administered in this fashion with NMA in the C. parvum + LPS model, massive necrosis ensued (Harbrecht et al., 1994). This five-fold increase in liver damage was prevented by coadministration of prostacyclin. These results suggest that ·NO participates with prostacyclin to maintain blood flow in low flow states in the liver, associated with inflammation. Certainly other functions for ·NO exist (i.e., cGMP-mediated functions), but in states of large ·NO release a function in liver hemodynamics seems certain.

Brune and Lapetina (1989) reported that ·NO could activate a platelet ADP-ribosyltransferase that resulted in the ribosylation of a 39 kDa protein. Subsequent work revealed that the protein was glyceraldehyde phosphate dehydrogenase (GAP-DH), and that ribosylation was associated with reduced GAP-DH activity (Dimmeler et al., 1992). In our collaboration with Molina et al., (1992), we have shown that GAP-DH activity is dramatically inhibited in C. parvum-treated rats and that this action is associated with both a ribosylation and nitrosylation of the enzyme. Such a marked inhibition of a glycolytic enzyme could explain some of the metabolic changes observed in the liver in sepsis.

In macrophages, ·NO has well-described antimicrobial actions, especially against intracellular pathogens. Nussler et al., (1991), investigating the role of liver ·NO in the development of the hepatic stage of malaria, showed that ·NO prevents malarial growth in HC. This inhibition requires only small amounts of ·NO formation and may be stimulated by the parasite itself. These authors have also proposed a role for liver nonparenchymal cell–HC interaction and IL-6 induction of ·NO synthesis in the liver as an antiparasite effector molecule. It is interesting to note that HC possess such a potent antimicrobial effect mechanism, a function previously thought to exist only in immune cells.

VI. SUMMARY

Rodent KC and HC, as well as human HC, express an inducible ·NO synthase under septic or inflammatory conditions. *In vivo* in endotoxemia, this expression is transient. Our *in vivo* data indicate that this induced ·NO serves a protective role in the liver and reduces hepatic injury in endotoxemia. This protective action may be mediated by the capacity of ·NO to neutralize oxygen radicals and prevent platelet adherence and aggregation. Our *in vitro* studies show that HC-derived ·NO can activate soluble guanylate cyclase. Other *in vitro* effects include the nonspecific suppression of protein synthesis and a small reduction in mitochondrial aconitase activity. The relevance of these *in vitro* actions to hepatic function *in vivo* remains to be determined.

REFERENCES

Arthur, M. J. P., Bentley, F. S., Tanner, A. R., Kowalski-Saunders, P., Millward-Sadler, G. H., and Wright, R. (1985). *Gastroenterology* **89,** 1114–1122.

Beckman, J. S., Beckman, T. W., Chen, J., Marshall, P. A., and Freeman, B. A. (1990). *Proc. Natl. Acad. Sci. U.S.A.* **87,** 1620–1624.

Billiar, T. R., Curran, R. D., Stuehr, D. J., West, M. A., Bentz, B. G., and Simmons, R. L. (1989a). *J. Exp. Med.* **169,** 1467–1472.

Billiar, T. R., Curran, R. D., Stuehr, D. J., Ferrari, F. K., and Simmons, R. L. (1989b). *Surgery* **106,** 364–372.

Billiar, T. R., Curran, R. D., West, M. A., Hofmann, K., and Simmons, R. L. (1989c). *Arch. Surg.* **124,** 1416–1420.

Billiar, T. R., Curran, R. D., Stuehr, D. J., Stadler, J., Simmons, R. L., and Murray, S. A. (1990a). *Biochem. Biophys. Res. Commun.* **168,** 1034–1040.

Billiar, T. R., Curran, R. D., Stadler, J., Kispert, P. H., Kim, R., Harbrecht, B., Ochoa, J., Williams, D., and Simmons, R. L. (1990b). *Surg. Forum* **41,** 64–65.

Billiar, T. R., Curran, R. D., Harbrecht, B. G., Stuehr, D. J., Demetris, A. J., and Simmons, R. L. (1990c). *J. Leukocyte Biol.* **48,** 565–569.

Billiar, T. R., Curran, R. D., Stuehr, D. J., Hofmann, K., and Simmons, R. L. (1990d). In "Nitric Oxide From L-Arginine," p. 275. Experta Medical, Amsterdam.

Billiar, T. R., Curran, R. D., Harbrecht, B. G., Stadler, J., Williams, D. L., Ochoa, J. B., Di Silvio, M., Simmons, R. L., and Murray, S. A. (1992). Am. J. Physiol. **262**, C1077-C1082.

Bredt, D. S., and Snyder, S. H. (1992). Neuron **8**, 3-11.

Bredt, D. S., Hwang, P. M., and Snyder, S. H. (1990). Nature (London) **347**, 768-770.

Brune, B., and Lapetina, E. G. (1989). J. Biol. Chem. **264**, 8455-8458.

Burnett, A. L., Lowenstein, C. J., Bredt, D. S., Chang, T. S. K., and Snyder, S. H. (1992). Science **257**, 401-403.

Cho, H. J., Xie, Q., Calaycay, J., Mumford, R. A., Swiderek, K. M., Lee, T. D., and Nathan, C. (1992). J.Exp. Med. **176**, 599-604.

Curran, R. D., Billiar, T. R., Stuehr, D. J., Hofmann, K., and Simmons, R. L. (1989). J. Exp. Med. **170**, 1769-1774.

Curran, R. D., Billiar, T. R., Stuehr, D. J., Ochoa, J. B., Harbrecht, B. G., Flint, S. G., and Simmons, R. L. (1990). Ann. Surg. **212**, 462-469.

Curran, R. D., Ferrari, F. K., Kispert, P. H., Stadler, J., Stuehr, D. J., Simmons, R. L., and Billiar, T. R. (1991). FASEB J. **5**, 2085-2092.

Dimmeler, S., Lottspeich, F., and Brune, B. (1992). J. Biol. Chem. **267**, 16771-16774.

Drapier, J.-C., and Hibbs, J. B., Jr. (1988). J. Immunol. **140**, 2829-2838.

Geller, D. A., Nussler, A. K., Di Silvio, M., Lowenstein, C. J., Shapiro, R. A., Wang, S. C., Simmons, R. L., and Billiar, T. R. (1993a). Proc. Natl. Acad. Sci. U.S.A. **90**, 522-526.

Geller, D. A., Lowenstein, C. J., Shapiro, R. A., Nussler, A. K., Di Silvio, M., Wang, S. C., Nakayama, D. K., Simmons, R. L., Snyder, S. H., and Billiar, T. R. (1993b). Proc. Natl. Acad. Sci. U.S.A. **90**, 3491-3495.

Green, L. C., Ruiz de Luzuriaga, K., Wagner, D. A., Rand, W., Istfan, N., Young, V. R., and Tannenbaum, S. R. (1981). Proc. Natl. Acad. Sci. U.S.A. **78**, 7764-7768.

Harbrecht, B. G., Billiar, T. R., Stadler, J., Demetris, A. J., Ochoa, J. B., Curran, R. D., and Simmons, R. L. (1992a). Crit. Care Med. **20**, 1568-1574.

Harbrecht, B. G., Billiar, T. R., Stadler, J., Demetris, A. J., Ochoa, J., Curran, R. D., and Simmons, R. L. (1992b). J. Leukocyte Biol. **52**, 390-394.

Harbrecht, B. G., Stadler, J., Demetris, A. J., Simmons, R. L., and Billiar, T. R. (1994). Am. J. Physiol. **266**, G1004-G1010.

Hatakeyama, K., Harada, T., Suzuki, S., Watanabe, Y., and Kagamiyama, H. (1989). J. Biol. Chem. **264**, 12660-12664.

Hibbs, J. B., Taintor, R. R., and Vavrin, Z. (1987a). Science **235**, 473-476.

Hibbs, J. B. Jr., Vavrin, Z., and Taintor, R. R. (1987b). J. Immunol. **138**, 550-565.

Hibbs, J. B., Jr., Taintor, R. R., Vavrin, Z., and Rachlin, E. M. (1988). Biochem. Biophys. Res. Commun. **157**, 87-94.

Ignarro, L. J., Buga, G. M., Wood, K. S., Byrns, R. E., and Chaudhari, B. (1987). Proc. Natl. Acad. Sci. U.S.A. **84**, 9265-9269.

Kaplan, S. S., Billiar, T. R., Curran, R. D., Zdziarski, U. E., Simmons, R. L., and Basford, R. E. (1989). Blood **74**, 1885-1887.

Kilbourne, R. G., and Belloni, P. (1990). J. Natl. Cancer Inst. **82**, 772-776.

Kispert, P., Curran, R. D., Billiar, T. R., and Simmons, R. L. (1990). *Surg. Forum* **41**, 80–83.
Kwon, N. S., Nathan, C. F., and Stuehr, D. J. (1989). *J. Biol. Chem.* **264**, 20496–20501.
Kwon, N. S., Nathan, C. F., Gilker, C., Griffith, O. W., Mathews, D. E., and Stuehr, D. J. (1990). *J. Biol. Chem.* **265**, 13442–13445.
Kwon, N. S., Stuehr, D. J., and Nathan, C. F. (1991). *J. Exp. Med.* **174**, 761–767.
Lancaster, J. R. (1992). *Am. Sci.* **80**, 248–259.
Lapoivre, M., Chenais, B., Yapo, A., Lemaire, G., Thelander, L., Tenu, J. P. (1990). *J. Biol. Chem.* **265**, 14143–14149.
Leone, A. M., Palmer, R. M. J., Knowles, R. G., Francis, P. L., Ashton, D. S., and Moncada, S. (1991). *J. Biol. Chem.* **266**, 23790–23795.
Mellion, B. T., Ignarro, L. J., Ohlstein, E. H., Pontecarvo, E. G., Hyman, A. L., and Kadowitz, P. J. (1981). *Blood* **57**, 946–955.
Molina y Vedia, L., McDonald, B., Reep, B., Brune, B., Di Silvio, M., Billiar, T. R., and Lapetina, E. G. (1992). *J. Biol. Chem.* **267**, 24929–24932.
Moncada, S., Palmer, R. M. J., and Higgs, E. A. (1991). *Pharmacol. Rev.* **43**, 109–142.
Murphy, M. E., Piper, H. W., Watanabe, H., and Sies, H. (1991). *J. Biol. Chem.* **266**, 19378–19383.
Nussler, A., Drapier, J.-C., Renia, L., Pied, S., and Miltgen, F. (1991). *Eur. J. Immunol.* **21**, 227–230.
Nussler, A., Di Silvio, M., Billiar, T. R., Hoffman, R. A., Geller, D. A., Selby, R., Madariaga, J., and Simmons, R. L. (1992a). *J. Exp. Med.* **176**, 261–264.
Nussler, A. K., Geller, D. A., Shapiro, R. A., Di Silvio, M., Wang, S. C., Simmons, R. L., and Billiar, T. R. (1992b). *J. Leukocyte Biol.* **S3**, 37.
Palmer, R. M. J., Ferrige, A. G., and Moncada, S. (1987). *Nature (London)* **327**, 524–526.
Palmer, R. M. J., Ashton, D. S., and Moncada, S. (1988). *Nature (London)* **333**, 664–666.
Radomski, M. W., Palmer, R. M. J., and Moncada, S. (1987). *Lancet* **2**, 1057–1058.
Stadler, J., Billiar, T. R., Curran, R. D., Stuehr, D. J., Ochoa, J. B., and Simmons, R. L. (1991a). *Am. J. Physiol.* **260**, C910–C916.
Stadler, J., Curran, R. D., Ochoa, J. B., Harbrecht, B. G., Hoffman, R. A., Simmons, R. L., and Billiar, T. R. (1991b). *Arch. Surg.* **126**, 186–191.
Stadler, J., Bentz, B. G., Harbrecht, B. G., Di Silvio, M., Curran, R. D., Billiar, T. R., Hoffman, R. A., and Simmons, R. L. (1992). *Ann. Surg.* **216**, 539–546.
Stadler, J., Harbrecht, B. G., Di Silvio, M., Curran, R. D., Jordan, M. L., Simmons, R. L., and Billiar, T. R. (1993). *J. Leukocyte Biol.* **53**, 165–172.
Stuehr, D. J., and Marletta, M. A. (1985). *Proc. Natl. Acad. Sci. U.S.A.* **82**, 7738–7742.
Stuehr, D. J., and Nathan, C. F. (1989). *J. Exp. Med.* **169**, 1543–1555.
Stuehr, D. J., Gross, S. S., Sakuma, I., Levi, R., and Nathan, C. F. (1989a). *J. Exp. Med.* **169**, 1011–1020.
Stuehr, D. J., Kwon, N. S., Gross, S. S., Thiel, B., Levi, R., and Nathan, C. F. (1989b). *Biochem. Biophys. Res. Commun.* **161**, 420–426.
Stuehr, D. J., Kwon, N. S., Nathan, C. F., Griffith, O. W., Feldman, P. L., and Wiseman, J. (1991a). *J. Biol. Chem.* **226**, 6259–6263.

Stuehr, D. J., Cho, H. J., Kwon, N. S., Weise, M. F., and Nathan, C. F. (1991b). *Proc. Natl. Acad. Sci. U.S.A.* **88,** 7773–7777.

Tayeh, M. A., and Marletta, M. A. (1989). *J. Biol. Chem.* **264,** 19654–19658.

Wagner, D. A., Young, V. R., and Tannenbaum, S. R. (1983). *Proc. Natl. Acad. Sci. U.S.A.* **80,** 4518–4521.

Werner-Felmayer, G., Werner, E. R., Fuchs, D., Hausen, A., Reibnegger, G., and Wachter, H. (1990). *J. Exp. Med.* **172,** 1599–1607.

West, M. A., Billiar, T. R., Curran, R. D., Hyland, B. J., and Simmons, R. L. (1989). *Gastroenterology* **96,** 1572–1582.

Wood, K. S., and Ignarro, L. J. (1987). *J. Biol. Chem.* **262,** 5020–5027.

7
Role of Nitric Oxide in Allograft Rejection

Rosemary A. Hoffman, Jan M. Langrehr, and
Richard L. Simmons
Department of Surgery
University of Pittsburgh
Pittsburgh, Pennsylvania 15261

I. INTRODUCTION

Intense effort has been applied to define and overcome obstacles encountered during the transplantation of histoincompatible tissue. Allograft rejection is normally initiated by the presentation to T lymphocytes of histocompatibility antigens by antigen presenting cells of bone marrow derived monocyte–macrophage lineage. Complex cellular interactions ensue, including synthesis of inflammatory mediators and cytokines by lymphocytes, macrophages, and other inflammatory cells which result in activation of host cells capable of destroying donor tissue. The central role that various T lymphocyte subsets play in allograft rejection has been defined by studies in which T helper cells (CD4$^+$) have been shown to produce cytokines that facilitate B cell antibody production as well as acquisition of antidonor cytolytic function by CD8$^+$ T cells.

The dramatic effects observed when nitric oxide ($\cdot N{=}O$) synthase is induced in cocultures of macrophages and tumor cells (Hibbs *et al.*, 1987; Stuehr and Nathan, 1989) as well as Kupffer cells and hepatocytes (Billiar *et al.*, 1989;

Curran et al., 1989) led us to examine whether $\cdot N{=}O^1$ synthase is also induced in macrophage–lymphocyte cocultures. While macrophages play an essential role in the immune response via antigen presentation in the context of class I or II antigen and provision of a costimulus, inhibition of lymphocyte proliferation by macrophages is also a well-described phenomenon. Investigators have shown that macrophage synthesis of reactive oxygen intermediates, prostaglandins, polyamine oxidase, thymidine kinase, and arginase can result in inhibition of lymphocyte proliferation (Calderon et al., 1974; Allison, 1978; Metzger et al., 1980). It is now clear that synthesis of $\cdot N{=}O$ is another mechanism whereby macrophages regulate lymphocyte proliferation. However, the mechanism of $\cdot N{=}O$ induced inhibition of lymphocyte activation remains to be elucidated.

II. NITRIC OXIDE PRODUCTION IN RAT SPLENOCYTE MIXED LYMPHOCYTE REACTION

Coculture of lymphocyte populations from two histoincompatible organisms and assessment of proliferation is used as the *in vitro* analogue of allograft rejection. In contrast to the mouse species, studies of immune reactivity in the rat species using splenocyte populations have been difficult to perform (Folch and Waksman, 1973; Weiss and Fitch, 1977; Holt et al., 1981). However, addition of NMA (N^G-monomethyl-L-arginine, an inhibitor of $\cdot N{=}O$ synthase) to rat splenic mixed lymphocyte cultures promotes proliferation and allospecific cytolytic T lymphocyte (CTL) induction coincident with a decrease in supernatant NO_2^-/NO_3^- and citrulline levels (Hoffman et al., 1990). The data imply that the alloimmune interaction of two histoincompatible-splenocyte populations will trigger $\cdot N{=}O$ synthase, presumably via production of cytokines. Cytokines able to induce T lymphocyte proliferation were present in the allostimulated culture supernatants, whether or not NMA was added to inhibit $\cdot N{=}O$ production. These data indicate that in these cultures, the T cell receptor has been engaged and that cytokine synthesis was initiated but that $\cdot N{=}O$ synthesis results in the inability of T cells to utilize the cytokines that are produced. In contrast, addition of NMA to the mouse splenocyte mixed lymphocyte reaction (MLR), in which proliferation and CTL induction were observed in the absence of NMA, resulted in variable enhancement of the response (Hoffman et al., 1990). Species differences in various functions of rat and mouse peritoneal macrophages have been observed by others (Fishman, 1980; Albina et al., 1989) and may account for the difference in $\cdot N{=}O$ produced in splenocyte cultures from the two species.

[1] Abbreviations used: APC, antigen presenting cell; CTL, cytolytic T lymphocyte; GvHD, graft versus host disease; MLR, mixed leukocyte reaction; NMA, N^G-monomethyl-L-arginine, $\cdot N{=}O$, nitric oxide

Other investigators have confirmed and extended these observations on the effect of NMA on macrophage mediated suppression of lymphocyte proliferation. Albina and Henry (1991) have shown that the presence of red blood cells in rat splenocyte cultures stimulated with the T cell mitogen, Concanavalin A (Con A), will promote proliferation because the ·N=O which is produced has a high affinity for hemoglobin. Mills (1991) has also demonstrated that the inhibition of T lymphocyte proliferation mouse splenocyte cultures obtained from *Corynebacterium parvum* injected animals is due to ·N=O production by the activated macrophages in the splenocyte population. Albina *et al.* (1991) have shown that the inhibition observed in mouse splenocyte–peritoneal macrophage cocultures stimulated by Con A is due to ·N=O production and that the ·N=O production can be reversed by the addition of antibody to interferon-γ (IFNγ) as well as by NMA. There is a pronounced species variation in the induction of macrophage ·N=O synthesis when splenocyte populations as well as peritoneal cell populations are compared in the rat and mouse species (Hoffman *et al.*, 1990; Albina *et al.*, 1989). However, prior immune stimulation with agents such as C. *parvum* will promote macrophage ·N=O synthesis in the mouse species.

III. *IN VITRO* NITRIC OXIDE SYNTHESIS BY CELLS RECOVERED FROM RAT SPONGE MATRIX ALLOGRAFTS

A. Spontaneous Nitric Oxide Production

To determine if the ·N=O produced during the *in vitro* mixed lymphoctye reaction would be applicable to *in vivo* systems, the sponge matrix allograft model in the rat was utilized. This model has been well characterized in the mouse species (Roberts and Häyry, 1976; Ascher *et al.*, 1983; bishop *et al.*, 1989; Ford *et al.*, 1990) and provides a system in which allograft infiltrating cells as well as the fluid surrounding the graft can be recovered for *in vitro* analyses. Early graft infiltrating cell populations (days 1–3) consist primarily of granulocytes with a gradual conversion to a mononuclear cell infiltrate consisting of macrophages and lymphocytes in both mouse and rat species (Hoffman *et al.*, 1988; Langrehr *et al.*, 1991a). Analysis of sponge graft fluid in the rat species on various days postgrafting for NO_2^-/NO_3^- content revealed that significantly higher levels were detected in allogeneic grafts compared to syngeneic grafts on day 6 postgrafting. However, experiments designed to determine the half-life of NO_3^- in sponge grafts revealed that by 30 min after injection of 200 nmol NO_3^- into the sponge graft, the levels of NO_3^- in the sponge fluid had returned to control levels. Thus, analysis of sponge graft fluid for NO_2^-/NO_3^- indicates increased ongoing production of ·N=O but since these end products do not accumulate at the graft site, these levels do not reflect total production (Langrehr *et al.*, 1991a).

Culture of cells recovered from sponge grafts on various days postgrafting revealed that coincident with the generation of an alloimmune response (as detected by the allospecific cytolytic activity of cells recovered from the graft site) spontaneous ·N=O production by cells infiltrating an allograft was higher than the levels of ·N=O produced by cells infiltrating a syngeneic graft. When L-arginine was omitted from the culture medium, the ·N=O production was inhibited, demonstrating as in other systems, that this amino acid is the substrate for ·N=O synthase. Additionally, the ·N=O production was inhibited when the competitive inhibitor of ·N=O synthesis, NMA, was added to the culture system (Fig. 1). These data indicate that, during the generation of an alloimmune response *in vivo*, ·N=O is produced and that late postgrafting (day 8 in the rat species) spontaneous ·N=O production has abated.

B. Nitric Oxide Production on Reexposure to Alloantigen

To determine whether ·N=O production modifies allospecific CTL effector function of the graft infiltrating cells, the CTL activity of these graft infiltrating cells was assessed after culture with irradiated syngeneic or allogeneic lymph node lymphocytes (Langrehr et al., 1992a). This procedure results in reexposure of the allosensitized graft infiltrating cell population to the donor antigen to which they have been sensitized *in vivo*. Exposure of the allograft infiltrating cells, but not the syngeneic graft infiltrating cells to alloantigen resulted in massive induction of ·N=O synthesis and the CTL activity of these cultured cells was undetectable. However, when NMA was added to the graft infiltrating cells plus alloantigen stimulated cultures, allospecific CTL activity was readily detected. The immunological specificity of the alloantigen induced ·N=O production was demonstrated by culturing allograft infiltrating cells with third party alloantigen. Supernatant NO_2^-/NO_3^- levels were not increased when graft infiltrating cells were cultured with third party alloantigen. We hypothesized that sensitized T cells respond to restimulation with the specific antigen by secreting cytokines which induce ·N=O synthase in macrophages.

C. T Cell Cytokine Production in Presence and Absence of N^G-Monomethyl-L-Arginine

A natural consequence of immune activation is T lymphocyte synthesis of cytokines such as interleukin-2 (IL-2) and subsequent IL-2 induced proliferation of the sensitized lymphocyte population. Examination of the supernatants of allograft infiltrating cells restimulated with alloantigen revealed, as in the rat splenocyte MLR, that IL-2 was produced. This was true in both the presence and absence of NMA. IL-2 was not present in the supernatants of allograft infiltrating cells cultured with syngeneic antigen, indicating that reexposure to the sensitizing alloantigen was necessary for IL-2 synthesis. Data depicted in Fig. 2 dem-

7 Role of Nitric Oxide in Allograft Rejection 241

FIGURE 1

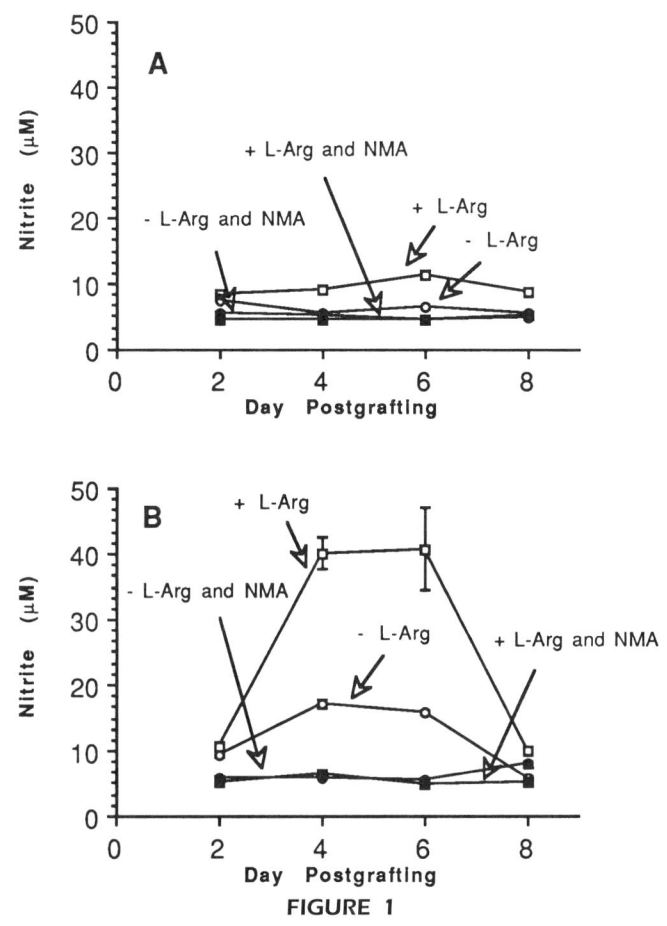

Effect of L-arginine and N^G-monomethyl-L-arginine (NMA) on spontaneous nitric oxide production by syngeneic and allogeneic sponge graft infiltrating cells. Syngeneic (A) and allogeneic (B) graft infiltrating cells were harvested on day 6 postgrafting and cultured in complete medium in the presence and absence of L-arginine (0.5 mM) and NMA (0.5 mM). After 48 hr of culture, supernatant nitrite levels were determined and are expressed as mean ± SD of triplicate cultures.

onstrate the time course of the detection of IL-2 in the supernatants of allograft infiltrating cells plus alloantigen in the presence and absence of NMA. In the absence of NMA, IL-2-like activity is present by 24 hr of culture and these cytokine levels remain constant at 48 and 72 hr of culture. Lymphocyte proliferation was not observed on any day of culture. In cultures carried out in the presence of NMA, IL-2 activity was present at 24 hr of culture but became undetectable at 48 and 72 hr of culture, coincident with the increase in ^{3}H-TdR

FIGURE 2

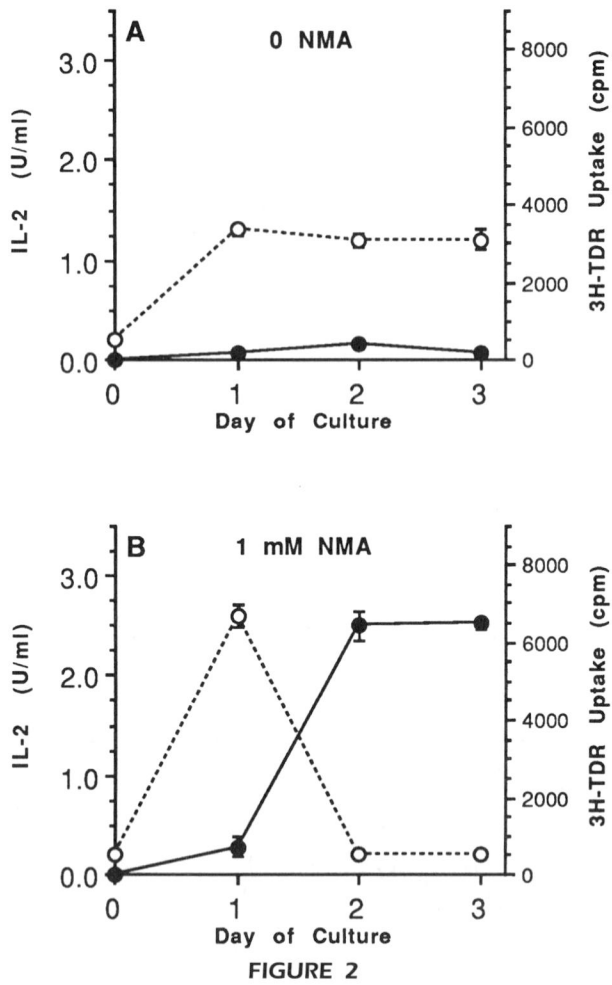

Effect of NMA on supernatant cytokine levels and ^{3}H-TdR uptake in cocultures of day 6 allograft infiltrating cells plus allogeneic lymph node lymphocytes. Lewis anti-ACI graft infiltrating cells were cultured with irradiated ACI lymph node lymphocytes in the presence of 0 mM (A) and 1.0 mM (B) NMA. Supernatant IL-2 levels (○) and ^{3}H-TdR uptake (●) were monitored on various days of culture and are expressed as mean ± SD of triplicate determinations. In cultures of allograft infiltrating cells plus syngeneic lymph node lymphocytes as well as syngeneic graft infiltrating cells plus syngeneic or allogeneic lymph node lymphocytes, supernatant IL-2 levels were undetectable in the bioassay and ^{3}H-TdR uptake was not above background levels (data not shown).

uptake of the cultured cells. We concluded that IL-2 production results from antigenic stimulation of the sensitized lymphocytes whether or not ·N=O was produced. A lymphocyte proliferative response to the IL-2, however, was prevented by ·N=O production and was only seen if ·N=O synthase activity was inhibited. We cannot explain, however, why IL-2 production is higher when NMA is present. It is possible that ·N=O partially inhibits IL-2 production which would be consistent with the effect of ·N=O on protein synthesis in other cells (e.g., hepatocytes) (Curran et al., 1989). Because IL-2 acts as a positive feedback mechanism in T cells by regulating its own receptors (Smith and Cantrell, 1985), quantitation of IL-2 production and IL-2R expression on T cells exposed to ·N=O should provide useful information.

D. Inhibition of Nitric Oxide Synthesis in Cocultures by FK506

The inhibitory effects of the immunosuppressive macrolide, FK506, on T cell cytokine mRNA synthesis have been described (Tocci et al., 1989). Inhibition of mitogen-stimulated as well as alloantigen-stimulated cultures can be achieved in the presence of FK506. Addition of FK506 to cultures of graft infiltrating cells stimulated with alloantigen resulted in complete inhibition of ·N=O synthesis. IL-2-like activity was not detected in the supernatants of cultures where FK506 was added (Langrehr et al., 1992a). Thus, it is likely that the mechanism of FK506-mediated inhibition of ·N=O synthesis is via prevention of T cell cytokine synthesis, one or more of which can activate macrophage ·N=O synthesis. Addition of IFNγ to graft infiltrating cells in the presence of FK506 resulted in induction of ·N=O synthesis, indicating that FK506 does not alter the induction of ·N=O synthesis in the macrophage but acts to prevent T cell activation in this system. The finding is analogous to other cell culture systems in which FK506 does not inhibit lymphocyte proliferation in the presence of an exogenous source of cytokines (Dumont et al., 1990).

E. Nitric Oxide Production by Adherent Graft Infiltrating Cells

The graft infiltrating cell population on days 6 to 8 postgrafting consists of approximately 50 to 60% cells of the monocyte–macrophage lineage, 20% granulocytes, and 20% lymphocytes, as determined by Wright Giemsa stain (Langrehr et al., 1993). Evaluation of ·N=O synthesis by the unseparated sponge graft infiltrating cells in the presence of various concentrations of IFNγ or lipopolysaccharide (LPS) revealed that enhanced ·N=O synthesis was observed in the presence of lower concentrations of these stimuli in the allograft, compared to the syngeneic graft infiltrating cells. IL-1β and TNFα induced allograft infiltrating cell ·N=O synthesis only at high concentrations of these cytokines while IL-6 was an ineffective stimulant. ·N=O synthesis by the unseparated graft infiltrating cell population in response to various agents is the result of complex cellular interactions. Therefore, the direct effect of cytokines on graft macrophage ·N=O synthesis was studied. Graft macrophage cultures were obtained by adherence to tissue culture wells. Macrophages obtained from allografts produced more ·N=O both spontaneously and in response to IFNγ and LPS than macrophages obtained from a syngeneic graft (Langrehr et al., 1993). Although unelicited rat peritoneal macrophages also produced ·N=O in response to IFNγ and LPS, when the nanomoles of NO_2^- per microgram adherent protein were compared, allograft macrophages produced significantly more ·N=O.

IV. IN VIVO NITRIC OXIDE SYNTHESIS DURING ALLOGRAFT REJECTION

A. Vascularized Organ Allografts

In order to determine whether serum NO_2^-/NO_3^- levels are elevated during allograft rejection, syngeneic and allogeneic liver and heart (Murase et al., 1990), small bowel (Langrehr et al., 1991b), skin (Ildstad et al., 1984), and sponge matrix grafts (Langrehr et al., 1991a) were performed in the rat. Serial assays of serum NO_2^-/NO_3^- levels revealed that elevated levels were detected before onset of clinical signs of rejection in the liver, heart, and small bowel grafts (Langrehr et al., 1992b). Serum NO_2^-/NO_3^- levels in animals that received a syngeneic graft remained within normal values at all times tested. Elevated serum NO_2^-/NO_3^- levels were not seen in animals that received a skin graft or subcutaneous sponge matrix graft. It is likely that the amount of ·N=O produced in the skin and sponge matrix graft models is low and the rapid renal excretion of NO_3^- may then prevent its detection in the serum.

B. Graft versus Host Reaction

The graft versus host (GvH) reaction is characterized by induction of immunosuppression and is associated with the production of various cytokines, including IFNγ. To determine whether elevated NO_2^-/NO_3^- levels were associated with this disease process, the GvH model in rats described by Markus et al. (1991) was utilized. Elevated NO_2^-/NO_3^- levels were detected before clinical symptoms of graft versus host disease (GvHD) were apparent and remained elevated until the animal became moribund. Recipients of a syngeneic bone marrow inoculum did not have elevated NO_2^-/NO_3^- levels at any time tested (Langrehr et al., 1992b).

The role of ·N=O in the inhibition of immune function tests using splenocytes from mice with GvHD has been examined (Hoffman et al., 1993). Compared to control mice, the agents Con A and LPS caused massive synthesis of ·N=O by splenocytes from animals with GvHD. Addition of NMA to the culture system reversed the production of ·N=O and enhanced the proliferative response to Con A. Addition of NMA to culture systems where GvHD splenocytes were added to a control culture, reversed the suppression of CTL generation and proliferation coincident with a decrease in supernatant NO_2^- levels. Peritoneal macrophages obtained from mice with GvHD synthesized ·N=O in response to LPS, an agent which does not induce ·N=O synthesis in peritoneal macrophages obtained from control mice. Thus, GvHD mice are primed for ·N=O synthesis and the enhanced levels of ·N=O detected in splenocyte cultures stimulated by mitogens are due to macrophage synthesis of ·N=O in response to cytokines, especially IFNγ produced by activated T cells.

C. Inhibition of Nitric Oxide Synthesis by FK506, Cyclosporin A

The serum NO_2^-/NO_3^- of the recipients of various organ and bone marrow–spleen cell allografts remained normal during therapy with cyclosporin A or FK506. This finding confirms our previous *in vitro* observation in which addition of FK506 to cultures of sponge graft infiltrating cells plus alloantigen completely inhibited ·N=O production, presumably due to inhibition of T cell cytokine production by FK506. However, when the immunosuppressive therapy was terminated and histologic evidence of recurring rejection of GvHD was ultimately detected (in the small bowel and the bone marrow–spleen cell transplants), elevated NO_2^-/NO_3^- levels were detected (Langrehr *et al.*, 1992b). In contrast, FK506-treated liver and heart allograft recipients (transplant models where graft acceptance is achieved with a brief course of immunotherapy) maintained normal serum NO_2^-/NO_3^- levels even after termination of the immunosuppression. Thus, these data suggest that the detection of elevated serum NO_2^-/NO_3^- levels in recipients of vascularized allografts correlates with the rejection reaction. Serum NO_2^-/NO_3^- levels may possibly prove to be an early marker of graft rejection, but the clinical utility of such an assay is, as yet, untested.

V. EFFECT OF NITRIC OXIDE SYNTHESIS IN DEFINED MACROPHAGE–LYMPHOCYTE COCULTURES

A. Comparison of P388D1 and RAW 264.7 as Allogeneic Antigen Presenting Cells

Previous studies on the effects of ·N=O production on the alloimmune response utilized bulk populations of responder and stimulator cells. In order to more clearly define the circumstances that induce ·N=O synthesis in allogeneic macrophage–lymphocyte cocultures, mouse splenocyte populations were depleted of accessory cells (>90% Thy 1.2$^+$) and cultured with mitomycin-C-treated macrophage cell lines, as the alloantigen presenting cells. The P388D1($H2^d$) and RAW 264.7 ($H2^d$) macrophage lines were selected because the former is a low ·N=O producer in response to LPS and cytokines while the latter is a high ·N=O producer. In the presence of increasing numbers of P388D1 plated per microwell, lymphocyte proliferation increased proportionally. However, in the presence of increasing numbers of RAW 264.7, lymphocyte proliferation was inhibited. Undetectable levels of NO_2^- were present in the supernatants of P388D1–lymphocyte cocultures while NO_2^- levels in RAW 264.7–lymphocyte cocultures increased in proportion to the number of RAW 264.7 plated. Addition of NMA inhibited ·N=O production and facilitated lymphocyte proliferation in RAW 264.7–lymphocyte cocultures. These data indicate that alloimmune stimulation can occur in the absence of ·N=O synthesis as is the case for P388D1–lymphocyte cocultures. However, if macrophages capable of ·N=O synthesis, such as RAW 264.7 are present in the system, ·N=O synthesis occurs with subsequent inhibition of lymphocyte proliferation (Hoffman et al., 1994).

B. Effect of Nitric Oxide Synthesis on RAW 264.7 Class II Antigen Expression and Antigen Presenting Cell Function

Initiation of an alloimmune response requires the expression of MHC (major histocompatibility) antigens on antigen presenting cell (APC). The most well-defined inducer of MHC class II antigen expression is IFNγ, produced by activated T cells. Because IFNγ is also a potent inducer of ·N=O synthesis in RAW 264.7 cells, we examined the induction of class II antigen on RAW 264.7 cells by IFNγ in the presence and absence of NMA. Class II antigen expression induced by IFNγ was similar whether or not NMA was present to prevent ·N=O synthesis. In addition, class I antigen expression was not altered by treating RAW 264.7 with IFNγ in the presence or absence of NMA. Analysis of the subsequent APC function of RAW 264.7 after treatment with IFNγ in the presence and absence of NMA revealed that, while IFNγ pretreatment of RAW 264.7 enhanced lymphocyte proliferation, treatment with IFNγ plus NMA resulted in similar induction of lymphocyte proliferation. Treatment of RAW 264.7 with IFNγ in the presence or absence of ·N=O production altered neither MHC expression nor APC function. Thus, the inhibitory effect of ·N=O of T lymphocyte responses does not seem to derive from a failure of the macrophage antigen presenting cell to express MHC antigen (Hoffman et al., 1994).

VI. EFFECT OF AUTHENTIC NITRIC OXIDE ON LYMPHOCYTE FUNCTION

A. Immediate Effect on Actively Proliferating Cells

The inhibition of lymphocyte proliferation and acquisition of cytotoxic T lymphocyte function in cultures where ·N=O is produced prompted an investigation of the effects of exposure of lymphocytes to authentic ·N=O gas ("authentic ·N=O") on subsequent function. When actively proliferating lymphocyte populations (previously stimulated by Con A, alloantigen, or IL-2) were exposed to authentic ·N=O, [3H] thymidine (^{3}H-TdR), uptake in the 4-hr period immediately postexposure was inhibited by approximately 50% when compared to deoxygenated media controls (R. A. Hoffman, J. M. Langrehr, and L. Simmons, unpublished observations, 1994). Exposure to authentic ·N=O was not cytotoxic, as assessed by trypan blue viability. Thus, regardless of the initial mitogenic stimulus, a brief exposure to ·N=O inhibits common proliferative pathways in lymphocytes, as it does in other cell systems (Hibbs et al., 1988).

B. Effect on Response to Interleukin-2, Antigen-Driven Proliferation

The ability of lymphocytes to recover proliferative function after exposure to authentic ·N=O was also assessed. The proliferative response of CTLL-2, a mouse T cell clone, in response to rmIL-2 24 hr after exposure to authentic ·N=O again revealed a 50% diminution of the ^{3}H-TdR uptake (Fig. 3). The ^{3}H-TdR uptake of a human T lymphocyte clone in response to alloantigen was also inhibited 48 hr after exposure to ·N=O. Whether the same mechanism responsible for decreased ^{3}H-TdR uptake of actively proliferating cells (described above) is operative in the system where lymphocytes are exposed to a mitogenic stimulus after ·N=O exposure remains to be determined.

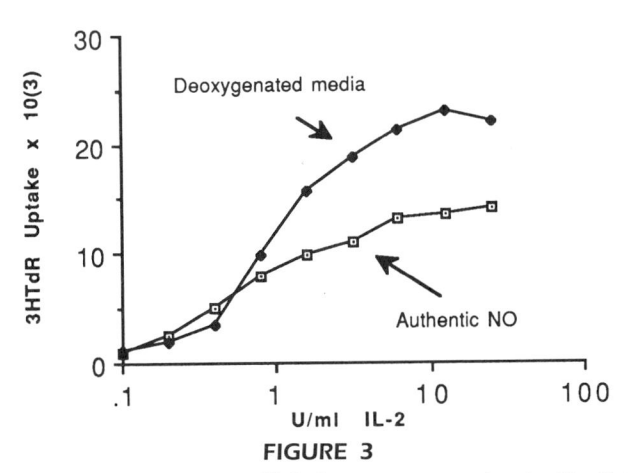

FIGURE 3

Proliferation of CTLL-2 in response to rmIL-2 after exposure to authentic ·N=O. CTLL-2 cells were exposed to 25% (of saturated solution) authentic ·N=O, washed, and reexposed to serial dilutions of rmIL-2 for 24 hr [^{3}H] Thymidine uptake was determined and is depicted relative to deoxygenated media controls.

C. Effect on Cytolytic Function and Motility

Exposure of lymphocytes to authentic ·N=O did not induce a global inhibition of lymphocyte function. The cytolytic activity of alloactivated mouse splenocytes as well as a murine allospecific cytolytic T lymphocyte clone were unaffected by exposure to ·N=O immediately prior to the cytotoxicity assay. Additionally, the motility of various lymphocyte populations, including Con A stimulated splenocytes, alloactivated mouse splenocytes, and a T lymphocyte clone were also unaffected by previous ·N=O exposure (R. A. Hoffman, J. M. Langrehr, and R. L. Simmons, unpublished observations, 1994). Thus, brief exposure to ·N=O does not affect T lymphocyte microfilament–microtubule function nor the enzymes required to induce target cell destruction.

VII. POSSIBLE MECHANISMS OF NITRIC OXIDE-INDUCED INHIBITION OF LYMPHOCYTE ACTIVATION

Although the precise biochemical nature of the inhibition of lymphocyte activation by ·N=O is unknown at this time, some speculations can be made with the knowledge available. In culture systems where ·N=O is produced, the temporal relationship between the synthesis of the cytokines that induce the macrophage to synthesize ·N=O and the actual initiation of ·N=O synthesis by the macrophage must be considered. In the culture systems we have studied thus far, IL-2 is detected in culture supernatants at 24 hr in the presence or absence of NMA. It can be assumed, therefore, that the T cell receptor has been engaged and the accessory cell costimulus has been provided in order for cytokine synthesis to be initiated. This process occurs in a relatively short time frame (several hours). Since, by definition, the inducible ·N=O synthase does not result in ·N=O synthesis until 4 to 6 hr after exposure to the cytokines which initiate the pathway, the initial events of allorecognition occur before ·N=O is synthesized. However, profound inhibition of lymphocyte proliferation and CTL activity is observed when ·N=O is produced, at a point after lymphocyte cytokine synthesis is initiated. Thus, lymphocytes are unable to proliferate when exposed to ·N=O even though cytokines able to induce proliferation are present. This information indicates that ·N=O can mediate inhibition of lymphocyte proliferation at a late stage in the activation process. However, lymphocyte exposure to ·N=O during the initial activation process, and examination of such parameters as intracellular second messengers, IL-2 receptor expression, cytokine mRNA levels, and so on, will provide valuable information concerning the effects of ·N=O on the early steps of lymphocyte activation.

A. Comparison to Lymphocyte Anergy

T lymphocyte anergy, the inability of a T cell to respond to secondary stimulation, is induced by presentation of antigen on chemically modified antigen presenting cells (APC) that cannot provide the appropriate costimulatory signal (Jenkins and Schwartz, 1987). In this system, lymphocyte proliferation is not observed upon reexposure of the anergized lymphocyte to an antigen presenting cell that provides an appropriate costimulus. Further analysis of this anergic state in lymphocytes demonstrated that on restimulation, the lymphocytes fail to produce IL-2, but IL-3, IFNγ, and IL-2R were partially induced (Jenkins et al., 1987). However, during the initial exposure to the chemically modified APC, an increase in intracellular calcium was detected and prevention of this increase by addition of EGTA prevented induction of the anergic state. In addition, protein kinase C (PKC) activation was detected in the unresponsive lymphocyte, indicating that a rise in intracellular calcium and PKC activation is not sufficient to activate IL-2 synthesis but can induce the anergic state. The costimulatory signal from a bone marrow-derived APC is necessary for IL-2 gene transcription and subsequent lymphocyte proliferation (Mueller et al., 1989). Whether exposure of lymphocytes to alloantigen in the presence of ·N=O will induce an anergic state is unknown at this time. This scenario would appear unlikely, since the lymphocytes recovered from the sponge allograft site, where exposure to ·N=O in vivo during contact with the alloantigen can be inferred by the elevated NO_2^-/NO_3^- levels in the fluid bathing the graft, proliferate and demonstrate enhanced CTL activity when reexposed to alloantigen in vitro in the presence of NMA. Thus, reexposure to ·N=O during the antigenic stimulus is necessary to prevent proliferation. However, the experiments described in Section VI.B where subsequent stimulation of lymphocytes with IL-2 after exposure to ·N=O results in half maximal stimulation demonstrates that there is an inhibitory mechanism that is not reversible within 24 to 48 hr after ·N=O exposure.

B. Nitric Oxide Effects on Intracellular Second Messengers

Endothelial cell-derived ·N=O has been shown to activate soluble guanylate cyclase and elevate tissue cGMP levels (Murad et al., 1979; Craven and DeRubertis, 1978; Gruetter et al., 1981) coincident with vascular smooth muscle relaxation. However, the mechanism by which elevations in cGMP levels cause smooth muscle relaxation is unknown. Subsequently, many culture systems where ·N=O is produced have documented a coordinate rise in cellular cGMP levels. Garg and Hassid (1989) have demonstrated an inhibitory effect on vascular smooth muscle cell proliferation by various ·N=O producing agents as well as by 8-bromo-cGMP. Conversely, other reports have documented a lack of correlation between the effects of ·N=O and elevated cGMP levels, including ·N=O induced decreases in cytosolic free calcium in BALB/c 3T3 fibroblasts (Garg and Hassid, 1991), ·N=O induced inhibition of bone-resorbing activity of the osteoclast (MacIntyre et al., 1991) and ·N=O induced ADP-ribosylation of a 39-kDa protein in platelets and other tissues (Brüne and Lapetina, 1989).

In lymphocytes as well, exposure to authentic ·N=O results in increased cGMP levels (R. A. Hoffman, J. M. Langrehr, and R. L. Simmons, unpublished observations, 1994). However, elevations in lymphocyte cGMP have been considered to positively, not negatively, affect T lymphocyte responses. T cell mitogens have been found to activate guanylate cyclase and to increase intracellular cGMP levels (reviewed in Hadden, 1988; Altman et al., 1990) in T cells. Moreover, addition of sodium nitroprusside to Con A stimulated human peripheral blood lymphocytes, while elevating cGMP levels did not inhibit ^{3}H-uridine incorporation, indicating that sodium nitroprusside does not influence RNA synthesis (Kaever and Resch, 1990). Alternatively, authentic ·N=O as well as ·N=O generating compounds, such as nitroprusside, have been demonstrated to inhibit calcium fluxes in various cell types including BALB/c 3T3 fibroblasts (Garg and Hassid, 1991) and ADP-stimulated human platelets (Morgan and Newby, 1989). An increase in intracellular calcium is associated with lymphocyte activation (reviewed in Altman et al., 1990) and disruption of this pathway may be a target for ·N=O mediated inhibition of lymphocyte function.

Work by Lander et al. (1993a) has focused on the immune-stimulatory properties of ·N=O through a cGMP-independent pathway. Human peripheral blood mononuclear cells exposed to sodium nitroprusside or S-nitroso-N-acetyl-penicillamine demonstrated enhanced membrane-associated protein tyrosine phosphatase, activated protein tyrosine kinase p56^{lck}, and induction of TNFα secretion and NF-κB binding activity. These investigators have also demonstrated that ·N=O activates G proteins in human lymphocytes (Lander et al., 1993b). Thus, although ·N=O itself is not mitogenic for lymphocytes, it obviously has stimulatory activity for several intracellular signaling pathways. The opposing effects of ·N=O, such as stimulation of some intracellular pathways and inhibition of the mitochondrial electron transport chain (Stuehr and Nathan, 1989), obviously need further clarification.

C. Inhibition of Ribonucleotide Reductase

Numerous cell culture systems where nitric oxide synthase is induced have the common feature of inhibition of proliferation or cytostasis in the cocultured cells, implying the existence of at least one common inhibitory mechanism in proliferating cells. One plausible common inhibitory mechanism is the reduction in enzymatic activity of ribonucleotide reductase (RR), a key enzyme for DNA synthesis. RR consists of two nonidentical subunits; one subunit contains the allosteric binding site for effector molecules while the other subunit contains a nonheme iron and possesses a tyrosyl free radical which is necessary for the reduction reaction (reviewed in Cory, 1989). The report of Lepoivre et al. (1990) using adenocarcinoma cells induced to make ·N=O, and that by Kwon et al. (1991) using tumor cells cocultured with ·N=O producing macrophages provide convincing evidence that ·N=O exerts an inhibitory effect on RR.

There is a strict correlation between the levels of RR activity and DNA synthesis. Low or undetectable levels are present in resting normal tissues (Takeda and Weber, 1981) while high levels are detected in regenerating liver (Larsson, 1969; King and Van Lancker, 1969) and in embryonic and newborn tissue (Elford, 1972). Additionally, RR activity is reported to be increased in human tumors relative to surrounding normal tissue (Gordon et al., 1970). Thus, in the experiments described in Section VI,A where exposure of actively proliferating lymphocytes to authentic ·N=O results in an immediate inhibition of ^{3}H-TdR uptake, inhibition of RR activity by ·N=O is a possible mechanism. However, exposure of freshly isolated human peripheral blood lymphocytes to authentic ·N=O also resulted in inhibition of subsequent ^{3}H-TdR uptake in response to phorbol myristate acetate (PMA) or anti-CD3 (R. A. Hoffman, J. M. Langrehr, and R. L. Simmons, unpublished observations, 1994). Because RR levels would be expected to be very low in normal peripheral blood lymphocytes, perhaps another mechanism of ·N=O induced cytostasis is operable in nontransformed cells.

D. Inhibition of Polyamine Synthesis

Metabolism of L-arginine via the arginase pathway results in production of L-ornithine, which is converted by ornithine decarboxylase to the polyamines putrescine, spermidine, and spermine which have been shown to be necessary for optimal DNA synthesis in lymphocytes (Fillingame et al., 1975). Speculation on the effects of diversion of L-arginine to the ·N=O pathway rather than the arginase pathway seems warranted. Our unpublished observations (1994) utilizing various lymphocyte clones (CTLL-2 and D10.G4.1) responding to recombinant cytokines (IL-2 and IL-4) ascertain that addition of NMA to these cultures does not affect subsequent proliferation, indicating that NMA does not affect synthesis of polyamines in the lymphocyte. In lymphocyte–macrophage coculture systems where ·N=O is being produced and L-arginine is metabolized via the oxidative pathway, the possibility exists that L-arginine is rate-limiting for lymphocyte polyamine synthesis. Addition of NMA to this culture system may result in increased arginine levels and thus facilitation of lymphocyte polyamine synthesis. Evaluation of the supernatants of these cultures for L-arginine and citrulline would provide evidence for this hypothesis.

Experiments utilizing D,L-α-difluoromethylornithine (DFMO), an irreversible inhibitor of ornithine decarboxylase, to determine its effect on mitogen- or alloantigen-induced lymphocyte proliferation and CTL induction have certain similarities to the inhibition seen in ·N=O-induced inhibition of lymphocyte function. Stimulation of both mouse and human T cells in the presence of DFMO results in partial inhibition of proliferation which can be reversed by addition of the polyamines putrescine, spermidine, and spermine (Bowlin et al., 1987; McCarthy et al., 1990). IL-2 levels in these lymphocyte culture supernatants as well as lymphocyte high affinity IL-2R expression were, however, unaffected in the DFMO-treated cultures compared to untreated cultures. This important finding is similar to the results found in our culture systems where IL-2 is detected but lymphocyte proliferation is not seen when ·N=O is produced. Additionally, the inhibitory effect of DFMO on CTL generation but not on proliferation in alloantigen stimulated cultures correlates with our results seen in rat splenic alloantigen-stimulated cultures. DFMO inhibition of CTL induction at a late, preeffector stage was proven by removal of $CD4^+$ cells from cells cultured for 3 days with DFMO (to remove possible source of help) and reculturing these cells in the presence of putrescine. Cytolytic activity was then detected after 2 days of cultures, indicating that polyamines are necessary for CTL induction (Schall et al., 1991). The distinction observed in DFMO-treated alloantigen stimulated culture between proliferation and CTL induction is also similar to our observation in rat splenic alloantigen-stimulated cultures where proliferation is detected in the presence of low concentrations of NMA (0.1 mM) but higher concentrations are needed (1–2 mM) to detect CTL activity (Hoffman et al., 1990). Work reported by Bowlin et al. (1988) indicates that CTL induction may be particularly sensitive to intracellular levels of the polyamine, spermine. Thus, if lymphocyte polyamine synthesis is affected by ·N=O, perhaps near total inhibition of ·N=O synthesis, achieved with 1–2 mM NMA may be necessary to observe CTL generation while proliferation may not have such stringent requirements. Assessment of lymphocyte ornithine decarboxylase activity as well as polyamine levels after stimulation in the presence of ·N=O would determine if ·N=O is inhibiting lymphocyte proliferation via inhibition of polyamine synthesis.

VIII. CONCLUSION

Ample evidence now exists that documents the role of cytokine induced macrophage synthesis of ·N=O as a potent cytostatic effector mechanism in cocultured cells. Since a multitude of cytokines is produced during an alloimmune reaction (some of which are potent inducers of macrophage ·N=O synthesis), it is not surprising that ·N=O synthase is induced during allograft rejection. Elucidation of the effects of ·N=O synthesis (i.e., graft destruction, inhibition of activation and/or effector function of activated cells) should provide insight into the complex cellular interactions that occur during an alloimmune response.

ACKNOWLEDGMENT

This work was supported by National Institutes of Health Grant AI-16869 (R. L. S.).

REFERENCES

Albina, J. E., and Henry, W. L., Jr. (1991). Suppression of lymphocyte proliferation through the nitric oxide synthesizing pathway. *J. Surg. Res.* **50**, 403–409.

Albina, J. E., Mills, C. D., Henry, W. L., Jr., and Caldwell, M. D. (1989). Regulation of macrophage physiology by L-arginine: Role of the oxidative L-arginine deaminase pathway. *J. Immunol.* **143**, 3641–3646.

Albina, J. E., Abate, J. A., and Henry, W. L., Jr. (1991). Nitric oxide production is required for murine resident peritoneal macrophages to suppress mitogen-stimulated T cell proliferation: Role of IFNγ in the induction of the nitric oxide-synthesizing pathway. *J. Immunol.* **147**, 144–148.

Allison, A. C. (1978). Mechanisms by which activated macrophages inhibit lymphocyte responses. *Immunol. Rev.* **40**, 3.

Altman, A., Coggeshall, K. M., and Mustelin, T. (1990). Molecular events mediating T cell activation. *Adv. Immunol.* **48**, 227–360.

Ascher, N. L., Chen, S., Hoffman, R. A., and Simmons, R. L. (1983). Maturation of cytotoxic T cells within sponge matrix allografts. *J. Immunol.* **131**, 617–621.

Billiar, T. R., Curran, R. D., Stuehr, D. J., West, M. A., Bentz, B. G., and Simmons, R. L. (1989). An L-arginine-dependent mechanism mediates Kupffer cell inhibition of hepatocyte protein synthesis *in vitro*. *J. Exp. Med.* **169**, 1467–1472.

Bishop, D. K., Jutila, M. A., Sedmak, D. D., Beattie, M. S., and Orosz, C. G. (1989). Lymphocyte entry into inflammatory tissues *in vivo*. Qualitative differences of high endothelial venule-like vessels in sponge matrix allografts vs. isografts. *J. Immunol.* **142**, 4219–4224.

Bowlin, T. L., McKown, B. J., Babcock, G. F., and Sunkara, P. S. (1987). Intracellular polyamine biosynthesis is required for interleukin-2 responsiveness during lymphocyte mitogenesis. *Cell. Immunol.* **106**, 420–427.

Bowlin, T. L., Davis, G. F., and McKown, B. J. (1988). Inhibition of alloantigen-induced cytolytic T lymphocytes *in vitro* with (2R, 5R)-6-heptyne-2, 5-diamine, an irreversible inhibitor of ornithine decarboxylase. *Cell. Immunol.* **111**, 443–450.

Brüne, B., and Lapetina, E. G. (1989). Activation of a cytosolic ADP-ribosyltransferase by nitric oxide-generating agent. *J. Biol. Chem.* **264**, 8455–8458.

Calderon, J., Williams, R. T., and Unanue, E. R. (1974). An inhibitor of cell proliferation released by cultures of macrophages. *Proc. Natl. Acad. Sci. U.S.A.* **71**, 4273–4277.

Cory, J. G. (1989). Role of ribonucleotide reductase in cell division. *In* "Inhibitors of Ribonucleoside Diphosphate Reductase Activity" (J. G. Cory and A. H. Cory, eds.), pp. 1–16. Pergamon, New York.

Craven, P. A., and DeRubertis, F. R. (1978). Restoration of the responsiveness of purified guanylate cyclase to nitrosoguanidine, nitric oxide, and related activators by heme and hemeproteins: Evidence for the involvement of the paramagnetic nitrosyl–heme complex in enzyme activation. *J. Biol. Chem.* **253**, 8433–8443.

Curran, R. D., Billiar, T. R., Stuehr, D. J., Hofmann, K., and Simmons, R. L. (1989). Hepatocytes produce nitrogen oxides from L-arginine in response to inflammatory products of Kupffer cells. *J. Exp. Med.* **170**, 1769–1774.

Dumont, F. J., Melino, M. R., Staruch, M. J., Koprak, S. L., Fischer, P. A., and Sigal, N. H. (1990). The immunosuppressive macrolides FK-506 and rapamycin act as reciprocal antagonists in murine T cells. *J. Immunol.* **144,** 1418–1424.

Elford, H. L. (1972). Functional regulation of mammalian ribonucleotide reductase. *Adv. Enzyme Regul.* **10,** 19–38.

Fillingame, R. H., Jorstad, C. M., and Morris, D. R. (1975). Increased cellular levels of spermidine or spermine are required for optimal DNA synthesis in lymphocytes activated by concanavalin A. *Proc. Natl. Acad. Sci. U.S.A.* **72,** 4042–4045.

Fishman, M. (1980). Functional heterogeneity among peritoneal macrophages. *Cell. Immunol.* **55,** 174–184.

Folch, H., and Waksman, B. H. (1973). Regulation of lymphocyte response *in vitro*. V. Suppressor activity of adherent and non-adherent rat lymphoid cells. *Cell. Immunol.* **9,** 12–24.

Ford, H. R., Hoffman, R. A., Wing, E. J., Magee, D. M., McIntyre, L. A., and Simmons, R. L. (1990). Tumor necrosis factor, macrophage colony-stimulating factor, and interleukin-1 production within sponge matrix allografts. *Transplantation* **50,** 460–466.

Garg, U. C., and Hassid, A. (1989). Nitric oxide-generating vasodilators and 8-bromocyclic guanosine monophosphate inhibit mitogenesis and proliferation of cultured rat vascular smooth muscle cells. *J. Clin. Invest.* **83,** 1774–1777.

Garg, U. C., and Hassid, A. (1991). Nitric oxide decreases cytosolic free calcium in BALB/c 3T3 fibroblasts by a cyclic GMP-independent mechanism. *J. Biol. Chem.* **266,** 9–12.

Gordon, H. L., Bardos, T. J., and Ambrus, J. L. (1970). Comparative study of ribonucleotide reductase activities in normal and neoplastic human tissue. *Res. Commun. Chem. Pathol. Pharmacol.* **1,** 749–756.

Gruetter, C. A., Gruetter, D. Y., Lyon, J. E., Kadowitz, P. J., and Ignarro, L. J. (1981). Relationship between cyclic guanosine 3':5'-monophosphate formation and relaxation of coronary arterial smooth muscle by glyceryl trinitrate, nitroprusside, nitrite and nitric oxide: Effects of methylene blue and methemoglobin. *J. Pharmacol. Exp. Ther.* **219,** 181–186.

Hadden, J. W. (1988). Transmembrane signals in the activation of T lymphocytes by mitogenic antigens. *Immunol. Today* **9,** 235–239.

Hibbs, J. B., Jr., Taintor, R. R., and Vavrin, Z. (1987). Macrophage cytotoxicity: Role for L-arginine deiminase and imino nitrogen oxidation to nitrite. *Science* **235,** 473–476.

Hibbs, J. B., Jr., Taintor, R. R., Vavrin, Z., and Rachlin, E. M. (1988). A cytotoxic activated macrophage effector molecule. *Biochem. Biophys. Res. Commun.* **157,** 87–94.

Hoffman, R. A., Ascher, N. L., Jordan, M. L., Migliori, R. J., and Simmons, R. L. (1988). Characterization of natural killer cell activity in sponge matrix allografts. *J. Immunol.* **140,** 1702–1710.

Hoffman, R. A., Langrehr, J. M., Billiar, T. R., Curran, R. D., and Simmons, R. L. (1990). Alloantigen-induced activation of rat splenocytes is regulated by the oxidative metabolism of L-arginine. *J. Immunol.* **145,** 2220–2226.

Hoffman, R. A., Langrehr, J. M., Wren, S. M., Dull, K. E., Ildstad, S. T., McCarthy, S. A., and Simmons, R. L. (1993). Characterization of the immunosuppressive effects of nitric oxide during graft-versus-host disease. *J. Immunol.* **151,** 1508–1518.

Hoffman, R. A., Langrehr, J. M., Dull, K. E., McCarthy, S. A., Jordan, M. L., and Simmons, R. L. (1994). Macrophage synthesis of nitric oxide in the mouse mixed leukocyte reaction. *Transplant. Immunol.* **2,** 313–320.

Holt, P. G., Warner, L. A., and Mayrhofer, G. (1981). Macrophages as effector of T suppression: T-lymphocyte-dependent macrophage-mediated suppression of mitogen-induced blastogenesis in the rat. *Cell. Immunol.* **63,** 57–70.

Ildstad, S. T., Wren, S. M., Sharrow, S. O., Stephany, D., and Sachs, D. H. (1984). In vivo and in vitro characterization of specific hyporeactivity to skin xenografts in mixed xenogeneically reconstituted mice (B10 + F344 rat → B10). *J. Exp. Med.* **160,** 1820–1835.

Jenkins, M. K., and Schwartz, R. H. (1987). Antigen presentation by chemically modified splenocytes induces antigen-specific T cell unresponsiveness in vitro and in vivo. *J. Exp. Med.* **165,** 302–319.

Jenkins, M. K., Pardoll, D. M., Mizuguchi, J., Chused, T. M., and Schwartz, R. H. (1987). Molecular events in the induction of a nonresponsive state in interleukin 2-producing helper T-lymphocyte clones. *Proc. Natl. Acad. Sci. U.S.A.* **84,** 5409–5413.

Kaever, V., and Resch, K. (1990). Role of cyclic nucleotides in lymphocyte activation. In "Current Topics in Membranes and Transport: Mechanisms of Leukocyte Activation" (S. Grinstein and O. D. Rotstein, ed.), Vol. 35, pp. 375–398. Academic Press, San Diego.

King, C. D., and Van Lancker, J. L. (1969). Molecular mechanisms of liver regeneration. VII. Conversion of cytidine to deoxytidine in rat regenerating livers. *Arch. Biochem. Biophys.* **129,** 603–608.

Kwon, N. S., Stuehr, D. J., and Nathan, C. F. (1991). Inhibition of tumor cell ribonucleotide reductase by macrophage-derived nitric oxide. *J. Exp. Med.* **174,** 761–767.

Lander, H. M., Sehajpal, P., Levine, D. M., and Novogrodsky, A. (1993a). Activation of human peripheral blood mononuclear cells by nitric oxide-generating compounds. *J. Immunol.* **150,** 1509–1516.

Lander, H. M., Sehajpal, P. K., and Novogrodsky, A. (1993b). Nitric oxide signaling: A possible role for G proteins. *J. Immunol.* **151,** 7182–7187.

Langrehr, J. M., Hoffman, R. A., Billiar, T. R., Lee, K. K. W., Schraut, W. H., and Simmons, R. L. (1991a). Nitric oxide synthesis in the in vivo allograft response: A possible regulatory mechanism. *Surgery* **110,** 335–342.

Langrehr, J. M., Hoffman, R. A., Banner, B., Stangl, M. J., Monyhan, H., Lee, K. K. W., and Schraut, W. H. (1991b). Induction of graft-versus-host disease and rejection by sensitized small bowel allografts. *Transplantation* **52,** 399–405.

Langrehr, J. M., Dull, K. E., Ochoa, J. B., Billiar, T. R., Ildstad, S. T., Schraut, W. H., Simmons, R. L., and Hoffman, R. A. (1992a). Evidence that nitric oxide production by in vivo allosensitized cells inhibits the development of allospecific CTL. *Transplantation* **53,** 632–640.

Langrehr, J. M., Murase, N., Markus, P. M., Cai, X., Neuhaus, P., Schraut, W., Simmons, R. L., and Hoffman, R. A. (1992b). Nitric oxide production in host-versus-graft and graft-versus-host reactions in the rat. *J. Clin. Invest.* **90,** 679–683.

Langrehr, J. M., White, D. A., Hoffman, R. A., and Simmons, R. L. (1993). Macrophages produce nitric oxide at allograft sites. *Ann. Surg.* **218,** 159–166.

Larsson, A. (1969). Ribonucleotide reductase from regenerating rat liver. *Eur. J. Biochem.* **11,** 113–121.

Lepoivre, M., Chenais, B., Yapo, A., Lemaire, G., Thelander, L., and Tenu, J. P. (1990). Alterations of ribonucleotide reductase activity following induction of the nitrite-generating pathway in adenocarcinoma cells. *J. Biol. Chem.* **265,** 14143–14149.

MacIntyre, I., Zaidi, M., Towhidul-Alam, A. S. M., Datta, H. K., Moonga, B. S., Lidbury, P. S., Hecker, M., and Vane, J. R. (1991). Osteoclastic inhibition: An action of nitric oxide not mediated by cyclic GMP. *Proc. Natl. Acad. Sci. U.S.A.* **88,** 2936–2940.

Markus, P. M., Cai, X., Ming, W., Demetris, A. J., Fung, J. J., and Starzl, T. E. (1991). FK 506 reverses acute graft-versus-host disease after allogeneic bone marrow transplantation in rats. *Surgery* **110,** 357–364.

McCarthy, M. A., Michalski, J. P., Sears, E. S., and McCombs, C. C. (1990). Inhibition of polyamine synthesis suppresses human lymphocyte proliferation without decreasing cytokine production or interleukin-2 receptor expression. *Immunopharmacology* **20,** 11–20.

Metzger, Z. V. I., Hoffeld, J. T., and Oppenheim, J. J. (1980). Macrophage-mediated suppression. I. Evidence for participation of both hydrogen peroxide and prostaglandins in suppression of murine lymphocyte proliferation. *J. Immunol.* **124,** 983–988.

Mills, C. D. (1991). Molecular basis of "suppressor" macrophages: Arginine metabolism via the nitric oxide synthetase pathway. *J. Immunol.* **146,** 2719–2723.

Morgan, R. O., and Newby, A. C. (1989). Nitroprusside differentially inhibits ADP-stimulated calcium influx and mobilization in human platelets. *Biochem. J.* **258,** 447–454.

Mueller, D. L., Jenkins, M. K., and Schwartz, R. H. (1989). Clonal expansion versus functional clonal inactivation: A costimulatory signalling pathway determines the outcome of T cell antigen receptor occupancy. *Annu. Rev. Immunol.* **7,** 445–480.

Murad, F., Arnold, W. P., Mittal, C. K., and Braughzer, J. M. (1979). Properties and regulation of guanylate cyclase and some proposed function for cyclic GMP. *Adv. Cyclic Nucleotide Res.* **11,** 175–204.

Murase, N., Kim, D. G., Todo, S., Cramer, D. V., Fung, J. J., and Starzl, T. E. (1990). Suppression of allograft rejection with FK506. I. Prolonged cardiac and liver survival in rats following short-course therapy. *Transplantation* **50,** 186–189.

Roberts, P. J., and Häyry, P. (1976). Sponge matrix allografts: A model for analysis of killer cells infiltrating mouse allografts. *Transplantation* **21,** 437–445.

Schall, R. P., Sekar, J., nee Tandon, P. M., and Susskind, B. M. (1991). Difluoromethylornithine (DFMO) arrests murine CTL development in the late, pre-effector stage. *Immunopharmacology* **21,** 129–144.

Smith, K. A., and Cantrell, D. A. (1985). Interleukin-2 regulates its own receptors. *Proc. Natl. Acad. Sci. U.S.A.* **82,** 864–868.

Stuehr, D. J., and Nathan, C. F. (1989). Nitric oxide: A macrophage product responsible for cytostasis and respiratory inhibition in tumor target cells. *J. Exp. Med.* **169,** 1543–1555.

Takeda, E., and Weber, G. (1981). Role of ribonucleotide reductase in expression of the neoplastic program. *Life Sci.* **28,** 1007–1014.

Tocci, M. J., Matkovich, D. A., Collier, K. A., Kwok, P., Dumont, F., Lin, S., DeGudicibus, S., Siekierka, J. J., Chin, J., and Hutchinson, N. I. (1989). The immunosuppressant FK506 selectively inhibits expression of early T cell activation genes. *J. Immunol.* **143**, 718–726.

Weiss, A., and Fitch, F. W. (1977). Macrophages suppress CTL generation in rat mixed leukocyte cultures. *J. Immunol.* **119**, 510–516.

8

Role of Nitric Oxide in Treatment of Foods[1]

Daren Cornforth
Department of Nutrition and Food Sciences
Utah State University
Logan, Utah 84322

I. INTRODUCTION

Nitric oxide (NO) is the compound responsible for the color and stability of cured meats. NO has also been identified in the headspace gases of swollen, canned green beans, and likely is involved in the antimicrobial action of nitrate added to some cheeses. Although patents have been issued for curing by direct exposure of meat to NO, this procedure has never been used commercially. In cured meats and cheese, NO is generated by reduction of added nitrate or nitrite. The literature on nitrate, nitrite, and NO in foods has direct application to the growing field of NO in living systems. For example, in the 1970s, it was recognized that nitrite was a precursor to carcinogenic nitrosamines in some cured meats. Subsequent studies revealed that various vegetables provide the largest source of dietary exposure to nitrate and nitrite (White, 1975). It was further shown that nitrate is endogenously produced in mammalian tissues, as demonstrated in studies on germfree rats on low nitrate/nitrite diets (Green *et al.*, 1981). Urinary nitrate levels were known to be higher in animals with bacterial infections. In a very satisfactory unification of studies in food chemistry, nutrition, and immunology, the origins of endogenous nitrate were finally determined to be from NO

[1] Dedicated to Arthur W. Mahoney (1939–1992)—colleague, mentor, friend.

produced by macrophages in the combat of infection (Hibbs et al., 1987; Granger et al., 1991).

II. HISTORICAL USE OF NITRATE AND NITRITE IN CURED MEATS

Nitrate is a common contaminant of desert and sea salt. The ancient practice of salt preservation of meat preceded the intentional use of nitrate as a curing salt by many centuries. The reddening effect of nitrate in preserved meats was not mentioned until Roman times. Wall saltpeter (calcium nitrate), found as an efflorescence on the walls of caves and stables, was used by ancient peoples in curing of meat (Binkerd and Kolari, 1975). Calcium nitrate is formed by nitrifying bacteria (Jensen, 1954). *The National Provisioner*, a meat industry publication, recommended in its 1894 handbook the following brine formula for 100 pounds of ham; 7.66 pounds salt, 2 pounds sugar, 0.33 pounds or 3300 parts per million (ppm) saltpeter in 4 gallons water (*National Provisioner*, 1952, as cited in Binkerd and Kolari, 1975).

The first intentional use of nitrite in meat curing brines is unknown. Many nineteenth century meat curing recipes recommended use of Sal Prunella to acquire good meat color. Sal Prunella is prepared from a fused mixture of sulfur and nitrate, and probably contained some nitrite. According to Binkerd and Kolari (1975), Polenske (1891) was the first to report the presence of nitrite in curing brines. He correctly concluded that the nitrite source was the bacterial reduction of nitrate. It was soon demonstrated that nitrite rather than nitrosylhemochromogen was responsible for cured meat color (Lehman, 1899; Kisskalt, 1899).

The first mention of nitric oxide in meat curing was by Hoagland (1908, as cited in Binkerd and Kolari, 1975). He concluded that the red pigment of cooked cured meats was the heat denatured product of nitrosohemoglobin. (We now recognize that denatured NO-myoglobin is the main cured meat pigment. Due to residual blood, some denatured NO-hemoglobin may also be present). Hoagland (1908) concluded that NO was the result of bacterial or enzymatic reduction of nitrate in curing brines to nitrite, nitrous acid, and nitric oxide.

A U.S. patent was issued in 1917 for the use of nitrite as a replacement for nitrate in curing brines (Doran, 1917, as cited in Binkerd and Kolari, 1975). Kerr et al., (1926) found that hams cured in a brine containing about 2000 ppm sodium nitrite were equivalent in flavor and color to hams cured with nitrate. The maximum nitrite level found in any part of the hams was 200 ppm. Based on these experiments, the U.S. Department of Agriculture (USDA) in 1925 authorized use of sodium or potassium nitrite in curing brines in federally inspected establishments, at 0.25 to 1 ounce per 100 pounds of meat, such that the finished product would contain no more than 200 ppm sodium nitrite (USDA, 1926, as cited in Binkerd and Kolari, 1975). The maximum level of nitrite permitted in cured meats remained at 200 ppm until the 1970s, when maximum levels were reduced to 120 ppm in bacon, and 156 ppm in other cured meats, to reduce nitrosamine levels in these products after cooking (USDA, 1975, 1978a,b). The use of nitrate salts in all meat and poultry products except dry cured or fermented sausage products was discontinued in 1975 (USDA, 1975).

The U.S. Food and Drug Administration (FDA) regulates the addition of nitrate and nitrite salts to fish products. Maximum permitted levels vary among products and types of fish, with up to 500 ppm residual sodium nitrate or 200 ppm sodium nitrite permitted as a preservative and color fixative in smoke cured sable fish, shad, or salmon (Code of Federal Regulations, 1981a,b; Committee on Nitrite and Alternative Curing Agents in Food, 1981).

III. NITRIC OXIDE IN MEAT CURING: AN OVERVIEW

Nitrate or nitrite salts are added to cured meats to provide cured meat color, and to enhance product stability. Curing salts have both antioxidant and antimicrobial effects. NO reactions with iron compounds are involved in all these beneficial effects. Nitrosohemochrome, consisting of NO, heme (Fe^{2+}), and the denatured globin of myoglobin, is the pink pigment of cured meats. Oxidation of meat lipids is catalyzed by both heme and nonheme iron. NO inhibits oxidative rancidity by (1) converting heme to nitrosoheme iron, which has antioxidant properties, (2) stabilizing heme iron to the effects of cooking, and (3) forming antioxidant complexes with free ionic iron. Finally, NO exerts antimicrobial effects against some anaerobic bacteria, including clostridia, by inactivating key iron–sulfur enzymes.

The major health concern regarding use of curing salts is the possibility of nitrosamine formation in the cured products. Nitrite ion appears to be the precursor compound required for nitrosamine formation, rather than NO. Inclusion of reductants such as ascorbate, now required in bacon, lowers nitrite level in the product and increases the level of NO available for cured meat color formation and stability.

IV. NITRIC OXIDE AND CURED MEAT COLOR

A. Enzymatic Reduction of Nitrate or Nitrite

Although nitrate was the traditional meat curing salt, Haldane (1901) demonstrated cured meat pigment development by addition of nitrite to hemoglobin. Hoagland (1908) concluded that bacterial or muscle tissue reduction of nitrate was the source of nitrite and NO in cured meats. Conclusive evidence of NO production in cured meat systems was provided by Walters and Taylor (1964). They demonstrated that fresh pig muscle minces were capable of anaerobic reduction of sodium nitrite to NO. Muscle tissue was not capable of reducing nitrate to NO, unless previously incubated at 30°C for 18 hr to encourage bacterial growth (Table 1). NO gas was identified by its characteristic insolubility in 0.9 N KOH, and solubility in alkaline sodium sulfite. Mass spectrometry of the gaseous products of the incubation of sodium nitrite with fresh pig muscle minces revealed a peak at mass/charge ratio of 30, attributed to NO. Infrared spectra of the dried gaseous products of incubation were initially inconclusive. The peak at 1870 cm^{-1} due to NO could not be recognized with certainty over the background absorption ascribed to small amounts of residual water. However,

TABLE 1
Production of Nitrosylmyoglobin from Sodium Nitrite or Potassium Nitrate[a]

Sample	Tissue preparation[b]	Anaerobic plate count (number/g)	Percent conversion of indigenous pigment to nitrosylmyoglobin in presence of NaNO$_2$	KNO$_3$
A	Fresh muscle mince	9.5 × 10^3	71	1.4
	Mince incubated at 30°C anaerobically for 18 hr	4.9 × 10^8	76	63.0
B	Fresh muscle mince	2.1 × 10^3	71	1.8
	Mince incubated at 30°C anaerobically for 18 hr	2.1 × 10^8	70	58.0
	Muscle minced after 48 hr at −20°C	—	64	5.6
C	Fresh muscle mince	4.7 × 10^3	70	0.0
	Mince incubated at 30°C anaerobically for 18 hr	1.3 × 10^9	79	71.0
	Muscle minced after 48 hr at −20°C	4.0 × 10^6	77	0.0

[a] After incubation at pH 6.0 with fresh or previously incubated (high microbial count) muscle minces.
[b] Manometric incubations under argon for 2 hr of 3.0-g samples of muscle minces in 3.0 ml of 0.20 M phosphate buffer, pH 6.0, containing antibiotic (chloromycetin, 10 mg%) with 0.5 ml of either 0.12% NaNO$_2$ or 0.17% KNO$_3$ initially in side arms of Thunberg tubes. Data from Walters and Taylor (1964).

in all tests involving nitrite, the admission of oxygen resulted in appearance of a peak at 1615 cm^{-1}, in accordance with the conversion of the weak absorber NO to the strong absorber NO$_2$. Control incubations of nitrite with whole citrated pig blood, glutathione, or NADH failed to produce NO, indicating that serum enzymes, hemoglobin, reduced glutathione, or NADH were not capable of directly reducing nitrite. Incubation of ascorbic acid with sodium nitrite at pH 6.0 was found manometrically to produce an appreciable NO fraction. However, fresh muscle minces did not contain appreciable amounts of ascorbate. Only fresh muscle minces had high activity for enzymatic reduction of nitrite. Heated controls had much less activity, as did minced tissue held 24 hr at 25°C.

Walters and Taylor (1964) concluded that the muscle cell respiratory system was responsible for the reduction of nitrite to NO in fresh tissue. Walters and co-workers (1967) went on to show that nitrite reduction by fresh muscle minces was due to the mitochondrial respiratory enzyme system. Nitrite in anaerobic systems acted as a terminal respiratory electron acceptor, with the formation of nitrosyl-ferricytochrome c. The NO was then transferred to metmyoglobin (MMb) by the action of NADH-cytochrome c reductase. Nitrosyl-metmyoglobin (NOMMb) was reduced to NO-myoglobin (NOMb) by mitochondrial enzyme systems, even in the presence of excess nitrite. In addition to its role as a terminal electron acceptor, nitrite in muscle systems acted as a pigment oxidant, causing rapid oxidation of myoglobin (Mb) and oxymyoglobin (MbO$_2$) to MMb.

A nonmitochondrial muscle diaphorase has also been described (Koizumi and Brown (1971), capable of forming NOMb in solutions containing nitrite and Mb, using methylene blue as a nonspecific electron carrier. NOMb formation by both the mitochondrial and nonmitochondrial diaphorase enzyme systems requires the presence of reduced NADH.

B. Nonenzymatic Reduction of Nitrate or Nitrite

Reith and Szakaly (1967a,b) studied the formation and stability of NOMb in both model systems and in canned meats. In model systems, optimum NOMb formation occurred at a molar ratio of 1 : 200 : 1 of Mb : sodium ascorbate : nitrite, respectively. At lower ascorbate concentrations, spectrophotometrically detectable levels of MMb remained in solution. When ascorbate was replaced with cysteine, a higher level of nitrite, to a molar ratio of 1 Mb : 3 NaNO$_2$ was required for optimum formation of NOMb. Solutions containing cysteine had markedly lower nitrite concentrations after 3 hr, accompanied by the smell of NO$_2$. Ferrous sulfate improved both the speed and degree of conversion to NOMb. The speed of conversion of reactants to NOMb was also increased by higher temperatures (10°–50°C) and lower pH (5.0 or 5.5 versus 6.0). However, at higher pH, NOMb was much more stable on exposure to light.

In canned meats, ascorbate also enhanced NOMb production (Reith and Szakaly, 1967b). Without ascorbate, a Mb:nitrite molar ratio of 1:5 was needed for optimum formation of NOMb. With ascorbate, a ratio of 1 Mb:3 nitrite was sufficient for maximum NOMb formation. Sodium erythorbate was equivalent to sodium ascorbate for NOMb formation. Canned meats formulated with potassium nitrate showed no formation of cured meat pigment.

Ascorbate, cysteine, hydroquinone, and NADH are capable of acting as reductants for NOMb formation in model systems containing sodium nitrite and Mb (Fox and Ackerman, 1968). Ascorbate, cysteine, and hydroquinone all form nitroso-reductant intermediates which released NO, forming a NO–MMb complex which was then reduced to NOMb. Release of NO from the reductant–NO complex was rate limiting in production of NOMb. For NADH as reductant, reduction of NOMMb to NOMb was the rate limiting step. In summary, two reduction steps were required, the reduction of nitrite (as nitrous acid or its anhydride, N_2O_3) to NO, and reduction of NOMMb to NOMb.

Neither NADH nor NADPH are capable of directly reducing nitrite to NO, under aerobic or anaerobic conditions (Koizumi and Brown, 1971). NADH–flavin systems are also incapable of reducing nitrite to NO. However, NADH enhances NOMb formation in systems containing nitrite, MMb, and flavins (flavin mononucleotide, flavin adenine dinucleotide, or riboflavin). Koizumi and Brown (1971) concluded that Mb participates in the reduction of nitrite to NO. In support of this hypothesis, they observed that in anaerobic systems containing only sperm whale Mb and nitrite, the visible absorption spectra after 60 min had features indicative of both MMb and NOMb. Apparently, the NADH–flavin system is capable of reducing redox active sites on the globin portion of NOMMb, somehow leading to formation of NOMb. It is known that both Mb (ferrous) and MMb (ferric) will combine with NO to yield NOMb. NOMMb is believed to "autoreduce with time via internal electronic rearrangement" (Giddings, 1977). Thus, Mb has dual roles as reductant in cured meat systems. First, ferrous heme of Mb is rapidly oxidized to ferric MMb by nitrite, and NO is generated. Second, sites on the globin portion of NOMMb may be oxidized in the formation of NOMb.

C. Ascorbate Accelerates Nitric Oxide Formation

In contrast to fresh muscle, meat has low levels of NAD (Madhavi and Carpenter, 1993). Thus, NAD-dependent enzymatic pathways for NOMb formation are relatively unimportant in meat curing. In commercial practice, nitrite is reduced to NO by nonenzymatic means, including use of reductants such as ascorbate and erythorbate. Although meat has sufficient reducing ability to obtain a slow conversion of nitrite to NO, ascorbate or its isomer, erythorbate, is commonly added to curing brines or sausage emulsions to obtain faster NO production and thus a more rapid development of cured meat color. Care must be taken in use of ascorbates in curing brines. If ascorbate-containing brines are held for long periods of time, especially at elevated temperatures or at acid pH, nitrite will be lost as NO gas, reducing the effectiveness of the brine for meat curing. Ascorbate-containing brines should be held for no longer than a day at less than 50°F, and at alkaline or very slightly acid pH (Rust, 1977). Phosphates are commonly used to maintain slightly alkaline conditions in curing brines, and have the additional advantage of increasing the water retention of the cooked meat products.

D. Characterization of Mononitrosylhemochromogen, the Cured Meat Pigment

NO, generated by heating sodium nitrite and ferrous sulfate in dilute sulfuric acid, can bind to both ferric hematin and ferrous heme. Keilin (1955) found that urohematin, isolated from the urine of a patient with congenital porphyria, reacts with NO in alkaline solution to form a pink pigment with two sharp absorption bands of equal intensity at 563 and 528 nm. On reduction of the pigment with a few milligrams of sodium dithionite, a single strong absorption band is observed at 543 nm. In absence of NO, reduction of urohematin forms dihydroxyuroheme with two well defined bands at 546 and 578 nm. However, Keilin (1955) did not describe the NO-hematin or NO-heme compounds as either mono or dinitrosyl complexes.

Cured meat pigment levels may be quantified based on extraction of nitrosoheme into acetone solution (Hornsey, 1956). Best extraction is at 80% acetone/20% water, taking into account the water content of the sample. The nitrosoheme pigment complex exhibits absorption peaks at 476, 535, and 563 nm. A similar curve is obtained when hematin is reduced with sodium hydrosulfite (dithionite), and a trace of sodium nitrite added, followed by dilution with four parts acetone. In the quantitation procedure, a single absorbance reading is taken at 540 nm. Model systems containing nitrosoheme fade rapidly in light, since they lack reductants present in extracts of meat. Addition of cysteine to such systems greatly prolongs color stability (Hornsey, 1956).

Tarladgis (1962) proposed that mononitrosyl heme would be paramagnetic, due to the presence of the single unpaired electron of NO, while dinitrosyl heme would be diamagnetic. Since paramagnetic signals were not observed in EPR studies, he concluded that the acetone extractable cured meat pigment was dinitrosyl heme. In support of this conclusion, Lee and Cassens (1976) found that heated samples of myoglobin contained twice as much labeled N (from ^{15}N-labeled nitrite) as unheated samples. They concluded that globin was likely detached from myoglobin by heating, making available two binding sites for NO on the heme ring.

More recent evidence indicates that the cured meat pigment is actually a mononitrosyl complex. Killday *et al.*, (1988) showed that dithionite reduction of chlorohemin in dimethyl sulfoxide, followed by exposure to NO gas in argon formed a red pigment [nitrosyl iron(II) protoporphyrin] that was identical to pigment extracted in acetone from cured corned beef. Both compounds had an R_f of 0.55 on thin layer chromatography plates developed with ethyl acetate–methanol (1:1). The infrared spectrum of the synthesized compound had a nitrosyl stretch at 1656 cm^{-1}, consistent with a bent NO-ligand state. Absorption bands corresponding to a second coordinated nitrosyl ligand were not present in the 1900-cm^{-1} region. Fast atom bombardment mass spectroscopy of the synthesized pigment showed fragments consistent with a molecular ion of 646 atomic mass units, corresponding to a mononitrosyl heme complex, as opposed to a dinitrosyl heme complex, which would show clusters at 30 amu greater. Killday et al. (1988) thus concluded that cured meat pigment was a mononitrosyl heme complex.

Killday et al. (1988) also provided evidence for internal autoreduction of ferric nitrosyl heme complexes, as previously proposed by Giddings (1977). Heating of chlorohemin(iron-III) dimethyl ester in dimethyl sulfoxide solution with imidazole and NO produced a product with an infrared spectra identical to that of nitrosyl iron(II) protoporphyrin dimethyl ester prepared by dithionite reduction. Both spectra clearly showed the characteristic nitrosyl stretch at 1663 and 1665 cm^{-1}. They thus proposed a mechanism for formation of cured meat pigment which includes internal autoreduction of NOMMb via globin imidazole residues. A second mole of nitrite is proposed to bind to the heat-denatured protein, possibly at a charged histidine residue generated in the previous autoreduction step.

V. NITRIC OXIDE AS AN ANTIOXIDANT IN CURED MEATS

"Warmed-over flavor" (WOF) is the term used to describe the stale or rancid flavor and odor that develops in cooked meats with refrigeration, or especially with reheating (Sato and Hegarty, 1971; Younathan, 1985). Unsaturated lipids, especially those of the membrane phospholipid fraction, are the compounds undergoing autoxidation (Igene and Pearson, 1979). Heme compounds such as hemin catalyze lipid autoxidation (Kanner et al., 1984; Tappel, 1955; Liu, 1970a,b) but the muscle pigments myoglobin and hemoglobin are much less active as catalysts of lipid oxidation (Sato and Hegarty, 1971; Love and Pearson, 1974). Metal chelating agents, sodium nitrite, and various reductants inhibit lipid oxidation when added to cooked ground beef, but iron powder increases oxidation as measured by thiobarbituric acid (TBA) values (Sato and Hegarty, 1971). It is now well established that in cooked meats, ionic iron or low molecular weight iron complexes are the main prooxidants (Sato and Hegarty, 1971; Igene *et al.*, 1979). Reductants such as ascorbate and cysteine may be prooxidants in meat homogenates or model systems containing ionic iron (Liu, 1970a,b), apparently by partial reduction of ferric iron to the ferrous form, which then catalyzes the degradation of lipid hydroperoxides (Kanner *et al.*, 1984). However, higher levels of ascorbate inhibit lipid oxidation. Lipid oxidation rates are highest when the ratio of ferrous to ferric iron is approximately one. The antioxidant effect of high ascorbate levels on iron-catalyzed lipid oxidation is due to complete reduction of iron (Miller and Aust, 1989).

Sodium nitrite is very effective in preventing WOF (Igene *et al.*, 1979; MacDonald *et al.*, 1980a). Igene *et al.* (1979) washed fresh meat to remove pigments, then cooked the meat with or without the pigment fraction, and with or without 156 ppm sodium nitrite. WOF was much less pronounced, and TBA values were lower, in beef or chicken meat cooked with nitrite (Table 2).

Igene *et al.* (1979) further showed that the free or nonheme iron in the washed pigment fraction was responsible for the increase in oxidation of cooked meat (Table 3). Cooking or treatment of the pigment fraction with hydrogen peroxide released iron from heme, increasing ionic iron levels. Addition of EDTA to chelate ionic iron resulted in lower TBA values in cooked samples (Table 3). Ionic iron released from heme pigments during cooking accelerated lipid oxidation and WOF. Nitrite apparently stabilizes heme pigments to heat degradation, due to formation of the NO–heme pigment complex. NO itself may have antioxidant properties, by its ability to act as a free radical acceptor (Sato and Hegarty, 1971). Nitrite or derivatives such as NO may also react with and stabilize unsaturated lipids (Woolford and Cassens, 1977; Frouin *et al.*, 1975; Zubillaga and Maerker, 1987).

Some studies indicate that nitrite or related compounds have antioxidant activity even in systems that do not contain heme pigments. MacDonald *et al.*

TABLE 2
Effect of Nitrite on Thiobarbituric Acid Values and Sensory Scores of Cooked Beef[a]

Treatment	TBA no.	Panel score
Beef with pigment, no nitrite	1.93[a]	2.42[a]
Beef with pigment, + nitrite	0.21[b]	4.42[b]
Beef without pigment, no nitrite	0.61[c]	3.50[b]
Beef without pigment, + nitrite	0.42[b,c]	4.31[b]

[a] Taste panel score was 1–5, where 1 corresponds to pronounced WOF and to 5 no WOF. Values in the same column with different letter superscripts were significantly different ($p < 0.05$). Adapted from Igene *et al.* (1979).

TABLE 3
Ionic Iron Levels in Beef Pigment Extracts, and Thiobarbituric Acid Values Obtained on Heating Extracts with Washed Beef Muscle[a]

Treatment	Nonheme Fe^{2+} (μg/g meat)	Mean TBA no.
Fresh meat pigment extract	1.80	5.00
$EDTA^b$ + fresh meat pigment extract	—	1.55
Cooked meat pigment extract	5.51	4.35
EDTA + cooked meat pigment extract	—	1.46
H_2O_2 oxidized pigment extract	13.59	6.02
EDTA + H_2O_2 oxidized extract	12.33	1.54

[a] For comparison, total iron (heme + nonheme) in fresh meat pigment extracts was 20.64 μg/g meat. Adapted from Igene et al. (1979).
[b] EDTA, Ethylenediaminetetraacetate, 2%.

(1980b) reported that linoleic acid oxidation catalyzed by Fe^{2+} or Fe^{2+}-EDTA was substantially inhibited by as little as 10 ppm sodium nitrite. S-Nitrosocysteine, a compound generated during curing of meat, has antioxidant activity in an aqueous linoleate model system (Kanner, 1979) and in cooked turkey meat (Kanner and Juven, 1980). Cysteine–Fe–NO is not capable of catalyzing oxidation of β-carotene in model systems (Kanner et al., 1984), although cysteine–Fe^{2+} is very active in the same system (Table 4).

Kanner et al. (1984) further showed that cysteine–Fe–NO is inhibitory to hemin or cysteine–Fe^{2+} catalyzed oxidation of β-carotene. They explained the antioxidant activity of cysteine–Fe–NO and hemin-NO as the simultaneous quenching of free radicals [Eq. (1)] and decomposition of hydroperoxides [Eq. (2)]. The free radicals formed during decomposition of hydroperoxides are

TABLE 4
Effect of Cysteine–Fe–NO Complex on Linoleate-β-Carotene Oxidation[a]

Treatment	β-Carotene oxidation (nmol/min)
Cysteine (5×10^{-4} M)	3.1 ± 0.8
Fe^{2+} (5×10^{-5} M)	9.3 ± 0.8
Cysteine–Fe^{2+}	13.4 ± 0.6
Cysteine–Fe–NO	0

[a] Data from Kanner et al., 1984.

quenched by NO complexes remaining in the system [Eq. (3)], inhibiting initiation reactions,

$$Cys-Fe^{2+}-NO\cdot + R\cdot \longrightarrow RNO\cdot + Cys-Fe^{2+} \qquad (1)$$

$$Cys-Fe^{2+} + ROOH \longrightarrow RO\cdot + OH^- + Cys-Fe^{3+} \qquad (2)$$

$$Cys-Fe^{2+}-NO\cdot + RO\cdot \longrightarrow Cys-Fe^{2+} + RONO, \qquad (3)$$

where $R\cdot$ is the alkyl radical, $RO\cdot$ is the alkoxy radical, and ROOH is hydroperoxide.

Thus, antioxidant effects of nitrite in cured meats appear to be due to the formation of NO. Kanner et al. (1991) also demonstrated antioxidant effects of NO in systems where reactive hydroxyl radicals ($\cdot OH$) are produced by the iron-catalyzed decomposition of hydrogen peroxide (Fenton reaction). Hydroxyl radical formation was measured as the rate of benzoate hydroxylation to salicylic acid. Benzoate hydroxylation catalyzed by cysteine–Fe^{2+}, ascorbate–EDTA–Fe^{2+}, or Fe^{2+} was significantly decreased by flushing of the reaction mixture with NO. They proposed that NO$\cdot$ liganded to ferrous complexes reacted with H_2O_2 to form nitrous acid, hydroxyl ion, and ferric iron complexes, preventing generation of hydroxyl radicals.

VI. NITRITE AND NITRIC OXIDE AS ANTIMICROBIAL AGENTS

A. Bacteriostatic Effects

Salt was long regarded as the agent responsible for the shelf stability of cured meats. Nitrate and later nitrite, were thought to only contribute to cured meat color. Tarr (1941a,b) showed that nitrite was an inhibitor of bacterial spoilage in fish, but recognition of the antibotulinal properties of nitrite and nitrate has been a relatively recent development. Steinke and Foster (1951) were apparently the first to demonstrate antibotulinal effects of nitrite in a processed meat product. They reported development of botulism toxin in 9–23 days in liver sausages containing 2.5% salt, 0.1% sodium nitrate, inoculated with 5000 spores/g, and incubated at 30°C. This time was markedly increased by a slight increase in salt concentration, a decrease in the incubation temperature, or a decrease in the spore inoculum number. The addition of 100–200 ppm sodium nitrite to the sausage enhanced the inhibitory effect of salt and nitrate (Table 5).

Sodium nitrite has multiple bacteriostatic effects, and is inhibitory to both aerobic and anaerobic organisms. Nitrite is most inhibitory to both aerobic (Tarr, 1941a,b; Castellani and Niven, 1955) and anaerobic (Shank et al., 1962) microbes over the pH range of 5.0–5.5, consistent with the hypothesis that undissociated nitrous acid ($pK_a = 3.4$) is the active compound. Shank et al. (1962) concluded that rapid nitrite loss due to conversion of nitrous acid to gaseous NO

TABLE 5
Effect of Salt, Nitrate, and Nitrite on Botulinal Toxin Formation in Liver Sausage Held at 30°C[a]

Treatment	First detection of toxin (+ or − at indicated days)
Salt, 2.00%	3, +
Salt, 2.25%	3, +
Salt, 2.50%	6, +
Salt, 2.75%	16, +
Salt, 3.00%	23, −
Salt, 3.25%	30, −
Salt, 2.50% + sodium nitrate, 0.1%	16, +
Salt, 2.5% + sodium nitrite, 200 ppm	30, +
Salt, 2.5% + 0.1% nitrate + 200 ppm nitrite	30, −

[a] Adapted from Steinke and Foster, 1951.

and NO_2 at pH values below 5.0 accounted for the decreased antimicrobial effects of nitrite at acid pH.

Castellani and Niven (1955) and Shank et al. (1962) both discussed the possibility that NO derived from nitrite reacted with heme enzymes of the cytochrome system, inhibiting aerobic respiration in gram-negative bacteria. Kucera and Skladal (1990) confirmed this possibility. They found nitrite reduction by dissimilatory nitrite reductase of *Paracoccus denitrificans* produced a substance, probably NO, that inhibited respiratory cytochrome oxidase of the same organism. This observation in part explains the sensitivity of even the denitrifying bacteria to rapid changes in nitrite concentration. However, Castellani and Niven (1955) concluded that this was not the sole mechanism of nitrite inhibition, since nitrite was most inhibitory under anaerobic conditions, where the cytochrome systems would not be important in the metabolism of the organisms. They also noted that some streptococci were strongly inhibited by nitrite, even though these bacteria are devoid of heme-containing cytochromes. They further cited the finding by Tarr (1941a) that aerobic respiration by a strain of *Achromobacter* was not inhibited by nitrite. Castellani and Niven (1955) observed that anaerobic nitrite inhibition that occurred when nitrite was autoclaved in media could be reversed by addition of sulfhydryl compounds such as cysteine or glutathione. They concluded that a complex was formed between nitrite and sulfhydryl compounds, limiting the availability of sulfhydryl compounds for microbial growth. They also suggested that nitrite may inactivate active sulfhydryl groups of pyruvate-metabolizing enzymes, as previously observed in *Vibrio comma* (Bernheim, 1943) and *Fusarium* (Nord and Mull, 1945). O'Leary and Solberg (1976) reported decreased activity of sulfhydryl enzymes and decreased concentration of membrane free-sulfhydryl groups in Clostridium perfringens cells inhibited with sodium nitrite. Hansen and co-workers have convincingly shown that nitrous acid modification of membrane sulfhydryl groups is responsible for the inhibition of spore outgrowth of Bacillus cereus, an aerobic sporeformer (Morris et al., 1984; Buchman and Hansen, 1987).

Nitrite inhibits anaerobic respiration, glucose and proline active transport, and oxidative phosphorylation in Pseudomonas aeruginosa, a nonfermenting, denitrifying bacteria (Rowe et al., 1979). Streptococcus faecalis and S. lactis were not inhibited, presumably because these species lack cytochromes, and glucose is not transported by an active transport mechanism. Yarbrough et al. (1980) extended this work to show that intracellular aldolase activity of Escherichia coli, P. aeruginosa, and S. faecalis were inhibited by addition of 10–100 mM (460–4600 ppm) nitrite. They suggested that S. lactis and S. faecalis were impermeable to nitrite, because these bacteria were nitrite-resistant, although aldolase from these bacteria was nitrite-sensitive. It was concluded that nitrite inhibition of active transport occurred via inhibition of the cytochrome chain, preventing formation of the proton gradient needed for active transport.

Shank et al. (1962) investigated the possibility that NO inhibition of heme enzymes of gram-negative bacteria was responsible for the predominance of gram-positive, catalase-negative bacteria on cured meat. Test organisms were absorbed onto filter discs, then exposed to NO gas in a vacuum dessicator for 30 min. The dessicator was evacuated and backfilled with oxygen-free nitrogen gas three times prior to the introduction of NO. Some samples had a small volume of air injected into the NO chamber, resulting in the formation of 0.18–0.36% NO_2. After treatment, sample discs were removed under nitrogen, and placed in sterile trypticase soy broth for enumeration. Pseudomonas fluorescens numbers decreased substantially due to NO treatment. Staphylococcus aureus, and strains of clostridia and Lactobacillus also had lower numbers after NO treatment, but Streptococcus durans was unaffected. All bacterial numbers were reduced by 90% or more after exposure to NO + 0.36% NO_2. Shank et al. (1962) noted that in the presence of oxygen, NO is oxidized to NO_2, which in turn forms nitric and nitrous acids, the latter having much more pronounced antimicrobial effects. They thus concluded that NO was relatively inert, and that nitrous acid was the primary antimicrobial agent in nitrite-cured meats. However, these experiments cannot rule out an antimicrobial role for NO itself. NO penetrating the microbial cell, or generated inside the cell, would certainly be expected to be more damaging than exposure of NO only at the cell surface. As pointed out by Shank et al. (1962), NO is poorly soluble in water (56.3 ppm at 25°C, 760 mm Hg). Some of the variable sensitivity reported by Shank et al. (1962) to NO gas may very well be due to differential permeability of microbial cells to NO, as well as differences among bacterial species in content of enzyme systems sensitive to NO.

B. Mechanism of Inhibition of Clostridia

Reports in the 1960s and early 1970s on the role of nitrite as a precursor to carcinogenic nitrosamines stimulated renewed interest in the mechanism of bacterial inhibition, and especially of botulinal inhibition by nitrite, toward the goal of perhaps developing nitrite substitutes in cured meats. A number of informative and comprehensive reviews were completed on this topic (Sofos et al., 1979; Benedict, 1980; Pierson and Smoot, 1982; Committee on Nitrite and Alternative Curing Agents in Food, 1981; Council for Agricultural Science and Technology, 1978).

Nitrite is a precursor to other inhibitors. Perigo et al. (1967) reported development of a clostridial inhibitor when nitrite was autoclaved with culture media. Iron was also a precursor (Huhtanen and Wasserman, 1975; Asan and Solberg, 1976; Moran et al., 1975). Roussins salt, nitrosocysteine, and cysteine–NO–Fe complexes have been suggested as responsible for "Perigo factor" inhibition. However, Perigo factor is neutralized by meat particles (Johnson et al., 1969) and requires higher temperatures than normally used in cured meat processing (Ashworth et al., 1973). Thus, Perigo factor is not thought to be responsible for the nitrite-induced inhibition observed in cured meats.

Most cured meats receive only moderate heat treatment (74°C or less). In such products, added ionic iron actually decreases nitrite inhibition of clostridia (Van Roon and Olsman, 1977; Tomplin et al., 1978a, 1979). Hemoglobin and myoglobin have similar effects (Tompkin et al., 1978b; Vahabzadeh et al., 1983; Miller and Menichillo, 1991). Benedict (1980) proposed that nitrite ties up iron, reducing its availability as a nutrient for bacterial growth. However, chelation with EDTA or EDTA plus carbon monoxide was not inhibitory to meat samples inoculated with clostridia (Tompkin et al., 1979; Vahabzadeh et al., 1983). Only meat samples formulated with nitrite were inhibitory. Iron addition probably stimulates botulinal growth by lowering residual nitrite level, as indicated by results of Ando (1974) and Lee et al. (1981) for ionic iron, and Kim et al. (1987) for addition of heme iron. Kim et al. (1987) further proposed that ionic iron addition may result in protein–cystyl–Fe–NO complexes which chelate NO more effectively than similar complexes without iron. In support of this proposal they cited the work of Van Roon and Olsman (1977), who found that S-nitrosocysteine inhibited clostridial growth, but a complex of cystyl–NO–Fe was less inhibitory.

Lee et al. (1978) investigated the possibility that sulfhydryl groups were required in the formation of the clostridial inhibitor in cured meats. Sulfhydryl groups of meat proteins were blocked by treatment with silver lactate, then the samples were cooked with sodium nitrite before inoculation with C. botulinum spores. Botulinal growth as measured by gas and toxin production was similar to controls without silver lactate treatment. They thus concluded that sulfhydryl groups were not required for the antibotulinal effects of nitrite in cured meats.

Tompkin et al. (1978a,b) were apparently the first to suggest the possibility that NO reacts with and inactivates an essential iron compound(s) within the newly germinated botulinal cell, preventing outgrowth. They further suggested that ferredoxin, an iron–sulfur protein cofactor for electron transport and energy production in clostridia, might be one such compound. Meyer (1981) reported that NO irreversibly inactivated the iron–sulfur nitrogenase of C. pasteurianum, but not the Mo–iron–sulfur form of nitrogenase. Woods et al. (1981) reported that pyruvate : ferredoxin oxidoreductase, an iron sulfur enzyme, was inactivated by NO derived from nitrite. This enzyme is part of the phosphoroclastic system responsible for ATP generation in clostridia, with generation of CO_2 and H_2 as by-products, and pyruvate as substate. Ferredoxin participates as an electron carrier in the reaction. Activity of the reconstituted phosphoroclastic system of C. sporogenes was measured by ferredoxin reduction of NAD. The enzyme-containing fraction from nitrite-treated cells had no activity, even when reconstituted with ferredoxin fron untreated cells (Table 6). The nitrite-treated ferredoxin fraction contained less protein but retained some capability for NAD reduction when reconstituted with enzyme fraction from untreated controls. Small amounts of NO were detected in the cell free system. Woods et al. (1981) noted that the 10,000-g fraction containing ferredoxin was much paler from nitrite-treated cells than controls, but concluded on the basis of activity in reconstituted systems that pyruvate–ferredoxin oxidoreductase was the iron–sulfur protein most sensitive to NO. They further pointed out that inhibitory effects were observed at nitrite levels (4.4 mM, 300 ppm) comparable to those used in meat processing, in contrast to the high nitrite levels (15 mM) used by O'Leary and Solberg (1976) to demonstrate inhibition of aldolase and glyceraldehyde-3-phosphate dehydrogenase in C. perfringens. Woods and Wood (1982) further demonstrated nitrite inhibition of the phosphoroclastic system in several strains of C. botulinum, as indicated by pyruvate accumulation in cell suspensions incubated at 37°C with 4.4 mM sodium nitrite.

TABLE 6
Activity of Reconstituted Phosphoroclastic System in *Clostridium sporogenes* Cells[a]

Protein fraction	Ferredoxin fraction	NAD reduction (nmol min^{-1} mg protein^{-1})	Total protein (mg)	Ferredoxin protein (mg)
Control	Control	0.86	71.5	1.0
Control	Nitrite-treated	0.16	71.0	0.5
Nitrite-treated	Control	0.00	58.0	1.0
Nitrite-treated	Nitrite-treated	0.00	57.5	0.5

[a] Cells were incubated with 4.4 mM nitrite at 37°C for 1 hr before disruption and separation of the ferredoxin and protein fractions. Data from Woods et al. (1981).

Reddy et al. (1983) demonstrated by electron spin resonance (ESR) spectroscopy the presence of iron–sulfur proteins in C. botulinum cells. Signals were observed at $g = 2.02$ in washed, sonicated cells, and at $g = 1.94$ in cell preparations reduced with sodium dithionite. Because a single iron–sulfur protein is not capable of producing signals at both $g = 2.02$ and $g = 1.94$, these results were indicative of at least two different iron–sulfur proteins in C. botulinum cells. Treatment of cells with 200 ppm sodium nitrite at 35°C for 45 min resulted in the appearance of a signal at $g = 2.035$, characteristic of iron–nitrosyl complexes, and disappearance of the signal at $g = 1.94$. The signal at $g = 2.035$ was intensified by inclusion of 500 ppm ascorbate in the incubation treatment. Ascorbate is known to enhance the antibotulinal effects of nitrite in cured meat products, perhaps in part due to the ability to chelate iron (Tompkin et al., 1978c). However, the primary effect of ascorbate is to rapidly reduce nitrous acid to NO (Kim et al., 1987; Payne et al., 1990a), as follows:

$$2\ HNO_2 + \text{ascorbate} \longrightarrow 2\ NO + \text{dehydroascorbate} + 2\ H_2O$$

Reddy et al. (1983) concluded that NO inactivation of iron–sulfur proteins was the probable mechanism of botulinal inhibition in nitrite-treated foods. In support of this conclusion, Carpenter et al. (1987) observed decreased activity of clostridial pyruvate–ferredoxin oxidoreductase and lower cytochrome c reducing ability by ferredoxin in extracts of cells treated with nitrite. NO treatment also inhibits yeast pyruvate decarboxylase (a non-iron–sulfur protein) and pyruvate–ferredoxin oxidoreductase from C. perfringens (McMindes and Siedler, 1988). They suggested that thiamine-dependent decarboxylation of pyruvate may be an additional site for antimicrobial effects of NO.

Payne et al. (1990a) also observed a signal at $g = 1.94$, characteristic of proteins such as ferredoxin, in the EPR spectra of normal C. sporogenes cells. Cells incubated with high levels (15 mM) of both nitrite and ascorbate were no longer viable. The EPR spectra of treated cells no longer showed the signal at $g = 1.94$, indicating that oxidation or destruction of the iron–sulfur clusters had occurred. The EPR spectra of these cells also had a prominent signal at $g = 2.03$ (characteristic of low spin Fe–NO complexes), a weak downward signal at $g = 2.0$, an unidentified broad signal at $g = 1.65$, and a signal at $g = 4.0$ indicative of high spin Fe–NO. With the exception of the signal at $g = 1.65$, the EPR spectra of nitrite-treated C. sporogenes were identical to that of nitrite-treated C. botulinum (Reddy et al., 1983). Payne et al. (1990a) further demonstrated inhibition of C. sporogenes cell growth by treatment with NO gas, or with addition to the growing cultures of both nitrite and ascorbate (Table 7).

Both hydrogenase and pyruvate–ferredoxin oxidoreductase activities were inhibited in nitrite-treated cells, and the inhibition was greater with addition of both nitrite and ascorbate (Payne et al., 1990a). However, they noted that enzyme inhibition was not complete. Meyer (1981) also reported incomplete in-

TABLE 7
Inhibition of Exponential Phase Cultures of *Clostridium sporogenes* by Nitrate, Nitrite, and Nitric Oxide[a]

Inhibitor	Cell yield (% of control)	Growth in fresh medium
1 mM NaNO₃	100	Yes
1 mM NaNO₂	100	Yes
4 mM NaNO₂	77	Yes
0.5 mM NaNO₂/4 mM ascorbate	64	Yes
0.5 mM NaNO₂/10 mM ascorbate	<3	No
0.186 mM nitric oxide	<3	No

[a] Cell yield was calculated from values of wet weights of harvested cells after growth in the presence or absence of inhibitors. Each inhibitor was added as a filter-sterilized solution after 4 hr of cell growth. Yield of control cells was 3 g/liter. Each culture was inoculated into fresh normal growth medium after inoculation with inhibitor, to see if growth recovery occurred. From Payne et al. (1990a).

activation of iron–sulfur enzymes by NO. He found that NO treatment of the nitrogenase system of *C. pasteurianum* rapidly and irreversibly inactivated the iron–sulfur enzyme, but not the molybdenum-iron-sulfur form of the enzyme. Payne et al. (1990a) observed that the EPR spectra of ferredoxin treated with nitrite and ascorbate still showed a signal at $g = 1.94$ for reduced (4Fe-4S) centers, substantiating the previous conclusion of Woods et al. (1981) that reduced ferredoxin is relatively stable to attack by nitrite.

Based on the incomplete inhibition of iron–sulfur proteins by nitrite, and the observation that low-spin Fe–NO EPR signals were observed by *C. sporogenes* cultures that recovered from nitrite treatment, Payne et al. (1990a) concluded that the antimicrobial effect of nitrite or NO cannot be explained by direct inhibition of preformed pyruvate–ferredoxin oxidoreductase or hydrogenase.

Payne et al. (1990b) in a companion paper reported that some iron–thiol–nitrosyl compounds, including the EPR silent [Fe₄S₃(NO)₇], were much more inhibitory to *C. sporogenes* than nitrite. Cells incubated with these complexes exhibited growth inhibition and yielded an EPR signal at $g = 1.65$, similar to cells inhibited by nitrite + ascorbate treatment. Payne et al. (1990b) proposed that sufficient iron and other precursors were present for formation of these compounds in cured meats. They noted that nitrite and Fe–S–NO compounds both have greater inhibitory activity when added to cell cultures than when assayed with purified enzyme systems. They suggested that iron–thiol–nitrosyl complexes may inhibit assembly of iron–sulfur clusters into proteins in growing cells, and that these compounds might also be responsible for the pigmentation observed in inhibited cells.

C. Antimicrobial Significance of Protein-Bound Nitrite

Tracer studies with bacon showed that about 25% of the ^{15}N of added nitrite is incorporated into muscle proteins (Woolford and Cassens, 1977). Kubberod et al. (1974) found that in model systems with high nitrite concentrations, high temperature, and low pH, it was possible for all of the sulfhydryl groups of myosin to react within a few minutes. However, under milder conditions comparable to meat curing, the reaction rate was much slower, and the authors concluded that the direct reaction between nitrite and myosin (the major meat protein) was responsible for only a small portion of the nitrite lost in the curing process. Tryptophan appears to be the major amino acid capable of reversible binding of nitrosyl groups (Ito et al., 1979; Nakai et al., 1978). Ito et al. (1983) monitored the increase in absorbance at 330 nm as a measure of tryptophan nitrosylation in nitrite-treated proteins. The reversible release of nitrosyl groups was demonstrated by the formation of nitrosylhemochrome when nitrosated lysozyme or bovine serum albumin were incubated with myoglobin in the presence of ascorbic acid. Mellet et al. (1986) reported that nitrosotryptophan acts as a reservoir of nitrite, progressively releasing nitrite (as nitrous acid) as the free nitrite concentration decreases in the medium. They concluded that the bound nitrite fraction should be classified as either reversibly or irreversibly bound, where nitrosotryptophan would be a major contributor to the reversibly bound fraction, and would be measured in the Association of Official Analytical Chemists (AOAC) nitrite assay (AOAC, 1980), where hot water is used to extract nitrite from foodstuffs.

Mirna (1974) described a procedure in which bound nitrite is measured as the content of nitrite detectable with Griess reagent (AOAC method) after cleavage of nitroso compounds with Hg^{2+}, minus the free nitrite. Mirna (1974) considered nitroso thiols to be the major form of protein bound nitrite. Olsman (1977) pointed out that protein bound nitrite should be considered as a reservoir for nitrosation, just as free nitrite. Van Roon and Olsman (1977) found that S-nitrosocysteine and related iron thiols had antibotulinal properties. Vahabzadeh et al. (1983) concluded that the antibotulinal effects of addition of washed, prenitrosylated myoglobin to a cooked pork product were due to release of nitrite from protein-bound sites. Addition of ionic iron to cured meat systems increases protein-bound nitrite levels, but reduces botulinal inhibition (Van Roon and Olsman, 1977; Olsman, 1977; Kim et al., 1987). Kim et al. (1987) thus suggested that NO release may be slower for protein bound Cys–NO–Fe complexes than from similar complexes without iron. Cys–NO complexes are known to readily release the nitroso group. Byler et al. (1983) described the equilibrium reaction between cysteine and nitrite ion in weakly acid conditions, as follows:

$$H^+ + NO_2^- + CysSH \longrightarrow CysSNO + H_2O$$

They observed formation of CysSNO at pH values as high as pH 5.5, and suggested that this reaction may occur with free sulfhydryls of nitrite treated foods, or with bacterial enzymes.

Malin et al. (1989) reported deamination of ε-amino groups of lysine in proteins treated at pH 3.0 with high levels of nitrite. They reported deamination of 13% of the ε-amino groups of poly-L-lysine at pH 5.5–6.4 during a 24-hr period, suggesting the possibility for some loss of available lysine in slightly acid, nitrite-treated foods. However, protein-lysyl groups are not a site for formation of protein-bound nitrite, since N_2 gas was released as a result of lysine deamination.

VII. DIRECT APPLICATION OF NITRIC OXIDE TO MEATS

Braddock and Dugan (1969) reported no differences in color development of nitrite-treated or NO-treated, freeze-dried beef. Cured color development after NO treatment was more intense for drier (1.7% moisture) samples, compared to samples containing 11.3% moisture. The higher moisture samples also turned brown faster during storage at 25°C. Harper and Kittle (1960) obtained a patent for improving color stability of packaged cured meats with NO. McBrady (1968) was issued a patent for curing of sliced meat products such as bacon by rapid exposure to NO. For optimum color stability, the slices on a conveyor were warmed to 125°–140°F and acidified to pH 4.0–5.0 by application of a lactic acid spray, then passed through a salt–sugar solution before entering the NO gas chamber. Rankin (1973) described a procedure requiring NO exposure for 5 min for bacon slices, but a 24-hr exposure for larger pieces (10 cm thick). Vahabzadeh et al. (1983) reported the presence of 150–160 ppm nitrite in meat samples blended for 30 sec in the presence of NO. The chamber was evacuated prior to introduction of NO, but apparently oxygen retained in the meat, or during air exposure after blending, allowed oxidation of NO to nitrite or nitrous acid. Vahabzadeh et al. (1983) concluded that NO would not be a suitable substitute for sodium nitrite in meat curing, since residual nitrite was still present in NO treated meats. Although the NO curing system described by McBrady (1968) and assigned to Swift and Co. (Chicago, IL) is technically feasible, the process has not received USDA approval for commercial use.

Cured meat pigment (nitrosyl ferrohemochrome) may be prepared by gaseous NO treatment of hemin (Shahidi et al., 1985). O'Boyle et al. (1990) used cured meat pigment produced by this method in the production of nitrite-free weiners, lowering the possibility for nitrosamine formation in these products during heat processing. This process also has not yet been incorporated into commercial practice.

VIII. NITRITE, NITRIC OXIDE, AND NITROSAMINE FORMATION

Concern first developed about nitrosamines in nitrite-cured foods when outbreaks of hepatotoxicosis in mink and sheep in Norway were traced to N-nitrosodimethylamine arising from addition of nitrite as a preservative to herring meal, followed by high temperature drying, and subsequent use in animal foods (Ender et al., 1964; Hansen, 1964; reviewed by Committee on Nitrite and Alternative Curing Agents in Food, 1981). Fazio et al. (1971) later demonstrated presence of this carcinogenic nitrosamine in nitrate or nitrite treated salmon and other fish used for human food. Bacon was the major red meat product of concern, due to formation of detectable levels of N-nitrosopyrrolidine after frying (Crosby et al., 1972). Crosby et al. (1972) stated that "the principal chemical route to N-nitrosamines is the reaction of secondary amines (such as proline) with nitrous acid in the presence of nitrite ion." Nitrous acid anhydride (N_2O_3) is the nitrosating agent for secondary amines, rather than nitrous acid itself, and the rate of nitrosamine formation is dependent on the concentration of amine and nitrous acid, as follows (Challis, 1981; Mirvish, 1975; reviewed by Committee on Nitrite and Alternative Curing Agents in Food, 1981):

$$\text{Rate} = k_1[\text{amine}][\text{HNO}_2]^2$$

Rates of nitrosation are thus fastest for most secondary amines at about pH 2.5–3.4 (nitrous acid $pK_a = 3.4$). Only the unprotonated amine substrate reacts with the nitrosating agent.

Inhibition of nitrosation by pH control has counteracting effects; at low pH the conjugated acid form of the amine substrates are unreactive, while the protonated nitrous acid is most reactive. At pH values above 6.0, nitrous acid is converted to inactive nitrite ion.

NO itself is not capable of acting as a nitrosating agent unless oxidized in air to NO_2 (Challis and Kyrtopoulos, 1978), or in presence of catalysts such as iodine (Challis and Outram, 1979) or monovalent silver ions (Challis and Outram, 1978). However, nitrosation by gaseous NO_2 does not require acid conditions, and is generally faster than for nitrous acid in aqueous conditions (Committee on Nitrite and Alternative Curing Agents in Foods, 1981).

Nitrosamine formation in foods is inhibited by ascorbic acid (Mirvish, 1981) and α-tocopherol (Fiddler et al., 1978; Mergens et al., 1978). Both inhibitors act by reducing nitrous acid or other nitrosating agents to NO. Current USDA regulations for bacon manufacture now require 120 ppm nitrite, down from the former level of 200 ppm, and incorporation of 500 ppm ascorbate, erythorbate, or α-tocopherol into the curing brine, to lower residual nitrite levels in the product after processing, lowering the potential for nitrosamine formation during frying.

IX. NITRATE PREVENTION OF "LATE GAS" DEFECT IN CHEESE

In some countries, including Germany, the Netherlands, Egypt, and South Africa (but not the U.S.A., France, Italy, or Greece), nitrate is used to prevent late-blowing gas defect during ripening of semihard Edam, Gouda, or Domiati type cheeses (Glaeser, 1989). The defect is related to contamination of cheese milk with spores of lactate-fermenting clostridia, especially *Clostridium tyrobutyricum* (Wasserfall and Teuber, 1979). Contamination originates mainly from silage fed to dairy cattle. The spores survive pasteurization, causing gas production during the later stages of cheese ripening. The active compound inhibiting germination of these spores is nitrite produced from nitrate by microbial or enzymatic reduction, or both (Wasserfall and Teuber, 1979). It is likely that NO derived from nitrite may be involved in the inhibition of *C. tyrobutyricum* in cheese, similar to the inhibition reported for other clostridia. However, a role for NO in inhibition of late-blowing defect has not yet been demonstrated. Early gas defects during cheese ripening are reportedly due to coliform growth. Galesloot and Hassing (1983) reported that growth of several coliforms, including *E. coli, Enterobacter aerogenes,* and *Citrobacter freundii* were actually slightly favored by nitrate in model experiments.

X. NITRIC OXIDE IN SWOLLEN CANS OF GREEN BEANS AND SPINACH

Baumgart *et al.* (1983) isolated a strain of *B. stearothermophilis* able to produce NO from the nitrate dissolved in the fluid of canned green beans, leading to soft spoilage and swelling or can blowing. Hawat and Achtzehn (1971) reported that NO, derived from nitrate via microbial nitrite assimilation (denitrification) caused development of blown cans of spinach. They developed an apparatus for NO measurement in spinach preserves via NO_2 using the Saltzmann method (Saltzmann, 1949). Elkins *et al.* (1969) described a gas chromatography method for determination of headspace gases, including NO, in canned foods, but noted that in analysis of several hundred cans as part of a study of nitrate detinning, NO was not found as a headspace gas constituent.

XI. SUMMARY

Nitric oxide (NO) is the compound responsible for the color and stability of cured meats, and may be involved in the anticlostridial action of nitrate added to some cheeses. NO has also been implicated as a major constituent in the headspace gases of swollen cans of spinach and green beans. Although patents have been issued for meat curing by direct exposure to NO, this procedure has never been used commercially. In cured meats, NO is generated by microbial reduction of nitrate, or by enzymatic or nonenzymatic reduction of nitrite. The three major advantages of nitrite use in cured meats are to develop pink cured meat color, to inhibit rancidity, and to inhibit growth of C. *botulinum* and other food pathogens or food spoilage organisms. NO–iron interactions are in some manner involved in all of these benefits. NO reacts with heme of heat-denatured myoglobin to form the cured meat pigment mononitrosylhemochrome. Interestingly, myoglobin itself is able to react with nitrite to form a ferrous heme–NO complex without aid from exogenous reductants, via ferrous heme iron reduction of nitrite to form a ferric heme–NO complex, followed by an internal autoreduction from an electron donor group on the myoglobin protein (perhaps a histidine ligand), generating the ferrous heme–NO complex. NO inhibits hemin or ionic iron catalysis of lipid oxidation by formation of ferrous heme–NO· radical, which has been shown to have antioxidant properties, and by formation of low molecular weight cysteine–Fe–NO· complexes with antioxidant properties. Formation of the heme–NO pigment complex may also lower the amount of ionic iron released from heme during cooking. NO reacts with and at least partially inactivates pyruvate–ferredoxin oxidoreductase, an important iron–sulfur enzyme of ATP generation in clostridia. Low molecular weight Fe–S–NO complexes formed during meat curing may also inhibit clostridial cell incorporation of iron–sulfur groups into functioning enzymes. Nitrous acid, rather than NO, has been shown to be the precursor to nitrosamine formation in nitrite-treated foods. Accordingly, nitrosamine formation can be lowered or eliminated by lowering the level of nitrite added, and by inclusion of reductants such as ascorbate or α-tocopherol to more rapidly reduce nitrous acid to NO. Ascorbate or its isomer, erythorbate, has also been shown to accelerate cured meat color development and to increase the antibotulinal effects of nitrite. Both effects are due to the rapid reduction of nitrite to NO.

REFERENCES

Ando, N. (1974). Some compounds influencing color formation. In "Proceedings of the International Symposium on Nitrite and Meat Production" (B. Krol and B. J. Tinbergen, eds.), p. 149. Pudoc, Wageningen, The Netherlands.

AOAC (1980). "Official Methods of Analysis," 13th Ed., p. 381. Association of Official Analytical Chemists, Washington, D.C.

Asan, T., and Solberg, M. (1976). Inhibition of *Clostridium perfringens* by heated combinations of nitrite, sulfur, and ferrous or ferric ions. Appl. Environ. Microbiol. **31**, 49–52.

Ashworth, J., Hargreaves, L. L., and Jarvis, B. (1973). The production of an antimicrobial effect in pork heated with sodium nitrite under simulated commercial pasteurization conditions. *J. Food Technol.* **8,** 477–484.

Baumgart, J., Hinrichs, M., Weber, B., and Kupper, A. (1983). Spoilage and blowing of canned green beans by *Bacillus stearothermophilis*. *Chemie Mikrobiologie Technologie der Lebensmittel.* **8,** 7–10.

Benedict, R. C. (1980). Biochemical basis for nitrite inhibition of *Clostridium botulinum* in cured meat. *J. Food Protection* **43,** 877–891.

Bernheim, F. (1943). The significance of the amino groups for the oxidation of various compounds by the *Cholera vibrio (V. comma)*. *Arch. Biochem.* **2,** 125–133.

Binkerd, E. F., and Kolari, O. E. (1975). The history and use of nitrate and nitrite in the curing of meat. *Food Cosmet. Toxicol.* **13,** 655–661.

Braddock, R. J., and Dugan, L. R., Jr. (1969). Moisture effect on the nitric oxide pigments in freeze-dried beef. *Food Technol.* **23,** 1085–1086.

Buchman III G. W., and Hansen, J. N. (1987). Modification of membrane sulfhydryl groups in bacteriostatic action of nitrite. *Appl. Environ. Microbiol.* **53,** 79–82.

Byler, D. M., Gosser, D. K., and Susi, H. (1983). Spectroscopic estimation of the extent of S-nitrosothiol formation by nitrite action on sulfhydryl groups. *J. Agric. Food Chem.* **31,** 523–527.

Carpenter, C. E., Reddy, D. S. A., and Cornforth, D. P. (1987). Inactivation of clostridial ferredoxin and pyruvate–ferredoxin oxidoreductase by sodium nitrite. *Appl. Environ. Microbiol.* **53,** 549–552.

Castellani, A. G., and Niven, C. F., Jr. (1955). Factors affecting the bacteriostatic action of sodium nitrite. *Appl. Microbiol.* **3,** 154–159.

Challis, B. C. (1981). The chemistry of formation of N-nitroso compounds. *In* "Safety Evaluation of Nitrosatable Drugs and Chemicals" (G. G. Gibson and C. Ioannides, eds.), pp. 16–55. Taylor & Francis, London.

Challis, B. C., and Kyrtopoulos, S. A. (1978). The chemistry of nitroso compounds. Part 12. The mechanism of nitrosation and nitration of aqueous piperidine by gaseous dinitrogen tetraoxide and dinitrogen trioxide in aqueous alkaline solutions. Evidence for the existence of molecular isomers of dinitrogen tetraoxide and dinitrogen trioxide. *J. Chem. Soc. Perkin Trans.* **2,** 1296–1302.

Challis, B. C., and Outram, J. R. (1978). Rapid formation of N-nitrosamines from nitric oxide in the presence of silver (I) salts. *J. Chem. Soc., Chem. Commun.*, 707–708.

Challis, B. C., and Outram, J. R. (1979). The chemistry of nitroso-compounds. Part 15. Formation of N-nitrosamines in solution from gaseous nitric oxides in the presence of iodine. *J. Chem. Soc. Perkin Trans.* **1,** 2768–2775.

Code of Federal Regulations (1981a). Title 21, Section 172.170. Office of the Federal Register, Washington, D.C.

Code of Federal Regulations (1981b). Title 21, Section 172.175. Office of the Federal Register, Washington, D.C.

Committee on Nitrite and Alternative Curing Agents in Food (1981). Historical perspective. *In* "The Health Effects of Nitrate, Nitrite, and N-Nitroso Compounds," Chap. 2. National Academy Press, Washington, D.C.

Committee on Nitrite and Alternative Curing Agents in Food (1981). The utility of nitrate and nitrite added to foods. In "The Health Effects of Nitrate, Nitrite, and N-Nitroso Compounds," Chap. 3. National Academy Press, Washington, D.C.

Committee on Nitrite and Alternative Curing Agents in Food (1981). The chemistry of nitrate, nitrite, and nitrosation. In "The Health Effects of Nitrate, Nitrite, and N-Nitroso Compounds," Chap. 4. National Academy Press, Washington, D.C.

Council for Agricultural Science and Technology (CAST). (1978). Nitrite in meat curing: Risks and benefits. Report 74. Iowa State University, Ames, Iowa.

Crosby, N. T., Forman, J. K., Palframan, J. F., and Sawyer, R. (1972). Estimation of steam-volatile N-nitrosamines in foods at the 1 µg/kg level. *Nature (London)* **238**, 342–343.

Doran, G. F. (1917). Art of curing meats. U.S. Patent 1,212,614.

Elkins, E. R., Kim, E. S., and Farrow, R. P. (1969). Nitrous oxide and other gases in headspace of canned foods. *Food Technol.* **23**, 1419–1421.

Ender, F., Havre, G., Helgebostad, A., Koppang, N., Masden, R., and Ceh, L. (1964). Isolation and identification of a hepatotoxic factor in herring meat produced from sodium nitrite preserved herring. *Naturwissenschaften* **51**, 637–638.

Fazio, T., Damico, J. N., Howard, J. W., White, R. H., and Watts, J. O. (1971). Gas chromatographic determination and mass spectrometric confirmation on N-nitrosodimethylamine in smoke-processed marine fish. *J. Agric. Food Chem.* **19**, 250–253.

Fiddler, W., Pensabene, J. W., Piotrowski, E. G., Phillips, J. G., Keating, J., Mergens, W. J., and Newmark, H. L. (1978). Inhibition of formation of volatile nitrosamines in fried bacon by the use of cure-solubilized α-tocopherol. *J. Agric. Food Chem.* **26**, 653–656.

Fox, J. B., and Ackerman, S. A. (1968). Formation of nitric oxide myoglobin: Mechanisms of the reaction with various reductants. *J. Food Sci.* **33**, 364–370.

Frouin, A., Jondeau, D., and Thenot, M. (1975). Studies about the state and availability of nitrite in meat products for nitrosamine formation. *Proc. 21st European Meeting Meat Research Workers, Bern*, 200.

Galesloot, T. E., and Hassing, F. (1983). Effect of nitrate and chlorate and mixtures of these salts on the growth of coliform bacteria. Results of model experiments related to gas defects in cheese. *Netherlands Milk Dairy J.* **37**, 1–9.

Giddings, G. G. (1977). The basis of color in muscle foods. *J. Food Sci.* **42**, 288–294.

Glaeser, H. (1989). Use of nitrate in cheese production. *Dairy Ind. Int.*, 19–23.

Granger, D. L., Hibbs, J. B., Jr., and Broadnax, L. M. (1991). Urinary nitrate excretion in relation to murine macrophage activation. Influence of dietary L-arginine and oral N^G-monomethyl-L-arginine. *J. Immunol.* **146**, 1294–1302.

Green, L. C., Tannenbaum, S. R., and Goldman, P. (1981). Nitrate synthesis in the germfree and conventional rat. *Science* **212**, 56–58.

Haldane, J. (1901). The red colour of salted meat *J. Hyg. (Cambridge)* **1**, 115–122.

Hansen, M. A. (1964). An outbreak of liver toxic injury in ruminants: Clinical observations and results of some hepatic tests in cattle and sheep. *Nord. Vet. Med.* **16**, 323–342.

Harper, R. H., and Kittle, R. W. (1960). Gettering of vacuum packages. U. S. Patent 2,925,346

Hawat, H., and Achtzehn, M. K. (1971). Formation and determination of nitric oxide in sterile spinach preserves. *Nahrung* **15**, 869–883.

Hibbs, J. B., Jr., Taintor, R. R., and Vavrin, Z. (1987). Macrophage cytotoxicity: Role for L-arginine deiminase and imino nitrogen oxidation to nitrite. *Science* **235,** 473–476.

Hoagland, R. (1908). The action of saltpeter upon color of meat. *In* "Bureau of Animal Industry 25th Annual Report. 1908," pp. 301–314. U.S. Department of Agriculture, Washington, D.C.

Hornsey, H. C. (1956). The colour of cooked cured pork. I. Estimation of the nitric oxide-haem pigments. *J. Sci. Food Agric.* **7,** 534–540.

Huhtanen, C. N., and Wasserman, A. E. (1975). Effect of added iron on the formation of clostridial inhibitors. *Appl. Microbiol.* **30,** 768–770.

Igene, J. O., King, J. A., Pearson, A. M., and Gray, J. I. (1979). Influence of heme pigments, nitrite, and non-heme iron on the development of warmed-over flavor (WOF) in cooked meat. *J. Agric. Food Chem.* **27,** 838–842.

Igene, J. O., and Pearson, A. M. (1979). Role of phospholipids and triglycerides in warmed-over flavor development in meat model systems. *J. Food Sci.* **44,** 1285–1290.

Ito, T., Cassens, R. G., and Greaser, M. L. (1979). Reaction of nitrite with tryptophyl residues of protein. *J. Food Sci.* **44,** 1144–1146 and 1149.

Ito, T., Cassens, R. G., Greaser, M. L., Lee, M., and Izumi, K. (1983). Lability and reactivity of nonheme protein-bound nitrite. *J. Food Sci.* **48,** 1204–1207.

Jensen, L. B. (1954). "Microbiology of Meats," 3rd Ed. The Garrard Press. Champaign, Illinois.

Johnson, M. A., Pivnick, H., and Samson, J. M. (1969). Inhibition of *Clostridium botulinum* by sodium nitrite in a bacteriological medium and meat. *Can. Inst. Food Technol. J.* **2,** 52–55.

Kanner, J. (1979). S-nitrosocysteine (RSNO), an effective antioxidant in cured meat. *J. Am. Oil Chem. Soc.* **56,** 74–76.

Kanner, J., and Juven, B. J. (1980). S-Nitrosocysteine as an antioxidant, color-developing, and anticlostridial agent in comminuted turkey meat. *J. Food Sci.* **45,** 1105–1108 and 1112.

Kanner, J., Harel, S., Shagalovich, J., and Berman, S. (1984). Antioxidative effect of nitrite in cured meat products: Nitric oxide–iron complexes of low molecular weight. *J. Agric. Food Chem.* **32,** 512–515.

Kanner, J., Harel, S., and Granit, R. (1991). Nitric oxide as an antioxidant. *Arch. Biochem. Biophys.* **289,** 130–136.

Keilin, J. (1955). Reactions of free haematins and haemoproteins with nitric oxide and certain other substances. *Biochem. J.* **59,** 571–579.

Kerr, R. H., Marsh, C. T., Schroeder, W. F., and Boyer, E. A. (1926). The use of sodium nitrite in the curing of meat. *J. Agric. Res.* **33,** 541–551.

Killday, B. K., Tempesta, M. S., Bailey, M. E., and Metral, C. J. (1988). Structural characterization of nitrosylhemochromogen of cooked cured meat: Implications in the meat-curing reaction. *J. Agric. Food Chem.* **36,** 909–914.

Kim, C., Carpenter, C. E., Cornforth, D. P., Mettanant, O., and Mahoney, A. W. (1987). Effect of iron form, temperature, and inoculation with *Clostridium botulinum* spores on residual nitrite in meat and model systems. *J. Food Sci.* **52,** 1464–1470.

Kisskalt, K. (1899). Bietrage zur Kenntnis der Ursachen des Rotwerdens des Fleisches biem Kochen, nebst einen Versuchen uber die Wirkung der schwefligen. Saure auf die Fleischfarbe. *Arch. Hyg. Bakt.* **35,** 11.

Koizumi, C., and Brown, W. D. (1971). Formation of nitric oxide myoglobin by nicotinamide adenine dinucleotides and flavins. *J. Food Sci.* **36,** 1105–1109.

Kubberod, G., Cassens, R. G., and Greaser, M. L. (1974). The reaction of nitrite with sulfhydryl groups of myosin. *J. Food Sci.* **39,** 1228–1230.

Kucera, I., and Skladal, P. (1990). Formation of a potent respiratory inhibitor at nitrite reduction by nitrite reductase isolated from the bacterium *Paracoccus denitrificans*. *J. Basic Microbiol.* **30,** 515–522.

Lee, M., and Cassens, R. G. (1976). Nitrite binding sites on myoglobin. *J. Food Sci.* **41,** 969–970.

Lee, M., Cassens, R. G., and Fennema, O. R. (1981). Effect of metal ions on residual nitrite. *J. Food Proc. Preserv.* **5,** 191–205.

Lee, S. H., Cassens, R. G., and Sugiyama, H. (1978). Factors affecting inhibition of *Clostridium botulinum* in cured meats. *J. Food Sci.* **43,** 1371–1374.

Lehman, K. (1899). Uber das Haemorrhodin, ein neues weit verbreitetes Blutfarbstoffderivat. *Sber. Phys.-Med. Ges. Wurzb.* **4,** 57.

Liu, H. P. (1970a). Catalysts of lipid peroxidation in meats. 1. Linoleate peroxidation catalyzed by MetMb or Fe(II)-EDTA. *J. Food Sci.* **35,** 590–592.

Liu, H. P. (1970b). Catalysts of lipid peroxidation in meats. 2. Linoleate oxidation catalyzed by tissue homogenates. *J. Food Sci.* **35,** 593–595.

Love, J. D., and Pearson, A. M. (1974). Metmyoglobin and nonheme iron as prooxidants in cooked meat. *J. Agric. Food Chem.* **22,** 1032–1034.

Madhavi, D. L., and Carpenter, C. E. (1993). Aging and processing affect color, metmyoglobin reductase and oxygen consumption of beef muscles. *J. Food Sci.* **58,** 939–942 and 947.

Malin, E. L., Greenberg, R., Piotrowski, E. G., Foglia, R. A., and Maerker, G. (1989). Deamination of lysine as a marker for nitrite–protein reactions. *J. Agric. Food Chem.* **37,** 1071–1076.

McBrady, W. J. (1968). Curing apparatus. U.S. Patent 3,393,629.

MacDonald, B., Gray, J. I., Stanley, D. W., and Usborne, W. R. (1980a). Role of nitrite in cured meat flavor: Sensory analysis. *J. Food Sci.* **45,** 885–888, 904.

MacDonald, B., Gray, J. I., and Gibbons, L. N. (1980b). Role of nitrite in cured meat flavor: Antioxidant role of nitrite. *J. Food Sci.* **45,** 893–897.

McMindes, M. K., and Siedler, A. J. (1988). Nitrite mode of action: Inhibition of yeast pyruvate decarboxylase (E.C. 4.1.1.1) and clostridial pyruvate:ferredoxin oxidoreductase (E.C. 1.2.7.1) by nitric oxide. *J. Food Sci.* **53,** 917–919 and 931.

Mellet, P. O., Noel, P. R., and Goutefongea, R. (1986). Nitrite–tryptophan reaction: Evidence for an equilibrium between tryptophan and its nitrosated form. *J. Agric. Food Chem.* **34,** 892–895.

Mergens, W. J., Kamm, J. J., Newmark, H. L., Fiddler, W., and Pensabene, J. (1978). Alpha-tocopherol: Uses in preventing nitrosamine formation. *In* "Environmental Aspects of N-Nitroso Compounds" (E. A. Walker, L. Griciute, M. Castegnaro, and R. E. Lyle, eds.), pp. 199–212. IARC Scientific Publication NO. 19. International Agency for Research on Cancer, Lyon, France.

Meyer, J. (1981). Comparison of carbon monoxide, nitric oxide, and nitrite as inhibitors of the nitrogenase from *Clostridium pasteurianum*. *Arch. Biochem. Biophys.* **210,** 246–256.

Miller, A. J., and Menichillo, D. A. (1991). Blood fraction effects on the antibotulinal efficacy of nitrite in model beef sausages. *J. Food Sci.* **56,** 1158–1160 and 1181.
Miller, D. M., and Aust, S. D. (1989). Studies of ascorbate dependent, iron-catalyzed lipid peroxidation. *Arch. Biochem. Biophys.* **271,** 113–119.
Mirna, A. (1974). Determination of free and bound nitrite. *In* "Proceedings of the International Symposium on Nitrite and Meat Production, (B. Krol and B. J. Tinbergen, eds.), pp. 21–29. Pudoc, Wageningen, The Netherlands.
Mirvish, S. S. (1975). Formation of N-nitroso compounds: Chemistry, kinetics, and *in vivo* occurrence. *Toxicol. Appl. Pharmacol.* **31,** 325–351.
Moran, D. M., Tannenbaum, S. R., and Archer, M. C. (1975). Inhibitor of *Clostridium perfringens* formed by heating sodium nitrite in a chemically defined medium. *Appl. Microbiol.* **30,** 838–843.
Morris, S. L., Walsh, R. C., and Hansen, J. N. (1984). Identification and characterization of some bacterial membrane sulfhydryl groups which are targets of bacteriostatic and antibiotic action. *J. Biol. Chem.* **259,** 13,590–13,594.
Nakai, H., Cassens, R. G., Greaser, M. L., and Woolford, G. (1978). Significance of the reaction of nitrite with tryptophan. *J. Food Sci.* **43,** 1857–1860.
National Provisioner. (1952). "The Significant Sixty," Section 2, p. 266. Chicago, Illinois.
Nord, F. F., and Mull, R. P. (1945). Recent progress in the biochemistry of fusaria. *Adv. Enzymol.* **5,** 165–205.
O'Boyle, A. R., Rubin, L. J., Diosady, L. L., Aladin-Kassam, N., Comer, F., and Brightwell, W. (1990). A nitrite-free curing system and its application to the production of weiners. *Food Technol.* **44 (5),** 88–104.
O'Leary, V., and Solberg, M. (1976). Effect of sodium nitrite inhibition on intracellular thiol groups and on the activity of certain glycolytic enzymes in *Clostridium perfringens*. *Appl. Environ. Microbiol.* **31,** 208–212.
Olsman, W. J. (1977). Chemical behavior of nitrite in meat products-1. The stability of protein-bound nitrite during storage. *In* "Proceedings of the Second International Symposium on Nitrite and Meat Production" (B. J. Tinbergen, and B. Krol, eds.), p. 101. Pudoc, Wageningen, The Netherlands.
Payne, M. J., Woods, L. F. J., Gibbs, P., and Cammack, R. (1990a). Electron paramagnetic resonance spectroscopic investigation of the inhibition of the phosphoroclastic system of *Clostridium sporogenes* by nitrite. *J. Gen. Microbiol.* **136,** 2067–2076.
Payne, M. J., Glidewell, C., and Cammack, R. (1990b). Interactions of iron-thiol-nitrosyl compounds with the phosphoroclastic system of *Clostridium sporogenes*. *J. Gen. Microbiol.* **136,** 2077–2087.
Perigo, J. A., Whiting, E., and Bashford, T. E. (1967). Observations on the inhibitor of vegetative cells of *Clostridium sporogenes* by nitrite which has been autoclaved in a laboratory medium, discussed in the context of sublethally processed cured meats. *J. Food Technol.* **2,** 377–397.
Pierson, M. D., and Smoot, L. A. (1982). Nitrate, nitrite alternatives, and the control of *Clostridium botulinum* in cured meats. *Crit. Rev. Food Sci. Nutr.* **17,** 141–187.
Polenske, E. (1891). Uber den Verlust, welchen das Rinkfleisch und Nahrwert durch das Pokein erleidet, sowie uber die Veranderungen saltpeterhaltiger Pokellaken. *Arb. K. GesundhAmt.* **7,** 471.

Rankin, M. D. (1973). A new method of curing. *Int. Food Sci. Technol. Proc.* **6,** 157–163.
Reddy, D., Lancaster, J. R., Jr., and Cornforth, D. P. (1983). Nitrite inhibition of *Clostridium botulinum:* Electron spin resonance detection of iron–nitric oxide complexes. *Science* **221,** 769–770.
Reith, J. F., and Szakaly, M. (1967a). Formation and stability of nitric oxide myoglobin. I. Studies with model systems. *J. Food Sci.* **32,** 188–193.
Reith, J. F., and Szakaly, M. (1967b). Formation and stability of nitric oxide myoglobin. II. Studies on meat. *J. Food Sci.* **32,** 194–196.
Rowe, J. J., Yarbrough, J. M., Rake, J. B., and Eagon, R. G. (1979). Nitrite inhibition of aerobic bacteria. *Curr. Microbiol.* **2,** 51–54.
Rust, R. E. (1977). "Sausage and Processed Meats Manufacturing," p. 24. American Meat Institute, Chicago, Illinois.
Saltzman, B. E. (1954). Colorimetric microdetermination of nitrogen dioxide in the atmosphere. *Anal. Chem.* **26,** 1949–1955.
Sato, K., and Hegarty, G. R. (1971). Warmed-over flavor in cooked meats. *J. Food Sci.* **36,** 1098–1102.
Shahidi, F., Rubin, L. J., Diosady, L. L., and Wood, D. F. (1985). Preparation of the cooked cured-meat pigment, dinitrosyl ferrohemochrome, from hemin and nitric oxide. *J. Food Sci.* **50,** 272–275.
Shank, J. L., Silliker, J. H., and Harper, R. H. (1962). The effect of nitric oxide on bacteria. *Appl. Microbiol.* **10,** 185–189.
Sofos, J. N., Busta, F. F., and Allen, C. E. (1979). Botulism control by nitrite and sorbate in cured meats: A review. *J. Food Protection* **42,** 739–770.
Steinke, P. K. W., and Foster, E. M. (1951). Botulinal toxin formation in liver sausage. *Food Res.* **16,** 477–484.
Tappel, A. L. (1955). Unsaturated lipid oxidation catalyzed by hematin compounds. *J. Biol. Chem.* **217,** 721–733.
Tarladgis, B. G. (1962). Interpretation of the spectra of meat pigments. II. Cured meats. The mechanism of colour fading. *J. Sci. Food Agric.* **13,** 485–491.
Tarr, H. L. A. (1941a). Bacteriostatic action of nitrites. *Nature (London)* **147,** 417–418.
Tarr, H. L. A. (1941b). The action of nitrites on bacteria. *J. Fish. Res. Bd. Can.* **5,** 265–275.
Tompkin, R. B., Christiansen, L. N., and Shaparis, A. B. (1978a). The effect of iron on botulinal inhibition in perishable canned cured meat. *J. Food Technol.* **13,** 521–527.
Tompkin, R. B., Christiansen, L. N., and Shaparis, A. B. (1978b). Causes of variation in botulinal inhibition in perishable canned cured meat. *Appl. Environ. Microbiol.* **35,** 886–889.
Tompkin, R. B., Christiansen, L. N., and Shaparis, A. B. (1978c). Antibotulinal role of isoascorbate in cured meat. *J. Food Sci.* **43,** 1368–1370.
Tompkin, R. B., Christiansen, L. N., and Shaparis, A. B. (1979). Iron and the antibotulinal efficacy of nitrite. *Appl. Environ. Microbiol.* **37,** 351–353.
U.S. Department of Agriculture, Bureau of Animal Industry. (1926). USDA, BAI Order 211, Amendment 4, 1925, p. 1 (revised).

U.S. Department of Agriculture (1975). Nitrates, nitrites and salt: Notice of proposed rule making. *Fed. Regist.* **40,** 52614–52616.
U.S. Department of Agriculture (1978a). Final report on nitrites and nitrosamines. Report to the Secretary of Agriculture by the Expert Panel on Nitrites and Nitrosamines, Food Safety and Quality Service, U.S. Department of Agriculture, Washington, D. C.
U.S. Department of Agriculture (1978b). Nitrates, nitrites and ascorbates (or isoascorbates) in bacon. *Fed. Regist.* 43,20992–20995.
Vahabzadeh, F., Collinge, S. K., Cornforth, D. P., Mahoney, A. W., and Post, F. J. (1983). Evaluation of iron binding compounds as inhibitors of gas and toxin production by *Clostridium botulinum* in ground pork. *J. Food Sci.* **48,** 1445–1451.
Van Roon, P. S., and Olsman, W. J. (1977). Inhibitory effect of some Perigo-type compounds on *Clostridium* spores in pasteurized meat products. In "Proceedings of the Second International Symposium on Nitrite and Meat Production" (B. J. Tinbergen and B. Krol, eds.), p. 53. Pudoc, Wageningen, The Netherlands.
Walters, C. L., and Taylor, A. M. (1964). Nitrite metabolism by muscle *in vitro*. *Biochim. Biophys. Acta* **86,** 448–458.
Walters, C. L., Casselden, R. J., and Taylor, A. M. (1967). Nitrite metabolism by skeletal muscle mitochondria in relation to haem pigments. *Biochim. Biophys. Acta* **143,** 310–318.
Wasserfall, F., and Teuber, M. (1979). Action of egg white lysozyme on *Clostridium tyrobutyricum*. *Appl. Environ. Microbiol.* **38,** 197–199.
White, J. W., Jr. (1975). Relative significance of dietary sources of nitrate and nitrite. *J. Agric. Food Chem.* **23,** 886–891.
Woods, L. F. J., and Wood, J. M. (1982). A note on the effect of nitrite inhibition on the metabolism of *Clostridium botulinum*. *J. Gen. Microbiol.* **52,** 109–110.
Woods, L. F. J., Wood, J. M., and Gibbs, P. A. (1981). The involvement of nitric oxide in the inhibition of the phosphoroclastic system in *Clostridium sporogenes* by sodium nitrite. *J. Gen. Microbiol.* **125,** 399–406.
Woolford, G., and Cassens, R. G. (1977). The fate of sodium nitrite in bacon. *J. Food Sci.* **42,** 586–589.
Yarbrough, J. M., Rake, J. B., and Eagon, R. G. (1980). Bacterial inhibitory effects of nitrite: Inhibition of active transport, but not of group translocation, and of intracellular enzymes. *Appl. Environ. Microbiol.* **39,** 831–834.
Younathan, M. T. (1985). Causes and prevention of warmed-over flavor. *Reciprocal Meat Conference Proceedings* **38,** 74–79.
Zubillaga, M. P., and Maerker, G. (1987). Antioxidant activity of polar lipids from nitrite-treated cooked and processed meats. *J. Am. Oil Chem. Soc.* **64,** 757–760.

9

The Enzymology and Occurrence of Nitric Oxide in the Biological Nitrogen Cycle

Thomas C. Hollocher
Department of Biochemistry
Brandeis University
Waltham, Massachusetts 02254

I. GLOBAL NITROGEN CYCLE[1]

Figure 1 represents the global nitrogen cycle (Clark and Rosswall, 1981; Sprent, 1987). Nitrogen enters the cycle through the reduction of N_2 to ammonia (NH_3) by nitrogen-fixing bacteria. The NH_3 is then converted into the proteins, nucleic acids, and aminosugars of the bacteria and the symbiotic plants with which the bacteria are associated. Nitrogen thus enters the biosphere where most of it remains at redox level 3^-. The oxidation of amino acids, aminosugars, and nucleotides by bacteria and animals ultimately regenerates NH_3, either directly or through the prior production of urea or uric acid. Some of the NH_3 so released to soil and water by these catabolic processes will be reassimilated, but some will

---

[1] Abbreviations: ATP, adenosine-5'-triphosphate; EPR, electron paramagnetic resonance; EXAFS, extended X-ray absorption fine structure; Hb, hemoglobin; Hb$^+$, oxidized (met-) hemoglobin; NADH, reduced form of nicotinamide-adenosine dinucleotide; PMS, phenazine methosulfate (methylsulfate salt of N-methylphenazonium cation); TMPD, N,N,N'N'-tetramethylphenylene-1,4-diaminium dication; SDS–PAGE, sodium dodecyl sulfate–polyacrylamide gel electrophoresis.

FIGURE 1

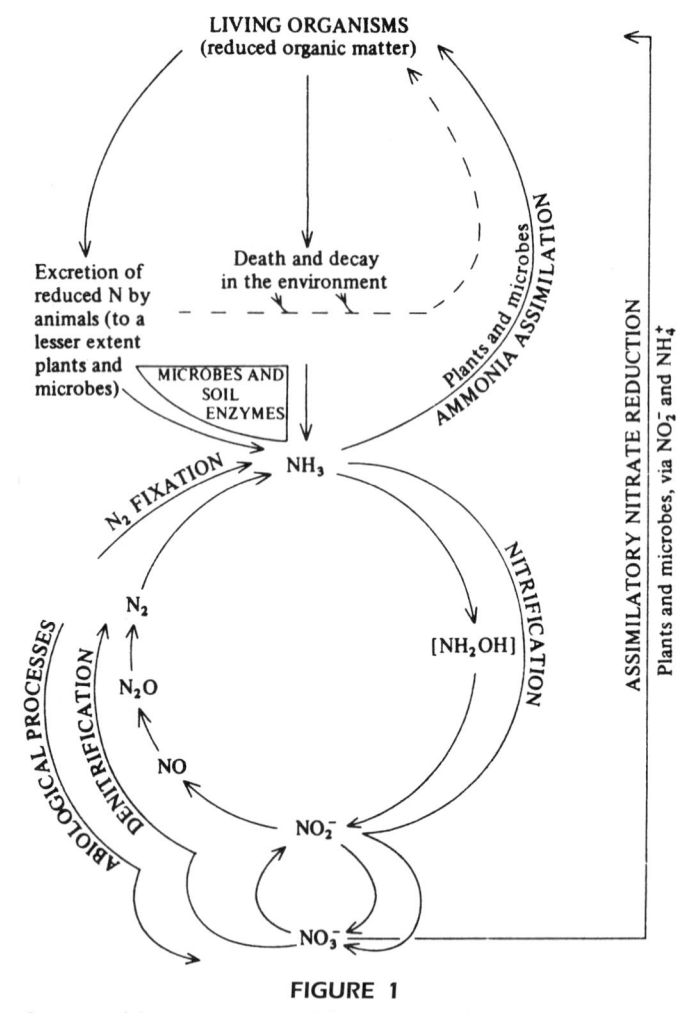

Summary of the major reactions of the nitrogen cycle. From Sprent (1987).

be aerobically oxidized by specialized bacteria, the nitrifying bacteria, to nitrite and nitrate.

For the most part, nitrification involves the activity of two very different classes of bacteria (Wallace and Nicholas, 1969; Kuenen and Robertson, 1987). The first step, the 6-electron oxidation of NH_3 to nitrite (NO_2^-), is carried out by a small number of genera of autotrophic bacteria, exemplified by *Nitrosomonas*; the second step, the 2-electron oxidation of nitrite to nitrate (NO_3^-) by a similarly small number of genera, exemplified by *Nitrobacter*. Nitrate is far more stable than nitrite and is a common environmental anion, whereas nitrite is not.

Some of the nitrite and nitrate produced by nitrification can be rereduced to ammonia by plants, fungi, and bacteria and returned to the biosphere. Nitrate can be reduced to nitrite by the assimilatory nitrate reductases of these organisms or (under anaerobic conditions) by the dissimilatory, energy-yielding[2] nitrate reductase of bacteria. Nitrite in turn can be reduced to ammonia by special nitrite reductases of the siroheme type (Hirasawa et al., 1987; Krueger and Siegel, 1982; Vega and Kamin, 1977; Vega et al., 1975; Zumft, 1972) or the hexaheme c type (Blackmore et al., 1986; Liu and Peck, 1981; Liu et al., 1983). These enzymes have the remarkable ability to reduce nitrite by six electrons without the apparent release of intermediates. These six-electron reductions of nitrite are for assimilation of nitrogen, detoxification of nitrite, or as an anaerobic sink for reducing equivalents and are not in general energy yielding (Tiedje, 1988). The processes that shuttle nitrogen among ammonia, nitrate, nitrite, and organic nitrogen are interesting and vital, but they do not succeed in returning nitrogen to the atmosphere as N_2 and thereby close the nitrogen cycle. A separate bacterial metabolism, termed denitrification, is the one that regenerates N_2 (Knowles, 1982; Payne, 1981; Tiedje, 1988). Because the anaerobic dissimilatory reduction of nitrate to nitrite (often referred to as nitrate respiration) is a capability widely distributed among bacteria, including many that cannot further reduce nitrite by a dissimilatory route, there is some disagreement whether denitrification should be considered to begin with nitrate or nitrite. The critical ability of denitrifying bacteria in either case is the reduction of nitrate and nitrite to N_2. In denitrifying bacteria, the process represents an anaerobic respiration which is linked to energy-conserving proton translocation and ultimately ATP synthesis. Very low partial pressures of O_2 are generally required for induction of the denitrification apparatus but there are some exceptions to this rule among aerobic denitrifiers. Most denitrifiers are facultative aerobes and can switch rather rapidly between O_2 and N-oxide respirations. Although most denitrifiers produce N_2 as the final product, some terminate at N_2O (Tiedje, 1988) and serve as one source for the N_2O that exists at very low concentrations in the atmosphere.

It is remarkable that the nitrogen cycle employs extreme oxidation–reduction states of nitrogen. Nitrification carries N from 3^- to 5^+, whereas denitrification reduces N from 5^+ to 0. A more direct cycle using only redox states of N between

---

[2] Most or all of the product of a dissimilatory pathway of nitrogen metabolism is lost to the environment and is not assimilated into cellular materials. A dissimilatory pathway is energy-yielding if the redox steps are mediated by components of the proton-pump-linked electron transport system of the cytoplasmic membrane. By energy-yielding, we imply that some of the negative free-energy of exergonic redox reactions is coupled to and thereby stored in endergonic processes, such as ATP synthesis and development of electrical charge and ion concentration gradients across the cytoplasmic membrane.

3⁻ and 0 seems not to exist. Redox extremes come about because nitrification employs a powerful oxidant, O_2, in the presence of which nitrate is stable. Nitrification and denitrification operate, respectively, in the aerobic (oxygenated) and anaerobic (or microaerobic) domains of the biosphere.

As will be discussed further in this chapter, there is now much evidence to suggest that NO is an obligatory intermediate in the denitrification pathway. Furthermore, there is evidence that NH_3 nitrifiers can synthesize the denitrification apparatus in addition to the nitrification apparatus and that the former system can produce NO and N_2O (also N_2 in at least one case) from nitrite under low partial pressures of O_2. It is possible therefore that NO may be an intermediate in the denitrification activity of nitrifiers and so arise as a secondary consequence of NH_3 oxidation. NO can also be produced by nondenitrifying organisms under certain conditions. For example, NO can be slowly produced by the anaerobic reduction of nitrite, but only in absence of nitrate, by a variety of enteric bacteria. Some of the NO can be further reduced to N_2O.

Because of the great toxicity of NO toward bacteria (Mancinelli and McKay, 1983), it can be expected that any pathway involving NO as an intermediate would be regulated so as to maintain NO at extremely small steady-state concentrations.

The rank order of size of the reservoirs of nitrogen on earth are atmospheric N_2, nitrate in sea water, and organic nitrogen of the biosphere and soils (Sverdrup et al., 1942). Nitrite and free ammonium are rare by comparison. Nitrate is also found at low concentrations in soils and fresh water. NO is extremely rare in the environment and when detected in sea water is typically at concentrations of less than 1 nM (Ward and Zafiriou, 1988).

II. DENITRIFYING BACTERIA

A. Diversity

The ability to denitrify is widely distributed among bacteria and much is now known of their taxonomy and global distribution (Tiedje, 1988; Gamble et al., 1977). They include hetero- or organotrophic, phototrophic, and lithotrophic organisms. The last group oxidizes NH_3, nitrite, H_2, or reduced sulfur compounds to support reduction of N-oxides, and because they also reduce CO_2 for synthesis of organic materials, they can be designated as autotrophic as well. Among the organotrophic denitrifiers are bacteria that are fermentative, nonfermentative, halophilic, thermophilic, spore-forming, magnetotactic, pathogenic, and N_2-fixing. The oxidation of organic substances drives denitrification most frequently both with respect to the number of taxonomic genera represented and the dominance of populations in nature. The most common denitrifiers in nature, particularly in soils, are species of *Pseudomonas* and the closely related *Alcaligenes*, both organotrophic. None of the true denitrifiers, bacteria that can reduce N-oxide at high rates and couple these reactions to proton translocation and ATP synthesis, is an obligate anaerobe. Certain species of *Propionibacterium* and *Chromobacterium* can reduce nitrite to N_2O, but in slow reactions that are not energy linked. Most denitrifiers lack a fermentative capability, and those few that can ferment, *Azospirillum*, *Chromobacterium*, and *Bacillus*, are considered to be only weakly fermentative compared, for example, to *Escherichia coli*.

Ten different families of eubacteria are represented among true denitrifiers: Rhodospirillaceae, Cytophagaceae, budding bacteria, Spirillasceae, Pseudomonadaceae, Rhizobiaceae, Halobacteriaceae, Neisseriaceae, Nitrobacteraceae, and Bacillaceae. Absent are obligate anaerobes, gram-positive organisms other than *Bacillus*, and Enterobacteriaceae. *Pyrobaculum aerophilum* sp. nov. represents the first example of a denitrifying archaebacterium (Völkl et al., 1993). This organism is also hyperthermophilic, with an optimum growth temperature of 100°C, and can grow organotrophically as a facultative aerobe. Anaerobic growth is supported by nitrate or nitrite which are reduced to N_2. Whether all steps in the reduction of possible intermediate N-oxides, such as NO, are enzymic at temperatures of 100°C remains to be determined. Previous to the characterization of *Py. aerophilum*, the only thermophilic denitrifying bacteria known were *Bacilli* (Gokce et al., 1989; Tiedje, 1988) with the possible exception of *Thermothrix thioparus* (Brannan and Caldwell, 1980) which seems to present an ambiguous phenotype with respect to denitrification (Hollocher and Kristjansson, 1992). In a broad sense, denitrifiers can be viewed as aerobes that can switch over to reduction of N-oxides under conditions of O_2 deficiency. Denitrification is thus an alternative style of respiration.

B. Pathway of Denitrification

The pathway for the reduction of nitrate by denitrifiers is generally recognized to involve four steps as shown in Eq. (1) (Betlach and Tiedje, 1981; Braun and Zumft, 1991; Goretski and Hollocher, 1990; Knowles, 1982; Payne, 1981). The N-oxides represented,

$$NO_3^- (5^+) \longrightarrow NO_2^- (3^+) \longrightarrow NO (2^+) \longrightarrow N_2O (1^+) \longrightarrow N_2(0) \quad (1)$$

nitrate, nitrite, nitric oxide, and nitrous oxide from left to right, undergo reductions of 2, 1, 1, and 2 equivalents, respectively, on the pathway to N_2. The numbers in parentheses in Eq. (1) are the formal oxidation states of nitrogen in the respective compounds. The intermediate N-oxides, nitrite, NO, and N_2O, are believed to be free (freely diffusible) intermediates. The pathway depends on the operation of four enzymes in succession: nitrate reductase, a Mo-containing enzyme associated with the cytoplasmic membrane (Carlson et al., 1982; Craske and Ferguson, 1986; Forget, 1971; Stewart, 1988); nitrite reductase, a heme (iron)- (Parr et al., 1976; Sawhney and Nicholas, 1978; Timkovich et al., 1982; Weeg-Aerssens et al., 1991; Yamanaka and Okunuki, 1963a–c) or Cu-containing (Denariaz et al., 1991; Fenderson et al., 1991; Godden et al., 1991; Iwasaki et al., 1975) enzyme which is generally soluble and localized in the periplasmic space of gram-negative bacteria; nitric oxide reductase, a heme-containing enzyme associated with the cytoplasmic membrane (Carr et al., 1989; Dermastia et al., 1991; Heiss et al., 1989); and nitrous oxide reductase, a soluble Cu-containing enzyme also localized in the periplasmic space (Coyle et al., 1985; Snyder and Hollocher, 1987; Teraguchi and Hollocher, 1989). The nitrous oxide reductase of *Flexibacter canadensis* is membrane-bound and unique in this regard (Jones et al., 1992). These enzymes of denitrification are terminal reductases in the sense that they are served reductively by the electron transport system. They are also termed dissimilatory reductases to indicate that the nitrogen is destined for the most part not to be assimilated into cellular material but rather to be discarded as N_2. The reduction of each of the above four N-oxides by the denitrification pathway is energy linked so that the process can support cell growth (Knowles, 1982; Koike and Hattori, 1975; Payne, 1981). It should be noted that the efficiency of free energy conservation in the reduction of NO by denitrifying bacteria, as measured by the member of protons translocated across the membrane per electron consumed in oxidant pulse experiments, is similar with most denitrifiers to that observed for the reduction of the other N-oxides (Boogerd et al., 1981; Garber et al., 1982; Kristjansson et al., 1978; Leibowitz et al., 1982; Shapleigh and Payne, 1985b). [There was some confusion in this regard in one report (Shapleigh and Payne, 1985b) between protons per N-oxide molecule reduced as opposed to protons per electron consumed.] There may be exceptions to this picture, because NO reduction has been reported (Vosswinkel et al., 1991) not to be linked to proton translocation in *Azospirillum brasilense* Sp. 7 and *Pseudomonas aeruginosa*.

Although the pathway of Eq. (1) is now based on much evidence (Section III) and is unambiguous in the case of at least one bacterium [*Pseudomonas stutzeri* strain Zobell (f. sp. *P. perfectomarina*)], there have been alternative hypothesis. One hypothesis, advanced by the Hollocher group (Garber and Hollocher, 1981; St. John and Hollocher, 1977), considered NO as a likely intermediate, but one that remained at least partly enzyme-bound and was not entirely free to diffuse. This view was based on the outcome of certain kinetic and isotope experiments which can be summarized as follows. When denitrifying bacteria were challenged simultaneously with [^{15}N]nitrite and ordinary NO, the cells reduced both compounds concomitantly to N_2 (or to N_2O in the presence of acetylene which is a specific inhibitor (Balderston et al., 1976; Yoshinari and Knowles, 1976) of nitrous oxide reductase). In the process, little ^{15}NO was generally detected in the gas phase pool of NO and there was relatively little isotopically mixed N_2O formed. That is, most of the ^{14}N and ^{15}N reduced to N_2O appeared as $^{14}N_2O$

and $^{15}N_2O$ and usually rather little as $^{14,15}N_2O$. Although there were some variability and exceptions, the overall inefficiency in trapping ^{15}NO in the NO pool and the small degree of isotopic mixing in N_2O implied that the reductions of nitrite and NO followed largely separate pathways that did not share a common intermediate. One way for this to occur was for two molecules of nitrite to be reduced to N_2O without release of NO from the enzyme, which was presumed to be nitrite reductase. In addition, it was observed with all denitrifying bacteria studied that V_{max} for nitrite reduction was always substantially greater than that for NO (Garber and Hollocher, 1981; St. John and Hollocher, 1977). This implied that if NO were a free intermediate, it should accumulate during reduction of nitrite. But it did not, and, in general, NO could not be detected during the steady-state reduction of nitrite by the means then available, at least in the case of bacteria capable of vigorous denitrification. NO was detected (Firestone et al., 1979) with *Pseudomonas aureofaciens* and *Pseudomonas chlorographis* through use of the short-lived isotope, ^{13}N, but the kinetic data were insufficient to decide whether NO was an intermediate or side product. The above ^{15}N tracer and kinetic results were later shown by Goretski and Hollocher (1990) to be artifactual. Because NO can partially inhibit its own reduction, even at low (40 μM) solution concentrations, the true V_{max} for NO reduction was at least twice that for nitrite. This result required, given the very low K_m (≤ 0.5 μM) for NO, that NO would not accumulate in the steady-state and that the steady-state concentration of NO would be $K_m/2$ or lower. The earlier ^{15}N isotope results were explained by a lack of homogeneity regarding where [^{15}N]nitrite and NO were reduced in solution. [^{15}N]Nitrite was reduced to N_2O in bulk solution where the nitrite resided and where the $^{15}NO/^{14}NO$ ratio was high. On the other hand, NO was reduced principally at the gas–aqueous interface where the ratio of $^{15}NO/^{14}NO$ was low. When this artifact of disequilibrium was remedied by lowering cell concentrations and allowing the physical mixing to keep pace with the biochemistry, it was demonstrated that free NO was kinetically and isotopically competent to be a free intermediate in the pathway (Goretski and Hollocher, 1990).

A second alternative hypothesis for the pathway of denitrification was advanced by Averill and Tiedje in 1982 [Eq. (2)]. In this scheme, the relevant enzyme for the reduction

$$NO_2^- + E \rightleftharpoons E \cdot NO_2^- \underset{\mp H_2O}{\overset{\pm 2H^+}{\rightleftharpoons}} E \cdot NO^+ \xrightarrow{NO_2^-}$$

$$E \cdot N_2O_3 \underset{-2H_2O}{\overset{+4H^+, 4e^-}{\rightarrow}} E + N_2O \quad (2)$$

of nitrite and synthesis of N_2O is presumed to be nitrite reductase and NO is nowhere involved. The enzyme serves first to activate nitrite for nitrosyl (NO^+) transfer, as, for example, by protonation or dehydration following protonation.

This activated species is then attacked by the N atom of a second nitrite anion acting as a nucleophile to form an enzyme-bound N_2O_3 which subsequently undergoes reduction by 4 equivalents (equiv), perhaps via enzyme-bound trioxodinitrate ($HN_2O_3^-$) with average redox state of N^{2+}. Note in this scheme that the initial N−N bond en route to N_2O is formed at the N^{3+} level. One distinctive feature is the avoidance of NO as an obligatory intermediate. Although there is evidence (Sections III and V,C) that nitrite reductases of denitrifying bacteria can activate nitrite for nitrosyl (NO^+) transfer reactions, the mechanism of this reaction is as yet unknown and its relevance to denitrification *in vivo* is still unclear. The weight of evidence indicates that both nitrite and nitric oxide reductases are essential and that nitrite reductase is unable alone to support denitrification. That is, the pathway of denitrification is as shown in Eq. (1). The evidence is discussed in Section III below.

The first product of nitrosyl transfer to nitrite in Eq. (2), $E \cdot N_2O_3$, contains N−N bonded N_2O_3 which is itself a well-known and powerful nitrosyl donor. It is reasonable to suppose therefore that nitrosyl transfer reactions with N- and O-nucleophiles could involve both $E \cdot NO^+$ (or $E \cdot HONO$) and $E \cdot N_2O_3$. In addition, the involvement of a second molecule of nitrite for denitrification would require that the substrate saturation curve should be sigmoidal to reflect a term second-order in nitrite concentration. No such effect has been reported to our knowledge. The use of bimolecular substrate kinetics in dilute solutions to generate an intermediate subject to solvolysis seems metabolically unwise but not impossible.

Four technical problems have historically impeded an understanding of the role of NO in denitrification. The first of these was the lack of a potent and highly specific inhibitor for nitric oxide reductase. The second was the difficulty of measuring the very low steady-state NO concentrations (≤ 60 nM) that apply during denitrification. The remaining problems concerned difficulties in purifying nitric oxide reductase and the lack of mutants in that enzyme. Because of the great toxicity of NO, it was generally assumed that all but temperature sensitive mutants in nitric oxide reductase would be lethal if cells were grown anaerobically on nitrate or nitrite. Although temperature sensitive mutants were never developed, the last of these problems is being resolved through the use of modern techniques in molecular biology. It should be noted that there are many reports in the literature of the detection of NO in denitrifying systems, among the first of which were the ^{15}N-isotope studies of Wijler and Delwiche (1954) with soil samples and the work of Baalsrud and Baalsrud (1954) with anaerobic cultures of *Thiobacillus denitrificans*. While the detection of NO is significant, proof that NO is a metabolic intermediate in denitrification depends on rigorous kinetic, enzymatic, and genetic criteria that have been met only relatively recently.

III. EVIDENCE FOR AND AGAINST NITRIC OXIDE AS AN INTERMEDIATE IN DENITRIFICATION PATHWAY

In this section we examine the evidence as to whether NO is an intermediate in denitrification. The weight of evidence suggests, in our opinion, that NO is in fact an obligatory, free intermediate in the major, if not sole, pathway.

All denitrifying bacteria have an active enzyme system capable of reducing NO to N_2O and that NO reduction is often coupled to proton translocation and other endergonic processes with an efficiency similar to that of other N-oxides (Boogerd et al., 1981; Garber et al., 1982; Kristjansson et al., 1978; Leibowitz et al., 1982; Shapleigh and Payne, 1985b). This activity is missing in aerobically adapted cells and is associated with biosynthesis of the denitrification apparatus. However, because of the lack of a specific inhibitor for either nitrite reductase or nitric oxide reductase and the partial inhibition of nitric oxide reductase by NO (see below and Section IV,A), it was not possible to decide from the above observations alone whether there were in fact two reductases operating in sequence as in Eq. (1) or whether nitrite reductase could function to reduce both nitrite and NO. A more successful approach was that of differential inactivation.

Shapleigh and Payne (1985a) showed for denitrifying bacteria which make use of a Cu-containing dissimilatory nitrite reductase (e.g., *Achromobacter cycloclastes*) that nitrite reductase could be inactivated in cell-free extracts by the Cu-chelator, diethyldithiocarbamate, with little or no effect on nitric oxide reductase activity. The use of diethyldithiocarbamate not only supported the existence of a highly active and separate nitric oxide reductase in this class of denitrifying bacteria, but has also been used as a means to differentiate among those denitrifiers that produce Cu- or heme-containing (cytochrome cd_1) nitrite reductases (Shapleigh and Payne, 1985a; Tiedje, 1988). The latter enzyme is not inactivated by chelating agents because of the very strong coordination of Fe within the macrocyclic ring system. The nonionic detergent, Triton X-100, has been used to differentially inhibit or inactivate nitric oxide reductase relative to nitrite reductase in cell-free systems (Shapleigh et al., 1987), showing again an independent functional existence for these two activities.

Because methods for the purification of both Cu- and heme-type nitrite reductases have been available for some time, it is possible to ask whether such dissimilatory enzymes produce NO or N_2O in the reduction of nitrite *in vitro*. Experiments addressing this question are not trivial, because of the great chemical reactivity of NO. It is essential to select a reducing system and appropriate conditions such that NO is not reduced chemically to N_2O, for otherwise confusion could arise regarding whether NO or N_2O was the primary product of the enzyme reaction. It is remarkable that many reports on nitrite reductases seem to have ignored this important consideration in the presentation of data. Several systems have been found that satisfy both the need for a rapid reduction of the enzyme and lack or virtual lack of chemical reduction of NO. Among the most widely used are NADH/PMS [reduced form of nicotinamide-adenosine dinucleotide/phenazine methosulfate (methylsulfate salt of N-methylphenazonium cation)], ascorbate/PMS and ascorbate/TMPD (N, N, N'N'-tetramethylphenylene-1,4-diaminium) for chemical systems and reduced cytochromes c or reduced small Cu-proteins of the azurin type for biochemically reconstructed systems. In the chemical systems, the redox dye, PMS or TMPD, serves as a redox shuttle or catalyst between the principal reducing agent and the enzyme. In the case of ascorbate at neutral pH, it is essential to have present phosphate salts or a chelating agent in order to suppress the reduction of NO by ascorbate that is catalyzed by contaminating iron salts (Zumft and Frunzke, 1982). Product analyses with both Cu- and heme-type nitrite reductases from a variety of denitrifying bacteria using the reducing systems alluded to above show that NO is the major primary product and that the enzyme is specific for nitrite, having little or no reactivity with NO (Denariaz et al., 1991; Gudat et al., 1973; Iwasaki and Matsubara, 1971, 1972; Kakutani et al., 1981; Kim and Hollocher, 1983, 1984; LeGall et al., 1979; Liu et al., 1986; Mancinelli et al., 1986; Masuko et al., 1984; Newton, 1969; Parr et al., 1976; Robinson et al., 1979; Sawada et al., 1978; Silvestrini et al., 1990; Timkovich et al., 1982; Weeg-Aerssens et al., 1991; Yamanaka and Okunuki, 1963a; Yamanaka et al., 1961; Zumft et al., 1987). N_2O is often detected but usually at the level of one to a few percent and there is often ambiguity whether this small amount of N_2O arose by enzymatic or chemical means. In one study with the heme-type enzyme from P. aeruginosa, it was shown that the rate of reduction of NO to N_2O by ascorbate/TMPD by all routes, enzymatic and chemical, was substantially smaller than the rate of formation of N_2O from nitrite by the enzyme (Kim and Hollocher, 1983), thus implying a direct enzymatic synthesis for the small amount (1–2%) of N_2O formed. In addition, the N_2O formed from a mixture of [^{15}N]nitrite and ordinary NO was sufficiently enriched in ^{15}N to suggest that N_2O was being formed from nitrite without equilibration with the NO pool. It would appear then that dissimilatory nitrite reductases of denitrifying bacteria produce NO as their chief product *in vitro*. Although there are a few reports to the contrary in which N_2O was alleged to be the major direct product (Sawhney and Nicholas, 1978; Wharton and Weintraub, 1980; Zumft and Vega, 1979), we know of no clear demonstration supported by the necessary array of chemical and enzymatic controls. Notwithstanding the above results, there are at least two novel means, discussed in the next few paragraphs, by which dissimilatory nitrite reductases can generate N_2O.

In the case of the Cu-type enzyme from A. cycloclastes, it has been shown (Averill and Tiedje, 1990; Jackson et al., 1991) that nitrite can be reduced to N_2O in the presence of NO in what resembles a self-catalyzed reaction. At the outset, NO was overwhelmingly the major product in the reduction of nitrite, but, as NO accumulated, there was a shift to N_2O production. At the same time, this enzyme, typical of Cu-type nitrite reductases in general, did not utilize NO as a substrate when it was presented alone. The chemistry behind the NO-induced reduction of nitrite to N_2O is not entirely clear, but it may reflect the operation of a variant of the scheme of Averill and Tiedje (1982) [Eq. (2)]. In this variant scheme [Eq. (3)] nitrite and NO act cooperatively to displace the equilibrium

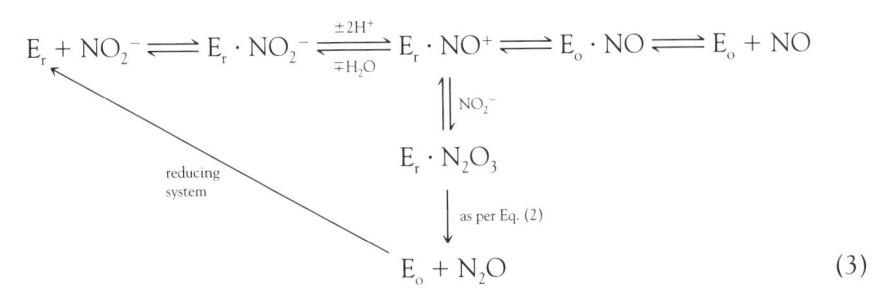

$$E_r + NO_2^- \rightleftharpoons E_r \cdot NO_2^- \underset{\mp H_2O}{\overset{\pm 2H^+}{\rightleftharpoons}} E_r \cdot NO^+ \rightleftharpoons E_o \cdot NO \rightleftharpoons E_o + NO$$

$$\Updownarrow NO_2^-$$

$$E_r \cdot N_2O_3$$

reducing system

$$\downarrow \text{as per Eq. (2)}$$

$$E_o + N_2O \qquad (3)$$

among nitrite, NO, and reduced (r) and oxidized (o) forms of the enzyme to favor the nitrosyl donor species, $E_r \cdot NO^+$, which can then react with a second nitrite anion (Goretski and Hollocher, 1991). The ultimate consequences of this nitrosyl transfer reaction under reducing conditions is the synthesis of N_2O when or if the concentration of NO becomes high. Other mechanistic possibilities exist, of course. NO-induced reduction of nitrite to N_2O has not yet been reported for a cytochrome cd_1 nitrite reductase.

^{18}O-exchange studies of Ye et al. (1991) support, we believe, the catalysis by nitrite reductase of redox reversibility between nitrite and NO as depicted in the first line of Eq. (3). They observed by analyzing the ^{18}O content of product N_2O that all eight strains of denitrifying bacteria studied could catalyze the exchange of ^{18}O between water and nitrite or NO by way of an electrophilic (nitrosyl donor) species of NO. The rates and extent of these exchange reactions depended on whether the bacterium made use of a heme- or Cu-type nitrite reductase. Contrary to the conclusions of Ye et al. (1991), we do not believe that this study otherwise informs about the pathway of denitrification or whether NO is an intermediate.

The work of Ye et al. (1991) was extended to include Tn5 mutants in *nir* genes for nitrite reductase of *Pseudomonas fluorescens* (Ye et al., 1992a) and *Pseudomonas* sp. strain G-179 (Ye et al., 1992b) which use, respectively, the cytochrome cd_1- and Cu-type nitrite reductase. The five mutants of *P. fluorescens* characterized not only lacked nitrite reductase and $NO/H_2O-^{18}O$ exchange activities, but also showed levels of nitric oxide reductase activity that were diminished roughly by a factor of two. The reason for the decrease in nitric oxide reductase activity of the mutants is not clear, but may represent a change in expression level of nitric oxide reductase that is linked to Tn5 mutations in nitrite reductase genes. In *P. stutzeri* and presumably also in other denitrifiers, the nitrite and nitric oxide reductase genes are located within a roughly 30-Kb denitrification gene cluster (Braun and Zumft, 1992). No such linked decrease in nitric oxide reductase activity was observed in the case of nitrite reductase Tn5 mutants of *P. stutzeri*, however (Zumft et al., 1988). Nir$^-$ mutants of strain G-179, that mapped in the nitrite reductase genes, lacked nitrite reductase and NO/H$_2$O–^{18}O exchange activities but were essentially normal in nitric oxide reductase and resembled *P. stutzeri* in this regard. These results and the resistance of nitric oxide reductase of strain G-179 to inactivation by the Cu-chelator, diethyldithiocarbamate, show that nitrite reductase does not normally participate in NO reduction, in spite of its catalysis of the NO/H$_2$O–^{18}O exchange. It would appear therefore that the equilibrium of the first line of Eq. (3) can occur with both heme- and Cu-type nitrite reductases, but reaction of the nitrosyl donor species with nitrite, as the first step in N$_2$O formation, is significant only with the Cu-type enzyme and then only at high concentrations of NO.

The second means of inducing nitrite reductase to produce N$_2$O also applied principally but not exclusively to the Cu-type enzyme and involves the use of hydroxylamine (NH$_2$OH) as reducing agent (Denariaz et al., 1991; Iwasaki and Matsubara, 1972; Iwasaki and Mori, 1958; Kim and Hollocher, 1984; Liu et al., 1986; Masuko et al., 1984; Matsubara, 1970; Zumft et al., 1987). The reaction catalyzed [Eq. (4)] is one in which one atom of N in N$_2$O arises from nitrite and the other from NH$_2$OH.

$$H^+ + NO_2^- + NH_2OH \longrightarrow N_2O + 2H_2O \qquad (4)$$

Thus, the reducing agent reacts covalently with nitrite as the nitrite is reduced by 2 equiv and the NH$_2$OH is oxidized by 2 equiv. This kind of reaction can best be described as a nitrosyl transfer from nitrite, presumably activated as the result of binding to the enzyme, to NH$_2$OH acting as a N-nucleophile (Garber and Hollocher, 1982; Kim and Hollocher, 1984). Analogous nitrosations can occur also with azide (N$_3^-$) and primary amines (Garber and Hollocher, 1982; Kim and Hollocher, 1984; Pichinoty et al., 1969; Ye et al., 1991). The chemical analog of this reaction, the reaction of nitrous acid, nitrosyl chloride, N$_2$O$_3$, or alkyl nitrite esters with NH$_2$OH, is well known and is thought to proceed by formation of *cis*- or *cis*- plus *trans*-hyponitrite (HON=NO), which yields N$_2$O by dehydration (Bonner et al., 1978, 1983; Bothner-By and Friedman, 1952; Cooper et al., 1970). The enzyme-catalyzed reaction between nitrite and NH$_2$OH is therefore probably an example of the general ability of dissimilatory nitrite reductases to catalyze nitrosyl transfer reactions between an activated, electrophilic form of nitrite and nucleophiles. These reactions are discussed in greater detail below and in Section V,C.

Although the nitrite reductases of denitrifying bacteria seem to be NO-producing reductases *in vitro*, we cannot so easily say that they similarly are NO-producing *in vivo*. It is not so rare for an enzyme to catalyze more than one reaction or be served by more than one substrate, but it is virtually without precedence for an enzyme to change its chemistry with the same substrate between *in vivo* and *in vitro* situations. The reduction of nitrite *in vivo* has been probed by several means which target the activity of either nitrite or nitric oxide reductase.

As discussed above, mutational probes have demonstrated that nitrite reductase can be eliminated without effect or with little effect on nitric oxide reductase activity. In a particularly elegant demonstration of independent function, Glockner et al. (1993) cloned the Cu-containing nitrite reductase of *P. aureofaciens* into a *Nir⁻ Nor⁻* double mutant MK3202 of *P. stutzeri* which lacked both the cytochrome cd_1 nitrite reductase and nitric oxide reductase. The Cu enzyme was functional in the host and produced NO as the product of reduction of nitrite. In addition to genetic manipulations, certain denitrifiers can be grown in such as way that nitrite reductase is fully induced but nitric oxide reductase is not (e.g., *P. stutzeri* grown under O_2-limiting conditions) (Frunzke and Zumft, 1986). Such induction-modified cells reduce nitrite to NO which can initially accumulate in large amounts and then subsequently be reduced slowly to N_2O and N_2 (Goretski and Hollocher, 1990). Judging from kinetics, nitrite reductase was fully active as a producer of NO in these modified cells but nitric oxide reductase was much diminished in activity. In extreme cases, the initial rates of production of NO and N_2O were as high as 10:1. Azide can differentially inhibit NO uptake more than nitrite uptake in certain denitrifying bacteria (Goretski and Hollocher, 1990), with the result again that NO is produced from nitrite in large amounts and accumulates. This behavior is in marked contrast to unmodified cells for which the steady-state concentration of NO during reduction of nitrite is extremely small [≤ 60 nM (Zafiriou *et al.*, 1989; Goretski *et al.*, 1990; Kalkowski and Conrad, 1991)]. In short, any modification that effectively cripples nitric oxide reductase activity in cells leads to accumulation of NO when nitrite is reduced. By this measure, nitrite reductase of both the Cu- and heme-types would appear to be NO-producing, not N_2O-producing, *in vivo*. The most dramatic and convincing demonstration of this view was reported for *P. stutzeri* by Braun and Zumft (1991). They cloned and deleted part of the structural gene of nitric oxide reductase of *P. stutzeri* and then inserted the modified gene into wild-type cells by means of a nonreplicative gene replacement plasmid, pSUP203. The resulting *Nor⁻* mutant colonies could grow by O_2 or N_2O respiration but not by nitrate or nitrite respiration. In fact, nitrate and nitrite were conditionally lethal substrates due to NO production. As one might predict, mutants constructed with deletions in both nitrite and nitric oxide reductase structural genes, *Nir⁻ Nor⁻*, could grow anaerobically on nitrate but not on nitrite. These double mutants could not produce NO and were therefore protected. The genetic experiments prove, at least for *P. stutzeri*, that nitrite reductase is NO-producing *in vivo*, nitric oxide reductase is an obligatory separate enzyme of the denitrification pathway, and NO is an obligatory intermediate.

Another general means to probe the *in vivo* denitrification apparatus is through the trapping of NO. As already pointed out, the steady-state concentrations of NO during denitrification have been found to be extremely low. This fact alone does not tell whether NO is an intermediate in the pathway, and kinetic information is required on the matter of flux into and from the NO pool. If NO were an intermediate, then the rates of both the injection of NO into the NO pool and its reductive removal would be large and in fact equal to the rate of uptake of nitrite. If NO were the product of side reactions, then the flux of N through the NO pool could be small and need not have any connection with the rate of nitrite reduction. A way to measure flux through the NO pool involves the partitioning of NO in the pool between the normal reductive reaction and some other process that will irreversibly remove NO from the cell suspension. If the rate of diversion of NO into the trapping branch and the partition ratio between the trapping process and the normal reductive path were known, then one could in principal calculate the flux of N through the NO pool. The trapping of NO has been accomplished qualitatively by the use of extracellular cytochrome *c* to form the oxidized cytochrome *c*-NO complex (Kucera, 1989) and of extracellular membrane particles containing an NADH–NO oxidoreductase system (Carr *et al.*, 1989). The chief problems with cytochrome *c* as a trapping agent are that the complex with NO is not formed irreversibly, and, in addition, there is a tendency under the anaerobic conditions required for denitrification for cytochrome *c* to become reduced. Quantitative trapping experiments were performed by Zafiriou *et al.* (1989) using a gas stripping method and by Goretski and Hollocher (1988) using extracellular hemoglobin (Hb) to trap NO. The HbNO complex is extremely stable, with a k_{on} exceeding 10^7 M^{-1} s^{-1} and a k_{off} less than 1 hr^{-1} (Cassoly and Gibson, 1975). In the gas stripping method, a dilute suspension of denitrifying bacteria was vigorously sparged with an inert gas and NO and N_2O in the gas stream were determined by a chemiluminescence NO_x detector and sensitive gas chromatographic methods, respectively. The concentrations of NO and N_2O in the gas stream were determined at different flow rates and the product of concentration times flow rate gave the rate at which these gases were being removed from the bacterial suspension. At sufficiently high flow rates, this product tended to become constant and this was taken as an indication that the stripping process was removing all of the NO and N_2O that appeared in the extracellular space. Similarly with the trapping experiments using Hb, the fraction of total nitrite N that appeared as HbNO tended to plateau at high Hb concentrations. The occurrence of a slow chemical reaction between nitrite and Hb [see discussion surrounding Eq. (5) below] required that the cell concentration be rather large to assure that the denitrification of nitrite would be rapid compared with the chemical reaction. Alternatively, nitrate was used to largely avoid the chemical reaction. In spite of the difference in cell concentration of some 1000-fold between the gas stripping and Hb methods, similar results were obtained. For most denitrifiers studied, upwards of 70% of the nitrite N could be trapped as extracellular NO. Why was the figure not 100%? A simple geometric argument can be made to the effect that the fraction of nitrite N trapped by extracellular processes should be about 50% and nothing like 100%. If the NO-producing nitrite reductase is in the periplasmic space (Alefounder and Ferguson, 1980; Coyne *et al.*, 1990; Page and Ferguson, 1989) and the nitric oxide reductase in the cytoplasmic membrane, as generally believed (Carr *et al.*, 1989; Dermastia *et al.*, 1991; Heiss *et al.*, 1989; Miyata *et al.*, 1969; Zumft, 1993), and if nitric oxide reductase has a cellular V_{max} greater than that of nitrite reductase (Goretski and Hollocher, 1990) and a high affinity for NO (Goretski *et al.*, 1990; Kalkowski and Conrad, 1991; Remde and Conrad, 1991; Zafiriou *et al.*, 1989), then half of the NO produced would tend to diffuse out of the periplasmic space into the extracellular fluid where the trapping process could remove it, whereas the other half would diffuse into the cytoplasmic membrane where it would be intercepted and reduced by nitric oxide reductase. A more sophisticated argument based on actual enzyme activities and diffusional halftimes for NO exiting a cell gave a similar result (Goretski and Hollocher, 1988). For a given bacterium, the fraction of N trapped was essentially independent of the cell concentration and the amount of nitrite or nitrate reduced. At low Hb concentrations, the fraction of N trapped depended on the Hb concentration and from this dependency it was possible to estimate the rate at which nitric oxide reductase could intercept NO (Goretski and Hollocher, 1988). Overall the trapping experiments were consistent with the notion that most or all of the nitrite N passed through a small pool of NO which was free to diffuse and partition between an extracellular trap and intracellular reduction by nitric oxide reductase. NO was thus an intermediate in denitrification by this criterion.

Although the isotope experiments of Goretski and Hollocher (1990) with *Paracoccus denitrificans* were not exactly trapping experiments, the fact that [^{15}N]nitrite and NO proceeded to N_2O via pathways which shared at least one mono-N intermediate in common strongly supported NO as a free intermediate. Because NO and N_2O represent 1- and 2-electron reduction products of nitrite, it is inescapable that NO must itself be the common intermediate.

If NO were an obligatory intermediate in the denitrification pathway, then there should exist a separate and specific enzyme to reduce NO to N_2O. This enzyme would need to keep pace with the flux of denitrification as set by nitrite reductase and would also need to maintain a low steady-state concentration of NO in order to suppress the inherent toxicity of NO (Mancinelli and McKay, 1983). These requirements, plus the need to intercept NO against diffusional loss to the environment, all suggested that nitric oxide reductase should have high activity and a low K_m. The enzyme has now been purified essentially to homogeneity from *P. stutzeri* (Heiss et al., 1989) and *Pa. denitrificans* (Dermastia et al., 1991) following its solubilization, respectively, with Triton X-100 or octyl glucoside. Partial purifications of the enzyme from *Pa. denitrificans* and *A. cycloclastes* were reported by Carr and Ferguson (1990) and Jones and Hollocher (1993), respectively, in procedures using dodecyl maltoside as the detergent. The enzyme was shown to be a unique cytochrome *bc* complex in all three organisms. It is particularly significant that nitric oxide reductase from *A. cycloclastes*, which makes use of a Cu-type nitrite reductase, is very similar in its physical and chemical properties and intracellular abundance to the enzyme from *P. stutzeri* and *Pa. denitrificans*, which make use of the cytochrome cd_1-type nitrite reductase. Thus, it would appear that the Cu nitrite reductase does not normally function as a nitric oxide reductase in *A. cycloclastes*. In addition to its purification, the structural genes for nitric oxide reductase of *P. stutzeri* have been identified by cloning (Braun and Zumft, 1991; Heiss et al., 1989). These results demonstrate together that nitric oxide reductase is unambiguously different from nitrite reductase and, furthermore, is abundant, exhibits high turnover and low K_m, is specific for NO, and produces N_2O as its sole product. *In vivo*, the apparent K_m for nitric oxide reductase was inferred to be less than or equal to 10 nM on the basis of steady-state NO concentrations among a variety of different denitrifying bacteria (Goretski et al., 1990; Kalkowski and Conrad, 1991; Remde and Conrad, 1991) and 400 nM by the gas-stripping method (Zafiriou et al., 1989). *In vitro*, the purified enzyme from *Pa. denitrificans* had been shown to have an apparent K_m of ≤ 0.5 μM in the ascorbate/PMS assay (Dermastia et al., 1991). A further discussion of this enzyme appears in Section IV.

The discussion in this section on pathway has so far marshaled arguments favoring NO as an intermediate in denitrification. Evidence to the contrary exists and this evidence, although circumstantial, needs to be examined. The first line of evidence, largely attributed to the work of Averill and co-workers, concerns experiments in which the denitrification of [^{15}N]nitrite to N_2O was placed in competition with the catalysis of nitrosyl transfer from [^{15}N]nitrite to a nucleophile (azide or NH_2OH) (Aerssens et al., 1986; Hulse and Averill, 1989; Weeg-Aerssens et al., 1987, 1988). Because denitrification yields $^{15}N_2O$ and nitrosyl transfer yields $^{14,15}N_2O$ plus $^{14}N_2$ with azide or $^{14,15}N_2O$ with NH_2OH, the two pathways could be distinguished by mass spectrometry. Catalysis of nitrosyl transfer from nitrite to NH_2OH by an *Alcaligenes* sp. and a Cu-type nitrite reductase was studied (Iwasaki and Matsubara, 1972; Iwasaki and Mori, 1958; Matsubara, 1970), but this reaction was not recognized as nitrosation until the more general studies of Garber and Hollocher (1982) with intact bacteria, and Kim and Hollocher (1984) with a purified cytochrome cd_1 nitrite reductase. More recent studies (Aerssens et al., 1986; Hulse and Averill, 1989; Weeg-Aerssens et al., 1987, 1988) with P. stutzeri and A. cycloclastes revealed additional properties of the system. The most important for our purposes were the observations that increasing the ratio of [^{15}N]nitrite to azide increased the ratio of $^{15}N_2O$ to $^{14,15}N_2O$ as products and increasing the concentration of nitrite (in absence of azide) suppressed the incorporation of ^{18}O from water into product N_2O. The incorporation of ^{18}O from water into nitrite and N_2O is thought to result from the catalysis by nitrite reductase of nitrosyl transfer from nitrite to water. In this exchange reaction water would behave as an O-nucleophile, just as azide can behave as an N-nucleophile in the nitrosyl transfer leading to N_2O. Although it is expected that increased azide concentration should favor the nitrosation branch by mass action, it was remarkable that nitrite should favor the branch that forms $^{15}N_2O$ ("denitrification" branch), because nitrite reductase has a low K_m for nitrite and was apparently saturated during the experiments under discussion. The results imply that a second molecule of nitrite can influence the partition ratio, and they were taken to support the Averill–Tiedje pathway of denitrification indicated in Eq. (2) or (3). By that scheme, the change in the partition ratio comes about because the denitrification pathway requires a second molecule of nitrite, acting as an N-nucleophile, to form the N–N dimeric species that is subsequently reduced to N_2O. Thus, denitrification and the related nitrosyl transfer to azide (or water) were thought to be reflections of different nitrosyl transfer reactions and simple competition among reactants would be expected. In this view of denitrification, the first N–N bond is formed by nitrite reductase (not nitric oxide reductase) at the N^{3+} redox level.

Other interpretations are, of course, possible. The action of nitrite on the partition ratio might occur by way of a low affinity effector site on nitrite reductase. It may also be relevant that most, if not all, of the in vivo and in vitro systems used to demonstrate nitrosyl transfer to azide were inhibited in or lacked nitric oxide reductase activity and therefore surely accumulated substantial amounts of NO (Goretski and Hollocher, 1990, 1991). NO is known to be an inhibitor of nitrite reductase, as well as nitric oxide reductase, and could conceivably modify the way in which nitrite reductase processed nitrite. Whether NO might modify the catalysis of nitrosation reactions was tested with P. stutzeri and A. cycloclastes, which contain the heme- and Cu-type nitrite reductase, respectively (Goretski and Hollocher, 1991). Cells were incubated with [^{15}N]nitrite and azide in a sidewell bottle in which acidic $CrSo_4$, a NO scavenger, occupied the sidewell. This system was effective in absorbing NO from the system, and a consequence of the absorption of NO was a dramatic suppression of nitrosyl transfer to azide. The study also demonstrated that the major initial product of the reduction of nitrite in azide-containing systems was NO, contrary to the expectations of Eq.

(2). The possible role of NO in facilitating the catalysis of nitrosyl transfer may be its ability, in cooperation with nitrite, to shift the distribution of enzyme species in favor of the nitrosyl donor species [Eq. (3)].

The second line of evidence that the dimer pathway of denitrification [Eq. (2)] may be significant comes from ^{15}N-isotope fractionation studies with *P. stutzeri* (Bryan et al., 1983; Shearer and Kohl, 1988). In an isotope fractionation experiment, nitrite at natural abundance is reduced by denitrifying bacteria and the reaction is terminated when about half of the nitrite has been reduced. The remaining nitrite is then converted chemically to N_2 and assayed by means of an isotope ratio mass spectrometer to see if it has become more enriched in ^{15}N. If there is an enzymatic kinetic isotope effect disfavoring ^{15}N in the reduction of nitrite and if the association/dissociation of nitrite with enzyme is rapid relative to the bond making/breaking (covalent) step, then the enzyme can discriminate against ^{15}N and an enrichment or fractionation in ^{15}N can occur in the unreacted nitrite. It was observed by Bryan et al. (1983) and Shearer and Kohl (1988) for both cells and cell-free systems that the extent of isotope fractionation depended on the nitrite concentration. This is a surprising result and implies that nitrite can modulate its own covalent modification. This is in principle possible if two molecules of nitrite were to cooperate in the relevant covalent reaction, as might occur, for example, in Eq. (2) (Shearer and Kohl, 1988). In this case, an increase in the concentration of nitrite might make the transfer of nitrosyl from nitrite to a second molecule of nitrite more rapid and thus affect the isotope fractionation factor. The observed isotope fractionation factor depends on the individual fractionation factors for the forward and reverse reactions at the step in question and therefore on the rate ratio for the forward and reverse reactions. Alternatively, if Eq. (1) applied, the binding of a second molecule of nitrite to an effector site might modify the structure of the transition state for nitrite reduction and change the degree of isotope fractionation in this way. Other possibilities may also exist, as, for example, an effect that would make the association of nitrite with enzyme more or less reversible. Loss of reversible binding would abolish isotope fractionation. Shearer and Kohl (1988) reported that the nitrite reductase of the test organism, *P. stutzeri* JM300, was "sticky" and that nitrite is committed to reduction once bound. This conclusion, which was based on the lack of nitrite/water ^{18}O-exchange, is puzzling because it should have obviated the N-isotope fractionation of nitrite.

In summary, a considerable body of enzymatic, genetic, and analytical data supports the view that the major, if not sole, pathway of denitrification involves NO as an obligatory intermediate and requires the action of nitric oxide reductase. On the other hand, the ability of nitrite to modify nitrosyl transfer ratios and the ^{15}N isotope fractionation factor during its reduction, are consistent with the reductive scheme of Averill and Tiedje (1982). It was suggested (Goretski and Hollocher, 1991) that because catalysis of nitrosyl reactions, including the putative transfer of nitrite nitrosyl to nitrite, may depend on or be controlled by NO, it is possible that Eq. (2) may serve as a metabolic escape route for reduction of nitrite when nitric oxide reductase is inhibited. Thus there may actually be two pathways, a major one with NO as an intermediate and a minor one that can proceed via nitrosyl transfer to nitrite under extraordinary conditions.

IV. NITRIC OXIDE REDUCTASE, NITRIC OXIDE-CONSUMING ENZYME OF DENITRIFICATION PATHWAY

A. Characteristics

This membrane associated enzyme has been highly purified first from *P. stutzeri* (Heiss *et al.*, 1989) and later from *Pa. denitrificans* (Dermastia *et al.*, 1991) with use of nonionic detergents. A partial purification of the enzyme from *Pa. denitrificans* was reported somewhat earlier by Carr and Ferguson (1990). More recently, the enzyme was partially purified (60–70% pure) from *A. cycloclastes*, a bacterium which uses a Cu-type nitrite reductase (Jones and Hollocher, 1993). The fact that the enzyme from *P. stutzeri* and *Pa. denitrificans* was pure or nearly so after a purification of about 100-fold suggests that it may comprise something of the order of 1% of the total membrane protein and perhaps 0.5% of the total cell protein. The enzyme, which is specific for NO and produces N_2O exclusively, is a tight complex of a cytochrome *b* and a cytochrome *c* with peptide M_r of 38 K and 17.5 K respectively by sodium dodecyl sulfate–polyacrylamide gel electrophoresis (SDS–PAGE). The cytochrome *c/b* mole ratio lies in the range of 1–1.7 for the three organisms studied to date. It is possible therefore that the functional unit is a heterodimer *in vivo*. The α bands of the reduced spectrum of the enzyme center at 552 and 559–560 nm (Fig. 2) for the cytochrome *c* and *b* components, respectively. There are about 3 mol of heme per 73,000 g [38,000 + 2(17,500)] for the enzyme from *Pa. denitrificans*, so that each subunit contains apparently only one heme. Although the iron content is about twice as great as the content of heme iron, the function of the nonheme iron is obscure at the moment. The electron paramagnetic resonance spectrum of the oxidized enzyme (Fig. 3) showed evidence for both high and low spin ferric heme centers but no evidence for nonheme iron centers. Because the enzyme from *P. stutzeri* on the one hand and *Pa. denitrificans* and *A. cycloclastes* on the other has been assayed by two different systems and at different pH values, it is difficult to make exact comparisons among the reported kinetic parameters. This difficulty is further compounded by the observations that NO can inhibit the enzyme at concentrations above about 10 μM, the enzyme is strongly inhibited by fatty acids, and the specific activity depends on which detergent is present to support its solu-

FIGURE 2

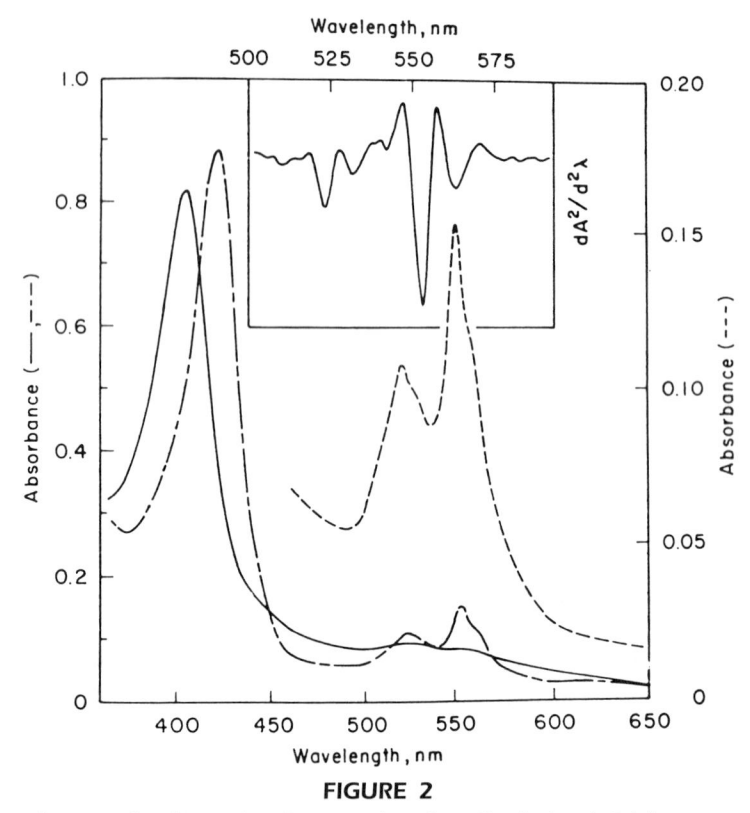

Spectra of nitric oxide reductase from *Paracoccus denitrificans* (batch eluate). Solid curve, enzyme as prepared (oxidized); dashed and dot-dashed curves, reduced with a small excess of dithionite. Inset solid curve shows the second derivative of the dashed curve. The protein concentration was about 200 μg/ml. From Dermastia et al. (1991).

bility. In any case, the preparations are reported to have specific activities ranging from 10–40 μmol min^{-1} mg^{-1} from pH 6.5 to 4.8. The preparation of Dermastia et al. (1991) from *Pa. denitrificans* showed a specific activity of 10 and 22 μmol min^{-1} mg^{-1} at pH 6.5 and 5.0 (the pH optimum), respectively, but these values were averages over progress curves that were modulated by inhibition of the enzyme by NO. The specific activities would be about three times greater if based on maximum rates. Maximum rate data taken at pH 6.5 gave a k_{cat} value of 36 sec^{-1}, assuming a M_r of 73,000 for the unit enzyme. The estimate of K_m was 0.4 μM ≤ K_m ≤ 1 μM. Thus, the effective second order rate constant, k_{cat}/K_m, may be 4–9 × 10^7 M^{-1} sec^{-1}. This value is only 10–100 smaller than the diffusion-controlled limit for small molecules in water and is of a magnitude similar to those found for diffusion-controlled encounters between proteins and

9 Enzymology and Occurrence in Biological Nitrogen Cycle

FIGURE 3

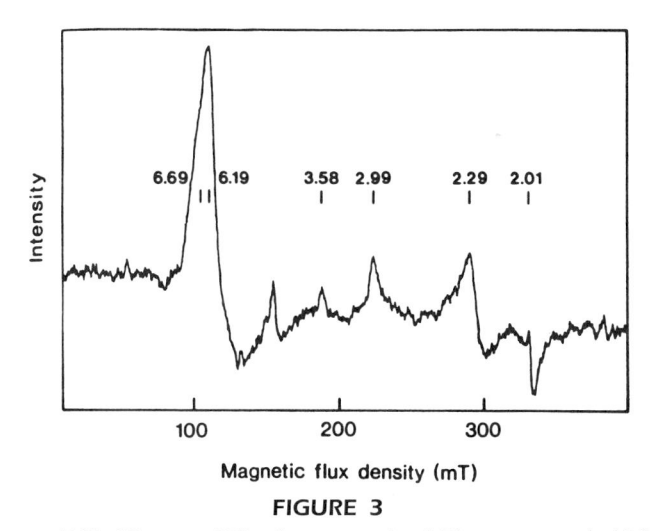

EPR spectrum at 12 K of *P. stutzeri* NO reductase as isolated. There was a total of 3.8 mg protein per ml of 20 mM Tris hydrochloride (pH 8.5)–0.1% Triton X-100–0.5% n-octyl-β-D-thioglucopyranoside. Conditions of recording: microwave frequency, 9.30 GHz; microwave power, 2 mW; modulation amplitude, 1 mT at 100 kHz; scan rate, 1.3 mT/s. From Heiss *et al.* (1989).

ligands or enzyme and substrate (Alberty and Hammes, 1958; Blacklow *et al.*, 1988; Schmitz and Schurr, 1972; Schurr and Schmitz, 1976). These kinetic parameters plus the great abundance of the enzyme in denitrifying cells imply that the enzyme should be a very active one *in vivo*. In fact, the *in vitro* kinetics of the purified enzyme come within a factor of about 2 of satisfying the *in vivo* activity of the NO uptake system, which is about 0.5 μmol min^{-1} mg^{-1} (Dermastia *et al.*, 1991; Goretski and Hollocher, 1990). The redox potentials for the cytochrome *b* and *c* components of the enzyme have been estimated to be +322 and −280 mV, respectively (Zumft, 1993). These values strongly suggest that the cytochrome *b* component provides the binding site for NO.

As far as we know, nitric oxide reductase has not been shown to catalyze the NO/H$_2$O–^{18}O exchange reaction.

The *in vitro* assays were not trivial to devise inasmuch as they must involve reducing systems that do not react chemically at significant rates with NO. The system used by Heiss *et al.* (1989) and Dermastia *et al.* (1991) was ascorbate linked to enzyme via PMS; that of Carr and Ferguson (1990) was isoascorbate linked via 2,3,5,6-tetramethylphenylene-1,4-diamine and horse heart cytochrome *c*. The assay system of Carr and Ferguson (1990) was particularly clever, because it included glucose, glucose oxidase, and catalase as an oxygen scavenger. Thus the assay mixture itself served to generate the anaerobic conditions required for the introduction of NO.

The genes for nitric oxide reductase activity in P. *stutzeri* are being dissected (Braun and Zumft, 1991, 1992), and the structural genes for nitric oxide reductase and nitrite reductase have been found to be part of a greater denitrification gene cluster, at least in P. *stutzeri* (Braun and Zumft, 1992; Jüngst et al., 1991a,b). NorC and norB are the structural genes, respectively, for the cytochrome c and b components of nitric oxide reductase and these have been sequenced for P. *stutzeri* (Zumft, 1993). These genes are contiguous and transcribed as a single operon in the sequence, norC, norB. From the gene sequence (Zumft, 1993; Zumft et al., 1994), the cytochrome b peptide is inferred to have a true M_r of 53,000 and thus must run anomalously fast on SDS–PAGE. It is inferred to have 12 membrane-spanning α-helices and to have both the N and C termini on the cytoplasmic side of the membrane. The histidine residues for axial coordination of heme iron may be on the periplasmic side. The cytochrome c component is inferred to have only a single membrane-spanning helix and to have the bulk of its mass along with the C terminus and heme covalent binding site on the periplasmic side of the membrane. This picture suggests that in gram-negative denitrifiers at least NO is produced in the periplasmic space by the periplasmic nitrite reductase and is reduced there by nitric oxide reductase without any need to cross over into the cytoplasmic space.

B. Mechanisms and Models

It was surprising that nitric oxide reductase should be comprised of cytochromes, because ferrous nitrosyl complexes tend to be extremely stable and do not readily decompose into ferric complexes and N_2O or NO^- (Bottomley, 1978; Eisenberg and Meyer, 1975; Enemark et al., 1975; McCleverty, 1979; Pandey, 1983; see also Chapter 2). Moreover, the reduction of Fe·NO complexes occurs generally at redox potentials lower than that of the standard hydrogen electrode and is virtually independent of the nature of the trans ligand (Olson et al., 1982). Nevertheless, there are ferrous systems capable of reducing NO to N_2O apparently via nitroxyl (nitrosylhydride: HNO, NO^-) as intermediate under mild conditions. One that has been studied extensively (Bonner and Pearsall, 1982; Pearsall and Bonner, 1982) is the aqueous ferrous ion in weakly acidic solutions, pH 4.6–6.0. The reduction of $^{15}N^{18}O$ by ferrous ion in presence of trioxdinitrate, a nitroxyl generator by virtue of its spontaneous decomposition in water, produced N_2O that arose by self-dimerization ($^{15}N_2^{18}O$) as well as $^{14}N^{15}N^{18}O$ and $^{15}N^{14}N^{16}O$ as the result of cross dimerization of the $^{14}N^{16}O^-$ from trioxodinitrate with something derived from $^{15}N^{18}O$. Because it is known that nitroxyl dimerizes rapidly (Bazylinski and Hollocher, 1985a), the result was interpreted to indicate that $^{15}N^{18}O^-$ was an intermediate in the reduction of $^{15}N^{18}O$. The rate law at pH 6 for NO reduction by ferrous ion suggested the involvement of the *gem*-dinitrosyl complex of ferrous ion [ON·Fe(II)·NO] because of a term that was first order in ferrous ion but second order in NO (Pearsall and Bonner, 1982).

But, because there was also a first-order term, reduction via a dinitrosyl complex may not be compulsory. It is doubtful that cytochromes could participate in NO reduction via dinitrosyl complexes, because of strong axial coordination of Fe by at least one protein ligand. It is of course possible that the nonheme iron of nitric oxide reductase is the actual site of reduction of NO.

Although relatively little is known about the mechanism of action of nitric oxide reductase, one mechanism that is attractive in its simplicity is the one-electron reduction of NO to nitroxyl which would then spontaneously and rapidly dimerize and dehydrate to form N_2O (Gordon et al., 1963; Bazylinski and Hollocher, 1985a). Nitroxyl can be trapped by a variety of reactive compounds, including $[Ni(CN)_4]^{2-}$ at high pH (Bonner and Akhtar, 1981) and oxidized cytochrome c (Doyle et al., 1988) or oxidized (met-) hemoglobin (Hb^+) at neutral pH (Bazylinski and Hollocher, 1985b), so there is some hope of using a nitroxyl trap to probe the mechanism of the enzyme. Under the same conditions whereby extracellular Hb was effective in trapping NO generated during the denitrification of nitrate or nitrite, Hb^+ was ineffective in trapping nitroxyl, as evidenced by a lack of formation of HbNO (Goretski and Hollocher, 1988). This negative result is not conclusive, however, because nitroxyl is expected to be ionized at pH 7.0 and so may not be permeant across the cell membrane. It is unlikely that Hb^+ could be useful as a nitroxyl rap for cell-free systems, inasmuch as the reducing agent needed to drive the enzymatic reduction of NO is likely to reduce Hb^+ to Hb which would then form HbNO from NO. An effective trap for nitroxyl under reducing conditions is the thiol group which can react rapidly with nitroxyl to form disulfide and hydroxylamine (Doyle et al., 1988) but only slowly with NO (Clancy et al., 1990; Pryor et al., 1982). There is evidence to suggest that nitroxyl-forming reactions may generate one or the other of two species of nitroxyl (Donald et al., 1986). One reacts very rapidly with O_2 to form peroxynitrite whereas the other does not. The former and latter species can be formed by the photolysis and spontaneous decomposition, respectively, of trioxodinitrate. These two species are thought to be triplet and singlet species, respectively, and these two spin states may have similar energy levels. This duality in nitroxyl necessarily complicates any attempt to trap nitroxyl in the enzymatic reaction, because a trapping reagent that might be effective toward one species of nitroxyl may not necessarily be effective toward the other. That nitroxyl may have singlet and triplet states of similar energy is perhaps not too surprising when one considers that NO^- is isoelectronic with O_2 which exists as a ground-state triplet.

Turk and Hollocher (1992) have attempted to trap nitroxyl using the dithiol, dithiothreitol, during the reduction of NO by purified nitric oxide reductase from *Pa. denitrificans*. Because the dithiol served in this system as both electron donor to the enzyme and trap for nitroxyl, the ratio of thiol groups oxidized to NO consumed would have been 1 if nitroxyl trapping did not occur and 3 if nitroxyl trapping were 100% efficient. The observed ratios were in the 1.3–2 range and thus support the idea that a significant fraction of the reaction proceeds with nitroxyl as an intermediate.

V. DISSIMILATORY NITRITE REDUCTASES, ENZYMES THAT GENERATE NITRIC OXIDE IN DENITRIFYING BACTERIA

Denitrifying bacteria produce one or the other of two usually soluble but very dissimilar nitrite reductases, heme- and Cu-containing types (Brittain et al., 1992; Coyne et al., 1989), both of which are found in the periplasmic space of gram-negative bacteria (Coyne et al., 1990).

A. Cytochrome cd$_1$

It is remarkable that this enzyme can accept both O_2 and nitrite as its oxidizing substrate, and in fact it was originally isolated as an oxidase rather than a nitrite reductase. Its role in nitrite reduction became clear when it was realized that its synthesis was induced only under anaerobic conditions and its oxidase activity was relatively small compared with the total O_2-reducing activity of cells (Lam and Nicholas, 1969; Yamanaka, 1964; Yamanaka and Okunuki, 1963a).

Although nitrite reductase of the cd_1 type has been purified from several bacteria, that from P. aeruginosa is perhaps the best understood (Gudat et al., 1973; Kuronen and Ellfolk, 1972; Parr et al., 1976; Yamanaka and Okunuki, 1963a–c; Yamanaka et al., 1962). It has been highly purified and the amino acid sequence is now known (Silvestrini et al., 1989). The primary sequence of the enzyme from P. stutzeri is now also known (Jüngst et al., 1991b; Smith and Tiedje, 1992) and is highly similar to that of the enzyme from P. aeruginosa. Structurally it is a cytochrome with two different heme groups, the covalently linked heme c and the noncovalently bound heme d_1. The latter is in reality an iron dioxoacryloisobacteriochlorin (Fig. 4) and not an iron porphyrin (Chang and Wu, 1986; Chang et al., 1986). The spectrum of the enzyme (Fig. 5) presents bands in the 630–660 nm region attributed to cytochrome d_1 and a split α band at 550 nm attributed to reduced cytochrome c. The enzyme exists in solution as a dimer (Gudat et al., 1973: Kuronen and Ellfolk, 1972) with subunit M_r of about 60,000 (Silvestrini et al., 1989), and there is one heme of each type per subunit. Each subunit is therefore a potential two-electron donor when reduced. Some structural data are now available from X-ray diffraction and electron microscopy (Akey et al., 1980; Berger and Wharton, 1980; Takano et al., 1979) but not yet at the atomic level. Polarization data on a single crystal of the enzyme suggested that the ring planes of heme c and d_1 are perpendicular to each other (Makinen et al., 1983). Several lines of evidence, including Mössbauer and electron paramagnetic resonance (EPR) data (Huynh et al., 1982), show that reduced enzyme

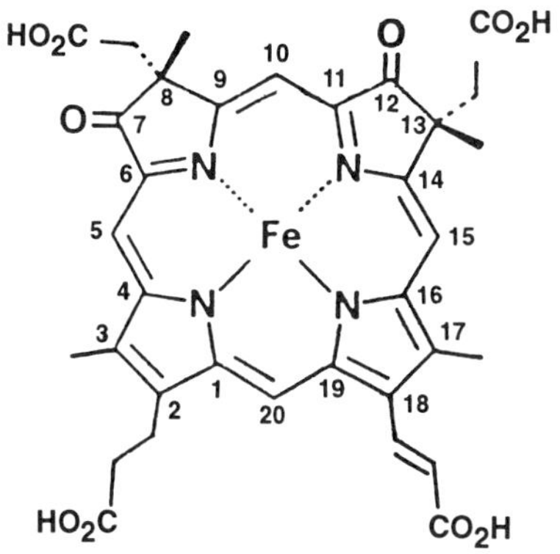

FIGURE 4
Structure of heme d_1. From Weeg-Aerssens et al. (1991).

contains low spin heme c and high spin heme d_1, whereas in oxidized enzyme both hemes are low spin. This result implies that the Fe of heme c is always six-coordinated, whereas that of heme d_1 is five-coordinated in the reduced state. The likely site of interaction of nitrite with reduced enzyme is therefore the heme d_1. Histidine N and methionine S have been assigned as axial ligands to heme c and two histidine N, to heme d_1 (Sutherland et al., 1986). The existence of a weak magnetic interaction between hemes c and d_1 (Sutherland et al., 1986) and calculations based on interheme electron-transfer rate (Schichmann and Gray, 1981) suggest that their iron atoms are rather distant (13–15Å) from each other. Among the kinetically most efficient reductants for cytochrome cd_1 are reduced, monoheme cytochrome c and reduced, small Cu-proteins such as azurin (Yamanaka and Okunuki, 1963b,c; Yamanaka et al., 1961), both of which are found in denitrifying bacteria. The results of a number of studies (Gudat et al., 1973; Horio et al., 1958; Parr et al., 1977; Wharton and Gibson, 1976; Wharton et al., 1973) indicate that inhibitors coordinate largely or entirely at heme d_1, which is therefore the likely binding site for O_2 or nitrite. Electron transfer between hemes c and d_1 has been studied by rapid kinetic methods including stopped flow (Greenwood et al., 1978; Parr et al., 1977; Wharton and Gibson, 1976; Wharton et al., 1973), rapid scanning spectrophotometry (Shimada and Orii, 1976), and temperature jump (Brunori et al., 1975). In general, the results seem consistent with the intramolecular electron transfer sequence $c \rightarrow d_1$ for both the reductive half-reaction of oxidized enzyme and the oxidative half-reaction of O_2 with reduced enzyme. On the other hand, the stopped-flow experiments of Greenwood et al. (1978) suggested a more complex scheme. They concluded that heme c underwent oxidation during the fastest, initial phase of a complex, three-phase process and that heme d_1 exhibited at that time a perturbed spectrum attributed

FIGURE 5

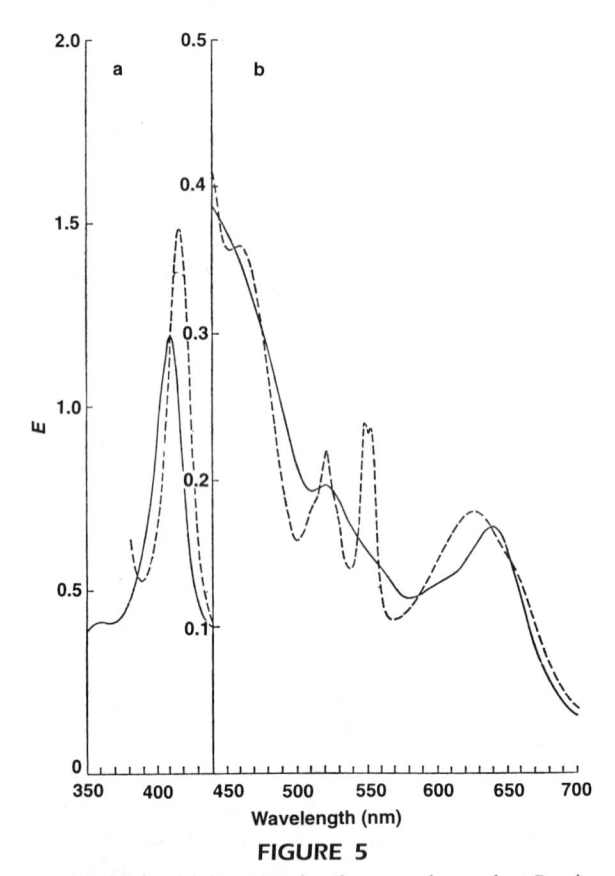

Absorption spectra of oxidized and dithionite-reduced nitrite reductase from *Pseudomonas aeruginosa*. Solid line, oxidized enzyme; dashed line, dithionite-reduced enzyme, in 40 mM potassium phosphate buffer, pH 6.9. The enzyme concentration was 8 μM and the spectra were recorded at room temperature (20°C) under N_2 in a Thunberg cuvette with a light-path of 1 cm. From Barber *et al.* (1976).

to a complex between heme d_1 and O_2 or a reduction intermediate of O_2. This study implied rapid internal electron transfer, $c \rightarrow d_1 \cdot O_2$, with rate constant of about 100 sec^{-1}.

Cytochrome cd_1 reduces nitrite by one electron to NO as the major product, providing a reducing system was used that does not reduce NO chemically (Kim and Hollocher, 1983, 1984; Mancinelli *et al.*, 1986; Parr *et al.*, 1976; Silvestrini *et al.*, 1979; Timkovich *et al.*, 1982; Yamanaka and Okunuki, 1963a). There is only one rapid kinetic study of cytochrome cd_1 with nitrite as the oxidant (Silvestrini *et al.*, 1990). This work, carried out at pH 8.0, indicated that there was intramolecular electron transfer, $c \rightarrow d_1$, as in the O_2 reaction, but that the enzyme underwent only a single turnover and terminated as the dead complex between reduced heme d_1 and NO at all concentrations of nitrite tried. In competition experiments with CO, there was no evidence that NO ever dissociated from an intermediate state before forming the final inactive complex. The reaction $c^{2+}d_1^{3+} \cdot NO \rightarrow c^{3+}d_1^{2+} \cdot NO$ was concluded to be rate limiting with $k = 1\ s^{-1}$. NO inhibition is released at lower pH, in spite of the still large binding constant between heme d_1 and NO. It may be that the reaction of previously reduced enzyme with nitrite is not a good model for events in the steady state. It is not clear how enzyme could turn over rapidly if NO complexation were significant after only one turnover. Nevertheless, it may well be that some and perhaps much of the enzyme exists as one or another NO complex during turnover under physiological conditions.

B. Copper-Type Enzymes

Dissimilatory nitrite reductases of the Cu-type have been purified from or detected in a variety of different denitrifying bacteria (see Denariaz et al., 1991 for a listing). It was concluded on the basis of antibody probes that although the majority of denitrifying isolates tested possessed the heme-type enzyme, the Cu-type was found over a wider range of phylogenetic groups (Coyne et al., 1989). The Cu enzyme from Achromobacter xylosoxidans (form. sp. Pseudomonas denitrificans) was originally reported to have a M_r of 70,000 with an α_2 structure and two type 1[3] Cu per molecule (Masuko et al., 1984), but more recent data based on X-ray scattering indicates that it has an α_3 structure (Grossmann et al., 1993). Ligands to Cu are both N and S (Sano and Matsubara, 1988). The enzyme from Alcaligenes faecalis S6 is reported to have an α_4 structure with four Cu per molecule, and the EPR spectral types are 1 and 2 (Kakutani et al., 1981). The Cu-type enzyme from P. aureofaciens is the first example of such an enzyme from the genus, Pseudomonas (Zumft et al., 1987). Although there are differences among Cu-type nitrite reductases with respect to M_r, spectral bands, amount of Cu, and perhaps EPR spectral properties, they seem to share antigenic similarities, to have subunits with M_r 30,000 to 40,000 and to be inactivated by the chelator, diethyldithiocarbamate (Coyne et al., 1989; Denariaz et al., 1991; Shapleigh and Payne, 1985a). Their primary amino acid sequences are similar and clearly indicate association within an homologous family. The sequences of enzyme from P. aureofaciens and Pseudomonas sp. G-179 are 64 and 78% identical, respectively, to that of the enzyme from A. cycloclastes, as inferred from the gene sequence (Glockner et al., 1993; Ye et al., 1993). The Cu nitrite reductases are also all

---

[3] Type 1 and 2 mononuclear copper centers in enzymes and proteins are distinguished by their different EPR spectra for Cu(II). Type 1 and 2 spectra exhibit comparatively narrowly and widely spaced hyperfine lines, respectively, for the electron–nuclear spin interaction. The more narrow the spacing, the weaker is the interaction. Type 1 cupric centers generally have an intensely blue color, whereas type 2 centers are virtually colorless.

inhibited by cyanide, CO, and mercurials, and the best reducing agents are monoheme cytochromes *c* or small azurin-like Cu proteins (Denariaz et al., 1991).

The enzyme from *A. cycloclastes* (Iwasaki and Matsubara, 1972; Iwasaki et al., 1975; Liu et al., 1986) is structurally the best defined of this family of enzymes. The primary sequence of the 340 amino acids of its subunit has been determined by direct sequencing of the protein (Fenderson et al., 1991) and a high resolution X-ray diffraction structure has been calculated (Godden et al., 1991). The enzyme is a trimer (Fig. 6) with a three-fold axis of symmetry. Each of the two domains of the subunit is similar in sequence and structure to domains in the Cu-containing oxidases, ascorbic acid oxidase and laccase, and in the small, blue Cu-proteins, such as plastocyanin. Each subunit in the trimeric unit contains 2 Cu atoms separated by approximately 12.5 Å. One Cu atom (type 1) is four-coordinated to two histidine N atoms, one cysteine S atom, and one methionine S atom within a Greek key β domain. The second Cu atom (type 2 or related) is located at a subunit–subunit interface and coordinated to one solvent O atom and three N atoms of histidine. One of these histidine imidazole groups is a component of the adjacent subunit. Because of this sharing of a Cu atom between pairs of subunits, it is likely that the monomer could not be active. This type 2 Cu site is the site of interaction of the enzyme with nitrite; the type 1 site presumably plays an electron transport role. Differential removal of Cu from the

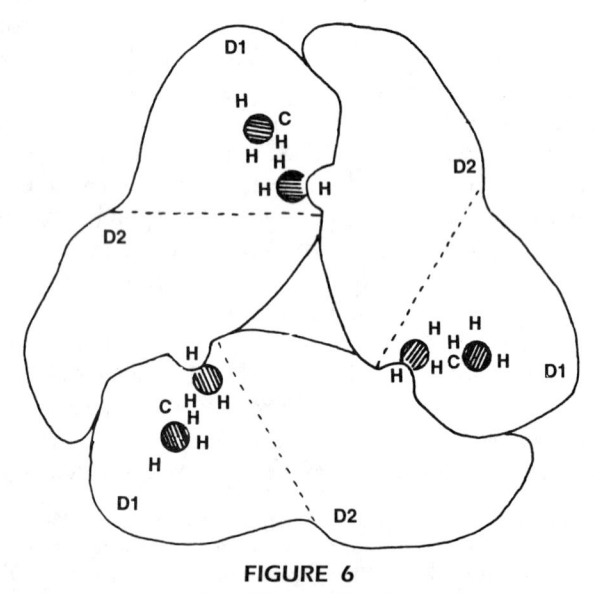

FIGURE 6

Arrangement of subunits in nitrite reductase from *Achromobacter cycloclastes*. Domains are denoted D1 and D2, copper sites are shaded spheres, and Cu ligands are denoted by one-letter abbreviations: C for cysteine sulfur and H for histidine imidazole nitrogen. From Fenderson et al. (1991).

nitrite reductase also showed that the type 2 site is the site of nitrite reduction (Libby and Averill, 1992).

The major product of nitrite reduction is NO (Denariaz et al., Iwasaki and Matsubara, 1972; Kakutani et al., 1981; Liu et al., 1986; Masuko et al., 1984; Sawada et al., 1978; Zumft et al., 1987) just as is the case with the cytochrome cd_1-type enzyme. As discussed previously in this chapter, the Cu-type enzyme from A. cycloclastes has been reported to reduce nitrite to N_2O in the presence of NO (Averill and Tiedje, 1990; Jackson et al., 1991), a catalytic capability apparently not exhibited by heme-type nitrite reductases.

Although the Cu-type family of nitrite reductase is comprised of soluble enzyme and localized in the periplasmic space in gram-negative bacteria, it has proved to be a membrane-bound enzyme in denitrifying Bacillus, which is gram positive and lacks an outer membrane and periplasmic space (Denariaz et al., 1991; Urata and Satoh, 1991; Ho et al., 1993).

C. Mechanisms and Models

A possible model for the cytochrome cd_1 type, NO-producing nitrite reductases is the reaction between nitrite and Hb (Doyle et al., 1981, 1984). This reaction, which produces Hb$^+$ and HbNO as products, can be written as occurring in two steps [Eq. (5)]. The second step is simply the complexation of NO by Hb.

$$Hb + H^+ + NO_2^- \longrightarrow Hb^+ + OH^- + NO$$

$$Hb + NO \longrightarrow HbNO \tag{5}$$

The first step is rate limiting and was shown to be first order in the concentrations of Hb and nitrite and to have a rate constant of 2.7 M^{-1} sec^{-1} at pH 7, 25°C. Although Hb$^+$ and HbNO were expected to be formed in a 50–50% mixture, that observed was about 70–30%, and this product distribution was independent of pH from 6 to 7, over which the rate changed by almost a factor of 10. The reason for the observed product distribution is not entirely clear, but the results suggest the involvement of an additional oxidant, presumably one derived from NO. The rate increased with decreasing pH and the plots of log k_2 versus pH were linear with a slope of about -0.9. This is in accord with the requirement for a proton in the reaction (specific acid catalysis) and suggested that nitrous acid (HONO) might be the true substrate. This idea was supported by the finding that ethylnitrite ester reacted with Hb in a way very similar to that of nitrite plus proton and with similar product ratios. At pH 7, 10°C, k_2 for ethylnitrite was 1.1×10^3 M^{-1} sec^{-1} and independent of pH from pH 6–7.5. This suggested that Eq. (6) applies and the alkoxyl group functions as an analog to the

$$Hb + EtONO \longrightarrow Hb^+ + EtO^- + NO \tag{6}$$

OH group of nitrous acid. The data, when extrapolated to 25°C, suggest that the reaction of ethylnitrite is about 1000 times faster at pH 7 than the reaction of nitrite and less than 10 times slower than the reaction of HNO_2. This is to be expected because water and ethanol have similar pK values near 16.

There are two major chemical problems in the reduction of nitrite to NO at pH 7. First is the problem of pushing additional negative charge onto an anion; second is the creation of a good leaving group for the N–O bond breaking step during dehydration. Protonation (or alkylation) of nitrite solves both problems in large part, because it quenches negative charge and makes the leaving group OH^- instead of the impossible O^{2-}. Accordingly, one would expect nitrite reductases to require protonation of nitrite in or prior to the transition state or an equivalent stabilization of the transition state by electrophilic catalysis through coordination of the leaving O-atom with a metal center. The Hb model suggests protonation, but the leaving of iron-coordinated OH^- or EtO^- cannot be ruled out as providing additional assistance. That is, both specific (or general) acid and electrophilic catalysis may occur. If enzyme-bound nitrite could resemble H_2ONO^+, one would predict that nitrite reductase should catalyze nitrosyl (NO^+) transfer reactions very effectively. Water should be a better leaving group than OH^- by many orders of magnitude (Jencks, 1969). In fact, as discussed in Section III of this chapter, nitrite reductases do catalyze nitrosyl transfer to a variety of nucleophiles, including NH_2OH, azide, and water under anaerobic conditions.

That cytochrome cd_1 catalyzes nitrosyl transfer reactions during its reduction of nitrite to NO suggests that formation of an iron–nitrosyl system ($Fe^{III}NO \rightleftharpoons Fe^{II}NO^+$) may be a normal event in the catalytic cycle (Garber and Hollocher, 1982 and references therein). Little is known however about the chemistry of heme d_1 and related Fe-chlorins in this regard. A relatively recent study (Ozawa et al., 1992) supports the view that $Fe^{II}NO^+$ species can be formed with Fe-chlorins. Ozawa et al. (1992) studied the chemical oxidation of $Fe^{II}NO$ complexes of octaethylchlorin and octaethylbacteriochlorin to yield $Fe^{II}NO$ π-cation radicals. These ring radicals could be valence isomerized to $Fe^{II}NO^+$ chlorin species by ligation of imidazole and other ligands to the sixth coordination position of the iron.

In spite of the above speculation, the actual mechanisms used by NO-producing nitrite reductases for reduction of nitrite and its activation for nitrosyl transfer are poorly understood. The fact that both the heme and Cu types of enzyme catalyze the reaction is remarkable in view of the fact that nitrosyl compounds of Fe complexes are well known, whereas Cu-nitrosyl compounds are rare [see Garber and Hollocher (1982) and Kim and Hollocher (1984) for further discussion]. On the other hand, both Fe and Cu can coordinate O and this might be more relevant for the activation of an O atom of nitrite for N–O bond breaking. In the case of the Cu-type nitrite reductase of *A. cycloclastes*, the

X-ray data suggest that nitrite coordinates the Cu^{II} type 2 site with one or both of its O atoms (Godden et al., 1991).

Cu(II) halides have been reported by Doyle et al. (1976) to react with NO to form reactive species which can oxidatively deaminate primary amines. The reactive species were inferred to be Cu–nitrosyl complexes [Cu(II)NO ⇌ Cu(I)NO$^+$] that serve as nitrosyl donors to amines in a reaction analogous to the Van Slyke reaction with HNO_2. Examples of Cu–nitrosyl compounds are otherwise so rare in the chemical literature [see the reviews in Carrier et al. (1992) and Tolman (1991)] that until very recently no mononuclear Cu compound had been isolated that contained a terminal NO ligand. The first such compound, {HB[3-(t-butyl)pyrazolyl]$_3$}Cu^{I}NO, has now been reported by Carrier et al. (1992). In the compound, Cu is coordinated to three pyrazole N and the N of NO, and the Cu–NO bond is short (1.76Å). This compound dissociates NO reversibly and can be thought of as an NO carrier rather than as a nitrosyl donor species. This family of pyrazolyl hydroborate Cu nitrosyls has been expanded to include 3-R, 5-R' disubstituted pyrazolyl groups (R = t-butyl; R' = H or phenyl), Fig. 7 (Ruggiero et al., 1993). The three pyrazolyl groups have a three-fold axis of symmetry through H–B and Cu–NO. The system comprises an 11-electron CuNO system (10 metal d and one NO π^*) and exhibits charge transfer bands [Cu(I)NO → Cu(II)NO$^-$] at 478 and 494 nm with spin pairing on NO$^-$. Although mononuclear Cu(II)NO ⇌ Cu(I)NO$^+$ complexes, which may have relevance to intermediates in nitrite reductases, have not yet been prepared, the existence of mononuclear Cu(I)NO complexes raises the possibility that N_2O production by Cu-type nitrite reductases in presence of nitrite and NO may involve Cu(I)NO and its dissociation into Cu(II) and NO$^-$. The NO$^-$ could then dimerize and dehydrate to form N_2O. In a related study, Tolman

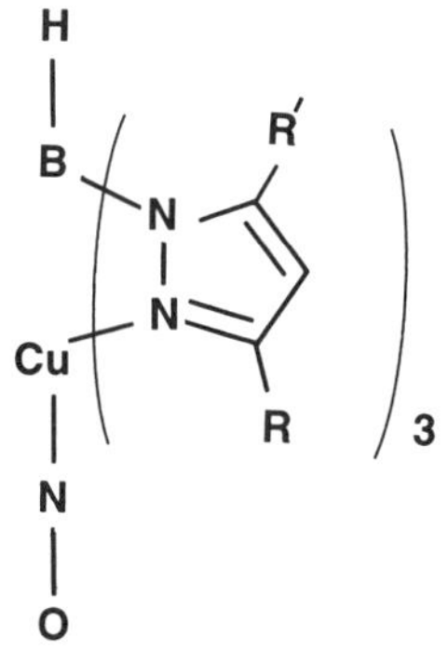

FIGURE 7

Structure of {HB[3,5-(disubstituted) pyrazolyl]$_3$}Cu^{I}NO, rare examples of mononuclear copper-nitrosyl complexes. After Ruggiero et al. (1993).

(1991) reported the structure of a Cu-nitrite compound, {HB[3-(t-butyl)pyrazolyl]$_3$}Cu^{II}NO$_2^-$, which would appear to be a reasonable model for the type 2 site of the nitrite reductase of *A. cycloclastes*. This compound contains a facially coordinating set of three pyrazoles that mimic the pyramidal array of histidyl imidazole groups coordinated to the type 2 Cu atom in the enzyme. In addition, the nitrite is nearly symmetrically coordinated to Cu in the model compound via its two O atoms, just as may be the case with the enzyme.

VI. GENOME FOR DENITRIFICATION

Studies by Zumft and co-workers have provided a detailed map for the organization of the denitrification genome in *P. stutzeri*. Many of the genes required for the biosynthesis of the cytochrome cd_1 nitrite reductase, nitric oxide reductase, and nitrous oxide reductase (*nir*, *nor*, and *nos* genes, respectively) are linked within a gene cluster extended over roughly 30 kb (Braun and Zumft, 1992; Jüngst et al., 1991a,b; Zumft, 1993). The *nir* and *nor* genes within this cluster are contiguous, with the *nor* genes lying downstream from the *nir* group. Although the linear sequence is *nir*Q, S, T, B, M, C, D followed by *nor*C, B, the *nir*Q gene is transcribed in the opposite direction from the other genes (right to left as written above). The functions encoded by these genes are as follows: *nir*Q, regulation of cytochrome cd_1 and nitric oxide reductase; *nir*S, structural gene for cytochrome cd_1; *nir*T, structural gene for a tetraheme cytochrome *c* which is presumed to have some electron transport function; *nir*B, structural gene for cytochrome c_{552} which exhibits peroxidase activity; *nir*M, structural gene for cytochrome c_{551}, the immediate electron donor to cytochrome cd_1; *nir*C, assembly or synthesis of cytochrome cd_1; *nir*D, conversion of protoporphyrin IX to the porphinedione of heme d_1; and, *nor*C and B, structural genes for cytochrome *c* and *b* components, respectively, of nitric oxide reductase. In *P. aeruginosa*, the sequence elucidated so far is *nir*Q followed by *nir*S, M, and C in that order (Zumft, 1993).

Deletion of the regulatory *nir*Q gene results simultaneously in a loss of the ability to reduce both nitrite and NO in *P. stutzeri* (Braun and Zumft, 1992). Nitrite reductase is synthesized and is active *in vitro* but nitric oxide reductase is synthesized in an inactive form. Although the exact function of the *nir*Q gene product is unknown, the gene encodes a protein homologous with the NtrC family of transcriptional activators (Zumft, 1993).

The transcription of *nir* and *nor* genes in *P. stutzeri* (Cuypers and Zumft, 1992, 1993) and *P. aeruginosa* (Zimmermann et al., 1991) would appear to be controlled in part by homologs of the Fnr protein of *E. coli* (Guest, 1992) which is a redox sensitive, transcriptional activator. Fnr boxes (binding sequences) are located upstream from *nor*C, *nir*M and *nir*S. It is likely therefore that Fnr-like proteins participate in a regulatory mechanism which derepresses the synthesis of nitrite and nitric oxide reductases under microaerobic conditions (Körner, 1993). Whether the concomitant and synergistic derepression caused by nitrite or nitrate, that is often observed with denitrifiers (e.g., Körner, 1993), works through Fnr proteins, remains to be determined.

VII. NITRIC OXIDE-REDUCING NITRITE REDUCTASES

Although the two types of dissimilatory nitrite reductases of denitrifying bacteria seem to produce NO as the overwhelmingly major product, there is now one example of a different type of nitrite reductase which can both produce NO from nitrite and also reduce NO to N_2O. This combined nitrite reductase–nitric oxide reductase is the hexaheme cytochrome c, as exemplified by the enzyme from *Wolinella succinogenes* (Costa et al., 1990). Hexaheme cytochrome c normally functions in *W. succinogenes*, *E. coli*, and other bacteria as an ammonia producing nitrite reductase. *Wolinella succinogenes* is an anaerobic bacterium which can grow using energy derived from the dissimilatory reduction of nitrate to NH_3, with nitrite as intermediate, and also from the reduction of N_2O to N_2 (Payne et al., 1982; Wolin et al., 1961; Yoshinari, 1980). Thus it has a nitrate reductase and nitrous oxide reductase in common with denitrifying bacteria, but lacks the cytochrome cd_1 or Cu-type nitrite reductase and a membrane-bound nitric oxide reductase. It was remarkable therefore that the bacterium could reduce NO to N_2O and N_2, albeit at rather modest rates (Payne et al., 1982). What enzyme catalyzed this reaction and what need has the organism for nitrous oxide reductase when the dissimilatory nitrite reductase seemed to be NH_3 producing? The answer to both questions followed from the observation (Costa et al., 1990) that the hexaheme c enzyme can produce either NH_3 or NO (six- and one-electron reductions, respectively) depending on the reducing system used and in particular on its redox potential. In general, low potential (strongly reducing) systems favored NH_3 production and higher potential ones favored NO production. The purified enzyme also has the ability to reduce NO to N_2O.

VIII. DENITRIFICATION BY CHEMOLITHOTROPHIC AMMONIA OXIDIZERS

NH_3 oxidizing bacteria (nitrifiers) gain energy for growth from the aerobic oxidation of NH_3 to nitrite (Kuenin and Robertson, 1987; Wallace and Nicholas, 1969). The prototype organism is *Nitrosomonas europaea*. These organisms cannot grow anaerobically by fermentation or by other means, such as heterotrophic denitrification. Nevertheless, it was reported (Blackmer et al., 1980; Goreau et al., 1980; Hynes and Knowles, 1984) that *Nitrosomonas* could produce N_2O, the production of which was stimulated by low partial pressures of O_2. Opinion was divided as to the significance of this observation, because it was possible that N_2O was formed by the reduction of a normal intermediate in the nitrification pathway from NH_3 to nitrite via NH_2OH or, alternatively, by a denitrification pathway (nitrite → N_2O) separate from the nitrification pathway. A major question for either route was what reducing agent would be used to reduce the precursor to N_2O. Because the utilization of carbon compounds was unlikely, the best candidate was NH_2OH, the immediate product of the oxidation of NH_3. NH_3 oxidation is catalyzed by ammonia monooxygenase (McTavish et al., 1993) and so has an absolute requirement for O_2. In the reaction, one atom of O from O_2 is inserted into NH_3 to form NH_2OH and the other is reduced to water (Hollocher et al., 1981). NH_2OH is itself the normal source of reducing equivalents for the reduction of this O atom and might also serve as the reducing agent in the reactions producing N_2O. It was possible then that NH_2OH would need to partition, depending on O_2 concentration, between electron donation to O_2, via the monooxygenase and the oxidase routes of nitrification, and to nitrite or some other intermediate compound to form N_2O. Because the redox states of N are 1^+ and 1^- in N_2O and NH_2OH, respectively, it is likely that electron donation from NH_2OH would need to be to some N compound with a redox state of greater than or equal to 2^+ in order to produce N_2O.

The weight of evidence strongly supports the existence of a separate denitrifying pathway in *Nitrosomonas* and presumably in related nitrifiers. The "soluble cytochrome oxidase" of *Nitrosomonas europaea* proved on purification (Miller and Nicholas, 1985; Miller and Wood, 1983) to be a Cu-type dissimilatory nitrite reductase similar in many ways, except perhaps in its ability to serve as an oxidase, to the Cu-type enzyme of denitrifying bacteria. One strain of *Nitrosomonas* was found that could reduce N_2O to N_2 (Poth, 1986), and Remde and Conrad (1990) reported that both NO and N_2O were produced from nitrite by cells using hydrazine (N_2H_4) as the electron donor. N_2H_4 is an analog of NH_2OH and can replace that intermediate in the part of the ammonia monooxygenase reaction involving the reduction of the second atom of O to water (Hollocher et al., 1981). The production of NO and N_2O were both found to increase (Anderson et al., 1993; Remde and Conrad, 1990) as the O_2 concentration decreased or cell density increased; furthermore chlorite (ClO_2^-) could block the production of NO and N_2O from nitrite but not the production of nitrite from NH_2OH. Evidently some NO is also released during NH_3 metabolism by *Nitrosomonas* (Anderson and Levine, 1986; Anderson et al., 1993).

Poth and Focht (1985) reported on a series of experiments with *N. europaea* in which [^{15}N]nitrite or [^{15}N]nitrate was mixed with NH_3 and the rates of formation and isotope compositions of nitrite and N_2O were followed. The systems were either well oxygenated (shaken) or poorly oxygenated (static). When shaken, no N_2O was produced by the cultures. With static cultures, N_2O but not N_2 was produced in the presence of nitrite but not nitrate, and nitrite was the preferred source of N for N_2O. The observed $^{14,15}N_2O/^{14}N_2O$ ratio was 0.25 rather than the value of 0.04 expected on the basis of the initial ratio of $^{15}NO_2^-$ to NH_3 and the total amount of N_2O formed. The results of analogous static experiments with harvested, washed cells were alleged to support the above results, but are difficult to rationalize in detail. The production of $^{15}N_2O$ and $^{14,15}N_2O$ terminated sequentially in that order well before the end of the experiment, whereas production of $^{14}N_2O$ proceeded unabated. This result implies that [^{15}N]nitrite was the preferred precursor for N_2O production early in the experiment when the system was well oxygenated but not later when the O_2 was depleted. On the other hand, only 4% of the [^{15}N]nitrite was converted to N_2O and the apparent size of the nitrite pool only doubled. That is, the total pool of $^{15}NO_2^-$ was not massively diluted with $^{14}NO_2^-$. Where then did the ^{15}N go and why was it apparently exhausted during the incubation? A major problem with static systems is that they can become stratified with respect to O_2, N_2O, and nitrite concentrations, and it is difficult to develop predictive kinetic models for such systems. It is likely in general that N_2O production occurred in strata of intermediate O_2 concentration that became depleted in ^{15}N and were sustained by diffusion of NH_3 from below. Notwithstanding these uncertainties, the isotope work lends support to the idea that low O_2 concentrations promote the denitrification of nitrite in *Nitrosomonas*. These experiments might well be repeated using controlled O_2/N_2 ratios so that known O_2 concentrations can be maintained homogeneously throughout the cell suspension.

It should be emphasized that denitrification by *Nitrosomonas* cannot normally occur under anaerobic conditions because of the critical role of O_2 in generating NH_2OH from NH_3 and the critical role of NH_2OH in reducing nitrite. This style of denitrification is sometimes referred to as aerobic denitrification. A number of probably analogous examples of aerobic denitrification can be cited (Robertson and Kuenen, 1984). One well documented example is that of *Thiosphaera pantotropha* (Bell and Ferguson, 1991; Robertson and Kuenen, 1990). The nitrite reductase of *T. pantotropha* is of the NO-producing cytochrome cd_1 class (Moir et al., 1993) in keeping with the relatively recent taxonomic transfer of this organism to *Pa. dentrificans* (Ludwig et al., 1993).

IX. DENITRIFICATION BY EUKARYOTIC MICROORGANISMS

It was long believed that bacteria were unique in their ability to denitrify. However, Shoun and Tanimoto (1991) and Shoun et al., (1989) demonstrated that the fungus, *Fusarium oxysporum*, could be induced to synthesize an enzyme system capable of the anaerobic reduction of nitrite to N_2O. Induction occurred under conditions of low oxygen concentrations in the presence of nitrate or nitrite. One and perhaps the only component of this nitrite reductase system is a unique, soluble cytochrome P-450 (P-450dNIR), which is more similar in its cDNA-inferred amino acid sequence to soluble, bacterial P-450 enzymes (especially to those of *Streptomyces*) than to the membrane-bound P-450s of eukaryotes (Kizawa et al., 1991). P-450dNIR has also been shown to be a highly active (turnover number of 500 sec^{-1}), NADH-dependent nitric oxide reductase (N$_2$O producing) (Nakahara et al., 1993). Although the denitrification-induced state of *F. oxysporum* can also reduce nitrate to N$_2$O, the enzyme responsible for nitrate reduction has not yet been identified. Overall, P-450dNIR would appear to reduce nitrite to N$_2$O in two steps, just as is the case with the bacterial denitrifying enzymes and with the hexaheme c nitrite reductase of *W. succinogenes* and other bacteria.

One curious feature of denitrification by *F. oxysporum* and other representatives of the *Fusarium* genus is "codenitrification" (Shoun et al., 1992; Tanimoto et al., 1992), which is the ability, as seen in ^{15}N-isotope experiments, for some product N$_2$O (or N$_2$ in some cases) to be isotopically mixed, with one N atom being derived from nitrite (or nitrate) and the other from another source. This other source can be unknown endogenous nitrogen compounds or added compounds, such as azide, salicylhydroxamic acid, and perhaps ammonia. The appearance of isotopically mixed species strongly suggests the occurrence of nonspecific nitrosyl transfer reactions from an enzyme-activated species of nitrite to N-nucleophiles. Alternatively (see Section X), the nitrosyl donor may be N$_2$O$_3$ or N$_2$O$_4$, arising from the spontaneous oxidation of intermediate NO by ambient O$_2$.

The fact that the cytochrome P-450 was induced even in the presence of NH$_3$, which is the end product of assimilatory N-oxide reductions, suggested that it might funciton in dissimilatory N-oxide reductions. Anaerobic growth experiments with induced cells showed that reduction of nitrate to nitrite was energy yielding in *F. oxysporum* but reduction of nitrite to N$_2$O was probably not (Shoun and Tanimoto, 1991).

The usual function of cytochromes P-450 is that of monooxygenation, the insertion of one O atom of O$_2$ into a substrate with concomitant two-electron reduction of the second O atom to water, and examples of P-450s acting as reductases are rare (Guengerich and MacDonald, 1984, 1990). Cytochrome P-450 enzymes can, under certain circumstances, reduce azo dyes and perhaps epoxides and reductively dehalogenate CCl$_4$. It is interesting that the inducible nitric oxide synthase of macrophages is a P-450-containing enzyme (White and Marletta, 1992) and P-450 monooxygenases of liver can convert the N^G-hydroxy-L-arginine intermediate in NO synthesis to citrulline and NO (Boucher et al., 1992).

X. NITRIC OXIDE PRODUCTION FROM NITRITE BY ENTERIC AND RELATED BACTERIA

It is known that certain bacteria, including *E. coli*, could use nitrite (but not nitrate) to nitrosate amines, such as morpholine or 2,3-diaminonaphthalene to yield N-nitrosomorpholine or 1[H]-naphthol[2,3-D] triazole, respectively (Calmels et al., 1987, 1988; Ralt et al., 1988). N-Nitrosomorpholine is a secondary nitrosamine and belongs to a family of compounds known to be potent carcinogens. The assays for the nitrosation reaction were carried out under aerobic conditions and high yields of nitrosated products could be obtained. Ralt et al. (1988) and Calmels et al. (1987, 1988) showed that the reaction was linked genetically to nitrate reductase genes and biochemically to the dissimilatory nitrate reductase, which is the enzyme responsible for energy-linked nitrate respiration. This connection was puzzling, because the normal substrate and product of this enzyme are nitrate and nitrite, respectively. Although the enzyme is not perfectly specific for nitrate and can, for example, reduce perchlorate, chlorate, chlorite, and hypochlorite (Hackenthal et al., 1964), nitrite was generally not thought to be a substrate. In pursuing this problem, Ji and Hollocher (1988a) demonstrated with E. coli that the nitrosation reaction could be separated into two steps. The first step was the slow enzymatic reduction of nitrite to NO which was stable under anaerobic conditions. The second step was the aerobic oxidation of NO chemically to yield short-lived but highly reactive nitrosating agents which were presumed to be N_2O_3 and/or N_2O_4. These workers also showed that the first step was strongly inhibited by nitrate and azide but only weakly by cyanide. These were characteristics expected of the respiratory nitrate reductase. Furthermore, the enzyme responsible for NO production was induced by nitrate, particularly when cell growth occurred under anaerobic conditions. Similar NO forming reactions were shown to occur in seven other genera of enteric bacteria (Ji and Hollocher, 1988b), *Serratia marcescens* (Anderson and Levine, 1986) and *Bacillus cereus* (Kalkowski and Conrad, 1991). It was later shown that purified nitrate reductase exhibited a relatively weak, NO-producing nitrite reductase activity that was also strongly inhibited by nitrate and azide (Ji and Hollocher, 1989). In contrast, purified Cu- and heme-type nitrite reductases (NO-producing) from *A. cycloclastes* and *Pa. denitrificans*, respectively, exhibited strong nitrite reductase activities which were not inhibited by nitrate or low concentrations of azide. The identification of nitrate reductase as the nitrosating enzyme of enteric bacteria was entirely consistent with the finding that the induction of the activity was inhibited by tungstate (Calmels et al., 1987). Tungstate is a homolog of molybdate, which is required for the synthesis of the Mo-cofactor at the active site of nitrate reductase.

Although the two-electron redox mechanism of the respiratory nitrate reductase is not known, there are chemical arguments and model systems to suggest (Berg and Holm, 1985) that it, and other mononuclear Mo-containing enzymes as well, may catalyze O atom transfer reactions to and from substrate via a $[Mo=O]^{4+}$ (Mo–oxo) species according to Eq. (7). The Mo atom in Mo=O can be thought of as having redox state IV or VI depending on whether the bonding electrons are viewed as metal- or O-centered, respectively. In any case, the system can be thought of an Mo-stabilized O atom which is somewhat analogous to the ferric–oxo systems of monooxygenases.

$$Mo^{4+} + NO_3^- \rightleftharpoons [Mo=O]^{4+} + NO_2^- \qquad (7)$$

Extended X-ray absorption fine structure (EXAFS) data suggest that nitrate reductase and analogous other Mo enzymes cycle between redox states Mo(IV) and Mo(VI) and that the Mo(VI) state contains one more short (double or oxo) bond than does the Mo(IV) state (Bordase et al., 1980; Cramer et al., 1981, 1984, 1985). Nitrite dehydrogenase, the enzyme of *Nitrobacter* that can catalyze a redox equilibrium between nitrite and nitrate and works physiologically by catalyzing the oxidation of nitrite (i.e., the exact reverse of nitrate reductase), is also a Mo-containing enzyme. Its mechanism was shown by DiSpirito and Hooper (1986) and Friedman et al. (1986) to be that of O atom transfer by use of a double labeling (^{15}N,^{18}O) method. When an anaerobic mixture initially containing [^{18}O]nitrate and [^{15}N]nitrite was incubated with *Nitrobacter*, [^{15}N,^{18}O]nitrate was formed to indicate ^{18}O transfer between the two substrates (Fig. 8). In addition, [^{15}N,^{18}O]nitrite was also formed at similar rates. The formation of this product can be explained if [^{15}N,^{18}O]nitrate, once formed at the active site, could rotate without dissociation and present an unlabeled O atom for return transfer to Mo(IV). Following O atom transfer (reverse of steps -2 and -3 of Fig. 8), the species that would dissociate would then be [^{15}N,^{18}O]nitrite.

For the one-electron reduction of nitrite to NO by nitrate reductase, an O atom transfer mechanism is unlikely, inasmuch as O atom transfer is inherently a two-electron mechanism.

XI. NITRIC OXIDE PRODUCTION BY HETEROTROPHIC NITRIFIERS

Heterotrophic nitrifiers can oxidize nitrogenous compounds, including NH_3, NH_2OH, amines, hydroxamic acids, and oximes, and utilize organic compounds as carbon and energy sources (Alexander, 1977; Alexander et al., 1960; Focht and Verstraete, 1977; Walker, 1978). Chief products are often nitrite and nitrate. They are thus distinguished from autotrophic nitrifiers which generate energy for growth solely from the oxidation of inorganic nitrogen compounds and reduce CO_2 for synthesis of cellular constituents. Heterotrophic nitrifiers generally oxidize their nitrogen-containing substrates at low rates (with some exceptions) as compared with the autotrophic nitrifiers. The heterotrophic nitrifiers include a variety of bacteria, including *Pseudomonas* (Amarger and Alexander, 1968; Castignetti and Hollocher, 1984), *Alcaligenes* (Anderson et al., 1993; Castignetti and Gunner, 1981), and *Arthrobacter* (Vestraete and Alexander, 1972), and fungi (Alexander, 1977). Now that NO is known to be a hormone in animals (Chap-

FIGURE 8

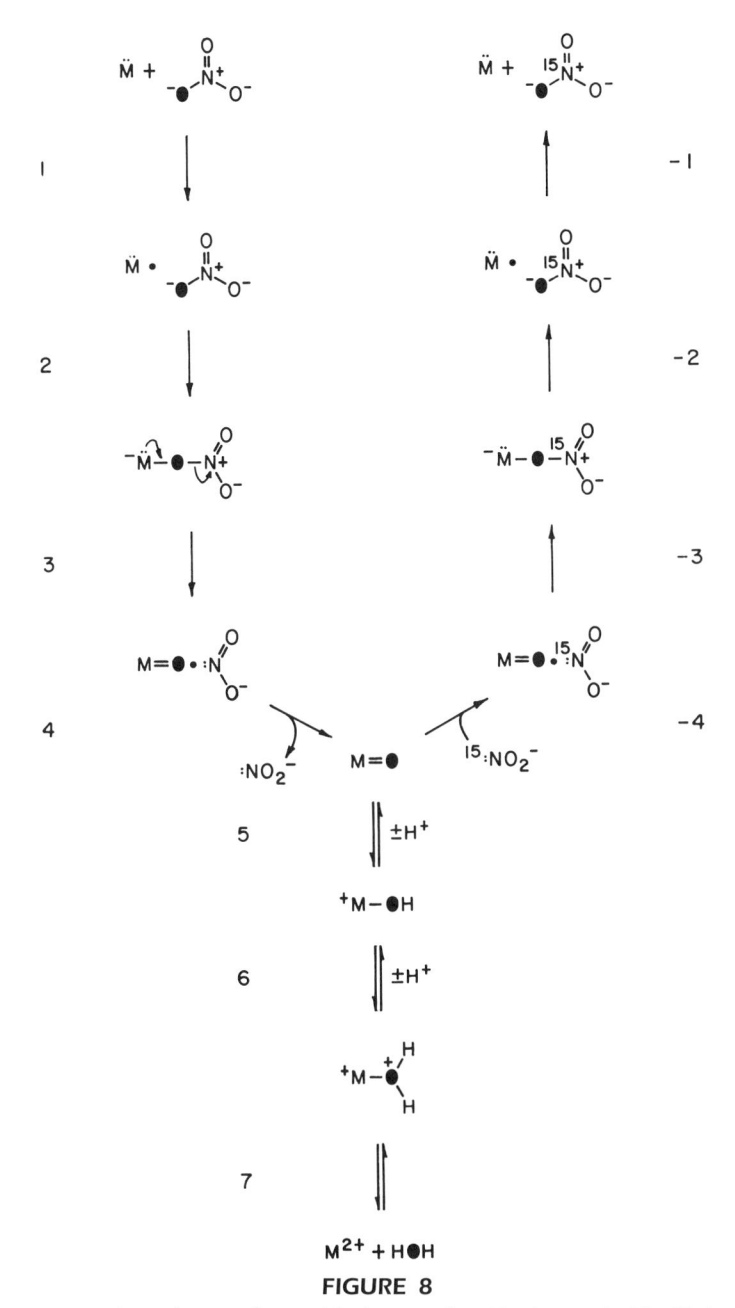

Oxygen atom transfer mechanism of nitrite dehydrogenase from *Nitrobacter agilis*. The filled-in oxygen atom represents [18]O. The lower pathway provides for O exchange with water, a process that competes with O atom transfer from nitrate to nitrite. M represents a molybdenum atom. From Friedman *et al.* (1986).

ters 3–7), the entire animal kingdom may need to be reclassified among heterotrophic nitrifiers. It is likely that a wide array of pathways contribute to the oxidation of nitrogen compounds. Whether NO may be a product or intermediate in the pathways of heterotrophic, nitrifying microorganisms seems to have received little attention as yet. Papen et al. (1989) reported that Al. faecalis could produce nitrite, nitrate, N_2O, and NO when growing aerobically in peptone–meat extract or a defined medium containing ammonium citrate as the nitrogen source. Tests were made for the presence of dissimilatory nitrate and nitrite reductases, but none was detected. It is possible therefore that production of NO and N_2O by Al. faecalis is not by way of an associated denitrification pathway. By use of oxidized cytochrome c as a trap for NO, Bell and Ferguson (1991) showed that both NO and N_2O are formed by T. pantotropha during reduction of nitrite and that both nitric oxide and nitrous oxide reductases are active. This unusual heterotrophic nitrifier thus has the additional ability to carry out aerobic denitrification.

As the result of a study in which NO and N_2O production was measured among several organisms (e.g., Al. faecalis) as a function of the partial pressure of O_2, Anderson et al. (1993) concluded that heterotrophic nitrification may well be a significant source of N_2O in microaerobic soils.

XII. TOXICITY OF NITRIC OXIDE TOWARD BACTERIA, A TOPIC JOINING BACTERIA WITH THE NITRIC OXIDE SYNTHASE OF ANIMAL CELLS

NO is well known to have antimicrobial activity and can be a potent bactericide at gas phase levels of 1 part per million (ppm) (Mancinelli and McKay, 1983). The mode of action of NO against bacteria is probably complex and varied. It has been shown to inactivate or complex with heme, iron–sulfur, and Cu proteins (Henry et al., 1991; Lancaster and Hibbs, 1990; Zumft, 1993), quench the tyrosine radical of the Fe-type ribonucleotide reductase (Lepoivre et al., 1991), nitrosylate (probably via a higher N-oxide) cysteinyl thiols of proteins (Stamler et al., 1992), deaminate the N-terminal and other amino groups of proteins (Moriguchi et al., 1992), and inactivate some enzymes lacking redox metals, for example, glyceraldehyde-3-phosphate dehydrogenase (Dimmeler et al., 1992). As might be expected by its ability to promote deamination reactions, NO has been shown to be mutagenic for Salmonella typhimurium and to promote C $\rightarrow$ T mutations (Wink et al., 1993). A deletion Nor⁻ mutant of P. stutzeri generates levels of NO from nitrite that are sufficiently great enough to make the mutation conditionally lethal (Braun and Zumft, 1991). Thus, even denitrifiers have the potential to be damaged or killed by the NO they generate as an intermediate in the pathway, if nitric oxide reductase is specifically inhibited or inactivated—a useful thought in the design of an inhibitor of denitrification for agricultural application.

It is likely that activated macrophages exert much of their bactericidal effect through the NO generated by the cytokine-inducible nitric oxide synthase. Certainly this effect is absent if macrophages lack arginine or if the synthase is inhibited by arginine analogs (Marlette et al., 1988, 1990). On the other hand, it has been difficult to determine whether the effect in aerobic environments is due to NO per se or, in part at least, to higher oxides of N arising from the spontaneous reaction of NO with O_2^- and H_2O_2, which are produced along with the NO, and ambient O_2. The coexistence with NO of reactive higher oxides of N, such as peroxynitrite ($ONOO^-$) (Radi et al., 1991) and the nitrosylating agents, N_2O_3 and N_2O_4, is plausible. Nevertheless, Mülsch et al. (1992) elegantly demonstrated that NO is the principal N-oxide to be both produced by macrophages in vitro and to find its way into the target organism (yeast). In the experiment, intracellular NO was trapped by reaction with Fe(II) diethyldithiocarbamate, which had been introduced into permeabilized yeast, and the resulting intracellular iron–nitrosyl complex was quantified by EPR. The apparent stoichiometry between L-arginine consumed and iron–nitrosyl formed was about 1:1. This result obviates the possibility that N-oxides other than NO are the primary reactive compounds within the yeast.

ACKNOWLEDGMENTS

The preparation of the manuscript for this chapter was supported in part by Grant DCB 88-16273 from the National Science Foundation.

REFERENCES

Aerssens, E., Tiedje, J. M., and Averill, B. A. (1986). Isotope labeling studies on the mechanism of N–N bond formation in denitrification. *J. Biol. Chem.* **261**, 9652–9656.

Akey, C. E., Moffat, K., Wharton, D. C., and Edelstein, S. J. (1980). Characterization of crystals of a cytochrome oxidase (nitrite reductase) from *Pseudomonas aeruginosa* by X-ray diffraction and electron microscopy. *J. Mol. Biol.* **136**, 19–43.

Alberty, R. A., and Hammes, G. G. (1958). Application of the theory of diffusion-controlled reactions to enzyme kinetics. *J. Phys. Chem.* **62**, 154–159.

Alefounder, P. R., and Ferguson, S. J. (1980). The location of dissimilatory nitrite reductase and the control of dissimilatory nitrate reductase by oxygen in *Paracoccus denitrificans*. *Biochem. J.* **192**, 231–240.

Alexander, M. (1977). "Introduction to Soil Microbiology." Wiley, New York.

Alexander, M., Marshall, K. C., and Hirsch, P. (1960). Autotrophy and heterotrophy in nitrification. *Proc. 7th Int. Conference of Soil Sci.*, Madison, WI **2**, 586–591.

Amarger, N., and Alexander, M. (1968). Nitrite formation from hydroxylamine and oximes by *Pseudomonas aeruginosa. J. Bacteriol.* **95**, 1651–1657.

Anderson, I. C., and Levine, J. S. (1986). Relative rates of nitric oxide and nitrous oxide production by nitrifiers, denitrifiers, and nitrate respirers. *Appl. Environ. Microbiol.* **51**, 938–945.

Anderson, I. C., Poth, M., Homstead, J., and Burdige, D. (1993). A comparison of NO and N_2O production by the autotrophic nitrifier *Nitrosomonas europaea* and the heterotrophic nitrifier *Alcaligenes faecalis*. *Appl. Environ Microbiol.* **59**, 3525–3533.

Averill, B. A., and Tiedje, J. M. (1982). The chemical mechanism of microbial denitrification. *FEBS Lett.* **138**, 8–12.

Averill, B. A., and Tiedje, J. M. (1990). Abstract INOR 103, Symposium on Inorganic and Biological Chemistry of Nitrogen, 200th ACS National Meeting American Chemical Society, Washington, D. C., Aug. 26–31.

Baalsrud, K., and Baalsrud, K. S. (1954). Studies on *Thiobacillus denitrificans*. *Arch. Microbiol.* **20**, 34–62.

Balderston, W. L., Sherr, B., and Payne, W. J. (1976). Blockage by acetylene of nitrous oxide reduction in *Pseudomonas perfectomarinus*. *Appl. Environ. Microbiol.* **31**, 504–508.

Barber, D., Parr, S. R., and Greenwood, C. (1976). Some spectral and steady-state kinetic properties of *Pseudomonas* cytochrome oxidase. *Biochem. J.* **157**, 431–438.

Bazylinski, D. A., and Hollocher, T. C. (1985a). Evidence from the reaction between trioxodinitrate(II) and ^{15}NO that trioxodinitrate decomposes into nitrosyl hydride and nitrite in neutral aqueous solution. *Inorg. Chem.* **24**, 4285–4288.

Bazylinski, D. A., and Hollocher, T. C. (1985b). Metmyoglobin and methemoglobin as efficient traps for nitrosyl hydride (nitroxyl) in neutral aqueous solution. *J. Am. Chem. Soc.* **107**, 7982–7986.

Bell, L. C., and Ferguson, S. J. (1991). Nitric and nitrous oxide reductases are active under aerobic conditions in cells of *Thiosphaera pantotropha*. *Biochem. J.* **273**, 423–427.

Berg, J. M., and Holm, R. H. (1985). A model for the active sites of oxo-transfer molybdoenzymes: Synthesis, structure, and properties. *J. Am. Chem. Soc.* **107**, 917–925.

Berger, H., and Wharton, D. C. (1980). Small-angle X-ray scattering studies of oxidized and reduced cytochrome oxidase from *Pseudomonas aeruginosa*. *Biochim. Biophys. Acta* **622**, 355–359.

Betlach, M. R., and Tiedje, J. M. (1981). Kinetic explanation for accumulation of nitrite, nitric oxide, and nitrous oxide during bacterial denitrification. *Appl. Environ. Microbiol.* **42**, 1074–1084.

Blacklow, S. C., Raines, R. T., Lim, W. A., Zamore, P. D., and Knowles, J. R. (1988). Triosephosphate isomerase catalysis is diffusion controlled. *Biochemistry* **27**, 1158–1167.

Blackmer, A. M., Bremner, J. M., and Schmidt, E. L. (1980). Production of nitrous oxide by ammonia-oxidizing chemoautotrophic microorganisms of soil. *Appl. Environ. Microbiol.* **40**, 1060–1066.

Blackmore, R., Roberton, A. M., and Brittain, T. (1986). The purification and some equilibrium properties of the nitrite reductase of the bacterium *Wolinella succinogenes*. *Biochem. J.* **233**, 547–552.

Bonner, F. T., and Akhtar, M. J. (1981). Formation of nitrosyltricyanonickelate (NiNO(CN)$_3^{2-}$) in a direct NO$^-$ displacement reaction. *Inorg. Chem.* **20,** 3155–3160.
Bonner, F. T., and Pearsall, K. A. (1982). Aqueous nitrosyliron(II) chemistry. 1. Reduction of nitrite and nitric oxide by iron(II) and (trioxodinitrato)iron(II) in acetate buffer. Intermediacy of nitrosyl hydride. *Inorg. Chem.* **21,** 1973–1978.
Bonner, F. T., Dzelzkalns, L. S., and Bonucci, J. A. (1978). Properties of nitroxyl as intermediate in the nitric oxide–hydroxylamine reaction and in trioxodinitrate decomposition. *Inorg. Chem.* **17,** 2487–2494.
Bonner, F. T., Kada, J., and Phelan, K. G. (1983). Symmetry of the intermediate in the hydroxylamine–nitrous acid reaction. *Inorg. Chem.* **22,** 1389–1391.
Boogerd, F. C., Van Verseveld, H. W., and Stouthamer, A. H. (1981). Respiration-driven proton translocation with nitrate and nitrous oxide in *Paracoccus denitrificans. Biochim. Biophys. Acta* **638,** 181–191.
Bordas, J., Bray, R. C., Garner, C. D., Gutteridge, S., and Hasnain, S. S. (1980). X-Ray absorption spectroscopy of xanthine oxidase. *Biochem. J.* **191,** 499–508.
Bothner-By, A. A., and Friedman, L. (1952). The reaction of nitrous acid with hydroxylamine. *J. Chem. Phys.* **20,** 459–462.
Bottomley, F. (1978). Electrophilic behavior of coordinated nitric oxide. *Acc. Chem. Res.* **11,** 158–162.
Boucher, J.-L., Genet, A., Vadon, S., Delaforge, M., Henry, Y., and Mansuy, D. (1992). Cytochrome P450 catalyzes the oxidation of N^G-hydroxy-L-arginine by NADPH and O$_2$ to nitric oxide and citrulline. *Biochem. Biophys. Res. Commun.* **187,** 880–886.
Brannan, D. K., and Caldwell, D. E. (1980). *Thermothrix thiopara*: Growth and metabolism of a newly isolated thermophile capable of oxidizing sulfur and sulfur compounds. *Appl. Environ. Microbiol.* **40,** 211–216.
Braun, C., and Zumft, W. G. (1991). Marker exchange of the structural genes for nitric oxide reductase blocks the denitrification pathway of *Pseudomonas stutzeri* at nitric oxide. *J. Biol. Chem.* **266,** 22785–22788.
Braun, C., and Zumft, W. G. (1992). The structural genes of the nitric oxide reductase complex from *Pseudomonas stutzeri* are part of a 30-kilobase gene cluster for denitrification. *J. Bacteriol.* **174,** 2394–2397.
Brittain, T., Blackmore, R., Greenwood, C., and Thomson, A. J. (1992). Bacterial nitrite-reducing enzymes. *Eur. J. Biochem.* **209,** 793–802.
Brunori, M., Parr, S. R., Greenwood, C., and Wilson, M. T. (1975). A temperature-jump study of the reaction between azurin and cytochrome *c* oxidase from *Pseudomonas aeruginosa. Biochem. J.* **151,** 185–188.
Bryan, B. A., Shearer, G., Skeaters, J. L., and Kohl, D. H. (1983). Variable expression of the nitrogen isotope effect associated with denitrification of nitrite. *J. Biol. Chem.* **258,** 8613–8617.
Calmels, S., Ohshima, H., Rosenkranz, H., McCoy, E., and Bartsch, H. (1987). Biochemical studies on the catalysis of nitrosation by bacteria. *Carcinogenesis (London)* **8,** 1085–1088.
Calmels, S., Ohshima, H., and Bartsch, H. (1988). Nitrosamine formation by denitrifying and non-denitryfying bacteria: Implications of nitrite reductase and nitrate reductase in nitrosation catalysis. *J. Gen. Microbiol.* **134,** 221–226.

Carlson, C. A., Ferguson, L. P., and Ingraham, J. L. (1982). Properties of dissimilatory nitrate reductase purified from the denitrifier *Pseudomonas aeruginosa*. *J. Bacteriol.* **151**, 162–171.

Carr, G. J., and Ferguson, S. J. (1990). The nitric oxide reductase of *Paracoccus denitrificans*. *Biochem. J.* **269**, 423–429.

Carr, G. J., Page, M. D., and Ferguson, S. J. (1989). The energy-conserving nitric-oxide-reductase system of *Paracoccus denitrificans*. *Eur. J. Biochem.* **179**, 683–692.

Carrier, S. M., Ruggiero, C. E., Tolman, W. B., and Jameson, G. B. (1992). Synthesis and characterization of a mononuclear copper nitrosyl complex. *J. Am. Chem. Soc.* **114**, 4407–4408.

Cassoly, R., and Gibson, Q. H. (1975). Conformation, co-operativity and ligand binding in human hemoglobin. *J. Mol. Biol.* **91**, 301–313.

Castignetti, D., and Gunner, H. B. (1981). Nitrite and nitrate synthesis from pyruvate oxime by an *Alcaligenes* sp. *Curr. Microbiol.* **5**, 379–384.

Castignetti, D., and Hollocher, T. C. (1984). Heterotrophic nitrification among denitrifiers. *Appl. Environ. Microbiol.* **47**, 620–623.

Chang, C. K., and Wu, W. (1986). The porphinedione structure of heme d_1. *J. Biol. Chem.* **261**, 8593–8596.

Chang, C. K., Timkovich, R., and Wu, W. (1986). Evidence that heme d_1 is a 1,3-porphyrindione. *Biochemistry* **25**, 8447–8453.

Clancy, R. M., Miyazaki, Y., and Cannon, P. J. (1990). Use of thionitrobenzoic acid to characterize the stability of nitric oxide in aqueous solutions and in porcine aortic endothelial cell suspensions. *Anal. Biochem.* **191**, 138–143.

Clark, F. E., and Rosswall, T. (eds.) (1981). "Terrestrial Nitrogen Cycles." Ecol. Bull., Stockholm, No. 33, Swedish Natural Science Research Council, Stockholm.

Cooper, J. N., Chilton, J. E., and Powell, R. E. (1970). Reaction of nitric oxide with alkaline hydroxylamine. *Inorg. Chem.* **9**, 2303–2304.

Costa, C., Macedo, A., Moura, I., Moura, J. J. G., LeGall, J., Berlier, Y., Liu, M.-Y., and Payne, W. J. (1990). Regulation of the hexaheme nitrite/nitric oxide reductase of *Desulfovibrio desulfuricans*, *Wolinella succinogenes*, and *Escherichia coli*. A mass spectrometric study. *FEBS Lett.* **276**, 67–70.

Coyle, C. L., Zumft, W. G., Kroneck, P. M. H., Körner, H., and Jakob, W. (1985). Nitrous oxide reductase from denitrifying *Pseudomonas perfectomarina*. Purification and properties of a novel multicopper enzyme. *Eur. J. Biochem* **153**, 459–467.

Coyne, M. S., Arunakumari, A., Averill, B. A., and Tiedje, J. M. (1989). Immunological identification and distribution of dissimilatory heme cd_1 and nonheme copper nitrite reductases in denitrifying bacteria. *Appl. Environ. Microbiol.* **55**, 2924–2931.

Coyne, M. S., Arunakumari, A., Pankratz, H. S., and Tiedje, J. M. (1990). Localization of the cytochrome cd_1 and copper nitrite reductases in denitrifying bacteria. *J. Bacteriol.* **172**, 2558–2562.

Cramer, S. P., Wahl, R., and Rajagopalan, K. V. (1981). Molybdenum sites of sulfite oxidase and xanthine dehydrogenase. A comparison by EXAFS. *J. Am. Chem. Soc.* **103**, 7721–7727.

Cramer, S. P., Solomonson, L. P., Adams, M. W. W., and Mortenson, L. E. (1984). Molybdenum sites of *Escherichia coli* and *Chlorella vulgaris* nitrate reductase: A comparison by EXAFS. *J. Am. Chem. Soc.* **106**, 1467–1471.

Cramer, S. P., Liu, C.-L., Mortenson, L. E., Spence, J. T., Liu, S.-M, Yamamoto, I., and Ljungdahl, L. G. (1985). Format dehydrogenase molybdenum and tungsten sites—observation by EXAFS of structural differences. *J. Inorg. Biochem.* **23**, 119–124.

Craske, A., and Ferguson, S. J. (1986). The respiratory nitrate reductase from *Paracoccus denitrificans*. Molecular characterization and kinetic properties. *Eur. J. Biochem.* **158**, 429–436.

Cuypers, H., and Zumft, W. G. (1992). Regulatory components of the denitrification gene cluster of *Pseudomonas stutzeri*. In "Pseudomonas: Molecular Biology and Biotechnology" (E. Galli, S. Silver, and B. Witholt, eds.), pp. 188–197. American Society of Microbiology, Washington, D. C..

Cuypers, H., and Zumft, W. G. (1993). Indentification and mutagenesis of the *fnr* homolog of *Pseudomonas stutzeri* (V114). *BioEngineering* **9**, 21.

Denariaz, G., Payne, W. J., and LeGall, J. (1991). The denitrifying nitrite reductase of *Bacillus halodenitrificans*. *Biochim. Biophys. Acta* **1056**, 225–232.

Dermastia, M., Turk, T., and Hollocher, T. C. (1991). Nitric oxide reductase. Purification from *Paracoccus denitrificans* with use of a single column and some characteristics. *J. Biol. Chem.* **266**, 10899–10905.

Dimmeler, S., Lottspeich, F., and Brüne, B. (1992). Nitric oxide causes ADP-ribosylation and inhibition of glyceraldehyde-3-phosphate dehydrogenase. *J. Biol. Chem.* **267**, 16771–16774.

DiSpirito, A. A., and Hooper, A. B. (1986). Oxygen exchange between nitrate molecules during nitrite oxidation by *Nitrobacter*. *J. Biol. Chem.* **261**, 10534–10537.

Donald, C. E., Hughes, M. N., Thompson, J. M., and Bonner, F. T. (1986). Photolysis of the N=N bond in trioxodinitrate: Reaction between triplet NO$^-$ and O_2 form peroxonitrite. *Inorg. Chem.* **25**, 2676–2677.

Doyle, M. P., Siegfried, B., and Hammond, J. J. (1976). Oxidative deamination of primary amines by copper halide nitrosyls. The formation of geminal dihalides. *J. Am. Chem. Soc.* **98**, 1627–1629.

Doyle, M. P., Pickering, R. A., DeWeert, T. M., Hoekstra, J. W., and Pater, D. (1981). Kinetics and mechanism of the oxidation of human deoxyhemoglobin by nitrites. *J. Biol. Chem.* **256**, 12393–12398.

Doyle, M. P., Pickering, R. A., daConceiçao, J. (1984). Structural effects in alkyl nitrite oxidation of human hemoglobin. *J. Biol. Chem.* **259**, 80–87.

Doyle, M. P., Mahapatro, S. N., Broene, R. D., and Guy, J. K. (1988). Oxidation and reduction of hemoproteins by trioxodinitrate(II). The role of nitrosyl hydride and nitrite. *J. Am. Chem. Soc.* **110**, 593–599.

Eisenberg, R., and Meyer, C. D. (1975). The coordination chemistry of nitric oxide. *Acc. Chem. Res.* **8**, 26–34.

Enemark, J. H., Feltham, R. D., Riker-Nappier, J., and Bizot, K. F. (1975). Stereochemical control of valence. IV. Comparison of the structures and chemical reactivities of five- and six-coordinate diarsine complexes of the cobalt nitrosyls {CoNO}^{8} group. *Inorg. Chem.* **14**, 624–632.

Fenderson, F. F., Kumar, S., Adman, E. T., Liu, M.-Y., Payne, W. J., and LeGall, J. (1991). Amino acid sequence of nitrite reductase: A copper protein from *Achromobacter cycloclastes*. *Biochemistry* **30**, 7180–7185.

Firestone, M. K., Firestone, R. B., and Tiedje, J. M. (1979). Nitric oxide as an intermediate in denitrification: Evidence from nitrogen-13 isotope exchange. *Biochem. Biophys. Res. Commun.* **91**, 10–16.

Focht, D. D., and Verstraete, W. (1977). Biochemical ecology of nitrification and denitrification. *Adv. Microbiol. Ecol.* **1**, 135–214.

Forget, P. (1971). Les nitrate-reductases bacteriennes. Solubilisation, purification, et properties de l'enzyme A de *Micrococcus denitrificans*. *Eur. J. Biochem.* **18**, 442–450.

Friedman, S. H., Massefski, W., and Hollocher, T. C. (1986). Catalysis of intermolecular oxygen atom transfer by nitrite dehydrogenase of *Nitrobacter agilis*. *J. Biol. Chem.* **261**, 10538–10543.

Frunzke, K., and Zumft, W. G. (1986). Inhibition of nitrous-oxide respiration by nitric oxide in the denitrifying bacterium *Pseudomonas perfectomarina*. *Biochim. Biophys. Acta* **852**, 119–125.

Gamble, T. N., Betlach, M. R., and Tiedje, J. M. (1977). Numerically dominant denitrifying bacteria from world soils. *Appl. Environ. Microbiol.* **33**, 926–939.

Garber, E. A. E., and Hollocher, T. C. (1981). ^{15}N tracer studies on the role of NO in denitrification. *J. Biol. Chem.* **256**, 5459–5465.

Garber, E. A. E., and Hollocher, T. C. (1982). ^{15}N, ^{18}O tracer studies on the activation of nitrite by denitrifying bacteria. *J. Biol. Chem.* **257**, 8091–8097.

Garber, E. A. E., Castignetti, D., and Hollocher, T. C. (1982). Proton translocation and proline uptake associated with reduction of nitric oxide by denitrifying *Paracoccus denitrificans*. *Biochem. Biophys. Res. Commun.* **107**, 1504–1507.

Glockner, A. B., Jüngst, A., and Zumft, W. G. (1993). Copper-containing nitrite reductase from *Pseudomonas aureofaciens* is functional in a mutationally cytochrome cd_1-free background (NirS$^-$) of *Pseudomonas stutzeri*. *Arch. Microbiol.* **160**, 18–26.

Godden, J. W., Turley, S., Teller, D. C., Adman, E. T., Liu, M.-Y., Payne, W. J., and LeGall, J. (1991). The 2.3 ångstrom X-ray structure of nitrite reductase from *Achromobacter cycloclastes*. *Science* **253**, 438–442.

Gokce, N., Hollocher, T. C., Bazylinski, D. A., and Jannasch, H. W. (1989). Thermophilic *Bacillus* sp. that shows the denitrification phenotype of *Pseudomonas aeruginosa*. *Appl. Environ. Microbiol.* **55**, 1023–1025.

Gordon, S., Hart, E. J., Matheson, M. S., Rabani, J., and Thomas, J. K. (1963). Reactions of the hydrated electron. *Disc. Faraday Soc.* **36**, 193–205.

Goreau, T. J., Kaplan, W. A., Wofsy, S. C., McElroy, M. B., Valois, F. W., and Watson, S. W. (1980). Production of NO_2^- and N_2O by denitrifying bacteria at reduced concentrations of oxygen. *Appl. Environ. Microbiol.* **40**, 526–532.

Goretski, J., and Hollocher, T. C. (1988). Trapping of nitric oxide produced during denitrification by extracellular hemoglobin. *J. Biol. Chem.* **263**, 2316–2323.

Goretski, J., and Hollocher, T. C. (1990). The kinetic and isotopic competence of nitric oxide as an intermediate in denitrification. *J. Biol. Chem.* **265**, 889–895.

Goretski, J., and Hollocher, T. C. (1991). Catalysis of nitrosyl transfer by denitrifying bacteria is facilitated by nitric oxide. *Biochem. Biophys. Res. Commun.* **175**, 901–905.

Goretski, J., Zafiriou, O. C., and Hollocher, T. C. (1990). Steady-state nitric oxide concentrations during denitrification. *J. Biol. Chem.* **265**, 11535–11538.

Greenwood, C., Barber, D., Parr, S. R., Antonini, E., Brunori, M., and Colosimo, A. (1978). The reaction of *Pseudomonas aeruginosa* cytochrome *c*-551 oxidase with oxygen. *Biochem. J.* **173,** 11–17.

Grossmann, J. G., Abraham, Z. H. L., Adman, E. T., Neu, M., Eady, R. R., Smith, B. E., and Hasnain, S. S. (1993). X-Ray scattering using synchrotron radiation shows nitrite reductase from *Achromobacter xylosoxidans* to be a trimer in solution. *Biochemistry* **32,** 7360–7366.

Gudat, J. C., Singh, J., and Wharton, D. C. (1973). Cytochrome oxidase from *Pseudomonas aeruginosa* I. Purification and some properties. *Biochim. Biophys. Acta* **292,** 376–390.

Guengerich, F. P., and MacDonald, T. L. (1984). Chemical mechanisms of catalysis by cytochromes P-450: A unified view. *Acc. Chem. Res.* **17,** 9–16.

Guengerich, F. P., and MacDonald, T. L. (1990). Mechanisms of cytochrome P-450 catalysis. *FASEB J.* **4,** 2453–2459.

Guest, J. R. (1992). Oxygen-regulated gene expression in *Escherichia coli*. *J. Gen. Microbiol.* **138,** 2253–2263.

Hackenthal, E., Mannheim, W., Hackenthal, R., and Becher, R. (1964). Die reduction von perchlorat durch bacterien. I. untersuchungen an intakten zellen. *Biochem. Pharmacol.* **13,** 195–206.

Heiss, B., Frunzke, K., and Zumft, W. G. (1989). Formation of the N–N bond from nitric oxide by a membrane-bound cytochrome *bc* complex of nitrate-respiring (denitrifying) *Pseudomonas stutzeri*. *J. Bacteriol.* **171,** 3288–3297.

Henry, Y., Ducrocq, C., Drapier, J.-C., Servent, D., Pellat, C., and Guissani, A. (1991). Nitric oxide, a biological effector. Electron paramagnetic resonance detection of nitrosyl-iron-protein complexes in whole cells. *Eur. Biophys. J.* **20,** 1–15.

Hirasawa, M., Shaw, R. W., Palmer, G., and Knaff, D. B. (1987). Prosthetic group content and ligand-binding properties of spinach nitrite reductase. *J. Biol. Chem.* **262,** 12428–12433.

Ho, T. P., Jones, A. M., and Hollocher, T. C. (1993). Denitrification enzymes of *Bacillus stearothermophilus*. *FEMS Microbiol. Lett.* **114,** 135–138.

Hollocher, T. C., and Kristjansson, J. K. (1992). Thermophylic denitrifying bacteria: A survey of hot springs in southwestern Iceland. *FEMS Microbiol. Ecol.* **101,** 113–119.

Hollocher, T. C., Tate, M. E., and Nicholas, D. J. D. (1981). Oxidation of ammonia by *Nitrosomonas europaea*. Definitive ^{18}O-tracer evidence that hydroxylamine formation involves a monooxygenase. *J. Biol. Chem.* **256,** 10834–10836.

Horio, T., Higashi, T., Matsubara, H., Kusai, K., Nakai, M., and Okunuki, K. (1958). High purification and properties of *Pseudomonas* cytochrome oxidase. *Biochim. Biophys. Acta* **29,** 297–302.

Hulse, C. L., and Averill, B. A. (1989). Evidence for a copper–nitrosyl intermediate in denitrification by the copper-containing nitrite reductase of *Achromobacter cycloclastes*. *J. Am. Chem. Soc.* **111,** 2322–2323.

Huynh, B. H., Liu, M.-C., Moura, J. J. G., Moura, I., Ljungdahl, P. O., Münck, E., Payne, W. J., Peck, H. D., DerVartanian, D. V., and LeGall, J. (1982). Mössbauer and EPR studies on nitrite reductase from *Thiobacillus denitrificans*. *J. Biol. Chem.* **257,** 9576–9581.

Hynes, R. K., and Knowles, R. (1984). Production of nitrous oxide by *Nitrosomonas europaea*: Effects of acetylene, pH, and oxygen. *Can. J. Microbiol.* **30,** 1397–1404.

Iwasaki, H., and Matsubara, T. (1971). Cytochrome *c*-557(551) and cytochrome *cd* of *Alcaligenes faecalis*. *J. Biochem. (Tokyo)* **69,** 847–857.

Iwasaki, H., and Matsubara, T. (1972). A nitrite reductase from *Achromobacter cycloclastes*. *J. Biochem. (Tokyo)* **71,** 645–652.

Iwasaki, H., and Mori, T. (1958). Studies on denitrification. III. Enzymatic gas production by reaction of nitrite with hydroxylamine. *J. Biochem. (Tokyo)* **45,** 133–140.

Iwasaki, N., Noji, S., and Shidara, S. (1975). *Achromobacter cycloclastes* nitrite reductase. The function of copper, amino acid composition, and ESR spectra. *J. Biochem. (Tokyo)* **78,** 355–361.

Jackson, M. A., Tiedje, J. M., and Averill, B. A. (1991). Evidence for an NO-rebound mechanism for production of N_2O from nitrite by the copper-containing nitrite reductase from *Achromobacter cyclolastes*. *FEBS Lett.* **291,** 41–44.

Jencks, W. P. (1969). "Catalysis in Chemistry and Enzymology," pp. 163–242. McGraw-Hill, New York.

Ji, X.-B., and Hollocher, T. C. (1988a). Mechanism for nitrosation of 2,3-diaminonaphthalene by *Escherichia coli*: Enzymatic production of NO followed by O_2-dependent chemical nitrosation. *Appl. Environ. Microbiol.* **54,** 1791–1794.

Ji, X.-B., and Hollocher, T. C. (1988b). Reduction of nitrite to nitric oxide by enteric bacteria. *Biochem. Biophys. Res. Commun.* **157,** 106–108.

Ji, X.-B, and Hollocher, T. C. (1989). Nitrate reductase of *Escherichia coli* as a NO-producing nitrite reductase. *Biochem. Arch.* **5,** 61–66.

Jones, A. M., and Hollocher, T. C. (1993). Nitric oxide reductase from *Achromobacter cycloclastes*. *Biochim. Biophys. Acta* **1144,** 359–366.

Jones, A. M., Hollocher, T. C., and Knowles, R. (1992). Nitrous oxide reductase of *Flexibacter canadensis*: A unique membrane-bound enzyme. *FEMS Microbiol. Lett.* **92,** 205–210.

Jüngst, A., Braun, C., and Zumft, W. G. (1991a). Close linkage in *Pseudomonas stutzeri* of the structural genes for respiratory nitrite reductase and nitrous oxide reductase, and other essential genes for denitrification. *Mol. Gen. Genet.* **225,** 241–248.

Jüngst, A., Wakabayashi, S., Matsubara, H., and Zumft, W. G. (1991b). The *nir* STBM region coding for cytochrome cd_1-dependentent nitrite respiration of *Pseudomonas stutzeri* consists of a cluster of mono-, di-, and tetraheme proteins. *FEBS Lett,* **279,** 205–209.

Kakutani, T., Watanabe, H., Arima, K., and Beppu, T. (1981). Purification and properties of a copper-containing nitrite reductase from a denitrifying bacterium, *Alcaligenes faecalis* strain S-6. *J. Biochem. (Tokyo)* **89,** 453–461.

Kalkowski, I., and Conrad, R. (1991). Metabolism of nitric oxide in denitrifying *Pseudomonas aeruginosa* and nitrate-respiring *Bacillus cereus*. *FEMS Microbiol. Lett.* **82,** 107–111.

Kim, C.-H., and Hollocher, T. C. (1983). ^{15}N tracer studies on the reduction of nitrite by the purified dissimilatory nitrite reductase of *Pseudomonas aeruginosa*. *J. Biol. Chem.* **258,** 4861–4863.

Kim, C.-H., and Hollocher, T. C. (1984). Catalysis of nitrosyl transfer by a dissimilatory nitrite reductase (cytochrome cd_1). *J. Biol. Chem.* **259,** 2092–2099.

Kizawa, H., Tomura, D., Oda, M., Fukamizu, A., Hoshino, T., Gotoh, O., Yarui, T., and Shoun, H. (1991). Nucleotide sequence of the unique nitrate/nitrite-inducible cytochrome P-450 cDNA from *Fusarium oxysporum*. *J. Biol. Chem.* **266,** 10632–10637.

Knowles, R. (1982). Denitrification. *Microbiol. Rev.* **46,** 43–70.

Koike, I., and Hattori, A. (1975). Energy yield of denitrification: An estimate for growth yield in continuous culture of *Paracoccus denitrificans* under nitrate-, nitrite-, and nitrous oxide-limiting conditions. *J. Gen. Microbiol.* **88,** 11–19.

Körner, H. (1993). Anaerobic expression of nitric oxide reductase from denitrifying *Pseudomonas stutzeri*. *Arch. Microbiol.* **159,** 410–416.

Kristjansson, J. R., Walter, B., and Hollocher, T. C. (1978). Respiration-dependent proton translocation and the transport of nitrate and nitrite in *Paracoccus denitrificans* and other denitrifying bacteria. *Biochemistry* **17,** 5014–5019.

Krueger, R. J., and Siegel, L. M. (1982). Spinach siroheme enzymes: Isolation and characterization of ferredoxin-sulfite reductase and comparison of properties with ferredoxin-nitrite reductase. *Biochemistry* **21,** 2892–2904.

Kucera, I. (1989). The release of nitric oxide from denitrifying cells of *Paracoccus denitrificans* by an uncoupler is the basis for a new oscillator. *FEBS Lett.* **249,** 56–58.

Kuenen, J. G., and Robertson, L. A. (1987). Ecology of nitrification and denitrification. *In* "The Nitrogen and Sulfur Cycles" (J. A. Cole and S. Ferguson, eds.), Symp. No. 42, pp. 161–218. Society of General Microbiology, Cambridge Univ. Press, London.

Kuronen, T., and Ellfolk, N. (1972). A new purification procedure and molecular properties of *Pseudomonas* cytochrome oxidase. *Biochim. Biophys. Acta* **275,** 308–318.

Lam, Y., and Nicholas, D. J. D. (1969). A nitrite reductase with cytochrome oxidase activity from *Micrococcus denitrificans*. *Biochim. Biophys. Acta* **180,** 459–472.

Lancaster, J. R., and Hibbs, J. B. (1990). EPR demonstration of iron–nitrosyl complex formation by cytotoxic activated macrophages. *Proc. Natl. Acad. Sci. U.S.A.* **87,** 1223–1227.

LeGall, J., Payne, W. J., Morgan, T. V., and DerVartanian, D. (1979). On the purification of nitrite reductase from *Thiobacillus denitrificans* and its reaction with nitrite under reducing conditions. *Biochem. Biophys. Res. Commun.* **87,** 355–362.

Leibowitz, M. R., Garber, E. A. E., Kristjansson, J. K., and Hollocher, T. C. (1982). Artifacts associated with the use of thiocyanate and valinomycin/K$^+$ as permeant ions in oxidant pulse experiments in denitrifying bacteria. *Curr. Microbiol.* **7,** 305–310.

Lepoivre, M., Fieschi, F., Coves, J., Thelander, L., and Fontecave, M. (1991). Inactivation of ribonucleotide reductase by nitric oxide. *Biochem. Biophys. Res. Commun.* **179,** 442–448.

Libby, E., and Averill, B. A. (1992). Evidence that the type 2 copper centers are the site of nitrite reduction by *Achromobacter cycloclastes* nitrite reductase. *Biochem. Biophys. Res. Commun.* **187,** 1529–1535.

Liu, M.-C., and Peck, H. D. (1981). The isolation of a hexaheme cytochrome from *Desulfovibrio desulfuricans* and its identification as a new type of nitrite reductase. *J. Biol. Chem.* **256,** 13159–13164.

Liu, M.-C., Liu, M.-Y., Payne, W. J., Peck, H. D., and LeGall, J. (1983). *Wolinella succinogenes* nitrite reductase: Purification and properties. *FEBS Lett.* **19,** 201–206.

Liu, M.-Y., Liu, M.-C., Payne, W. J., and LeGall, J. (1986). Properties and electron transfer specificity of copper proteins from the denitrifier *Achromobacter cycloclastes*. *J. Bacteriol.* **166,** 604–608.

Ludwig, W., Mittenhuber, G., and Friedrich, C. G. (1993). Transfer of *Thiosphaera pantotropha* to *Paracoccus denitrificans*. *Int. J. Syst. Bacteriol.* **43,** 363–367.

McCleverty, J. A. (1979). Reactions of nitric oxide coordinated to transition metals. *Chem. Rev.* **79,** 53–76.

McTavish, H., Fuchs, J. A., and Hooper, A. B. (1993). Sequence of the gene coding for ammonia monooxygenase in *Nitrosomonas europaea*. *J. Bacteriol.* **175,** 2436–2444.

Makinen, M. W., Schichman, S. A., Hill, S. C., and Gray, H. B. (1983). Heme–heme orientation and electron transfer kinetic behavior of multisite oxidation–reduction enzymes. *Science* **222,** 929–931.

Mancinelli, R. L., and McKay, C. P. (1983). Effects of nitric oxide and nitrogen dioxide on bacterial growth. *Appl. Environ. Microbiol.* **46,** 198–202.

Mancinelli, R. L., Cronin, S., and Hochstein, L. I. (1986). The purification and properties of a cd-cytochrome nitrite reductase from *Paracoccus halodenitrificans*. *Arch. Microbiol.* **145,** 202–208.

Marletta, M. A., Yoon, P. S., Iyengar, R., Leaf, C. D., and Wishnok, J. S. (1988). Macrophage oxidation of L-arginine to nitrite and nitrate: nitric oxide is an intermediate. *Biochemistry* **21,** 8706–8711.

Marletta, M. A., Tayeh, M. A., and Hevel, J. M. (1990). Unraveling the biological significance of nitric oxide. *BioFactors* **2,** 219–225.

Masuko, M., Iwasaki, H., Sakurai, T., Suzuki, S., and Nakahara, A. (1984). Characterization of nitrite reductase from a denitrifier, *Alcaligenes*. sp. NCIB 11015. A novel copper protein. *J. Biochem. (Tokyo)* **96,** 447–454.

Matsubara, T. (1970). Studies on denitrification. VII. Gas production from amines and nitrites. *J. Biochem. (Tokyo)* **67,** 229–235.

Miller, D. J., and Nicholas, D. J. D. (1985). Characterization of a soluble cytochrome oxide/nitrite reductase from *Nitrosomonas europaea*. *J. Gen. Microbiol.* **131,** 2851–2854.

Miller, D. J., and Wood, P. (1983). The soluble cytochrome oxidase of *Nitrosomonas europaea*. *J. Gen. Microbiol.* **129,** 1645–1650.

Miyata, M., Matsubara, T., and Mori, T. (1969). Studies on denitrification XI. Some properties of nitric oxide reductase. *J. Biochem. (Tokyo)* **66,** 759–763.

Moir, W. B., Baratta, D., Richardson, D. J., and Ferguson, S. J. (1993). The purification of a cd_1-type nitrite reductase from, and the absence of a copper-type nitrite reductase from, the aerobic denitrifier *Thiosphaera pantotropha*. *Eur. J. Biochem.* **212,** 377–385.

Moriguchi, M., Manning, L. R., and Manning, J. M. (1992). Nitric oxide can modify amino acid residues in proteins. *Biochem. Biophys. Res. Commun.* **183,** 598–604.

Mülsch, A., Vanin, A., Mordvintcev, P., Hauschildt, S., and Busse, R. (1992). NO accounts completely for the oxygenated nitrogen species generated by enzymic L-arginine oxygenation. *Biochem. J.* **288,** 597–603.

Nakahara, K., Tanimoto, T., Hatano, K., Usuda, K., and Shoun, H. (1993). Cytochrome P-450 55A1 (P-450dNIR) acts as nitric oxide reductase employing NADH as the direct electron donor. *J. Biol. Chem.* **268,** 8350–8355.

Newton, N. (1969). The two-haem nitrite reductase of *Micrococcus denitrificans*. *Biochim. Biophys. Acta* **185**, 316–331.

Olson, L. W., Schaeper, D., Lancon, D., and Kadish, K. M. (1982). Characterization of several novel iron nitrosyl porphyrins. *J. Am. Chem. Soc.* **104**, 2042–2044.

Ozawa, S., Fujii, H., and Morishima, I. (1992). NMR studies of iron(II) nitrosyl π-cation radicals of octaethylchlorin and octaethylisobacteriochlorin as models for reaction intermediate of nitrite reductase. *J. Am. Chem. Soc.* **114**, 1548–1554.

Page, M. D., and Ferguson, S. J. (1989). A bacterial c-type cytochrome can be translocated to the periplasm as an apoform: The biosynthesis of cytochrome cd_1 (nitrite reductase) from *Paracoccus denitrificans*. *Mol. Microbiol.* **3**, 653–661.

Pandey, K. K. (1983). Transition metal nitrosyls in organic synthesis and in pollution control. *Coord. Chem. Rev.* **51**, 69–98.

Papen, H., von Berg, R., Hinkel, I., Thoene, B., and Rennenberg, H. (1989). Heterotrophic nitrification by *Alcaligenes faecalis*: NO_2^-, NO_3^-, and NO production in exponentially growing cultures. *Appl. Environ. Microbiol.* **55**, 2068–2072.

Parr, S. R., Barber, D., Greenwood, C., Phillips, B. W., and Melling, J. (1976). A purification procedure for the soluble cytochrome oxidase and some other respiratory proteins from *Pseudomonas aeruginosa*. *Biochem. J.* **157**, 423–430.

Parr, S. R., Barber, D., Greenwood, C., and Brunori, M. (1977). The electron-transfer reaction between azurin and the cytochrome c oxidase from *Pseudomonas aeruginosa*. *Biochem. J.* **167**, 447–455.

Payne, W. J. (1981). "Denitrification." Wiley, New York.

Payne, W. J., Grant, M. A., Shapleigh, J., and Hoffman, P. (1982). Nitrogen oxide reduction in *Wolinella succinogenes* and *Campylobacter* species. *J. Bacteriol.* **152**, 915–918.

Pearsall, K. A., and Bonner, F. T. (1982). Aqueous nitrosyliron(II) chemistry. 2. Kinetics and mechanism of nitric oxide reduction. The denitrosyl complex. *Inorg. Chem.* **21**, 1978–1985.

Pichinoty, F., Bigliardi-Rouvier, J., and Rimassa, M. R. (1969). La denitrification bacteriene I. Utilisation des amines aromatiques comme donneuses d'électrons dans la réduction du nitrite. *Arch. Microbiol.* **69**, 314–329.

Poth, M. (1986). Dinitrogen production from nitrite by a *Nitrosomonas* isolate. *Appl. Environ. Microbiol.* **52**, 957–959.

Poth, M., and Focht, D. D. (1985). ^{15}N kinetic analysis of N_2O production by *Nitrosomonas europaea*: An examination of nitrifier denitrification. *Appl. Environ. Microbiol.* **49**, 1134–1141.

Pryor, W. A., Church, D. F., Govindan, C. K., and Crank, G. (1982). Oxidation of thiols by nitric oxide and nitrogen dioxide: Synthetic utility and toxicological implications. *J. Org. Chem.* **47**, 156–159.

Radi, R., Beckman, J. S., Bush, K. M., and Freeman, B. A. (1991). Peroxynitrite oxidation of sulfhydryls. The cytotoxic potential of superoxide and nitric oxide. *J. Biol. Chem.* **266**, 4244–4250.

Ralt, D., Wishnok, J. S., Fitts, R., Tannenbaum, S. R. (1988). Bacterial catalysis of nitrosation: Involvement of the *nar* operon in *Escherichia coli*. *J. Bacteriol.* **170**, 359–364.

Remde, A., and Conrad, R. (1990). Production of nitric oxide in *Nitrosomonas europaea* by reduction of nitrite. *Arch. Microbiol.* **154**, 187–191.

Remde, A., and Conrad, R. (1991). Metabolism of nitric oxide in soil and denitrifying bacteria. *FEMS Microbiol. Ecol.* **85,** 81–93.

Robertson, L. A., and Kuenen, J. G. (1984). Aerobic denitrification—Old wine in new bottles? *Antonie Van Leenwenhoek* **50,** 525–544.

Robertson, L. A., and Kuenen, J. G. (1990). Combined heterotrophic nitrification and aerobic denitrification in *Thiosphaera pantotropha* and other bacteria. *Antonie Van Leenwenhoek* **57,** 139–152.

Robinson, M. K., Markinkus, K., Kennelly, P.J., and Timkovich, R. (1979). Implications of the integrated rate law for the reactions of *Paracoccus denitrificans* nitrite reductase. *Biochemistry* **18,** 3921–3926.

Ruggiero, C. E., Carrier, S. M., Antholine, W. E., Whittaker, J. W., Cramer, C. J., and Tolman, W. B. (1993). Synthesis and structure and spectroscopic characterization of mononuclear copper nitrosyl complexes: Models for nitric oxide adducts of copper proteins and copper exchange zeolites. *J. Am. Chem. Soc.* **115,** 11285–11298.

St. John, R. T., and Hollocher, T. C. (1977). Nitrogen-15 tracer studies on the pathway of denitrification in *Pseudomonas aeruginosa*. *J. Biol. Chem.* **252,** 212–218.

Sano, M., and Matsubara, T. (1988). Structural change in the one-electron oxidation–reduction at the copper site in nitrite reductase. Evidence from EXAFS. *Inorg. Chem. Acta Bioinorg. Chem* **152,** 53–54.

Sawada, E., Satoh, T., and Kitamura, H. (1978). Purification and properties of a dissimilatory nitrite reductase of a denitrifying phototrophic bacterium. *Plant Cell Physiol.* **19,** 1339–1351.

Sawhney, V., and Nicholas, D. J. D. (1978). Sulphide-linked nitrite reductase from *Thiobacillus denitrificans* with cytochrome oxidase activity: Purification and properties. *J. Gen. Microbiol.* **106,** 119–218.

Schichmann, S. A., and Gray, H. B. (1981). Kinetics of the anaerobic reduction of ferricytochrome cd_1 by Fe(EDTA)$^{2-}$. Evidence for bimolecular and intramolecular electron transfers to the d_1 hemes. *J. Am. Chem. Soc.* **103,** 7794–7795.

Schmitz, K. S., and Schurr, M. J. (1972). The role of orientation constraints and rotational diffusion in bimolecular solution kinetics. *J. Phys. Chem.* **76,** 534–545.

Schurr, M. J., and Schmitz, K. S. (1976). Orientation constraints and rotational diffusion in bimolecular solution kinetics. A simplification. *J. Phys. Chem.* **80,** 1934–1936.

Shapleigh, J. P., and Payne, W. J. (1985a). Differentiation of cd_1 cytochrome and copper nitrite reductase production in denitrifiers. *FEMS Microbiol. Lett.* **26,** 275–279.

Shapleigh, J. P., and Payne, W. J. (1985b). Nitric oxide-dependent proton translocation in various denitrifiers. *J. Bacteriol.* **163,** 837–840.

Shapleigh, J. P., Davies, K. J. P., and Payne, W. J. (1987). Detergent inhibition of nitric-oxide reductase activity. *Biochim. Biophys. Acta* **911,** 334–340.

Shearer, G., and Kohl, D. H. (1988). Nitrogen isotope fractionation and ^{18}O exchange in relation to the mechanism of denitrification of nitrite by *Pseudomonas stutzeri*. *J. Biol. Chem.* **263,** 13231–13245.

Shimada, H., and Orii, Y. (1976). Oxidation-reduction behavior of the heme c and heme d moieties of *Pseudomonas aeruginosa* nitrite reductase and the formation of an oxygenated intermediate at heme d. *J. Biochem. (Tokyo)* **80,** 135–140.

Shoun, H., and Tanimoto, T. (1991). Denitrification by the fungus *Fusarium oxysporum* and involvement of cytochrome P-450 in the respiratory nitrite reduction. *J. Biol. Chem.* **266,** 11078–11082.

Shoun, H., Suyama, W., and Yasui, T. (1989). Soluble, nitrate/nitrite-inducible cytochrome P-450 of the fungus *Fusarium oxysporum*. *FEBS Lett.* **244,** 11–14.

Shoun, H., Kim, D.-H., Uchiyama, H., and Sugiyama, J. (1992). Denitrification by fungi. *FEBS Microbiol. Lett.* **94,** 277–282.

Silvestrini, M. C., Colosimo, A., Brunori, M., Walsh, T. A., Barber, D., and Greenwood, C. (1979). A re-evaluation of some basic structural and functional properties of *Pseudomonas* cytochrome oxidase. *Biochem. J.* **183,** 701–709.

Silvestrini, M. C., Galeotti, C. L., Gervais, M., Schinina, E., Barra, D., Bossa, F., and Brunori, M. (1989). Nitrite reductase from *Pseudomonas aeruginosa*: Sequence of the gene and protein. *FEBS Lett.* **254,** 33–38.

Silvestrini, M. C., Tordi, M. G., Musci, G., and Brunori, M. (1990). The reaction of *Pseudomonas* nitrite reductase and nitrite. A stopped-flow and EPR study. *J. Biol. Chem.* **265,** 11783–11787.

Smith, G. B., and Tiedje, J. M. (1992). Isolation and characterization of a nitrite reductase gene and its use as a probe for denitrifying bacteria. *Appl. Environ. Microbiol.* **58,** 376–384.

Snyder, S. W., and Hollocher, T. C. (1987). Purification and some characteristics of nitrous oxide reductase from *Paracoccus denitrificans*. *J. Biol. Chem.* **262,** 6515–6525.

Sprent, J. I. (1987). "The Ecology of the Nitrogen Cycle." Cambridge Univ. Press, Cambridge.

Stamler, J. S., Simon, D. I., Osborne, J. A., Mullins, M. E., Jaraki, O., Michel, T., Singel, D. J., and Loscalzo, J. (1992). S-Nitrosylation of proteins with nitric oxide: Synthesis and characterization of biologically active compounds. *Proc. Natl. Acad. Sci. U.S.A.* **89,** 444–448.

Stewart, V. (1988). Nitrate respiration in relation to facultative metabolism in enterobacteria. *Microbiol. Rev.* **52,** 190–232.

Sutherland, J., Greenwood, C., Peterson, J., and Thomson, A. J. (1986). An investigation of the ligand-binding properties of *Pseudomonas aeruginosa* nitrite reductase. *Biochem. J.* **233,** 893–898.

Sverdrup, H. U., Johnson, M. W., and Fleming, R. H. (1942). "The Oceans. Their Physics, Chemistry and General Biology," pp. 165–227. Prentice-Hall, Englewood Cliffs, New Jersey.

Takano, T., Dickerson, R. E., Schichman, S. A., and Meyer, T. E. (1979). Crystal data, molecular dimensions and molecular symmetry in cytochrome oxidase from *Pseudomonas aeruginosa*. *J. Mol. Biol.* **133,** 185–188.

Tanimoto, T., Hatano, K., Kim, D., Uchiyama, H., and Shoun, H. (1992). Co-denitrification by the denitrifying system of the fungus *Fusarium oxysporum*. *FEMS Microbiol. Lett.* **93,** 177–180.

Teraguchi, T., and Hollocher, T. C. (1989). Purification and some characteristics of a cytochrome c-containing nitrous oxide reductase from *Wolinella succinogenes*. *J. Biol. Chem.* **264,** 1972–1979.

Tiedje, J. M. (1988). Ecology of denitrification and dissimilatory nitrate reduction to ammonia. In "Biology of Anaerobic Microorganisms" (A. J. B. Zehnder, ed.). pp. 179–244. Wiley, New York.

Timkovich, R., Dhesi, R., Martinkus, K. J., Robertson, M. K., and Rea, T. M. (1982). Isolation of *Paracoccus denitrificans* cytochrome cd_1: Comparative kinetics and other nitrite reductases. Arch Biochem. Biophys. **215**, 47–58.

Tolman, W. B. (1991). A model for the substrate adduct of copper nitrite reductase and its conversion to a novel tetrahedral copper(II) triflate complex. Inorg. Chem. **30**, 4877–4880.

Turk, T., and Hollocher, T. C. (1992). Oxidation of dithiothreitol during turnover of nitric oxide reductase: Evidence for generation of nitroxyl with the enzyme from *Paracoccus denitrificans*. Biochem. Biophys. Res. Commun. **183**, 983–988.

Urata, K., and Satoh, T. (1991). Enzyme localization and orientation of the active site of dissimilatory nitrite reductase from *Bacillus firmus*. Arch. Microbiol. **156**, 24–27.

Vega, J. M., and Kamin, H. (1977). Spinach nitrite reductase. Purification and properties of a siroheme-containing iron–sulfur enzyme. J. Biol. Chem. **252**, 896–909.

Vega, J. M., Garrett, R. H., and Siegel, L. M. (1975). Siroheme: A prosthetic group of the *Neurospora crassa* assimilatory nitrite reductases. J. Biol. Chem. **250**, 7980–7989.

Vestraete, W., and Alexander, M. (1972). Formation of hydroxylamine from ammonium by N-oxygenation. Biochim. Biophys. Acta **261**, 59–62.

Völkl, P., Huber, R., Drobner, E., Rachel, R., Burggraf, S., Trincone, A., and Stetter, K. O. (1993). *Pyrobaculum aerophilum* sp, nov., a novel nitrate-reducing hyperthermophilic archaeon. Appl. Environ. Microbiol. **59**, 2918–2926.

Vosswinkel, R., Neidt, I., and Bothe, H. (1991). The production and utilization of nitric oxide by a new, denitrifying strain of *Pseudomonas aeruginosa*. Arch. Microbiol. **156**, 62–69.

Walker, N. (1978). On the diversity of nitrifiers in nature. In "Microbiology-1978" (D. Schlessinger, ed.), pp. 346–347. American Society for Microbiology, Washington, D. C.

Wallace, W., and Nicholas, D. J. D. (1969). The biochemistry of nitrifying microorganism. Biol. Rev. **44**, 359–391.

Ward, B. B., and Zafiriou, O. C. (1988). Nitrification and nitric oxide in the oxygen minimum of the eastern tropical north pacific. Deep Sea Res. **35**, 1127–1142.

Weeg-Aerssens, E., Tiedje, J. M., and Averill, B. A. (1987). The mechanism of microbiol denitrification. J. Am. Chem. Soc. **109**, 7214–7215.

Weeg-Aerssens, E., Tiedje, J. M., and Averill, B. A. (1988). Evidence from isotope labeling studies for a sequential mechanism for dissimilatory nitrite reduction. J. Am. Chem. Soc. **110**, 6851–6856.

Weeg-Aerssens, E., Wu, W., Ye, R. W., Tiedje, J. M., and Chang, C. K. (1991). Purification of cytochrome cd_1 nitrite reductase from *Pseudomonas stutzeri* JM 300 and reconstitution with native and synthetic heme d_1. J. Biol. Chem. **266**, 7496–7502.

Wharton, D. C., and Gibson, Q. H. (1976). Cytochrome oxidase from *Pseudomonas aeruginosa*. IV. Reaction with oxygen and carbon monoxide. Biochim. Biophys. Acta **430**, 445–453.

Wharton, D. C., and Weintraub, S. T. (1980). Identification of nitric oxide and nitrous oxide as products of nitrate reduction by *Pseudomonas* cytochrome oxidase (nitrite reductase). *Biochem. Biophys. Res. Commun.* **97**, 236–242.

Wharton, D. C., Gudat, J. C., and Gibson, Q. H. (1973). Cytochrome oxidase from *Pseudomonas aeruginosa*. II. Reaction with copper protein. *Biochim. Biophys. Acta* **292**, 611–620.

White, K. A., and Marletta, M. A. (1992). Nitric oxide synthase is a cytochrome P-450 type hemoprotein. *Biochemistry* **31**, 6627–6631.

Wijler, J., and Delwiche, C. C. (1954). Investigation on the denitrifying process in soil. *Plant Soil* **5**, 155–169.

Wink, D. A., Darbyshire, J. F., Nims, R. W., Saavedra, J. E., and Ford, P. C. (1993). Reactions of the bioregulatory agent nitric oxide in oxygenated aqueous media: Determination of the kinetics for oxidation and nitrosation by intermediates generated in the NO/O_2 reaction. *Chem. Res. Toxicol.* **6**, 23–27.

Wolin, M. J., Wolin, E. A., and Jacobs, N. J. (1961). Cytochrome-producing anaerobic vibrio, *Vibrio succinogenes* sp. n. *J. Bacteriol.* **81**, 911–917.

Yamanaka, T. (1964). Identity of *Pseudomonas* cytochrome oxidase with *Pseudomonas* nitrite reductase. *Nature (London)* **204**, 253–255.

Yamanaka, T., and Okunuki, K. (1963a). Crystalline *Pseudomonas* cytochrome oxidase. I. Enzymatic properties with special reference to biological specificity. *Biochim. Biophys. Acta* **67**, 379–393.

Yamanaka, T., and Okunuki, K. (1963b). Crystalline *Pseudomonas* cytochrome oxidase. II. Spectral properties of the enzyme. *Biochim. Biophys. Acta* **67**, 394–406.

Yamanaka, T., and Okunuki, K. (1963c). Crystalline *Pseudomonas* cytochrome oxidase. III. Properties of the prosthetic groups. *Biochim. Biophys. Acta* **67**, 407–416.

Yamanaka, T., Ota, A., and Okunuki, K. (1961). A nitrite reducing system reconstructed with purified cytochrome components of *Pseudomonas aeruginosa*. *Biochim. Biophys. Acta* **53**, 294–308.

Yamanaka, T., Kijimoto, S., Okunuki, K., and Kusai, K. (1962). Preparation of crystalline *Pseudomonas* cytochrome oxidase and some of its properties. *Nature (London)* **194**, 759–760.

Ye, R. W., Toro-Suarez, I., Tiedje, J. M., and Averill, B. A. (1991). $H_2^{18}O$ isotope exchange studies on the mechanism of reduction of nitric oxide and nitrite to nitrous oxide by denitrifying bacteria. *J. Biol. Chem.* **266**, 12848–12851.

Ye, R. W., Arunakumari, A., Averill, B. A., and Tiedje, J. M. (1992a). Mutants of *Pseudomonas fluorescens* deficient in dissimilatory nitrite reduction are also altered in nitric oxide reduction. *J. Bacteriol.* **174**, 2560–2564.

Ye, R. W., Averill, B. A., and Tiedje, J. M. (1992b). Characterization of Tn5 mutants deficient in dissimilatory nitrite reduction in *Pseudomonas* sp. strain G-179, which contains a copper nitrite reductase. *J. Bacteriol.* **174**, 6653–6658.

Ye, R. W., Fries, M. R., Bezborodnikov, S. G., Averill, B. A., and Tiedje, J. M. (1993). Characterization of the structural gene encoding a copper-containing nitrite reductase and homology of this gene to DNA of other denitrifiers. *Appl. Environ. Microbiol.* **59**, 250–254.

Yoshinari, T. (1980). N_2O reduction by *Vibrio succinogenes*. *Appl. Environ. Microbiol.* **39**, 81–84.

Yoshinari, T., and Knowles, R. (1976). Acetylene inhibition of nitrous oxide reduction by denitrifying bacteria. *Biochem. Biophys. Res. Commun.* **69,** 705–710.

Zafiriou, O. C., Hanley, Q. S., and Snyder, G. (1989). Nitric oxide and nitrous oxide production and cycling during dissimilatory nitrite reduction by *Pseudomonas perfectomarina*. *J. Biol. Chem.* **264,** 5694–5699.

Zimmermann, A., Reimmann, C., Galimand, M., and Haas, D. (1991). Anarobic growth and cyanide synthesis of *Pseudomonas aeruginosa* depend on *anr*, a regulatory gene homologous with *fnr* of *Escherichia coli*. *Mol. Microbiol.* **5,** 1483–1490.

Zumft, W. G. (1972). Ferredoxin:nitrite oxidoreductase from *Chlorella*. Purification and properties. *Biochim. Biophys. Acta* **276,** 363–375.

Zumft, W. G. (1993). The biological role of nitric oxide in bacteria. *Arch. Microbiol.* **160,** 253–264.

Zumft, W. G., and Frunzke, K. (1982). Discrimination of ascorbate-dependent nonenzymatic and enzymatic, membrane-bound reduction of nitric oxide in denitrifying *Pseudomonas perfectomarinus*. *Biochim. Biophys. Acta* **681,** 459–468.

Zumft, W. G., and Vega, J. M. (1979). Reduction of nitrite to nitrous oxide by a cytoplasmic membrane fraction from the marine denitrifier *Pseudomonas perfectomarinus*. *Biochim. Biophys. Acta* **548,** 484–499.

Zumft, W. G., Braun, C., and Cuypers, H. (1994). Nitric oxide reductase from *Pseudomonas stutzeri*. Primary structure and gene organization of a novel bacterial cytochrome bc complex. *Eur. J. Biochem.* 219, 481–490.

Zumft, W. G., Döhler, K., Körner, H., Löchelt, S., Viebrock, A., and Frunzke, K. (1988). Defects in cytochrome cd_1-dependent nitrite respiration of transposon Tn5-induced mutants from *Pseudomonas stutzeri*. *Arch. Microbiol.* **149,** 492–498.

Zumft, W. G., Gotzmann, D. J., and Kroneck, P. M. H. (1987). Type 1, blue copper proteins constitute a respiratory nitrite-reducing system in *Pseudomonas aureofaciens*. *Eur. J. Biochem.* **168,** 301–307.

Index

Acidification, nitrite, 27-28
Aconitase
   iron–sulfur cluster function, 93
   islet mitochondria, IL-1 effects, 189-190
   mitochondrial, hepatocyte-derived NO effects, 229-230
Alloantigen presenting cells, P338D2 vs RAW 264.7 macrophage lines, 245-246
Allograft, rejection
   authentic NO effects on lymphocyte function, 246-248
   in vitro NO synthesis in sponge matrix allograft cells, 239-244
   in vivo NO synthesis during, 244-245
   mechanisms of NO-induced inhibition lymphocyte activation, 248-252
   nitric oxide synthesis effects in macrophage–lymphocyte cocultures, 245-246
Aminoguanidine, inhibition of inducible NO synthase, 205
Ammonia oxidizing bacteria, denitrification by, 321-323
Amyotrophic lateral sclerosis, superoxide dismutase mutation role, 63
Anergy, lymphocyte, 248-249
Antimicrobials
   nitrite and NO, 269-271, 328-329
      bacteriostatic effects, 269-271
      Clostridia inhibition, 272-275
      protein-bound nitrite, 276-277

Antioxidants
   nitric oxide in cured meats, 266-269
   nitric oxide role, 60-62
α_1-Antiprotease, inactivation by peroxynitrite, 67
L-Arginine
   nitric oxide biosynthesis, N^G-hydroxyl-L-arginine intermediate, 118, 143
   role in macrophage NOS subunit assembly, 163-164
Ascorbate, use in curing brines, 264-265
Ascorbic acid, effects on nitrosamine formation, 278
Autoantigens, β-cell specific, 179
Autoimmune diabetes, *see* Diabetes type I
Autoimmunity, in insulin-dependent diabetes mellitus, 178-180

Bacteria
   ammonia-oxidizing, denitrification by, 321-323
   chemolithotropic NH_3 oxidizers, 321-323
   denitrifying, *see* Denitrification
   gram-negative, NO inhibition of heme enzymes, 271
   nitric oxide production from nitrite, 324-326
   nitrifying
      denitrification pathway, 321-323
      nitric oxide production, 326-328
Beta cells, islet
   damage by free radicals, 198

Beta cells, islet (*Continued*)
  destruction during diabetes type I development
    macrophage role, 178-180
    models, 180
    T cell role, 178-180
  function, interleukin 1 effects
    early studies, 181-182
    mitochondrial, 185
    mRNA transcription, 182-185
    protein translation, 182-185
  interleukin 1-induced dysfunction, mechanism, 198-200
  interleukin 1-induced NO formation, 192-194
  nitric oxide-mediated dysfunction and destruction, 187-191
  nitric oxide production in human tissue, 202-204
Bioassay cascade method, similarity between NO and endothelium-derived relaxing factor, 113-114
Bioassays
  limitations, 34
  sensitivity, 34
Biological inactivation, rate for NO, 10-13
Blood flow, laminar, NO role, 7-8
Botulism toxin
  formation in cured meats, effects of salt, nitrate, and nitrite, 269-271
  nitric oxide inactivation of iron–sulfur proteins in nitrite-cured meats, 274
Buffer anions, effects on peroxynitrite decomposition, 54-55

Calmodulin, role in electron transfer control, 160-161
Cheese, nitrate effects on late gas defect, 279
Chemiluminescence
  nitric oxide detection, 35-36
  nitric oxide and endothelium-derived relaxing factor similarity, 114-115

Clostridium perfringens
  clostridial inhibitor in cured meats, 272-275
  free-sulfhydral groups, sodium nitrite effects, 271
Cocultures
  Kupffer cell–hepatocyte, 222-225
  macrophage–lymphocyte, NO synthesis effects, 245-246
Codenitrification, in *Fusarium* genus, 324
Colitis, peroxynitrite-induced, 68
Concentration, NO, information relayed by, 7-10
Contaminants, trace metals, NO reactions, 33-37
Copper
  binuclear center of molluscan hemocyanin, 97
  binuclear center of oxidized cytochrome oxidase, 97
  Cu(II)–NO complexes, 90, 97
  nitrite reductases in nitrifying and denitrifying bacteria, 315-317
  states in aqueous solution and proteins, 97
Copper, nitrosyl complexes, 97
Corpus cavernosum, smooth muscle relaxation, 119, 128
Corrosion, inhibition by nitrite, 279
Cyclic GMP
  glutamate-linked formation in cerebellum, NO mediation, 116-117
  role in IL-1 inhibitory effects on insulin secretion, 190-191
Cycloheximide, effects on IL-1-induced inhibition of insulin secretion, 182, 184, 190-191
Cyclooxygenase, in islet inflammation, 202-203
Cyclosporin A, inhibition of NO synthesis spleen cell allografts, 245
Cytochrome *c*, hexaheme, *Wollinella succinogenes*, 321
Cytochrome cd_1, in denitrifying bacteria, 312-315

Cytochrome oxidase
　nitric oxide complex, 90
　nuclear Cu(II), NO binding, 97
Cytochrome P450
　nitrosyl complexes, 91
　oxygen activation model, 155-156
　soluble, in *Fusarium oxysporum*, 323-324
Cytokines
　role in diabetes development, 180
　T cell production, N^G-monomethyl-L arginine effects, 240-243
Cytolytic T lymphocytes, allospecific activity, NO production effects, 240
Cytotoxic activated macrophages
　cell injury by, 141-142
　cytostatic/cytotoxic properties, 144
　and NO formation, 143

Denitrification
　by chemolithotrophic NH_3 oxidizers, 321-323
　codenitrification, 323-324
　diversity of denitrifiers, 292-293
　by eukaryotic organisms, 323-324
　genome, 320-321
　hexaheme cytochrome c, 321
　nitric oxide as intermediate in, 297-307
　nitric oxide reductase
　　characterization, 307-310
　　mechanisms, 310-312
　　models, 310-312
　nitrite reductase
　　Cu-type enzymes, 315-317
　　cytochrome cd_1, 312-315
　　mechanisms, 317-320
　　models, 317-320
　pathways, 293-296
Desaturase, nitrosyl complexes, 96
Detection methods, 32-33
　absorption into hydrophobic tubing, 36-37
　actions with trace metal contaminants, 33-37
　cheletrophic spin trapping, 39

Detection methods (*Continued*)
　chemiluminescence, 35-36
　electrochemical detection, 37-38
　electron paramagnetic spin resonance, 38-39
Dexamethasone, inhibition of inducible NO synthase, 206
Diabetes, autoimmune, *see* Diabetes type I
Diabetes type I
　animal models
　　low dose streptozotocin, 200
　　nonobese diabetic mouse, 201
　autoimmune nature, 178-180
　interventions, 204-206
　models, 180
Diazonium intermediates, formation, 25
Diffusion distances, hydroxyl radical and peroxynitrite, 43-44
Dimerization
　nitric oxide, 4
　nitric oxide–nitrogen dioxide reactions, 29-30
Dinitrogen dioxide, dimer characterization, 4
Dinitrogen tetroxide
　characterization, 29-30
　formation, 29
　isomeric forms, 29
Dinitrogen trioxide
　isomeric forms, 30
　oxidant actions, 30
　reactions, 30
Dinitrosylirondithiol complexes
　electron paramagnetic resonance, 144-146
　and nonheme iron–sulfur nitrosylation, 146-148
3,4-Dioxygenase–nitrosyl adducts, EPR spectrum, 95-96

Electrochemical oxidation, NO detection by, 37-38
Electronegativity
　nitrogen, 18-19

Electronegativity (*Continued*)
  nitrogen compared to oxygen and carbon, 19
Electron paramagnetic resonance
  dinitrosylirondithiol complexes, 144-146
  g tensor
    2.04 signals and nonheme iron–sulfur nitrosylation, 146-148
    properties, 99-101
  hyperfine coupling, 101-102
  iron–sulfur clusters, 91-95
  nitric oxide reductase spectrum in *Pseudomonas stutzeri*, 309
  nitrosyl complexes of hemoglobin and myoglobin, 89-91
  nonheme iron protein–NO complexes, 95-96
  relaxation, 102-103
  saturation, 102-103
  $S = \frac{1}{2}$ and $S > \frac{1}{2}$ systems, 103-107
  temperature effects, 102-103
  Zeeman splitting, 99-101
Electron paramagnetic spin resonance
  nitric oxide detection, 38-39
  nitric oxide properties as paramagnetic ligand, 84-86
Electron transfer
  calmodulin role, 160-161
  nitric oxide synthase flavins in, 158-159
Electrostatic fields, nitric oxide, molecular oxygen, and nitrosyldioxyl radical, 15
Endonuclease III, iron–sulfur cluster, 93
Endothelial cells, peroxynitrite production, 68
Endothelium-derived relaxing factor
  bioassay behavior, 31
  history of research, 11-12
  identification with NO, 113-115, 143
  nitrosothiols as, 31
  nitroxyl anion link, 24
Endotoxemia, *in vivo* role of NO, 232-233

Ferredoxins, NO complexes, 91-95
Ferritin, nitrosyl complexes, 96

Ferroheme–NO complexes
  characterization, 86-88
  in proteins, 88-91
Ferrous tetraphenylporphyrin 1-methylimidazole, nitrosyl complexes, 87
Ferrous tetraphenylporphyrin pyridine, nitrosyl complexes, 87
FK506, inhibition of NO synthesis
  in graft infiltrating cells, 243
  in spleen cell allografts, 245
Flavins, NO synthase
  reduction levels, 158-159
  role in electron transfer, 158-159
Florescence-activated cell sorting, NO production by β cells, 192-194
Formaldehyde, formation from DMSO oxidation, 48-50
Fusarium oxysporum, codenitrification, 324

Genome, denitrification in *Pseudomonas stutzeri*, 320-321
Glutamate
  neurotoxiciy, NO mediation, 125-126
  in synaptic function, link with NO, 124-125
Glutathione, reduced, and hepatocyte NO synthesis, 225-226
Graft infiltrating cells
  adherent, NO production, 243-244
  nitric oxide synthesis, inhibition by FK506, 243
Graft *versus* host reaction, NO synthesis *in vivo*, 244-245
Gram-negative bacteria, heme enzymes, NO inhibition, 271
Griess reaction, nitrite assay, 35
Guanylate cyclase
  activation by NO and protoporphyrin IX, 119-123
  ferrous heme iron, NO binding, 2
  inactivation after NO binding, 15-18
  soluble, activation in hepatocytes, 227-228

Haber–Weiss reaction
  hydroxyl radical production, 42-43
  vs metal-catalyzed nitration by peroxynitrate, 53
Heme
  electron transfer, flavin role, 158-159
  in macrophage NOS subunit assembly, 163-164
  in NO synthesis, 153-157
Hemocyanin, copper center, nitrosyl complexes, 97
Hemoglobin, nitrosyl complexes, 89-91
Hepatocytes
  in vitro NO actions
    activation of soluble guanylate cyclase, 227-228
    effects on mitochondrial aconitase, 229-230
    suppression of protein synthesis, 228-229
  nitric oxide production, 223
  nitric oxide synthase expression, 223-225
  nitric oxide synthesis, biochemistry, 227
Highest occupied molecular orbital, unpaired electron of NO, 3-4
High performance liquid chromatography, methods for nitrite and nitrate, 35
Horseradish peroxidase, nitrosyl complexes, 90
N-Hydroxyguanidine
  peracid reaction, 156-157
  reaction with one-electron oxidants, 157
N^G-Hydroxyl-L-arginine, as intermediate in NO biosynthesis from L-arginine, 118
Hydroxyl radical
  diffusion distance, 43-44
  formation from peroxynitrite, 48-49
  nitric oxide as scavenger of, 60-62
  production by iron-catalzyed Haber–Weiss reaction, 42-43
Hyperfine coupling, in EPR spectra, 101-102

Immunology
  autoimmune diabetes, *see* Diabetes type I
  studies prior to 1987, 141-142
Inactivation, NO under physiological conditions, 118
Inflammation, islet, cyclooxygenase effects, 202-203
Insulin-dependent diabetes, *see* Diabetes type I
Insulin secretion
  glucose-stimulated, interleukin 1 effects, 181-185, 198-200
  role of constitutive NO synthase, 196-197
Intercellular messengers, *see also* Second messengers; Signaling systems
  NO role, 6-7
Interferon γ, induction of NO synthesis in hepatocytes, 223-224
Interleukin 1
  effects on islet function
    early studies, 181-182
    mediation
      by nitric oxide, 186-187
      by oxygen radicals, 185-186
    mitochondrial, 185
    mRNA transcription, 182-185
    protein translation, 182-185
  induction of islet NO formation, 191-192
    in β cells, 192-194
    in macrophages, 194-195
  induction of NO synthase, signaling mechanism, 195-196
  induction of NO synthesis in hepatocytes, 223-224
Iron
  ferric, NO reactions, 6
  ferrous, NO reactions, 6
Iron complexes
  Cu(II)–NO, 90
  in cured meats, NO reactions, 273-275
  formation, 87-88

Iron complexes (Continued)
  iron–nitrosyl, IL-1-induced in islets, 187-188
Iron–sulfur proteins
  $g = 2.04$ dinitrosylirondithiol signal, 146-147
  nitric oxide complexes, 91-95
  nitric oxide inactivation
    and botulinal inhibition, 274
    in nitrite-treated foods, 274-275
Iron–thiol–nitrosyl compounds, inhibition of *Clostridium sporogenes* in cured meats, 275
Ischemia/reperfusion, peroxynitrite generation in vascular compartment, 40-41
Islets of Langerhans
  function, activated macrophage effects, 198
  inflammation, NO role, 202-203
  interleukin 1 effects
    initial studies, 181-182
    mitochondrial function, 185
    mRNA transcription, 182-185
    protein translation, 182-185
  interleukin 1 effects, mediation
    by nitric oxide, 186-187
    by oxygen radicals, 185-186
  interleukin 1-induced NO formation, 191-192
    in β-cells, 192-194
    in macrophages, 194-195
  nitric oxide-mediated dysfunction and destruction, 187-191
  nitric oxide production in human tissue, 202-204

Kramer's theorem, 85
Kupffer cells, hepatocyte–Kupffer cell cocultures, 222-223

Late gas defect, in cheese, 279
Lewis dot diagrams
  nitric oxide compared to nitrosonium ion and molecular nitrogen, 3

Lewis dot diagrams (Continued)
  oxidation states of simple N containing molecules, 20
  oxygen reductive chemistry, 5
Ligands, paramagnetic, 84-86
Lipopolysaccharide
  effects on protein synthesis in Kupffer cell–hepatocyte cocultures, 222-223
  induction of NO synthesis in hepatocytes, 223-224
Lipoxygenase, soybean, nitrosyl complexes, 95
Liver
  in vitro effects of NO on cell function, 227-230
  in vivo actions of NO, 231-233
  sepsis and inflammation, 221-225
Low density lipoproteins, oxidation by peroxynitrite, 68

Macrophages
  islet, IL-1-induced NO formation, 194-195
  nitric oxide synthase subunit assembly, 163-164
  nitric oxide synthesis and release, 119
  peroxynitrite production, 64-66
Magnetic resonance spectra, nitrosyl complexes, 86
Mars, Viking mission, 45-46
Meat curing
  antioxidant effects of NO, 266-269
  bacteriostatic effects of nitrate and NO, 269-271
  color
    ascorbate effects, 264-265
    enzymatic reduction of nitrate/nitrite, 261-263
    mononitrosohemochromogen characterization, 265-266
    nonenzymatic reduction of nitrate/nitrite, 263-264
  direct application of NO, 277
  nitrate use, history, 260-261

Index

Meat curing (Continued)
  nitric oxide use, overview, 261
  nitrite use, history, 260-261
Messenger RNA
  inducible NO synthesis in hepatocytes, 224-225
  preproinsulin, IL-1 effects, 182-185
Metallothionein, nitrosyl complexes, 96
N-Methyl-D-aspartate, activation, link with NO formation, 124, 126
N-Methyl-D-aspartate receptor, role in neuronal NO synthesis, 9-10
Mitochondria, islet, interleukin 1 effects, 185, 189-190
Mixed lymphocyte reaction, splenocyte, NO production, 238-239
Molecular nitrogen, Lewis dot diagram, 3
Molecular oxygen, unpaired electrons on, 5
N^G-Monomethyl-L-arginine
  effects on IL-1-induced inhibition of islet insulin secretion, 186-187
  effects on splenocyte mixed lymphocyte reaction, 238-239
Mononitrosohemochromogen, characterization, 265-266
Myoglobin, nitrosyl complexes, 89-91
Myohemoglobin, reaction with NO to produce nitrate, 2

NADH dehydrogenase, iron–sulfur clusters, 93
Neuroblastoma, N1E-115 cells, cytosolic NO synthase purification, 117
Neurons, NO synthesis, 9-10
Neurotoxicity, glutamate-induced, NO mediation, 125-126
Neutrophils
  nitric oxide modulatory role, 119-120
  peroxynitrite production in human, 68
Nitrate
  in cured meats
    enzymatic reduction, 261-263
    history of use, 260-261
    nonezymatic reduction, 263-264

Nitrate (Continued)
  Griess reaction assay, 35
  and late gas defect in cheese, 279
  production
    by macrophages, 65-66
    by NO reaction with oxyhemoglobin/oxymyoglobin, 2
  radiation damage, 45-46
  reduction by denitifiers, 293-296
Nitrate reductase, denitrifying bacteria, 300-301
Nitration, tyrosine, pathological implications, 53
Nitric acid, dimerization reactions with nitrogen dioxide, 29-30
Nitric oxide
  antioxidant role, 60-62
  bacterial production
    by heterotrophic nitrifiers, 326-328
    from nitrite, 324-326
  biological half-life, 2, 10-18
  chemistry, 22-25
  in cured meats
    antimicrobial effects, 269-271
    direct application, 277
    and nitrosamine formation, 278
  detection, see Detection methods
  endothelium-derived, biosynthesis, 115-118
  information communication by, 7-10
  intercellular messenger role, 6-7
  as labile nitroso precursor, 115
  Lewis dot diagram, 3
  metal reactions, 6
  neuronal messenger role, 124-129
  oxygen reactions, 4-6
  as paramagnetic ligand, 84-86
  as physiological mediator, 119-120
  reaction with superoxide to produce peroxynitrite anion, 3
  superoxide reactions, chemistry, 39-41
  in swollen cans, green beans and spinach, 279
Nitric oxide complexes
  copper complexes, 97

Nitric oxide complexes (Continued)
  ferroheme–nitrosyl model complexes, 87-88
  ferrohemes in proteins, 88-91
  iron–sulfur proteins, 91-95
  nonheme iron proteins, 95-96
  in proteins, 91-95
Nitric oxide reductase, denitrification
  characterization, 307-310
  mechanisms, 310-312
  models, 310-312
Nitric oxide synthase
  catalysis of odd-electron oxidation, 160
  composition, 150-153
Nitric oxide synthase (Continued)
  constitutive
    expression, 149
    and insulin secretion, 196-197
  conversion of L-arginine to NO plus L-citrulline, 116
  cytosolic, from N1E-115 neuroblastoma cells, 117
  dimeric structure, 163-164
  electron transfer
    calmodulin role, 160-161
    flavin role, 158-159
  inducible
    expression, 149-150
    in hepatocytes, 223-224
    inhibitors, 205-206
  interleukin 1-induced expression in β cells, 194
    signaling mechanism, 195-196
  isoforms
    cloning and sequencing, 151-153
    cytosolic and membrane-bound, 117-118
    prosthetic groups, 152
    purification, 150
  isolation, purification, and characterization, 116-117
  macrophage, subunit assembly, 163-164
  properties, 150-153
  reaction catalyzed by, 148-149
  tetrahydrobiopterin role, 161-163
Nitric oxide synthesis
  from L-arginine, 118, 143
  enzymology, 148-164
  in graft infiltrating cells, inhibition by FK506, 243
  heme role, 153-157
  in vitro by sponge matrix allograft cells, 239-240
  in vivo during allograft rejection
    graft versus host reaction, 244-245
    inhibition by cyclosporin A, 245
    inhibition by FK506, 245
  iron enzyme dysfunction, 143-144
  mammalian, 142-143
  mechanism, 149
  stoichiometry, 149
Nitrifying bacteria
  denitrification pathway, 321-323
  nitric oxide production, 326-328
Nitrite
  acidification, 27-28
  as corrosion inhibitor, 279
  in cured meats
    antimicrobial effects, 269-271
    effects on thiobarbituric values, 266-267
    enzymatic reduction, 261-263
    history of use, 260-261
    and nitrosamine formation, 278
    nonezymatic reduction, 263-264
  Griess reaction assay, 35
  nitric oxide production by enteric bacteria, 324-326
  production by macrophages, 65-66
  protein-bound, antimicrobial effects, 276-277
  ultraviolet spectrum at acidic pH, 28
Nitrite reductase
  denitrifying bacteria
    Cu-type enzymes, 315-317
    cytochrome cd_1, 312-315
    mechanisms, 317-320
    models, 317-320
    nitric oxide-reducing, 321
  ferroheme–NO complexes, 91

Nitrogen
　common names and IUPAC names, 21
　global cycle, 289-292
　molecular, Lewis dot diagram, 3
Nitrogenase, iron–molybdenum cofactor, 93
Nitrogen cycle, major reactions, 289-292
Nitrogen dioxide, dimerization reactions with NO, 29-30
Nitrogen oxides, *see also specific oxide*
　chemistry, 18-21
　command names and recommended IUPAC names, 21
　inorganic, mammalian synthesis, 140
Nitrosamines, formation in foods, 278
Nitrosomas europaea, denitification pathway, 321-323
Nitrosonium ion
　Lewis dot diagram, 3
　reactions, 25
Nitrosothiols
　as endothelium-derived relaxing factors, 31
　formation, 25
　spontaneous reaction with NO, 32
　synthetic preparation, 32
Nitrosoylmyoglobin, production in various tissue, 262
Nitrosyl complexes
　formation, 6
　iron–nitrosyl, IL-1-induced formation by islets, 187-188
　magnetic resonance spectra, 86
Nitrosyldioxyl radical, *see* Peroxynitrite radical
Nitrous acid, concentration, and nitrosamine formation in foods, 278
Nitroxyl anion
　chemistry, 23-24
　as endothelium-derived relaxing factor, 24
　peroxynitrite anion formation, 24
　singlet, 23
　triplet state, 24

Nomenclature, nitrogen oxides and related oxygen radical species, 21
Nonadrenergic–noncholinergic neurotransmitter, nitric oxide role in, 119, 127
Nuclear magnetic resonance spectroscopy, ^{15}N-labeled peroxynitrite, 58-59

Oxidation states
　nitrogen compounds, 19
　nitrogen at pH 7.0, 22
Oxygen, reductive chemistry, 4-6
Oxygen radicals
　liver damage, NO protective actions, 232
　mediation of IL-1 inhibitory effects on islet insulin-secretion, 185-186
　nomenclature, 21
　production by islet macrophages, 198
Oxyhemoglobin, reaction with NO to produce nitrate, 2

P388D1 macrophage cell lines, as allogeneic antigen presenting cells, 245-246
Peracids, reaction with *N*-hydroxyguanidine, 156-157
Peroxynitrite
　anion formation, 24
　buffer anion effects, 54-55
　cis conformation, 56-60
　decomposition rate, reactant concentration effects, 46-47
　diffusion distance, 43-44
　formation routes, 66-67
　generation in vascular compartment, ischemia/reperfusion-related, 40-41
　historical studies, 44-45
　oxidative pathways, 46-48
　　chemistry at neutral pH, 40
　　hydroxyl radical-like, 48-52
　　nitronium-like species formation, 52-53
　　pH dependence, 55-56
　　oxidative reactions at alkaline pH, 50-51

Peroxynitrite (Continued)
  production by macrophages, 64-66
  production via superoxide–NO reaction, 3
  reaction with superoxide dismutase, 62-63
  redox potentials of derived oxidants, 47-48
  sulfhydral oxidation, 54
  superoxide dismutase probe for, 63-64
  toxicity, 66-67
Peroxynitrite radical
  formation, 14
  formation via unpaired electrons, 5
  reduction potential at pH 7.0, 48
  stability, 14-15
pH effects
  peroxynitrite oxidation, 55-56
  products from peroxynitrous acid attack on DMSO, 50-51
Phosphoroclastic system, in *Clostridium sporogenes* cells, 273-274
Pigment, in cured meat, 265-266
Plastocyanin, nitrosyl complexes, 97
Platelet aggregation, in endotoxemia, NO effects, 232-233
Polyamine synthesis, lymphocyte, NO effects, 251-252
Protein synthesis
  hepatocyte
    in vitro suppression, 228-229
    lipopolysaccharide effects in cultures, 222-223
  islet, interleukin 1 effects, 182-185
  in Kupffer cell–hepatocyte cocultures, lipopolysaccharide effects, 222
Pseudomonas stutzeri
  denitrification genome, 320-321
  nitric oxide reductase EPR spectrum, 309
Pulmonary surfactant, peroxynitrite effects, 67
Pyruvate:ferredoxin oxidoreductase, inactivation by nitrite-derived NO, 273

Raman spectroscopy, peroxynitrite
  [14]N- and ^{15}N-labeled, 58-59
  vibrational modes, 58-59
RAW 264.7 macrophage cell lines
  as allogeneic antigen presenting cells, 245-246
  nitric oxide synthesis effects, 246
Reduction potential, NO, peroxynitrite, and derived oxidants, 48
Ribonucleotide reductase, inhibition, and NO-induced inhibition of lymphocyte activation, 250-251
Rubredoxins, characterization, 92

Second messengers, *see also* Signaling systems
  nitric oxide effects in lymphocytes, 249-250
  nitric oxide properties, 124-129
Sepsis, Kupffer cell–hepatocyte interactions, 222-225
Signaling systems, *see also* Second messengers
  intercellular, NO role, 6-7
  interleukin 1-induced expression of NO synthase, 195-196
  intracellular, NO effects in lymphocytes, 249-250
Signal transduction
  activation of cytosolic guanylate cyclase, 120-123
  nitric oxide activation of soluble guanylate cyclase, 6
SIN-1 nitrovasodilator, peroxynitrite production, 68
Sodium channels, amelioride-sensitive, peroxynitrite effects, 67
Sodium nitrate, as antioxidant in cured meats, 267
Sodium nitrite, bacteriostatic effects in cured meats, 269-271
Solubility, nitric oxide in water, 33-34
Soybean lipoxygenase, nitrosyl complexes, 95
Splenocytes, mixed lymphocyte reaction, NO production, 238-239

Sponge matrix allograft, in vitro NO synthesis, 239-240
Succinate dehydrogenase, mitochondrial, iron–sulfur clusters, 93-94
Sulfhydryl groups, oxidation by peroxynitrite, 54
Sulfite reductase, ferroheme–NO complexes, 91
Superoxide
 assay, cytochrome c method, 26
 buffer, NO destruction, 13-14
 nitric oxide as scavenger of, 60-62
 reaction with NO
  peroxynitrite anion production, 3
  rate of, 14
 toxicity to vascular targets, 42-43
Superoxide dismutase
 in amyotrophic lateral sclerosis, 63
 bacterial, iron-containing, 96
 peroxynitrite reactions, 62-63
 probe for peroxynitrite, 63-64
 reduction of tissue injury, 41-43
Synaptic function, NO–glutamate interrelationships, 124-125
Synaptic plasticity, NO role, 8-10

Tetrahydrobiopterin
 and hepatocyte NO synthesis, 225-226
 in hydroxylation reactions, 162
 in nitric oxide synthase, 161-163
 role in macrophage NOS subunit assembly, 163-164
Tetranitromethane, nitration of tyrosine residues, 53
Thiobarbituric acid, in cured meats, nitrite effects, 266-267
T lymphocytes
 activation, NO-induced inhibition mechanisms, 248-252
  inhibition of polyamine synthesis, 251-252

T lymphocytes (Continued)
  inhibition of ribonucleotide reductase, 250-251
  intracellular messengers, 249-250
 authentic NO effects
  actively proliferating cells, 246-247
  antigen driven proliferation, 247
  cytolytic function and motility, 248
  response to Il-2, 247
 cytolytic, see Cytolytic T lymphocytes
 role in diabetes development, 179-180
Toxicity
 nitric oxide toward bacteria, 328-329
 peroxynitrite, 66-67
Trace metals, NO reactions, 33-37
Transition metals
 catalysis of nitronium-like species from peroxynitrite, 52-53
 ferric iron complexes, 6
 ferrous iron complexes, 6
Tumor necrosis factor, induction of NO synthesis in hepatocytes, 223-224
Tyrosinase, copper pair, NO binding, 97
Tyrosine, nitration, pathological implications, 53

Ultraviolet radiation, damage to nitrate, 45-46
Ultraviolet spectra, nitrite at acidic pH, 28

Vascular smooth muscle, nitric oxide modulatory role, 119-120
Viking mission, 45-46

Warmed-over flavor, in cured meats, 266-269
Wollinella succinogenes, hexaheme cytochrome c, 321

Zeeman splitting, 99-101